BIRDS OF WESTERN AUSTRALIA

Plate 1.—Some Honeyeaters of the Genus *Meliphaga*.

Singing Honeyeater	Purple-gaped Honeyeater
Yellow-plumed Honeyeater	Grey-headed Honeyeater
Yellow-fronted Honeyeater	White-plumed Honeyeater

BIRDS

OF

WESTERN AUSTRALIA

By D. L. SERVENTY and H. M. WHITTELL

WITH ILLUSTRATIONS BY OLIVE SEYMOUR
H. O. WEBSTER
P. SLATER, G. E. BINSTED and A. THORPE

1976

UNIVERSITY OF WESTERN AUSTRALIA PRESS
PERTH, W.A.

First published in December 1948 by
Paterson Brokensha Pty Ltd

Second edition May 1951
Third Edition May 1962
Fourth edition May 1967

Fifth edition November 1976 published by
University of Western Australia Press
Nedlands, Western Australia

Eastern States of Australia, New Zealand, Papua and New Guinea:
Melbourne University Press, P.O. Box 278, Carlton South, Vic. 3053
United Kingdom, the Middle East, Africa and the Caribbean:
Prentice-Hall International, International Book Distributors Ltd,
66 Wood Lane End, Hemel Hempstead, Hertfordshire, England

United States of America, its territories and possessions, Canada:
International Scholarly Book Services, Inc.,
P.O. Box 555, Forest Grove, Oregon, 97116, U.S.A.

National Library of Australia
Cataloguing-in-Publication data:
Serventy, Dominic Louis
Birds of Western Australia.
Index.
ISBN 0 85564 101 0.
1. Birds—Western Australia—Identification.
I. Whittell, Hubert Massey, joint author.
II. Title.
598.29941

CONTENTS

Page

INTRODUCTION 1
Plan of the Book. Names. Native Names. Description of Species. Distribution. Nesting and Habits. Migration. Nesting Seasons. Introduced Species. Summary. Acknowledgements.

SECTION 1

THE HISTORY OF WESTERN AUSTRALIAN ORNITHOLOGY 14

I—The Period of the Early Travellers (1618-1700) 15
II—The Period of the Great Navigating Expeditions (1791-1826) 19
III—The Period of John Gilbert and the Exploration of the Colony (1829-1887) 28
(a) The Pioneer Observers (1829-1839) 28
(b) The Visit of John Gilbert (1839-1843) 33
(c) The Barren Years (1843-1887) 38
IV—The Modern Period (1887 onward) 40

SECTION II

BIRD GEOGRAPHY 47

SECTION III

THE BIRDS 65

Emus and Cassowaries: Order Casuariiformes 65
Emus: Family Dromaiidae 65
Penguins: Order Sphenisciformes 71
Penguins: Family Spheniscidae 71
Tubinares: Order Procellariiformes 75
Albatrosses: Family Diomedeidae 75
Petrels and Mutton-birds: Family Procellariidae 81
Storm-Petrels Family Hydrobatidae 99
Diving-Petrels: Family Pelecanoididae 103
Gannets, Cormorants and their Allies: Order Pelecaniformes 104
Frigate-birds: Family Fregatidae 105
Tropic-birds: Family Phaethontidae 104
Pelicans: Family Pelecanidae 107
Gannets and Boobies: Family Sulidae 108
Cormorants: Family Phalacrocoracidae 111
Grebe-like Birds: Order Podicipediformes 119
Grebes: Family Podicipedidae 119
Heron-like Birds: Order Ciconiiformes 123
Herons, Egrets and Bitterns: Family Ardeidae 123
Storks: Family Ciconiidae 131
Ibises and Spoonbills: Family Threskiornithidae 132
Duck-like Birds: Order Anatiformes 136
Swans, Geese and Ducks: Family Anatidae 136
Diurnal Birds of Prey: Order Falconiformes 157
Eagles, Goshawks, etc.: Family Accipitridae 157
Ospreys: Family Pandionidae 171
Falcons and Kestrels: Family Falconidae 172
Fowl-like Birds: Order Galliformes 177
Mound-builders: Family Megapodiidae 177
Pheasants and Quails: Family Phasianidae 180

Cranes, Rails, etc.: Order Gruiformes ... 182
Bustard Quails: Family Turnicidae ... 182
Cranes: Family Gruidae ... 184
Rails, Crakes and Water-hens: Family Rallidae ... 185
Bustards: Family Otidae ... 193
Waders, Gulls and Terns: Order Charadriiformes ... 194
Waders: Suborder Charadrii ... 194
Jacanas: Family Jacanidae ... 196
Painted Snipes: Family Rostratulidae ... 197
Oystercatchers: Family Haematopodidae ... 198
Plovers and Dotterels: Family Charadriidae ... 199
Curlews, Sandpipers and Snipes: Family Scolopacidae 209
Stilts and Avocets: Family Recurvirostridae ... 224
Stone-Curlews: Family Burhinidae ... 227
Pratincoles and Coursers: Family Glareolidae ... 228
Gulls and Terns: Suborder Lari ... 229
Skuas: Family Stercorariidae ... 229
Gulls and Terns: Family Laridae ... 231
Pigeons and Doves: Order Columbiformes ... 256
Pigeons and Doves: Family Columbidae ... 256
Parrots and Cockatoos: Order Psittaciformes ... 267
Parrots and Cockatoos: Family Psittacidae ... 267
Cuckoo-like Birds: Order Cuculiformes ... 293
Cuckoos: Family Cuculidae ... 293
Owls: Order Strigiformes ... 300
Hawk-Owls: Family Strigidae ... 300
Barn-Owls: Family Tytonidae ... 302
Frogmouths and Nightjars: Order Caprimulgiformes ... 304
Frogmouths: Family Podargidae ... 304
Owlet-Nightjars: Family Aegothelidae ... 305
Nightjars: Family Caprimulgidae ... 306
Swifts: Order Apodiformes ... 307
Swifts: Family Apodidae ... 307
Roller-like Birds: Order Coraciiformes ... 308
Kingfishers: Family Alcedinidae ... 308
Bee-eaters: Family Meropidae ... 312
Rollers: Family Coraciidae ... 314
Perching or Song-birds: Order Passeriformes ... 314
Scrub-birds: Family Atrichornithidae ... 314
Larks: Family Alaudidae ... 321
Swallows and Martins: Family Hirundinidae ... 321
Pipits: Family Motacillidae ... 327
Cuckoo-Shrikes: Family Campephagidae ... 328
Thrushes: Family Turdidae ... 332
Quail-Thrushes and Babblers: Family Timaliidae ... 332
Old World Warblers: Family Sylviidae ... 337
Australian Warblers: Family Maluridae ... 343
Old World Flycatchers: Family Muscicapidae ... 373
Fantails: Family Rhipiduridae ... 381
Monarch Flycatchers: Family Monarchidae ... 383
Whistlers and Shrike-Thrushes: Family Pachycephalidae ... 385
Shrike-Tits: Family Falcunculidae ... 389
Nuthatches: Family Neosittidae ... 395
Australian Tree-creepers: Family Climacteridae ... 396
Flower-peckers: Family Dicaeidae ... 398

Diamond-birds: Family Pardalotidae 399
Silvereyes: Family Zosteropidae 402
Honeyeaters: Family Meliphagidae 404
Australian Chats: Family Ephthianuridae 427
Finches and Buntings: Family Fringillidae 428
Grass-Finches: Family Estrildidae 429
Old World Sparrows: Family Passeridae 436
Starlings: Family Sturnidae 437
Magpie-Larks: Family Grallinidae 439
Wood Swallows: Family Artamidae 441
Australian Magpies: Family Cracticidae 447
Bower-birds: Family Ptilonorhynchidae 454
Crows and Ravens: Family Corvidae 456

INDEX 460

ILLUSTRATIONS

COLOURED PLATES

Plate

1. Some Honeyeaters of the genus *Meliphaga* ... Frontispiece
2. Races of the Western Rosella ... Facing page 281
3. Left: Masked Owl. Right: Barn Owl ... " " 302
4. Left: Banded Blue Wren. Right: Turquoise Wren ... " " 343
5. The chestnut-shouldered Wrens ... " " 349
6. Slate-backed Thornbill, Broad-tailed Thornbill and chestnut-rumped Thornbill ... " " 356
7. Yellow-rumped Thornbill, Samphire Thornbill and Western Thornbill ... " " 358
8. Upper: Southern Emu-Wren. Lower: Red-tipped Diamond-bird ... " " 400
9. Upper: Spinebill. Lower: Tawny-crowned Honeyeater ... " " 417

TEXT FIGURES

Figure | Page

1\. Distribution of Australian Faunas ... 47
2\. Distribution of Western and Eastern Spinebills ... 50
3\. Distribution of the Yellow Robins ... 51
4\. Distribution of majority of Bassian species ... 53
5\. Main distribution of the Smoker Parrot ... 54
6\. Southern and south-western boundaries of various honey-eaters ... 56
7\. Distribution of species which avoid the South-West corner 57
9\. Distribution of species along the southern corridor ... 58
9\. Distribution of species along the northern corridor ... 59
10\. Black-browed Albatross ... 77
11\. Yellow-nosed Albatross ... 78
12\. Fairy Prion ... 85
13\. Thin-billed Prion ... 87
14\. Medium-billed Prion ... 88
15\. Little Shearwater ... 90
16\. Fluttering Shearwater ... 92
17\. Kerguelen Petrel ... 98
18\. Crested Grebe ... 119
19\. Little Grebe ... 121
20\. Hoary-headed Grebe ... 122
21\. Royal Spoonbill ... 135
22\. Magpie Goose ... 137
23\. Cape Barren Goose ... 141
24\. Maned Goose, male ... 152
25\. Black-shouldered Kite ... 157
26\. Letter-winged Kite ... 158
27\. Whistling Eagle ... 162
28\. Australian Little Eagle ... 166
29\. Spur-winged Plover, Masked Plover, Banded Plover ... 200
30\. Oriental Dotterel ... 203
31\. Large Sand-Dotterel ... 204
32\. Mongolian Dotterel ... 205

33.	Japanese Snipe, Chinese Snipe, tail patterns	216
34.	Curlew Sandpiper	219
35.	Sharp-tailed Sandpiper	221
36.	Kelp Gull, Pacific Gull, beaks	236
37.	Kelp Gull, Pacific Gull, tail patterns	236
38.	European Common Tern, Roseate Tern, Arctic Tern	242
39.	Sooty Tern	248
40.	Bridled Tern	250
41.	Common Noddy	252
42.	Lesser Noddy	254
43.	Peaceful Dove	259
44.	Diamond Dove	260
45.	Flock Pigeon	266
46.	Little Corella	274
47.	Long-billed Corella	274
48.	Boobook Owl	300
49.	Noisy Scrub-bird	315
50.	Spotted Scrub-Wren	362
51.	Striated Grass-Wren	369
52.	Western Grass-Wren	369
53.	Pied Honeyeater	407
54.	Black Honeyeater	407
55.	White-fronted Honeyeater	419
56.	New Holland Honeyeater	421
57.	White-cheeked Honeyeater	422
58.	Spiny-cheeked Honeyeater	426
59.	White-breasted Wood-Swallow	442
60.	Masked Wood-Swallow	442
61.	Black-faced Wood-Swallow	445
62.	Dusky Wood-Swallow	445
	Map of Western Australia	end paper

INTRODUCTION

The need for some such book as this one was impressed on the authors by their own early experiences in Western Australian ornithology. Perhaps, however, they flatter themselves unduly for presenting to the bird-loving public a book which they like to feel is of the type they wished had been available to them when beginning the study of local birds.

Whatever opinions may be held on the merits or otherwise of the present work, most ornithologists are agreed that there are few Australian books that cater conveniently for the ordinary field worker who desires a compact summary of what is known of the birdlife of the area in which he is working. The standard reference works are both beyond the means of the ordinary person and in many respects are out of date, and the more popular books, covering the whole of Australia, can necessarily deal very superficially and quite inadequately with the birds of particular areas. In our opinion the time has arrived for the preparation of regional handbooks for so large and diversified a country as Australia, usable books that will not be overloaded with accounts of birds which the reader is hardly ever likely to come across and whose inclusion in an identification book will only contribute to instead of dispelling confusion.

The scope of such a book must be a matter of compromise; it should, of course, be based on natural regions rather than be limited by political boundaries. At the same time its compass should not be too narrow and should cover an area which may be conveniently though loosely described as the reader's "home ground"—with which, even though he may not have opportunities of knowing it all at first-hand, yet by virtue of business dealings, daily reading and so on, he feels he is reasonably familiar and intimate.

Both practical convenience and consideration of zoo-geography have impelled us to limit the scope of this book to that portion of the State south of the Kimberley border, namely about Latitude 19° or 20° South. The ornithological reasons for this decision are given in detail in Section II of this book; briefly, the Kimberley bird fauna is quite a distinctive one compared with that further south. The Kimberley Division is divided from the remainder of the State by a desert barrier and its monsoonal climate, productions and general problems mark it off from the country further south, which from the Pilbara to the Leeuwin is continuously settled, has freer communications and in general regards itself as one compared with the "far north." Eastwards, though we have in general limited ourselves to the State boundary, the book may have use in Central Australia. The naturalist at Alice Springs, for instance, will find in its pages almost all the birds he is likely to meet with there. For marine species, waders and many other aquatic groups, the book may be found useful over most of Australia.

PLAN OF THE BOOK

The first two sections deal with the history of Western Australian ornithology and a discussion of the bird geography of the State. Though complete in themselves both are supplementary to the third section, the detailed treatment of the species of birds occurring in our area. The species are arranged generally, as in our former editions and in the order given in *An Index of Australian Bird Names,* published by the Commonwealth Scientific and Industrial Research Organization in 1969. The scientific names therein are generally followed. In some cases both arrangement and names have been modified as a result of new knowledge. Recently various lists of Australian bird names have been proposed, in books and checklists, but we have avoided extreme lumping and adopted a taxonomic treatment we feel meets the practical needs of field observers. In order each species is discussed as follows:

NAMES

The accepted vernacular name is followed, in brackets, by all the local names known to us to be in use in Western Australia.

NATIVE NAMES

The native names given under certain species have been drawn primarily from ornithological authors and are quoted in the original spellings used by them. The main sources were given in our earlier editions.

Some native names, not many unfortunately, have passed into use as vernacular names, either locally or generally. Such include "Gnow" for the Mallee Fowl, "Berrin-berrin" for the Bee-eater, "Weero" for the Cockatoo Parrot and, slightly corrupted, "Chock-a-lock" for the Major Mitchell Cockatoo and "Wockengull" for the Smoker Parrot. Others seem to have had only a temporary currency, Examples are "Tilyabit" for the Magpie Lark at York (*vide* F. A. Grigson) and "Buan," once used in the Brookton district for the Dusky Wood-Swallow (compare this with Gilbert's rendering of the Aboriginal name: "Be-wo-wen," and Fletcher Moore's "Biwoen".

A few native bird names have survived in the components of place names. Some may be enumerated: Wagin and Wagerup (meaning the abode or resort of the Emu), Gnowangerup (abode of the Gnow or Mallee Fowl), Coolup (contraction of Koolyinyinup or Koolinup, place of the Bustard or Wild Turkey), Jandakot (place of the Whistling Eagle), Cowaramup (abode of the Purple-crowned Lorikeet) and Dowerin (cf. Towerin, a native name for the Twenty-eight Parrot). One interpretation of Karrakatta is the "hill of the Red-tailed Black Cockatoo" ("Karrak") (J. C. Hassell), though another version has it as the "hill of the crab" ("Karri").

The general name for a small bird in the South-West (Bibbulman tribe) was "Jida"; for a bird's nest, "Jida-mya" or "Jee-da-mya" and "Man-ga"; "Nur-do" was the name for an egg and "I-dal-ya" for feathers. At the Warburton Range the respective words are: bird, "Tjitati"; egg, "Ngampu." At Mt. Margaret egg is "Tjantji".

DESCRIPTION OF SPECIES

An outline description only is attempted, so that the reader may visualise the appearance of a bird, primarily for the purpose of field identification. Complete "feather by feather" descriptions are available in such works as A. J. North's *Nests and Eggs of Birds found Breeding in Australia and Tasmania* (1901-1914) and Gregory M. Mathews' *The Birds of Australia* (1910-1927). Concise modern descriptions are available in J. D. Macdonald's *Birds of Australia.* For extended information on marine birds readers are referred to *The Handbook of Australian Sea-Birds,* by D. L. Serventy, V. Serventy and J. Warham. All of the descriptions have been prepared by ourselves from local material, except in a few instances where no specimens were available to us. Essential diagnostic features in the description have been italicised to enable the reader to pick out at once the basic points for identification. Field marks, for ready recognition of the bird in life, utilising characters, voice or behaviour, have been given wherever possible. No uniform simple method of giving relative size is available and though there are objections to such a measure as "total length" (taken in inches in a straight line from the tip of the beak to the end of the tail feathers, with the specimen gently stretched) it seems to be the best way of conveying an impression of size to the ordinary reader. In some cases, particularly in larger birds, we have added wing span and the weight as well, though for some species these particulars are not accurately known or the range of weight variation has yet to be determined. In the case of the petrels, which the average ornithologist will mostly encounter as storm-drifted specimens on the ocean beaches, and of some other birds, we have also included the wing and beak lengths, in millimetres. Such measurements, particularly in the case of incomplete remains, often enable a satisfactory identification of species otherwise rather difficult to determine. The beak length so given is the "chord of the culmen," and is best taken with a pair of dividers by placing one point at the base of the beak, where the feathering begins, and the other point at the tip of the beak even though this be sharply decurved. The wing length is measured, from the carpal joint to the tip of the longest feather, by pressing the wing on to a ruler until it is quite flat. The wing span is taken from tip to tip of the fully extended wings by *gently* stretching the bird.

We have refrained in this edition from adopting *in toto* the metric system for linear dimensions and weights, as most people who will be using this book still think in terms of the old imperial units. As in previous editions fine measurements and weights, particularly of material that will be examined in the hand, such as beach-drifted seabirds, are given in metric units.

DISTRIBUTION

Only the Western Australian distribution is given in this book —the wider Australian occurrence of any species can be obtained from our *Systematic List of the Birds of Western Australia* published by the Western Australian Museum, and the general works mentioned on p. 3. The geographical distribution in this State has been given in as much detail as is possible at the present time and we have been deliberately positive in many of the statements made, both in this instance as well as in many of the other portions of the accounts of the species. Through this treatment the reader will be able to assess the significance of any observation he may make, either on distribution or behaviour, and so be in a position to know whether it represents an addition to knowledge and is therefore worthy of being placed on record. The statements made in the various editions of this book report contemporary knowledge, and no one realises more than we that much of it is destined to be amended or extended. The information published in the earlier editions has elicited more exact records from various observers and these have been incorporated in the present edition. However, much information has still to be uncovered or made known.

NESTING AND HABITS

Following the outline of distribution is a paragraph on nesting and after this a summary of what is known of general habits, migration, song and other pertinent details. The egg measurements are given in millimetres (1 in. = 25.4 mm.) but it was considered that the dimensions of very large eggs could be more conveniently recorded in inches. Except in special cases details of nesting are not given for species which do not breed in Western Australia.

MIGRATION

Western Australia shows remarkably little of the classical type of bird migration, as known in Europe and North America, and there is less regular migration than in the eastern States. We have very few consistent migrants in the large family of song-birds. The White-winged Triller is perhaps the best known of such migrants in the South-West and others are the Brown and Rufous Songlarks. The Reed Warbler appears to be migratory in the southern portion of the district only. The Welcome Swallow, Rufous Whistler and the Black-faced Cuckoo-Shrike, which are partly migratory in south-eastern Australia, are not so with us. Apart from the song-birds just mentioned, the only regular migrants among the breeding birds in the southern part of the State are the two Bronze Cuckoos, Pallid Cuckoo, Bee-eater and Sacred Kingfisher. Most of these birds move only into the northern part of the State in the winter months and do not travel beyond its borders.

The real, large-scale migrants in Western Australia are the Asiatic-breeders which visit us in the summer months: the 28 species of wading birds, the two northern Skuas and the Fork-tailed Swift. The Sooty Storm-Petrel, the Chinese Little Bittern, the Corncrake, the White-winged Black Tern and European Common and Arctic Terns are, on present evidence, only occasional migrants from the north. The Christmas Island race of the White-tailed Tropic-bird (the Golden Bosun-bird) ranges occasionally into northern waters. The Greater Frigate-bird, from the same island, wanders into our waters more extensively. The Kimberley Division receives additional occasional migrants from the north, including the Malay Banded Crake *(Rallina fasciata)*, the Blue-winged Pitta *(Pitta moluccensis)*, the eastern race of the Northern Hemisphere Barn Swallow *(Hirundo rustica)* and the Yellow Wagtail *(Motacilla flava)*. In contrast to the above a group of southern-breeding birds, all sea-birds, nesting in the Antarctic and Sub-Antarctic regions, comprising 28 species of petrels and albatrosses and the Southern (Great) Skua all visit local seas in the winter. The White-faced Storm-Petrel and the Fleshy-footed Shearwater are migratory, local-nesting petrels which make a trans-equatorial migration into the northern Indian Ocean waters in the winter months.

The Short-tailed Shearwater and the White-capped Albatross wander into the western Bight from south-eastern Australia, and the Australian Gannet is an autumn and winter migrant from Bass Strait and New Zealand. Two other species also visit us from New Zealand, the Fluttering Shearwater and the Double-banded Dotterel.

Most of our land and fresh water birds can be classed, broadly, as residents. Quite a few, however, make regular but local and partial migrations. Such include the Tree-Martin, Grey Fantail, Red-tipped Diamond-bird, Spotted Diamond-bird, Red Wattle-bird and several other honeyeaters, and the White-tailed Black Cockatoo, whose seasonal movements are referred to in greater detail in the accounts of the particular species. Other birds are irregular nomads, whose movements are controlled by local and varying shifts in the food supply, such as the flowering of the eucalypts in the case of the Purple-crowned Lorikeet, and the incidence of drought and good rains on some of the bird inhabitants of the inland and northern areas. Recorded data are lacking for much of this local migration and irregular wandering, which should provide a fruitful study for observers in the future.

Occasional exceptional seasons provide opportunities for the study of these movements. Thus the year 1952 was noteworthy for the irruption into the southern part of the State of many species characteristic of the Kimberley Division, central and eastern Australia. These included the Brolga, White Ibis, Pied Goose and Letter-winged and Black Kites. The situation appeared to be due to the populations of these species being built up to super-normal levels by good seasons, to be followed immediately by the unusually severe drought of 1951-52. This drastically reduced the carrying capacity of the northern country, compelling the excess populations of the nomadic species to seek survival by temporary emigration.

The phenomenal irruption of 1952 also highlighted another aspect of bird movement in Western Australia: the probable existence of a greater amount of regular seasonal movement between the Kimberley Division (and of eastern Australia) and the southern parts of the State than had been realised in the past. Some of the species which were concerned in the 1952 invasion, such as the Pelican, Straw-necked Ibis, White Egret, White-necked Heron, and possibly the Black Swan, appear each winter in varying numbers in the south supplementing local populations. The numbers of these birds were swelled considerably in the winter of 1952, which fact suggests that the origin of the visiting birds in normal years may also be the Kimberley Division. It would appear that in the winter, which is the dry or lean period in the Kimberleys, a certain proportion of the excess population, particularly of water birds, moves south, a partial migration which varies in its incidence according to the severity of environmental conditions in the far north. It is possible, however, that many individuals of the Straw-necked Ibis, the Black-tailed Native Hen and of the White-necked Heron may originate in the eastern States.

NESTING SEASONS

The nesting activity of Western Australian birds is very closely linked with the rainfall. So, in general, birds in the winter rainfall belt, the South-West, nest in the late winter and spring. Though the winter rain-bearing systems may extend north at times to the Pilbara and beyond, the predominant rains in the North-West are summer and autumn rains. The rainfall is irregular, however, and there may be prolonged droughts.

Like the other regularly recurring phenomena in the lives of birds, such as migration, reproduction takes place at the time of the year when it is biologically most advantageous for the species to do so. Therefore, it is nicely adjusted to the outside environment, so that the young are hatched out when food is plentiful and the external conditions most favourable for their development.

An essential organ involved in the intricate timing mechanism between the external environment and the breeding organs is the secretes being the chemical regulator of the ovary or testes, as the case may be. The question which primarily interests field ornithologists is what causes the pituitary gland itself to begin secreting periodically at the appropriate time? Experiments have shown that the external stimulus which most generally operates to activate the pituitary is *light,* the eye being the receptor, whence the stimulus is conveyed to the pituitary gland via the brain. It was the *increasing* amount of light in spring which in most cases set the pituitary mechanism in action. Captive birds have been brought into a breeding state "out of season" by artificially lighting up their cages to simulate spring conditions. A few species of birds reproduce while the light is *decreasing.* There are autumn and winter breeders such as the Emperor Penguin *(Aptenodytes forsteri)* and the New Zealand Kea Parrot *(Nestor notabilis).*

In Western Australia, too, we have several species of sea-birds that breed in winter. The Little Shearwater and the Great-winged Petrel are uncomplicated cases of winter-breeding in the southern parts of the State and there are others in the north. Several sea-birds show the interesting phenomenon of double-breeding, there being in the one year both autumn-breeding and summer-breeding. Among such species are the Lesser Frigate-bird, Masked Gannet, Brown Gannet, Pied Cormorant (the Black Cormorant, which is predominantly a fresh water species in this State, also exhibits this phenomenon), Silver Gull, Pacific Gull, Caspian Tern, Crested Tern, Fairy Tern, Bridled Tern, Sooty Tern and Common Noddy. In many cases both spring- and autumn-breeding occurs on the same island, with sometimes a low level of breeding during the intervening winter. An example is the Silver Gull at the Safety Bay and other local islands. Generally speaking, autumn (and winter) breeding is more characteristic of the northern islands, from Shark Bay northwards, and spring breeding is the rule in the southern islands. All but one colony of the Pied Cormorant are autumn and winter breeders, the exception being the population inhabiting the Abrolhos Islands, where nesting takes place in spring. All the Wedge-tailed Shearwater populations are spring and summer breeders, and at all the Pelican colonies, until recently, only autumn and winter breeding was known. In the case of the Silver Gull, Miss C. A. Nicholls has demonstrated that in that species, at least, the phenomenon may be due to *double-broodedness,* i.e. breeding birds produce two clutches of eggs in the year, the interval between successive layings being about three months. Double-broodedness does not occur among Silver Gulls in eastern Australia. Double-broodedness cannot, however, explain the situation in such species as the Pied Cormorant, where separate populations breed only in the spring or autumn. The predominance of autumn and winter breeding in the north may be correlated with the fact that, for nesting purposes, the winter period there, unlike the south, is biologically the more favourable season (more tranquil seas, drier weather). It could be assumed, as a reasonable working hypothesis, that photoperiodicity is the factor which triggers off the breeding season in these cases but it should be emphasised that no experimental proof exists as yet.

In any case, light is by no means the only probable activator and among land and fresh water birds other external factors are capable of stimulating the pituitary; among those which have been suggested are temperature, humidity, rainfall and water levels, and food supply. It is difficult to isolate the effective factor out of a number which might be acting simultaneously. Different species of birds and individuals of the same species vary in the speed of their response to a particular external stimulus and may not all come into breeding activity at exactly the same time. Natural selection is the final arbiter in the matter and modifies the response for the end that has to be served, namely that the young should be hatched at the most favourable period. Any departures from this optimum tend to be eliminated by natural selection.

In the South-West most birds nest in the spring and hence *appear* to behave as if the pituitary gland is stimulated by increasing hours of daylight. But it is clear that the situation cannot be so simply explained. Many of our birds start breeding very early in the season, some even before the winter solstice, namely when the hours of daylight are still decreasing. Obviously the light factor alone cannot be the operative external factor here. Opportunities of isolating the factor responsible come when there is some upset in the usual local environmental conditions which cause corresponding upsets in the local nesting regime. Field observers should be always on the alert to take full advantage of such occasions.

One opportunity presented itself to us in March 1937, in the Kalgoorlie area, following exceptionally heavy rains which filled all the temporary lakes and clay pans, quite transforming the country. Normally the birds in this area nest in the spring, but on this occasion it was found that many birds had developing breeding organs and nesting was reported the following month, April, by a local observer. The species concerned included the Grey Teal, Black-eared Cuckoo, Broad-tailed and Samphire Thornbills, Whiteface and Crested Bell-bird. Further south, in the Norseman and Salmon Gums districts, where the rains had not penetrated, there was no indication whatever of breeding.

It is obvious that the breeding cycle must be initiated before the material benefits of the rainy season become fully manifest and the question is what is the particular external stimulus which acts on the bird? It might be the sight of drenching rain itself or the coming greenness, acting through the eye as the receptor organ, or in some species which show a later response, the richer insect and other food becoming available, may set up internal reactions. The factor, whatever one it is, acts as part of a trigger mechanism to start the endocrine machinery going so that breeding will take place at the right period in the interests of the young. The process is quite involuntary and outside the control of the will of the individual. The nature of the environment probably determines which type of external factor is the key one for particular species —whether it be light, temperature, rain and associated phenomena, or food. The identification of whichever of these factors is the operative one in pressing the trigger in our various areas and for the different species is one of the many interesting field studies of the future.

It is a problem which offers a richer field for investigation in the North-West than perhaps in the South-West owing to the greater irregularity in its environmental conditions. The nesting of the North-West birds was first studied in any detail by Thomas Carter and F. Lawson Whitlock, and the latter considered that the nesting cycle agreed in essentials with that in the South-West in normal seasons but showed far more variation in consequence of variations in rainfall. An analysis of the status of the breeding birds of the North-West was attempted by I. C. Carnaby, based on observations by him in the upper Gascoyne and upper Ashburton pastoral areas over a period of ten years. including a prolonged and severe drought period.

Carnaby recognised three categories of nesting behaviour in the North-West; birds which nested immediatly after good rains, irrespective of the time of the year; birds which nested at a particular season each year, but only when conditions were suitable; and birds which nested regularly, irrespective of the conditions.

In recent years there has been a good deal of investigation into the subject, including close studies of the effects of unseasonal rains on breeding behaviour in various parts of the country. It is deduced that a combination of stimulating and inhibiting factors produces the breeding pattern within the area treated in this book. Rain and its associated effects are the primary stimulators for breeding in most of our land and fresh water birds. H. J. Frith has further shown by studies in the eastern States that water levels in rivers and billabongs control nesting times in ducks and some other waterfowl, even though the rain causing the rise in river flow may have fallen a considerable distance away in the watershed. The stimulus of rain may, however, be inhibited by unfavourable factors, of which cold and excessive wetness are the more important. The combination of both of these retard the effective start of nesting in the South-West until late July or August. In mild winter species like Banded Plover, Yellow-rumped Thornbill and Magpie Lark may begin nesting in May with the first rains, but usually the nests remain unattended until spring.

In the North-West, following summer rains, there is a general nesting among a variety of species in the autumn and early winter.

This slackens off markedly during the winter (May to July) owing to low night temperatures. There is an upsurge of nesting again in the spring, provided there have been adequate earlier rains. During drought nesting may be suppressed altogether or restricted to small "thunderstorm patches" where local rains fell. Characteristic of the summer climate of the North-West are the tropical cyclones or "willy-willys". These destructive rain-bearing storms originate in the Timor Sea, cross the coast usually between Broome and Onslow and gradually recurve south and south-east. Sometimes they bring unseasonal late summer and autumn rains as far south as the Wheatbelt and the South-West and lead to unusual autumn nesting, more marked in the former than in the latter region. This stimulus to breeding is not, however, so generally effective in producing autumn nesting in the south as it is in the North-West and many birds show only a partial response in breeding display and gonad development which fail to lead on to actual egg-laying or raising of young. Years in which a notable amount of autumn nesting *did* occur in the Wheatbelt and Eastern Goldfields were 1922, 1937, 1953 and 1955.

Studies of the breeding patterns in Western Australian birds show that Carnaby's categories, previously mentioned, are soundly based. A great many Western Australian bird species have divorced themselves from the tyranny, as it were, of rigid, regular breeding seasons (controlled by some constant factor such as increasing or decreasing daylight) and have found it more biologically advantageous to evolve a response to rainfall, or its effects with a more elastic timing of nesting periods. Thus they have become, as it were, opportunistic breeders taking advantage of favourable breeding conditions created by the generally erratic rainfall of most parts of the State. Three of the outstandingly successful opportunistic breeders are the Grey Teal, Budgerygah and Zebra Finch—species superbly adapted to Australian conditions and which have become among Australia's most widespread and successful bird species. The Zebra Finch will produce many successive clutches while conditions remain favourable—in excess of a year without gonad re-organization—and young birds are capable of reproducing at two months of age. Inhibiting factors are cold and extreme damp in winter and dehydration during summer drought. The Grey Teal has so telescoped its epigamic display that birds may produce shelled eggs 10 days after stimulating rains have fallen.

The land and fresh water species which we know are able to breed in response to an adequate rain stimulus, as shown by their capacity to nest indifferently in autumn or spring, are the following: Little Grebe, Little Pied and Little Black Cormorants, White-faced Heron, Black Swan, Mountain Duck, Black Duck, Grey Teal, Blue-winged Shoveler, Maned Goose, Blue-billed Duck, Whistling Eagle, Kestrel, Stubble Quail, Little Quail, Dusky Moorhen, Coot, Australian

Bustard, Banded Plover, Red-capped and Black-fronted Dotterels, White-headed Stilt, Avocet, Australian Dotterel, Peaceful and Diamond Doves, Common Bronzewing, Crested Pigeon, Red-plumed Pigeon, Weero, Smoker Parrot, Mulga Parrot, Bourke Parrot, Budgerygah, Black-eared Cuckoo, Narrow-billed Bronze Cuckoo, Owlet Nightjar, Red-backed Kingfisher, Tree-Martin, White-backed Swallow, Australian Pipit, Black-faced Cuckoo-Shrike, Ground Cuckoo-Shrike, White-winged Triller, Grey-crowned and White-browed Babblers, Cinnamon Quail-Thrush, Crimson, Orange and White-fronted Chats, Blue-and-white Wren, Variegated Wren, Banded Blue Wren, Rufous-crowned Emu-Wren, Whiteface, Banded Whiteface, Slate-backed Thornbill, Broad-tailed Thornbill, Samphire Thornbill, Yellow-rumped Thornbill, Weebill, Redthroat, Field Wren, Brown and Rufous Songlarks, Hooded and Scarlet Robins, Brown Flycatcher, Willy Wagtail, Rufous Whistler, Western Shrike-Thrush, Crested Bell-Bird, Wedgebill, Black-capped Sittella, White-browed and Rufous Tree-creepers, Red-browed and Red-tipped Diamond-birds, Spinebill, Brown Honeyeater, Singing Honeyeater, Yellow-fronted Honeyeater, Purple-gaped Honeyeater, Grey-headed Honey-eater, White-plumed Honeyeater, New Holland Honeyeater, White-cheeked Honeyeater, Yellow-throated Miner, Red and Little Wattle-birds, Spiny-cheeked Honeyeater, Painted Finch, Zebra Finch, Star Finch, Black-faced Wood-Swallow, Little Wood-Swallow, Magpie Lark, Little Crow, Australian Crow, and Spotted Bower-bird. Further observations will augment this list considerably.

There are many species, again, which despite opportunity, have not been observed to depart from a regular nesting season. These include the Emu (egg-laying takes place always in the autumn, throughout the State), most of the hawks (including the Spotted Harrier, Little Eagle, Wedge-tailed Eagle, Black-breasted Buzzard, Brown Hawk, Little Falcon, Grey Falcon), most of the parrots (including the Little Corella, Galah and Twentyeight), Bee-eater, Western Silvereye, the Butcher-birds and the Magpies. One may tentatively suggest that some of these species may still retain an ancestral response to photo-stimulation in the control of their breeding seasons. However, this has, as yet, been experimentally proved for only one species. Dr. J. A. Keast has demonstrated that the Silvereye in south-eastern Australia is photo-periodically controlled in its nesting cycle and this has recently been shown to be so for the Western Silvereye as well.

INTRODUCED SPECIES

There are four introduced species well established in this State—Spotted and Senegal Turtledoves, the Kookaburra and the Goldfinch. They are included in this book, together with the Domestic Pigeon, White Swan and Red-browed Finch which, though less extensively distributed, maintain themselves by local breeding. The House Sparrow is also included, though it is no longer a resident in the

State. There are three introductions which are not treated more particularly here as they can hardly be considered to have become "wild" birds. These are the Peafowl *(Pavo cristatus)* and Pheasant *(Phasianus colchicus),* which have been set free on Rottnest Island, and the Mallard *(Anas platyrhyncha)* which lives free in some of the metropolitan parks (Queen's Gardens, Hyde Park). It shows no indication of spreading or hybridising freely with the native Black Duck. The Indian or Ceylon Crow *(Corvus splendens)* is not infrequently transported to Fremantle in ships, but though individuals may remain at large for some time they have not succeeded in obtaining a permanent footing. The Starling *(Sturnus vulgaris)* has occasionally attempted to colonise from South Australia and has been included.

SUMMARY.

This book deals with 400 species of birds which may be grouped in the following categories: breeding land and fresh water species, 265; breeding sea-birds, 24; visiting land and fresh water species which do not breed south of the Kimberley Division in Western Australia, 57 (the 33 species of Asiatic wading birds are included here); visiting sea-birds which do not breed locally. 45; introduced species, 9; In the Kimberley Division there occur an additional 76 species. Thus the political division of Western Australia contains 467 naturally occurring and 9 introduced species. Australia as a whole contains 715 naturally occurring species, but it should be remembered that different authorities have varying ideas as to species limits, whether a particular form is to be regarded as a full species or only a sub-species.

ACKNOWLEDGMENTS

Though the information given is based on our own observations we have drawn extensively on the writings of ornithologists who have worked in the State, together with numerous unpublished field notes generously made available to us by our colleagues. We acknowledge, in particular, aid received from Mrs. Rosemary Balmford, Mrs. Rica Erickson, Mrs. Pauline Reilly, Mrs. Minilya Sharpe, Dr. Mary E. Gillham, Misses C. A. Nicholls and Olive Seymour, Drs. A. A. Burbidge, S. J. J. F. Davies, G. M. Dunnet, K. Immelmann, J. A. Keast, Ernst Mayr, G. F. Mees, Graeme Smith, Robert Stein, and G. M. Storr, Messrs. R. A. Aitken, D. G. Bathgate, M. G. Brooker, K. G. Buller, T. E. Bush, W. H. Butler, J. H. Calaby, I. C. Carnaby, G. S. Chapman, A. H. Chisholm, H. T. Condon, P. Congreve, A. M. Douglas, M. Ellis, J. A. Estbergs, J. R. Ford, P. J. Fuller, R. Garstone, L. Glauert, K. A. Hindwood, T. Jasper, C. F. H. Jenkins, A. Kalnins, E. Le Souef, E. Lindgren, W. H. Loaring, G. A. Lodge, J. Long, A. G. Mathews, Bro. Matthew Amedy, E. McCrum, V. McDougall, A. R. McEvey, R. C. Mitchell, T. K.

Moriarty, C. L. E. Orton, S. A. Parker, D. H. Perry, H. Tarlton Phillipps, A. H. Robinson, F. N. Robinson, P. T. Sandland, D. A. Saunders, E. H. Sedgwick, L. E. Sedgwick, V. N. Serventy, P. Slater, T. Spence, N. E. Stewart, R. H. Stranger, M. Tarburton, A. Thorpe, A. R. Tomlinson, H. O. Webster, S. R. White, F. Lawson Whitlock, J. Warham, H. M. Wilson and Rev. A. R. Wilbrey. We are indebted also to the leaders of the Harold Hall Expedition of the British Museum (Natural History), Mr. J. D. Macdonald and Mrs. B. P. Hall, and Messrs. G. Cowles and D. Goodwin of their staff. Dr. Ernst Mayr has been consulted on general taxonomic and nomenclatural problems, Dr. Finn Salomonsen on the nomenclature and classification of the honeyeaters, and Dr. H. J. Frith on the duck family.

For the illustrations we are indebted to Miss Olive Seymour for Plates 2, 3, 4, 5, 6, 7 and 9 and Figs. 10, 11, 24, 28, 30, 31, 32, 35, 39, 40, 41, 42, 46, 47, 48, 49 and 55; Miss C. A. Nicholls for Figs. 34 and 35; Mr. Hargreaves O. Webster for Plates 1 and 8 and Figs. 12 to 21, 27, 43, 44, 50, 55, 56 and 57; Mr. Peter Slater for Figs. 29, 59, 60, 61 and 62; Mr. G. E. Binsted for Figs. 22, 23 and 45; Dr. G. M. Dunnet for Fig. 38, and Mr. Arthur Thorpe for Figs. 25, 26, 51, 52, 53, 54, 63 and 64.

Mr. P. Stanley, Lands and Surveys Department, supervised the final preparation of the maps for the original edition.

SECTION I

THE HISTORY OF WESTERN AUSTRALIAN ORNITHOLOGY

In studying the ornithology of a country it is always of interest to learn the circumstances in which the information which we now possess came to be obtained, to know something about those who provided it, and of the incidents which led to their being there. Here something will be said of those circumstances, those persons and those incidents, which, in historical review, may be grouped into four fairly well-defined periods.

The first was the era of the early voyagers, from the first discovery of the western coast of Australia to the close of the seventeenth century. The earliest voyagers were all Dutch, none of them professed naturalists and their ornithological records were casual anecdotes. However, at the end of the century the first Englishman to reach the western coast, William Dampier, left us more substantial information. Following him there was nearly a century when almost nothing was written about our birds. Then came the period (1791-1826) of the great navigators, French and British; their cruises were not trading ventures as had been the voyages of the earlier visitors, but were properly organised exploring expeditions, and staffed by men who were trained in the functions of their appointments. Naturalists were invariably included in the personnel and sometimes they were also the medical officers of the expeditions. It was a period, consequently, when more tangible results were obtained, and the real foundations of modern knowledge were laid down. Proper collections of the modern type were made, and the specimens so preserved were, when they arrived in Europe, described on scientific (Linnean) principles.

These great voyages of discovery came to an end by the time of the founding of the Swan River Colony in 1829, and with a population settled, albeit sparsely, on the land, the opportunity was now afforded for the study of the land-bird fauna; hitherto all observations had been limited to the coastline. These resident observers were, however, relatively untrained persons but with a leaning towards natural history. On the whole their contributions to ornithology were spasmodic and had in most cases little scientific merit. However, the period was notable for the two visits to the colony of John Gilbert, whose work in Western Australia was done so methodically and enthusiastically, and to such good effect, that his achievements constitute the basis of our present-day knowledge of the bird-life of the southern part of the State. Therefore the period from the foundation of the colony to the middle of the century may justifiably be termed the period of John Gilbert.

After Gilbert's final departure from the colony in 1843 there was a long period during which little was added to knowledge of the birds of Western Australia. From the time of his leaving until the 'eighties, practically the only contributions to ornithology resulted from the Austin Exploring Expedition of 1854, Francis Thomas Gregory's two expeditions of 1858 and 1861, and the two collecting visits, in 1863 and 1868, of George Masters of the Australian Museum, Sydney, who collected in the vicinity of King George's Sound.

But in the 'eighties began a revival of interest in local bird study, an interest which has continued without interruption to the present day. During these last ninety years much investigation has been made into the bird-life of the State, not only by competent resident ornithologists, but in field-work undertaken by experienced naturalists who have visited Western Australia specifically for the purpose of studying its birds. This may be termed "The Modern Period."

I—THE PERIOD OF THE EARLY TRAVELLERS (1618-1700)

As in the case of the first geographical and other records of the west coast, the first mention of Western Australian birds is contained in letters written by commanders of ships of the Dutch East India Company. The first actual reference to birds is in a letter of **Haevick Claeszoon van Hillegom** of June 24, 1618, in which he describes the voyage of the ship *Seewolf* from South Africa to Java. Van Hillegom wrote that on May 5, 1618, he was in Latitude 28° 26′ South, namely, a little north of Geraldton, where he saw numbers of birds, many of which seemed to be land-birds, such as a white tern and a few ternlets. On May 11 he was off North-West Cape and observed land-birds up to the 16th degree.

Nine years later Supercargo **J. van Roosenbergh** wrote to the directors of the Dutch East India Company an account of the voyage of the *Het Wapen van Hoorn* to Java, in which letter he mentioned that on September 7, 1627, they sighted "the land of d'Eendracht, near Dirck Hartoch's reede" and saw a black bird with a white tail, having white streaks here and there under the wings, "a bird it seems of rare occurrence." Three or four days previously they had seen a number of sandpipers.

These early voyagers had, however, been merely passers-by while on uninterrupted voyages, but on June 4, 1629, **Francisco Pelsaert** was wrecked on the Abrolhos Islands and in the book describing his tragic misadventures is the first positive reference to a Western Australian land-bird; an observation of a number of "grey turtle doves" on the Wallaby Islands. These were doubtless the Brush Bronzewing Pigeons which still occur there and so are of considerable historical interest to naturalists. As he apparently did not visit the Southern and Easter Groups (according to the researches of Mrs. Henrietta Drake-Brockman) it is not strange that he omitted any reference to the great nesting colonies of terns and other sea-birds which make the southern Abrolhos islands so noteworthy to ornithologists. The first reference to these had to wait almost a century.

The next mention of birds is that by **Wollebrand Geleynzoon de Jongh** of the ship *Amsterdam* who, in 1635, when sailing past Dirk Hartog Island, recorded that on May 25 he saw a "small saturn-gull, and not above 6 or 7 other gulls." In the following year, on July 5, **Antonie Caen,** skipper of the ship *Banda,* when near Bernier Island in the same general area, described seeing two birds which were almost certainly Black Swans—the first time these birds were sighted by Europeans. On April 28, 1656, the ship *De Vergulde Draeck* was wrecked on a reef a little to the south of Jurien Bay and some of the survivors managed to keep alive by eating gulls and seal meat. The crew of the *Emeloort,* when searching for the wrecked vessel, saw, on March 15, 1658, many "gulls," entirely black but of small size—probably Wedge-tailed Shearwaters. On March 17 they saw several "wagtails."

Willem de Vlamingh's *Journal Wegens een Voyagie na het onbekende Zuid-lande en wyders na Batavia,* provides much more substantial information about the bird-life of the new land than do the usual captains' letters. From this publication of 1701 we learn that on December 29, 1696, his ship the *Geelvinck* anchored near the mouth of the Swan River and early in 1697 Vlaming landed first on Rottnest Island and then on the beach near Cottesloe. Four Black Swans were caught and two of them taken alive to Batavia. So this bird holds a premier place in the annals of Western Australia; it gave its name to the river on which our capital stands, and it furnishes the emblem of the State. The discovery of such a singular bird as a *black* swan aroused much interest throughout Europe and the Royal Society of London published the news in the form of a letter in its *Philosophical Transactions* in 1698, three years before Vlaming's own *Journal* appeared in print. The *Journal* further recorded that plenty of parrots, some cormorants, geese and divers were seen and a bird heard singing which was referred to as a "Nachtegaal" and was, perhaps, the Reed-Warbler, though it is quite conceivable that in those days the Noisy Scrub-bird may have lived on the banks of the Swan River. Pelicans and cockatoos (probably Long-billed Corellas) are also mentioned, and when Vlaming next landed, this time at a bay which was probably Jurien Bay, the footprint of an Emu or, as he called it, the Kasuarius, was observed.

Despite their historical interest these are, at best, incidental references to birds recorded casually by the early Dutch voyagers and it was left to an Englishman to furnish us with more satisfying information of the bird-life of the Western Australian coast in the seventeenth century. This man, **William Dampier** (1652-1715), not only recorded his observations in more detail than had the early Dutch merchant captains but he gave us also, when he published in 1703 the classic *Voyage to New Holland,* the first pictures of our birds.

In January 1688 the buccaneering ship *Cygnet* anchored in a bay on the coast of the Kimberley Division, now known as Cygnet

Bay, but the only mention of birds by Dampier who was a member of the crew, is that there were a few small land-birds, but none bigger than a Blackbird, and but few "sea-fowls." However, in his account of his second visit to New Holland, in the ship *Roebuck* in 1699, Dampier not only wrote extensively of the birds but also enhanced his book with six illustrations of them. On July 30, when approaching Houtman's Abrolhos, he saw what were doubtless Caspian Terns: "a Sort of Fowls the like of which we had not seen in the whole Voyage all the other Fowls having now left us. These were as big as Lapwings; of a grey Colour, black about their Eyes, with red sharp Bills, long Wings, their Tails long and forked like Swallows; and they flew flapping their Wings like Lapwings." On August 1, "as we were standing in we saw several large Sea-fowls, like our Gannets on the Coast of England, flying 3 or 4 together," and, what were probably Crested Terns, "A Sort of white Sea-Mews, but black about the Eyes, and with forked Tails."

Landing on the shores of Shark Bay on August 6, "There were but few Land-Fowls; we saw none but Eagles, of the larger Sorts of Birds; but 5 or 6 Sorts of small Birds. The biggest Sort of these were not bigger than Larks; some no bigger than Wrens, all singing with great Variety of fine shrill Notes; and we saw some of their Nests with young Ones in them. The Water-Fowls are Ducks (which had young Ones now, this being the Beginning of the Spring in these Parts); Curlews, Galdens, Crab-catchers, Cormorants, Gulls, Pelicans; and some Water-Fowl, such as I have not seen any where besides. I have given the Pictures of 4 several Birds of this Coast." The sketches referred to constitute the first illustrations ever published, not only of birds from this State, but from anywhere in Australia and, although crude, are easily identifiable. One is of the Avocet and a note added to the sketch states "The head & greatest part of ye neck of this bird is red & herein differs from the Avosetta of Italy." Another is a sketch of the Pied Oyster-catcher, or as Dampier terms it "the Crab-catcher," with the caption, "The Bill and Leggs of this Bird are of a Bright Red." There are also sketches of the Bridled Tern (referred to as "A Noddy of N. Holland") and of the Common Noddy. The "Galden" is not, however, illustrated and some years ago considerable discussion took place as to its identity. Apparently Dampier used the name for a species of Heron, perhaps the Reef-Heron, or alternatively the rarer Mangrove-Heron, which is still to be seen in the small tracts of mangroves fringing portion of the shores of Shark Bay.

On August 22, Dampier landed on one of the group of islands now known as the Dampier Archipelago and he named the particular one "Rosemary Island" because "There grow here 2 or 3 Sorts of Shrubs, one just like Rosemary. . . . It grew in great Plenty here, but had no smell." It has been pointed out, however, that the actual island on which Dampier landed was probably the one now known as Lewis Island, adjacent to the Rosemary Island of the present-day charts. Of the bird-life he noted "some Cormorants, Gulls, Crab-catchers, &c, a few small Land-Birds, and a sort of White Parrots, which flew a great many together." The white "Parrots" were Little Corellas, which still frequent the islands off the coast in this vicinity. On August 28 he "saw also some Boobies, and Noddy-birds; and in the Night caught one of these last. It was of another Shape and Colour than any I had seen before. It had a small long Bill, as all of them have, flat Feet like Ducks Feet; its Tail forked like a Swallow, but longer and broader, and the Fork deeper than that of the Swallow, with very long Wings; the Top or Crown of the Head of this Noddy was Coal-black, having also small black Streaks round about and close to the Eyes; and round these Streaks on each Side, a pretty broad white Circle. The Breast, Belly, and under part of the Wings of this Noddy were white; and the Back and upper part of its Wings of a faint black or smoak Colour. See a Picture of this and of the Common one." Dampier is referring to the Bridled Tern. He continues: "The Land-fowls that we saw here were Crows (just such as ours in England), small Hawks, and Kites; a few of each sort; But here are plenty of small Turtle-Doves, that are plump, fat and very good Meat. Here are 2 or 3 sorts of smaller Birds, some as big as Larks, some less; but not many of either sort. The Sea-Fowl are Pelicans, Boobies, Noddies, Curlews, Sea-pies, &c. and but few of these neither."

The first knowledge of the interesting sea-bird fauna of the Abrolhos Islands was gained by the shipwrecked people on the Dutch East Indiaman *Zeewyk* which grounded on the reef off Pelsart Island in the Southern Group in June 1727. They lived on Gun Island, in the shelter of the reef, until March the following year, when they sailed for Batavia in a vessel they built from salvaged materials. In the journal of the second mate, **Adriaen van der Graeff** it was stated that birds first appeared on the night of August 20 "which we have never seen by day, flying close to the ground and at about a man's height above ground and in every direction, like bats do in our country." These were Wedge-tailed Shearwaters and many were caught subsequently each night for food. Well over a thousand birds must have been taken according to the records until the birds ceased to be conspicuous after late October. The crew landed on Pelsart Island also, but no sea-birds other than the shearwaters are mentioned in the journal.

These early voyagers made no collections, or, at least, no specimens of their collecting survive except some botanical material brought back by Dampier and which presumably had been pressed in his journals and thus preserved. These specimens are now in the University Herbarium, Oxford. The art of preserving bird specimens was not then developed and it was not until the close of the eighteenth century that Western Australian birds began to be seriously investigated on scientific lines. But in the interim, for about one hundred years nothing new was recorded about our birds, either by ordinary travellers or by naturalists.

II—THE PERIOD OF THE GREAT NAVIGATING EXPEDITIONS (1791-1826)

In 1791 the British Government sent an expedition to Nootka Sound in British Columbia in the vessels *Discovery* and *Chatham*. **George Vancouver** (c. 1758-1798) was in command and **Archibald Menzies** (1754-1842) was naturalist. The instructions issued to Menzies by the Admiralty required him to investigate the whole of the natural history of the countries visited, paying attention to the nature of the soil, to the prospects of growing European grains, fruits, etc.; he was to dry specimens of all native trees, shrubs, plants, grasses and ferns that he should think worthy of being brought home, and he was to bring home either seeds or living plants of as many kinds as possible; he was to search for ores, metals and minerals, and give attention to beasts, birds and fishes; inquiries were to be made into the manners, customs, languages and religion of the natives. He was to keep a regular journal of all occurrences, which, together with all specimens collected, was to be delivered to the Admiralty on his return.

This journal, which is in the British Museum, shows Menzies to be an important figure in ornithology; in Western Australia he was the first to collect the Musk Duck, the Sooty Oyster-catcher, the Boobook Owl and the King and Rosella Parrots, and in California he was the collector of the type specimens of the California Condor and the California Quail. A perusal of his journal discloses that not only was he an accomplished botanist, in which branch of natural history he obtained considerable celebrity during his life, but that he was also a competent zoologist.

During the period September 28 to October 11, 1791, the expedition was at King George's Sound, for the discovery of which the expedition was responsible, Menzies was indefatigable in his exploration of the country and he made many interesting records in his journal. He noted that "The number of Birds which we saw about the Sound was likewise very inconsiderable, besides they were so wild & shy that we had opportunity to examine but few of them. Of land birds we saw eagles, hawks, owls, parrots, parrakeets, Quails & two or three other small birds. The Water birds consisted of Swans, Ducks, Penguins, Gulls, Terns, Plovers, Oyster-catchers & Sandpipers." Vancouver in 1798 published an account of the voyage in three volumes and in that work he made somewhat similar remarks regarding the bird-life encountered at King George's Sound, "Of the birds that live in or resort to the woods, the vultures may be said to be the most common, as we saw several of this species, or at least birds that were so considered." This is an error for the Wedge-tailed Eagle, as noted correctly by Menzies, but to those who have seen vultures similarly soaring at high altitudes, the comparison is not inapt. Vancouver continued: "Hawks of the falcon tribe with several others of that genus; a bird much resembling the English crow, parrots, parroquets, and a variety of small birds, some of which sing very melodiously, were those which attracted our attention the most; but all were so excessively wild and watchful that few specimens could be procured. Of the water-fowl, the black swan seemed as numerous as any other species of aquatic birds in the neighbourhood of Oyster Harbour, but they were seen in no other place. There were also black and white pelicans of a large sort, seen at a distance; and though ducks were in great numbers, we were very unsuccessful in taking them."

As already mentioned Menzies collected a specimen of the Boobook Owl and, in his journal, correctly classed it as the same species which J. R. Forster had obtained in New Zealand whilst on the *Resolution* with Captain Cook. Specimens of the King Parrot and Western Rosella were also obtained, and Menzies' remarks on the two Oyster-catchers are interesting: "On the sea shores & insulated rocks about the Sound were frequently seen what I considered as two distinct species of the genus *Haematopus* or Oyster-catcher. One of them is I believe common to many parts of the world in either of the temperate zones especially in the Southern, though not yet noticed as far as I know in general arrangements—the other I have seen nowhere but here and it appears to differ so much from that described in the *British Zoology* [the well-known work of Thomas Pennant (1726-1798) of which Menzies probably had a copy on board] that I have ventured to make it a new species, whether in this I am right or not Ornithologists must hereafter determine." He then gave a scientific description of the Sooty Oyster-catcher to which he applied the name *Haematopus fuliginosus,* and the Pied Oyster-catcher, to which he gave the name *Haematopus Novae Hollandiae.* Regarding the last-named he mentioned in the journal, "I conceive this to be the Bird Dampier figures in his *Voyage,* vol. 3, tab. p. 85, f. 4 but the Bill in his figure is too thick & not sufficiently streight, nor is the whiteness of the rump & bottom of the Tail expressed, in other respects the representation agrees pretty well."

Another bird which this visit of Menzies to King George's Sound was to bring to scientific knowledge was the Musk Duck. On October 1, Menzies recorded in his journal that he "came in sight of some small lakes on which we saw some curious ducks. I shot one of a dark grey colour with a wattle pendulous from the under side of its bill which induced me to name it *Anas caruncu-lata.*" Vancouver also mentioned this discovery: "A very peculiar one was shot, of a darkish-grey colour, with a bag like that of a lizard hanging under its throat, which smelt so intolerably of musk that it scented nearly the whole ship." Menzies' name for the bird was, however, never published and, when in 1796 the specimen was described by George Shaw, in *The Naturalist's Miscellany,* it received the name *Anas lobata.*

Further notes by Vancouver were: "There were many grey curlews and sea-pies; of the latter we procured a few which were excellent eating. The aquatic birds before enumerated, with shags, the common gull, two or three sorts of tern, and a few small penguins of a bluish colour, included the whole of the feathered tribe in the vicinity of the shores." It is interesting to note that two localities in the Sound, Quail Cove and Curlew Island, were named after birds. The ships sailed out of the Sound and towards the east on October 11 and the next day Menzies recorded in his journal: "I shot a kind of Petrel which flew about the vessels and which I had not before examined. It was the *Procellaria melanopus* of Linnaeus"—doubtless the Great-winged Petrel. Outside Esperance, Vancouver noticed more coastal and oceanic birds than had been seen on any other part of the coast; recording also that, in addition to gannets and two or three different sorts of terns, albatrosses and petrels, particularly the black and sooty, were in great abundance.

While Vancouver was in King George's Sound a French expedition consisting of the *Recherche* and the *Esperance* under the command of Admiral Bruny d'Entrecasteaux, left France for the purpose of ascertaining the fate of the expedition under La Perouse which had not been heard of since it left Botany Bay in 1788. In 1792 the French ships, on which were **Jacques-Julien Houton de Labillardiere** (1755-1834), the botanist, and **Claude Antoine Gaspard Riche** (1762-1797), the zoologist, sighted the coast of Western Australia at a point which was named Point D'Entrecasteaux, not far from Chatham Island. Several places on the south coast were charted and named, including the Archipelago of the Recherche. Labillardiere published in 1800 an account of the voyage, and in 1808 the official report was issued by **E. P. E. de Rossel** (1765-1829), one of the naval officers with the expedition. The collection of natural history specimens which had been made was confiscated by the Dutch when the expedition had broken up at Java in 1794, and handed over to the British. It was, however, later restored to the French and found its way to the Museum of Natural History in Paris and the specimens were described by French zoologists in various French scientific journals.

The vessels rounded the Leeuwin on December 9, 1792, and remained for a short period—only a few days—in Esperance Bay, so there are not many references to Western Australian birds in the records. However, Labillardiere noted that, when near the Leeuwin, "tropical birds" with red streaks and some albatrosses of a darker colour than those commonly seen at the Cape of Good Hope, flew around the ships. The first named could only have been the Red-tailed Tropic-bird, while the albatrosses mentioned were possibly one of the Sooty species, or even the Giant Petrel. Bad weather, no doubt, accounted for the sparseness of the records.

On December 12, 1792, Labillardiere landed on a small island, named Observatory Island, in the Archipelago of the Recherche, where there was a numerous flock of "swans" several of which allowed themselves to be taken by hand. "This new species is somewhat smaller than our wild swan, and of an ash-coloured grey, a little lighter on the belly. The bill is blackish, with a tumour of a sulphur-yellow at its base. The legs are slightly tinged with red." Labillardiere is referring to the Cape Barren Goose which was given the name *Cereopsis novae-hollandiae* by John Latham in 1801, from specimens the origin of which is not known, and this name was the first published. However, it is of interest to note that this was the first record of the species for Western Australia. In the official account of the voyage by de Rossel there is mention that the zoologist, Riche, believing it to be new to science, proposed to give the bird the scientific name of *Anas Terrae Leeuwin,* while in 1818, after the zoological collections at last reached France, Vieillot, also believing it to be new, named it *Anser griseus.*

On another small island, Labillardiere "saw two seagulls, a male and female, of the species called by Buffon, *bourgmestre,* and by Linnaeus, *larus fuscus,* perched upon the heights at a small distance from us." These two Pacific Gulls were unfortunate, as Labillardiere continues: "The female having been shot, the male, frightened by the noise of the explosion, took flight; but presently returned, and being determined not to abandon his mate was killed by her side." On December 15 he landed on yet another small island and stayed the night. "Some birds came to our retreat to quench their thirst. There killed the charming yellow turtle dove, remarkable for six or eight golden feathers towards the bases of its wings." This was the Brush Bronzewing, still abundant on the islands. He notes that he had previously found it at Recherche Bay in Tasmania. On the island there were many Little Penguins "concealed in very deep holes in the rocks from which it was frequently difficult to expel them. . . . The cold was sufficiently keen to induce us to kindle a fire, and as we had not much provisions, after I had chosen among the birds I had shot those which I wished to preserve for my collection, I gave the rest to my companions to be broiled." In the official account there is an extract from the journal of the commander Bruny d'Entrecasteaux, who wrote: "Penguins occur in great numbers on all the islands, I did not eat them, but our crew lived on them during our stay at this anchorage, and found them better than seals."

Riche, who had a longer spell on shore than the others, owing to having wandered away from a party which landed on the mainland, and not being found for a couple of days, saw emus, "cazoards" and a large white parrot with a red crest—probably the Major Mitchell Cockatoo—and, on the shore, several waders, including the Oyster-catcher. Labillardiere, while searching for his companion scientist, did not meet with much bird-life on the mainland and wrote that the region "affords but little food for birds; and accordingly I found none in this excursion except two species, a *muscicapa,* which I afterwards met with in the Moluccas, and a fine species of red-crested cockatoos, *psittacus moluccensis,* which are met with in the same islands, in flocks of many hundreds. When I attempted to approach them, they always removed to a greater distance, flying rapidly, with sudden starts, and emitting loud and very disagreeable shrieks."

In October 1800 a further French expedition under the command of **Nicolas Baudin** (1750-1803), left France with the object of exploring more fully the coast of New Holland. Baudin was captain of the corvette *Geographe,* on which the naturalists included **Francois Peron** (1775-1810), zoologist, **Stanislas Levillain** (?-1801), zoologist, **Leschenault de la Tour,** botanist and **Charles-Alexandre Lesueur** (1778-1846), natural history painter. The corvette *Naturaliste* was commanded by Emmanuel Hamelin, and had also a complement of naturalists. On the return of the expedition to France, Peron commenced the publication of the official account of the voyage, but owing to his death, in 1810, the work was completed by Louis Claude de Saulces de Freycinet (1779-1842), lieutenant on the *Naturaliste,* who commanded the schooner *Casuarina,* purchased by the expedition at Sydney.

The two vessels reached the south-west coast in May 1801, and Cape Naturaliste and Geographe Bay were named. Landing where Busselton now stands, Peron saw several flocks of Black Swans in the vicinity of the Vasse River, which then received its name. Unfortunately that is the only reference to the birds of the Busselton district in the portion of the account written by Peron. A storm drove the vessels out of Geographe Bay and they became separated, the *Geographe* going on to Shark Bay, and the *Naturaliste* to Rottnest Island.

Levillain landed on Garden Island, where "in the scrub we saw several birds of prey of the buzzard family and some very large crows. Along the shore there were several types of water birds, such as oyster-catchers with black bills and red feet and eyes, and black-and-white grebe, which continually strutted around the rocks" (namely the Pied Cormorant, still conspicuous there). After exploring a portion of the island, which they named "Ile aux Ours" from the number of seals (or sea-bears) they found on it, they sailed for the mainland and were driven on shore. Here Levillain saw a large number of red-bellied green parrots (possibly the King Parrot) and "a large black vulture-like bird of prey" the size of a turkey (the Wedge-tailed Eagle) besides many other small song-birds. He mentioned in his report that he found an enormous bird's nest in a fig tree. "This nest was made of roots and sticks and was some 6 to 7 feet from the ground. What bird's could it have been, I wonder." It was, apparently, the White-breasted Sea-eagle, but which, however, no longer nests in the district. Finally he recorded that he brought back with him to the ship some birds, "such as parrots, wagtails, etc., etc."

At the entrance to the Swan River, Freycinet saw "a prodigious multitude" of Pelicans, but could procure only one, and he recorded that great flocks of land birds, especially elegant parrots, flying among the trees, enlivened, by their presence, those unknown, wild, and desolate regions. These would be the Twenty-eight Parrots, still colourful denizens of the Perth bushland. Rowing into Perth Water, the party met with more flocks of Pelicans and shot two. Above Heirisson Island they came upon Black Swans for the first time and killed several. It would appear that then, as at the present day, Black Swans were more plentiful in the shallows above Heirisson Island than in the lower reaches of the Swan Estuary. Freycinet noticed that shortly after death the beak lost its beautiful red colour and became black. On Giraud Reef, near Carnac Island, a great number of sea-birds was seen, and on Garden Island he observed "partridges," which were probably Bronzewing Pigeons, and crows smaller than those of Europe, "but of a delicious taste."

Peron landed from the *Geographe* on Bernier Island, but was disappointed not to find more animal life. "If we except certain useless and disagreeable kinds of which we shall not stop to speak, all animals are rare on the sterile soil with which we are dealing; the class of birds, for example, is only represented by melancholy cormorants, by several species of noddies, petrels, gulls, sea-eagles and oyster-catchers, which, far from man and his works, multiply on these arid rocks. The division of land-birds is only represented by flycatchers and shrikes. We found, however, a beautiful species of tit with a blue breast, which deserves more particular mention." The only other place where Peron landed before the *Geographe* went on to Timor, was on Depuch Island, but all he could say on the bird-life was that "Birds are reduced to some species of flycatchers and shore-birds."

Meanwhile Freycinet had come to Shark Bay in the *Naturaliste*. Passing a night on a small, isolated and sterile island in Freycinet Estuary, he found "a prodigious number" of different sea-birds which, as soon as he landed, flew around screaming for a long time. His party killed several and collected some fresh eggs but "neither appeared to us good."

Leaving Timor again on November 13, the two vessels sailed south well out into the Indian Ocean, so their records are limited to birds observed at sea. Thus, on December 11, Peron recorded seeing a Cape Petrel in Latitude 20° South, and also a Tropic-bird and commented that therefore he saw together at the same place two birds, one of which, inhabiting exclusively antarctic seas, was at home amidst cold, fog and storms, whilst the other, the Tropic-bird, loved the calm of the tropics and their high temperature. Cape Petrels were observed all along the Western Australian coast and even in Geographe Bay. Peron further noted that he saw "Grey Petrels" (*Procellaria grisea* Linnaeus) for several days at the end of December between Latitude 32° and 33° South. The "great Equinoctial Petrel (*Procellaria equinoxialis* Linnaeus)" was recorded as having been within sight from Latitude 21° South to immediately south of King George's Sound, when it disappeared, while a little further on to the east, south of Bremer Bay, they saw for the first time a bird which Peron recorded as the "Chocolate-coloured Albatross (*Diomedea spadicea*)." This was most likely the Sooty Albatross, *Phoebetria fusca,* which had been named *spadicea* by the French ornithologist Lesson. They had met the common Albatross, *Diomedea exulans,* Linnaeus, east of Cape Leeuwin—two varieties, white and brown, which Peron considered to represent two species. Great Skuas were noted on January 9—an unusual record for this time of the year—together with several species of terns. Of these he distinguished three species—one he called *Sterna obscura* Linnaeus; to the second, which he did not recognise and considered to be a new species, he gave the name *Sterna melanosoma* on account of the black colour of the body; the third was also considered to be new and, being like a Caspian Tern, was named *Sterna caspioides.*

In December 1801, **Matthew Flinders** (1774-1814), in command of the *Investigator,* reached Cape Leeuwin. With him was the well-known botanist, **Robert Brown** (1773-1858), and the artist, **Ferdinand Bauer** (1760-1826). The *Investigator* remained three weeks at King George's Sound and Brown collected some birds. Birds collected on the voyage were divided between the Linnean Society of London and the British Museum and there is in the British Museum a list by Brown of the specimens of birds obtained on the voyage.

Of the specimens that reached the Linnean Society, the parrots were reported on by Heinrich Kuhl in 1820 and the parrots and pigeons by C. J. Temminck in 1821, while in 1827 N. A. Vigors and T. Horsfield included all species in a paper they wrote on the specimens of Australian birds belonging to the society. Those specimens in the British Museum were never written up and they have since disappeared. However, the specimens which reached the Linnean Society were later transferred to the British Museum where at least some of them are still to be found. In the early days, specimens of birds were usually mounted for exhibition and so deteriorated rapidly.

Among the birds collected by Brown was the type-specimen of the Brown Honeyeater, obtained at King George's Sound. It is unfortunate that Brown himself has left few notes on the birds he observed during the time he was in Australia and that Flinders made only scanty references to birds in his account of the voyage. He noted that he observed parroquets and Black Swans at King George's Sound and that "amongst the animal productions of the district the kangaroo and cassowary hold the first ranks . . . the cassowary showed nothing distinguishable at a distance from the same animal at that place [Port Jackson]; both were shy, as were the ducks, swans, and all birds." Near Point Possession "were found two nests of extraordinary magnitude. They were built upon the ground, from which they rose about two feet, and were of vast circumference and great interior capacity, the branches of trees and other matter of which each nest was composed, being enough to fill a small cart. If the magnitude of the constructor be proportionate to the size of the nest, Terra Australis must be inhabited by a species of bird little inferior to the condor of the Andes."

At Lucky Bay, near Esperance, "upon a rock on the side of the hill I found a large nest, very similar to those seen at King George's Sound. There were in it several masses resembling those which contain the hair and bones of mice and are disgorged by the owls in England after the flesh is digested. These masses were larger, and consisted of the hair of seals and of land animals, of the scaly feathers of penguins, and the bones of birds and small quadrupeds. Possibly the constructor of the nest might be an enormous owl; and, if so, the cause of the bird being never seen, while the nests were not scarce, would be from its not going out until dark; but from the very open and exposed situations in which the nests were found, I should rather judge it to be of the eagle kind, and that its powers are such as to render it heedless of any attempts from the natives upon its young.

"Geese and ducks were found here [Lucky Bay] and, not being very shy, some of them were killed by the shore parties. The goose was also found upon the islands; and is the same bird as I found [spoken of in the introduction, page cxxxv, as resembling the bernacle goose] frequenting Furneaux's Island in Bass Strait. This goose is described by M. Labillardiere . . . as a new species of swan." "Some of the little blue penguins, like those of Bass Strait, harboured under the bushes on Goose Island; and amongst the grass and upon the shore were a number of the bernacle geese, of which we killed nine, mostly with sticks."

On January 16, 1818, Lieutenant (later Admiral) **Phillip Parker King** (1791-1856) arrived at the Archipelago of the Recherche in H.M.S. *Mermaid* on a mission to complete the survey of the coasts of Australia which had been begun by Flinders. With him was **Allen Cunningham** (1791-1839), the botanist. King published in 1827 a general account of all the surveying trips in which he had been engaged on the coasts of Australia, but most of the Western Australian section of the survey was carried out off the coasts of what is now the Kimberley Division and there are not many references to the birds of the portion of the State with which we are here concerned. Further, the vessel was too small to permit of natural history specimens being collected. King's references to birds at the Archipelago of the Recherche are limited to a remark that on Goose Island the Cape Barren Goose appeared to be abundant. No landing, however, was made there, but on January 20 a landing was made on Seal Island in King George's Sound, where the party took some penguins and noted Cape Barren Geese and gulls on the island.

Regarding the bird-life of the vicinity of the Sound, King noted that the shoals at the mouth of the Kalgan River, which were very extensive, were covered with large flights of water-fowl; among which curlews and teals were abundant, and again: "Excepting the sea-fowl which consisted of geese, wild ducks, teals, curlews, divers, sea-pies, gulls, and terns, very few birds were seen, and those chiefly of the parrot and cockatoo tribe; a species of the latter was noticed of a rich black plumage, and very like the black cockatoo of New South Wales. . . . A little without the east entrance of the harbour we saw one of those prodigious large nests which Captain Flinders observed near Point Possession; it was built on the summit of an almost inaccessible rock, exposed to the S.W. winds; it measured four feet in diameter at the top, and nearly seven feet at the base; it appeared to have been deserted for some time, as the branches and seaweed, with which it was made, were strewed about the rock. Captain Flinders thought it probable that the inhabitant was an eagle; but on our subsequent visit to King George's Sound in 1821, we saw the same nest ocupied by a hawk of moderate size." This would evidently have been the Osprey.

After leaving King George's Sound the *Mermaid* proceeded to the area of North-West Cape, Depuch Island, Exmouth Gulf, the Dampier Archipelago and Nickol Bay. King put little on record regarding the bird-life of that part of Western Australia, but it is interesting to read that Pelicans, and particularly Curlews were numerous on the coast where Onslow now stands, and that King named a small river there, one of the outlets of the Ashburton, the Curlew River. Some shoals between Legendre and Gidley Islands were covered with immense flights of Pelican and other water-fowl, and on Lewis Island there was a large Osprey's nest. The only other note was that not a single Emu was seen, other than a footprint of one. The few birds collected during these survey voyages went to the Linnean Society of London.

A French expedition in the ship *Uranie* under the command of Louis de Freycinet, who had previously visited Western Australia in the *Naturaliste,* anchored in Shark Bay from September 12 to 27, 1818. Two famous French naturalists were on this expedition, **Jean Rene Constant Quoy** (1790-1869) and **Joseph Paul Gaimard** (1793-1835) and these two after their return to France wrote the zoological account of the voyage, which was published in 1825. Owing to a shipwreck, most of the zoological specimens were lost, including the Australian birds they had collected, so the account of the zoology was not very complete. They recorded, however, that in Shark Bay they met with new birds. These included the Western Grass-Wren, which they called "Merion natte," found on the Peron Peninsula, and the Black-and-White Wren from Dirk Hartog Island. Figures of these two species were given in the volume of plates accompanying the work, which de Freycinet published from 1825 onwards after his return to France, and they were scientifically described by L. C. H. Dumont in articles in the journal *Dictionnaire des Sciences naturelles.*

Quoy and Gaimard were again in Western Australian waters from 1826 to 1829 with Dumont d'Urville's expedition around the world in the *Astrolabe* (formerly the *Coquille*). The ship was at King George's Sound from October 7 to 25, 1826, and on the return to France the two naturalists published in 1830 the "Zoologie" volumes of the account of the voyage. This account was issued in magnificent form in fourteen volumes of text and seven volumes of maps and plates. To this work we owe the first description of the Twenty-eight Parrot, the Western Yellow Robin, the White-breasted Robin, the Banded Blue-Wren, and the Red-eared Firetail.

III—THE PERIOD OF JOHN GILBERT AND THE EXPLORATION OF THE COLONY (1829-1887)

(a) THE PIONEER OBSERVERS (1829-1839)

With the establishment of the Swan River Colony under Lieut.-Governor Sir James Stirling, in 1829, a new era began in the accumulation of information regarding the birds of the western portion of the Australian continent. As already described, all former collecting had been done by members of the staffs of survey ships, French and British—now the land-birds were to receive attention. Actually Stirling's preliminary survey of the Swan River in March 1827 yielded some interesting natural history information, recorded in the report of his companion, **Charles Fraser** (-1831), Colonial Botanist of New South Wales. Fraser observed the large flocks of the Long-billed Corella on the Perth foreshores, now extinct locally, and expressed his wonder at the abundance of waterfowl, particularly the Black Swans, one assemblage of which, estimated at over 500 birds, rose at once, "exhibiting a spectacle which, if the size and colour of the bird is taken into account, and the noise and rustling occasioned by the flapping of their wings previous to rising, is quite unique in its kind."

At the time of the actual founding of the settlement Captain C. H. (afterwards **Admiral Sir Charles) Fremantle** (1800-1869), of H.M.S. Challenger, made several natural history observations in his diary. On May 3, 1829, on the Swan River near Point Heathcote he saw 10 Black Swans and "an amazing number of a black species of Shag", undoubtedly the Little Black Cormorant. At Carnac Island he noted the existence of the colony of hair seals, of which his boat's crew killed 13, chiefly young ones.

The earliest explorations of the newly-formed colony were those carried out by some of the officials and officers of the garrison troops. These included **Dr. T. B. Wilson,** R.N., Ensign **Robert Dale** and Lieutenant **Archibald Erskine,** both of the 63rd Regiment, **John Septimus Roe** (1797-1878), Surveyor-General, and Surgeon **Alexander Collie,**

F.L.S., the first Government Resident of King George's Sound. All these took note of the bird-life they encountered during their travels, and there are many references to birds in the reports they handed in on their return from the trips. Certain of these reports, in some cases in the form of letters, were published in the local and London newspapers and most of them reprinted in London in 1833 in a small book, published by J. Cross, entitled *Journals of several expeditions made in Western Australia during the years 1829, 1830, 1831 and 1832.* These accounts, taken together, give us an insight into the kinds and numbers of birds noted by travellers in the early days of the colony and so make interesting comparison with the character of the bird-life as it is today. It is not possible, in this short account of the history of Western Australian ornithology, to repeat all the information contained in the journals, but one particular item may be mentioned as showing the interest and value of the early reports. Being at the mouth of the King River on June 4, 1831, Collie took advantage of a boat going to Coffin Island to look for seals and mutton-birds. He reported that upwards of five hundred "sooty petrels" were caught by three persons in three days. This mention of a "sooty" mutton-bird, together with that of the date on which the birds were observed nesting, proves that they were Great-winged Petrels. Collie's account is the first record of the species for Western Australia, though it was long overlooked and it was not realised until 1905 that the bird occurred in the State. Another interesting find, by Roe, was the first discovery of the nesting mounds of the Mallee-Fowl.

Lieutenant **Matthew Friend** of the Royal Navy, and a Fellow of the Royal Society of London, arrived at the Swan River Colony on January 30, 1830, in command of the *Wanstead* carrying immigrants. He, with his wife, remained in the Colony until March 19, when he sailed for England via Tasmania. He took with him a small collection of mammals and birds and it was exhibited in London on November 22, 1831, at a meeting of the Zoological Society. We have no information as to the species of birds in the collection and it would appear that at least two of the mammals actually came from Tasmania.

John Gould (1804-1881), the famous British ornithologist, had turned his attention to Australian birds in 1833, when he read a paper on the subject before the Zoological Society of London. From then, until he himself sailed for Australia in 1838, he described a number of "new" birds from this country, sent to him or to the Zoological Society by various travellers. In this way he described in 1836 the Black-tailed Native-hen, the Blue-billed Duck, the Red-winged Wren, the Western Shrike-Tit, the Black-capped Sittella, the Western Spinebill and the Yellow-plumed Honeyeater. The exact locality where these birds were obtained, and by whom, is unknown for most of them, but we do know that the Native-hen and the Duck were supplied by Lieutenant Breton and Captain Beete. **Henry**

William Breton, a Lieutenant in the Royal Navy, spent five weeks in Western Australia in October and November 1829 and, besides a paper on the Musk Duck read before the Zoological Society in 1831, published in 1833 a book *Excursions in New South Wales, Western Australia and Van Diemen's Land, during the years 1830, 1831, 1832 and 1833.* Captain (later Major) **John Picton Beete** (the name was mis-spelled "Brete" in some of the early records) was in Swan River Colony at the period, and was Lieut.-Governor from May 11 to 24, 1834. Breton gave a number of items concerning birds in Western Australia in his book.

William Nairne Clark (1804-1854), who achieved minor fame in various activities in the colonial life of the time, also enters our record as the author of the first bird-list for the Colony, published in his newspaper, *The Swan River Guardian* in 1836. In 1840 he published in *The Inquirer* newspaper accounts of his exploratory trips by sea along the coast to the west of Albany and interesting remarks on the birds met with were included.

Lieutenant (later Captain Sir) **George Grey** (1812-1898) of the 83rd Regiment, after an expedition in the Kimberley Division in 1837, did some exploratory work in the area of the Gascoyne River, using Bernier Island as his base and eventually reaching Perth overland after a trying journey. He published in 1841 an account of these expeditions, including a list of the species of birds which had come under the notice of Gould as inhabiting the western coast. The previous year Grey had sent to the British Museum sixty skins of birds he had collected on his trips; most of these passed through the hands of Gould who wrote scientific descriptions of them. Grey, after being for a short time Government Resident at Albany, went on to government service in South Australia, New Zealand and South Africa. Himself an ornithologist and philologist, Grey corresponded with Gould, especially from South Australia, and through publication of this correspondence and papers published by Gould, Grey's work affords a notable contribution to our knowledge of the birds of Western Australia and South Australia, his original information on the breeding habits of the Mallee-Fowl for instance, being outstanding.

Gould's energy and enthusiasm were undoubtedly responsible for the interest taken by many officials and settlers of the new Swan River Colony in its bird-life and this enterprise was to reach its culmination when, in 1839, he despatched **John Gilbert** from Tasmania to collect in the colony. Gilbert's work in Western Australia was of so remarkably brilliant a character that his activities in the new colony overwhelmingly obscure the work done by any other collector during the century. Before detailing what John Gilbert achieved in Western Australian ornithology, the work of contemporaries, who in a minor way added to knowledge of the bird-life, will be reviewed.

First should be mentioned **Francis Fraser Armstrong** (c. 1812-1896). The Armstrong family had arrived in the colony in December

1829 and, in December 1834, Francis was appointed Native Interpreter. He appears to have had a distinct taste for natural history in which his contacts with the aboriginal population no doubt greatly assisted him; and, doubtless influenced by the presence in the colony of two professional collectors, Preiss and Gilbert, he made use of his personal collecting ability to supplement the meagre wage he received in his colonial appointment. Writing to the Governor in 1848, he stated that, his salary having been reduced to sixty pounds a year, "I employed my spare time successfully in collecting the finest specimens of ornithology the colony produced." He advertised in the press that he had collections of preserved birds for sale and one collection advertised included "Variegated, Crested Grey, Musical or Plaide Wing, Green, Platycerus and Nanodes Parrots, Diamond and Indigo Birds, Watchmen, Night Screamer, Barking Owl (white), Tricoloured or Pink Cockatoo, Red Tree-climbers, Crested Butcher Birds, Painted Quails, Golden Cuckoos, Crimson Parrot, Black Eagle, large Australian Bustard, Crested Grebe, Black Swan, Nectar Suckers." Another collection included "Painted Snipe."

The ultimate destination of specimens collected by Armstrong is not known, but it is more than likely that some of them were acquired by **Dr. Johann August Ludwig Preiss** (1811-1883), a German botanist who arrived in the Swan River Colony in 1838, a few months before the arrival of John Gilbert. Preiss travelled extensively in the colony collecting all types of natural history specimens for museums in Germany and purchasing those items which he was unable to collect personally. In this way his collections amounted to 200,000 plant specimens, 200 species of insects, 181 species of birds and 60 to 80 species of reptiles. When Preiss returned to Germany in 1842 his collections were dispersed to various museums in that country and as late as 1937 some twenty-two of the birds brought back by him were seen by one of us in the Zoological Museum at Hamburg. These were lost when the museum was destroyed by bombing in World War II. However other specimens survive in several German and Austrian museums; these include nine in the Zoological Museum, Berlin; 165 in the Municipal Museum, Halberstadt; and 13 in the Natural History Museum, Vienna.

A third person who has left some interesting notes on the bird-life of this State in its earliest days was **George Fletcher Moore** (c.1800-c.1887), the Advocate-General. As early as 1834 a small volume containing extracts from his letters and journals was published in London. This contained no remarks on birds, but in 1842 he issued a *Descriptive Vocabulary of the Language in Common Use amongst the Aborigines of Western Australia,* which included the aboriginal names applied to many birds and remains a valuable source of information. The vocabulary was founded on a similar work published in 1839 by Grey, but included additions which Moore acknowledges to have been supplied by Francis Armstrong, Charles Symmons and John Bussell. His main observations were not published until 1884, in a book entitled *Diary of ten years' eventful life of an early settler in Western Australia.*

An early settler who assisted Gilbert in the collecting of Western Australian birds was **Lockier Clere Burges** (1813-1886), who was then farming with his two brothers, William and Samuel Burges, at "Tipperary," near York. From him Gilbert obtained the type-specimen of the Grey Falcon, also a specimen of the Pied Honeyeater, both coming from the York district. Gould, in his *Birds of Australia,* when referring to the Pied Honeyeater, wrote: "One of Mr. Gilbert's specimens had been given to him by Mr. Lock Burgess, to which gentleman I am also indebted for several other rare species kindly presented by him to Mr. Gilbert for me, whereby the interest and value of this work is much enhanced." L. C. Burges later did much travelling in the North-West and, in 1864, was manager of the Roebuck Bay Pastoral Company.

From 1837 to 1843 H.M.S. *Beagle* was engaged in surveying the coasts and rivers of Australia, first with Captain J. C. Wickham as Commander and later in charge of Captain (later Admiral) **John Lort Stokes.** The vessel had previously visited King George's Sound from March 6 to 14, 1836, when under the command of Captain Robert Fitzroy, and with the eminent biologist Charles Darwin, then only a young man of 27 years, on the staff. No notes on birds observed during that first visit have been recorded. During April and May 1840 the vessel was engaged on a survey of the Abrolhos Islands and, in his book, *Discoveries in Australia, with an account of the coasts and rivers explored and surveyed during the voyage of H.M.S. Beagle in the years 1837, 38, 39, 40, 41, 42, 43,* published in 1846, Stokes recorded that "the soil of Rat Island, and the south-west side of West Wallaby Island is filled with burrows of the sooty petrel or mutton bird, so that it forms rather troublesome walking. There was a flat in the centre of North Island, covered with coarse grass, where a great many quails were flushed, affording good sport. Three small flat islands just between the Wallaby Islands were called Pigeon Islands, the common bronze-wing pigeon being found there in great numbers. It may be remarked that the birds met with on Houtmans Abrolhos, with the exception of one, resembling in shape and colour a small quail (*Hemipodius scintillans,* Gould), were known and common on the mainland. The aquatic species were also familiar to us, but the habit of one kind, of a sooty-black colour, generally called noddies, was quite new—that of building their nests, which are constructed of sea-weed and contain only one egg, in trees." The bird Stokes mentions is the Lesser Noddy, and he is not quite accurate in his account of the Bronzewing Pigeon; it is the Brush, and not the Common Bronzewing, which occurs on these islands.

Benjamin Bynoe (c.1803-1865), an assistant surgeon of the Royal Navy, was in the *Beagle* and was very active in collecting specimens of natural history. He sent many birds, animals and plants to England. Among these were the type-specimens of at least two new birds from northern Australia, and he is particularly important to the history of ornithology of Western Australia, as covered in this *Handbook,* in that he collected on June 9, 1840, on Depuch Island, the type-specimen of the Painted Finch, which remained for a number of years the only specimen known to science. Gould, in his *Handbook,* mentions that Bynoe found the Plumed Pigeon in the country between Cape Hotham and Depuch Island.

(b) THE VISIT OF JOHN GILBERT (1839-1843)

In 1837 John Gould commenced the publication of his *Synopsis of the Birds of Australia,* and shortly afterwards *The Birds of Australia and the Adjacent Islands,* a folio work in which he intended to illustrate every known Australian bird. The latter enterprise had not got far (only two parts had been issued) when Gould decided that his own and other available material was quite inadequate for the formidable task he had set himself. There was no alternative other than that he himself should proceed to Australia to obtain the necessary specimens. With his coming Western Australian birds were to receive intensive study, and practically all the species inhabiting the southern portion of Western Australia were to be "discovered." Furthermore, for the first time specimens were to be so preserved that, at the present day, more than a hundred years later, their condition is as when they were first prepared by their collector.

To Tasmania went John Gould, accompanied by a professional taxidermist and collector, John Gilbert, and, Western Australia not falling into the collecting itinerary Gould had set himself, he deputed Gilbert to carry out the Western Australian portion of the enterprise. Gilbert came to Swan River Colony in the *Comet,* which arrived at Fremantle on March 6, 1839.

During his first visit most of Gilbert's collecting was done in the vicinity of Perth and inland as far as York. The colony was only ten years old and there was always danger from the Aborigines for any individual strolling about the country in the way that naturalists have. In consequence Gilbert was compelled to limit the extent of his travels. However, he made friends with Francis Armstrong within a few days of his arrival in the colony and the two made several trips together in the environs of Perth; among skins in the British Museum collected by Gilbert there is one of the Yellow-rumped Thornbill obtained on Mount Eliza on March 14, 1839. Fortunately, letters Gilbert wrote to Gould from Perth are in existence, also the lists he made out of specimens sent to England. He obtained the type-specimen of the Rufous Tree-creeper in the vicinity of Perth and also collected specimens of the Red-eared Firetail and of the Southern Emu-Wren —not one of these birds is today, owing to despoliation of the environment, found where he obtained them. Of all the birds, however, which Gilbert collected on this first visit to Western Australia the most interesting is, perhaps, the Western Bristle-bird, of which he obtained not only several skins, but also a nest and eggs. These he got in the Perth area, a locality where the bird has never since been recorded.

In addition to making friends with Armstrong, Gilbert was fortunate in his association with **James Drummond** (c.1784-1863), the famous botanical collector. Drummond had arrived in Western Australia in 1829 and taken up land in the Toodyay district, where he established himself at the farm still known as "Hawthornden." He also pastured sheep at the Moore River. From "Hawthornden" Drummond made a great many expeditions in search of botanical specimens all over the southern portion of the Colony and Gilbert joined him on several of these trips. Between August 18 and 22, 1839, Gilbert went with Drummond to Rottnest Island. Another member of the party was Preiss, the German collector. Besides visiting Rottnest Island, Gilbert twice visited the York district in 1839, on the second occasion going on with Drummond to the Wongan Hills, Drummond's youngest son, Johnston, also accompanying them. **Johnston Drummond** (1820-1845) worked on his father's property, "Hawthornden," and employed some of his time in collecting birds, which it appears he was ready to dispose of. Thus, writing to Sir J. D. Hooker in May 1839, James Drummond mentioned that his son Johnston had already sold £20 worth of bird-skins to Preiss. These and other transactions no doubt led Gilbert to complain, in a letter to Gould: "There are a number of Persons here who collect Birds but I fear I shall not be able to make any dealings with them, for their whole aim appears to [be] to make the most they can of their specimens. I have seen several very rare Birds. I mean considered rare here for which they ask the most extravagant prices and so vastly independent with it. There is a German here collecting for several Continental Museums. He is at present about 60 miles in the country; but I am given to understand he gives large prices for almost every thing in the shape of Natural History in the Town, for he is a general Collector. This will I fear act as a little drawback to my exertions & our wish to complete a collection by purchasing specimens; but I hope when we meet we shall be able to understand each other and if possible be of mutual Benefit to our Interests. I have been particular in impressing on the minds of Persons the importance of lending me assistance in a National point of view; and I am happy to say they generally appear to think I should be supported in preference to the Foreigner." Many specimens eventually did go to Gould, through Gilbert, and, in exchange, it was arranged that Drummond should receive a copy of Gould's forthcoming book *The Birds of Australia.* It was from Johnston Drummond that Gilbert obtained the type-specimen of the Scarlet-chested Parrot, which was named *Neophema splendida* by Gould in 1841.

When Gilbert again came to Western Australia in 1842 he made several more trips with James Drummond when the son was also present but, shortly after Gilbert finally left the Colony in 1843, Johnston Drummond was murdered by an Aboriginal. The earlier parts of Gould's *Birds of Australia,* which had been sent to Johnston Drummond in return for the assistance he gave Gilbert in collecting specimens of birds, are still in possession of members of the Drummond family.

Hoping to join Gould, who was collecting in New South Wales, Gilbert sailed from Fremantle for Sydney on January 29, 1840, and at King George's Sound the vessel on which he was travelling, the *Caledonia,* stayed for a fortnight. Here he seized the opportunity of joining William Nairne Clark and others in a trip to Kojonup. The trip covered altogether twelve days and among the natural history specimens obtained by Gilbert was a rare marsupial, Gilbert's Rat Kangaroo, *Potorus gilbertii,* which was then brought to scientific knowledge, and is now, unfortunately, extinct. Writing to Gould, he mentions that the party saw an enormous herd of kangaroos, over five hundred in one mob, on the Gordon Plains.

When leaving King George's Sound at the end of his first visit to Western Australia, Gilbert had as fellow-passenger, as far as Adelaide, George Grey, who had been acting for a short period as Government Resident at Albany, and was on his way to England. Writing from Sydney to Gould, Gilbert mentions that Grey had offered him £1 each for the whole collection of skeletons he had made. Besides the skeletons, Gilbert had made a large general collection of natural history specimens—birds and their eggs, marsupials, bats, reptiles, shells, plants. Excluding what he obtained while on the trip from King George's Sound to Kojonup and back, he had collected over 750 skins of birds, of over 175 species, besides eggs of 80 species; 18 species of quadrupeds, besides a large number of botanical specimens, crania of Aborigines, various native implements, geological specimens, insects and fishes. Writing to him Gould said: "Your collections from Western Australia do you much credit and I am perfectly satisfied that on your part, no exertions have been spared to accomplish the object of your mission." When he reached Sydney Gould had already left for England, so Gilbert went collecting at Port Essington, at the very north of what is nowadays known as Northern Territory. Thence he went on to England via Singapore.

Believing that much in the way of natural history still remained to be discovered in Western Australia, Gould persuaded Gilbert to return here for a second visit and he arrived at Fremantle in July 1842. Among the historical documents in the Queensland Museum is a small book in clear and elegant writing—probably that of E. C. Prince, who acted as secretary to Gould—which contains the instructions by Gould as to what particular specimens of natural history Gilbert was to pay attention. It contains many beautiful illustrations of birds, mainly of heads. Gould was busy on his book on the kangaroos of Australia, so was particularly anxious to obtain information about these animals. He noted: "Gain every possible information respecting the group. Collect adult males and females together with young of every species and if practicable ascertain the weight of each and the difference of the weight in the sexes. Procure also as many crania as possible labelling each with the name of the species and the sex. Procure the wallaby from Garden and Rottnest Islands. Dissect as many females as you can and learn if the young are found in the uterus and pouch at the same time; state the size of the young in the pouch. Specimens and crania of [Potorus] Gilbertii particularly wanted. For birds of Western Australia see the list. Find examples of the nests and eggs of all and also duplicates of the eggs. See after a second Emu in W. Australia. Collect Emus in every locality, ascertain the sex and procure if possible adult birds. Eggs of Leipoa and Megapodius in brine or new preservative. Ascertain about the Harriers."

Soon after his arrival on this second visit, Gilbert went inland to Toodyay and joined James Drummond on a visit to the Moora lakes, travelling via Culham, Bolgart, and the site of what was to become New Norcia. Then followed a visit to the Wongan Hills where Gilbert studied the nesting habits of the Gnow or Mallee-Fowl and sent Gould a long letter, dated September 28, 1842, giving a detailed account of these observations. He was a little early for much success in finding eggs in nesting mounds, but managed to procure five eggs in four different nests.

On his return to Perth, Gilbert investigated a report that the Mallee-Fowl nested in thickets between the sandhills close to the beach near Perth and Armstrong had even told him he remembered seeing a mound on Mt. Eliza (= King's Park) when Perth was first settled. Gilbert failed to substantiate these accounts though he gave them some credence. However, during the search he made the important discovery that the Western Whip-bird occurred in the dense coastal thickets and was able to identify, by hearing its strange distinctive notes, the self-same call he had heard a few weeks earlier at Wongan Hills.

Then followed immediately another trip with Drummond which was to have the highly important result of the discovery of the Noisy Scrub-bird. This was an excursion down the coast from Fremantle to Augusta and an attempt to strike east from Augusta to Mount Barker. At Drakesbrook, just within the Darling Range, on November 3, Gilbert got the type-specimen of the Noisy Scrub-bird. Archdeacon J. R. Wollaston, who was then at Picton, near Bunbury, has given some information about this trip in one of his celebrated diaries. The naturalists called on him on their way south on November 21, 1842, and Wollaston mentioned in his diary that they travelled about ten miles a day, walking chiefly to save their horses, which carried packs. Drummond was suffering from sore feet. On reaching Augusta, after visiting Captain Molloy at the Vasse, the two naturalists attempted to go on to Mount Barker but, in the words of Archdeacon Wollaston: "They found it impossible to penetrate the country with so small a force and badly provisioned as they were. It took them four days to do a distance of twenty miles, and they travelled along one continuous swamp to the Augusta. They very wisely turned back; for they were very jaded and somewhat tattered in their habiliments. They were loaded with specimens of plants and birds." However, they succeeded in getting as far east as Scotts River and, while at Augusta, visited the small islands lying off Cape Leeuwin, where Gilbert discovered the White-faced Storm-Petrel and the Fleshy-footed Shearwater breeding and obtained skins and eggs.

Doubtless encouraged by the reports made by the officers of the *Beagle* regarding the fauna of the Abrolhos Islands, Gilbert seized an opportunity to visit those islands in a boat which was being sent there by Captain Daniel Scott, the Harbour Master at Fremantle, to salvage what was possible from the barque *Ocean Queen,* wrecked on Pelsart Island on September 13, 1842. Leaving Fremantle on January 6, 1843, Gilbert was away over two months on this trip, which he afterwards referred to in a letter to Gould as "my late perilous trip to the Abrolhos, from which I had such a miraculous escape." Bad weather delayed the return of the party to Fremantle and the authorities were becoming anxious at the non-return of the vessel. The Harbour Master was on the point of setting out himself with a search party when the missing voyagers returned. Gilbert gave a very complete account of the bird-life of the islands in a letter to the local newspaper, *The Inquirer,* and in letters to Gould which were published in the *Proceedings of the Zoological Society of London.*

Next, Gilbert undertook a journey from Perth to King George's Sound, via Williams and Kojonup. On this trip he was very successful in collecting small animals, and in a letter to Gould he commented that he was indebted for some of the specimens to a domestic cat belonging to the military garrison at Williams which had captured them in the night. An interesting discovery he made while at King George's Sound during this visit was that the Noisy Scrub-bird occurred there. Back again in Perth by July 6, 1843, Gilbert made a final visit to York and on his return to Perth despatched a consignment of natural history specimens to Gould by H.M.S. *Beagle,* which ship had called at Fremantle and left for England on September 13, 1843. Leaving a further consignment for despatch by the *Napoleon,* which sailed for England on January 14, 1844, Gilbert left Western Australia for good on December 20, 1843, for Launceston via Adelaide. From Launceston he went on to Sydney and eventually joined Ludwig Leichhardt's famous overland expedition to Port Essington, during which he was promoted to the post of second-in-command. He was speared to death by natives on June 28, 1845, when only just over half the journey had been completed.

(c) THE BARREN YEARS (1843-1887)

In the decade after Gilbert's departure nothing important was added to the information on the bird-life of the State; in fact there is no work in ornithology to record at all for this period. In 1854, however, owing to the encouragement of the Colonial Secretary, **William Ayshford Sanford** (1818-1902), when Assistant Surveyor **Robert Austin** made his exploratory trip from Northam via Cowcowing to Geraldton, a small collection of bird-skins was made and information obtained on the birds of the inner regions of the State. On this trip was obtained a parrot which, although referred to as a Ground Parrot in the appendix of Austin's *Report,* which Sanford wrote on the bird-skins collected, was apparently the type specimen of that always rare bird, the Night Parrot. Sanford on his return to England became a well-known naturalist, specialising in palaeontology, particularly that of Somerset.

In 1858 **Francis Thomas Gregory** (1821-1888) led an expedition to examine the country east and north of the upper Murchison River, and although he does not appear to have brought back any specimens, Gregory made a number of notes in his journal on the various birds met with on the trip. On a second expedition in 1861, which was inland from Nickol Bay, Gregory collected some bird-skins which went to the British Museum, among them being the type-specimens of the Red-plumed Pigeon and the western form of the Spotted Bower-bird. Gregory's data added much to the knowledge of the bird-life of this portion of the State; notes which he supplied to John Gould were printed in the latter's *Handbook to the Birds of Australia,* 1865, and notes which he recorded in his journal appeared in 1862 in the *Journal of the Royal Geographical Society of London.* A collected account of his expeditions was published in Brisbane in 1884.

In the 'sixties, the Trustees of the Australian Museum, Sydney, sent over to King George's Sound their collector, **George Masters** (1837-1912) on two collecting trips, the first in 1866, when he collected at Albany, and the second, a longer visit, in 1868-69, when he collected as far inland as the upper reaches of the Pallinup or Salt River. From the first visit Masters took back to Sydney 281 bird-skins, among which was a specimen of the Noisy Scrub-bird and two of the Western Bristle-bird. The collection made during the second visit included no fewer than six specimens of the Noisy Scrub-bird, ten of the Bristle-bird, and eight of the Western Whipbird. Masters was one of the few naturalists who left published observations on the habits of the Noisy Scrub-bird, and references to other birds obtained by him are to be found in *Nests and Eggs of Birds found breeding in Australia,* by A. J. North, assistant in ornithology at the Australian Museum, published between 1901 and 1914.

Dr **Frederick Matthew Rayner** (?-1873) was surgeon on H.M.S. *Herald* from 1857 to 1861, during which period the ship did some surveying off the coast of Western Australia. In May 1858, he was in Shark Bay. Bird-skins collected by Rayner are in the British Museum.

From May 1879, and during the early months of 1880, **James Cockerell** was in Western Australia. He had arrived in South Australia from New South Wales to join Samuel White who was preparing to leave for the Aru Islands on a collecting trip. As the vessel in which the trip was to be made was not ready, Cockerell came over to Western Australia to collect birds for a few months. Skins collected by him while on this visit are in the British Museum.

From 1862 Western Australia had one who may be called her first resident naturalist—**William Webb** (c.1834-1897), an Englishman, born in Staffordshire, who settled at Albany in that year. A collector of seeds and plants, G. Maxwell (1804-1880), was then in business in Albany and Webb appears to have been in some sort of partnership with him and eventually succeeded to the business. Several plants were named after Webb by Baron von Mueller. At this period W. S. Macleay was forming the famous "Macleay Collection" in Sydney and Webb provided him with many specimens, including bird-skins. Thus in 1875 and 1876 Macleay's journal shows that Webb supplied no fewer than 60 specimens of mammals and 180 bird-skins, and in 1879 two further consignments are recorded. Webb is noteworthy as having been the collector of at least seven of the twenty specimens of the Noisy Scrub-bird now in existence and he has left us some field-notes on that rare bird in an article he wrote in the local newspaper—a number of articles by him on natural history subjects are to be found in *The Albany Mail,* which, however, commenced publication only in 1883. Webb supplied a case of 100 mounted birds (62 species) which formed part of the Western Australian exhibit at the Colonial and Indian Exhibition held in London in 1886.

Webb also sold bird specimens, and various natural history curios, to passengers on ships calling at Albany. The English botanical artist, Marianne North, in 1880 purchased from him "a pair of lovely green ground-parrots with spread tails" and mentions in her book of reminiscences that another collector at Albany sold to P. & O. passengers live cockatoos at two guineas each, his stock being supplied by Aborigines. There was a general interest in those days in display collections of colourful birds and the State was visited by various natural history dealers. One such was Frederick G. Ayres, of South Australia, who collected in Western Australia between 1888 and 1891, in between mining and other occupations. He also made good bird-skins and some were purchased from him by the National Museum, Melbourne.

Generally, however, the period between Gilbert's departure in 1843 until well towards the end of the century, was a barren one in local ornithology. The second generation settlers showed very little of the interest their fathers evinced in the natural history of the country and there was no native son to carry on the work of Gilbert and the other visitors.

IV—THE MODERN PERIOD (1887 ONWARD)

With the 'eighties, however, there opened a period of active work in ornithology in Western Australia. Early in 1887 there landed at Carnarvon, to settle in the colony, a young Englishman who was, for the next thirty-three years, to devote all his spare time to studying and collecting the birds of Western Australia, first in the Gascoyne district, then at Point Cloates, near North-West Cape, and later at Broomehill on the Great Southern Railway. Intermittently he was also to make observations and collections in the lower South-West, at Busselton southwards and in the vicinity of Albany. We refer to **Thomas Carter** (1863-1931). Of Yorkshire extraction, Carter had already studied birds in the north of England and had made an ornithological trip to Iceland before he came to Australia. Starting first as a "jackaroo" on Boolathanna Station, on the Gascoyne River, his first notes on the birds of Western Australia appeared in the English natural history magazine *The Zoologist*. In the year of his arrival in Western Australia he gained a knowledge of the bird-life of the area when he assisted to drive a mob of 2,000 sheep overland from Boolathanna to Perth. Following on a visit to Busselton, he worked on Minilya Station for a period and, in November 1889, settled at Point Cloates, where he was to remain for the next fourteen years carrying out pastoral pursuits on a 135,000-acre property. During this period Carter published many notes on birds observed and collected, and to him is attributable the discovery of the Rufous-crowned Emu-Wren and the Spinifex-bird. In 1903 Carter left Western Australia on a holiday to England, taking with him a collection of over 500 bird-skins, which he disposed of to Lord Rothschild for the famous "Tring Collection," and a series of birds' eggs, which passed to the British Museum. Returning to Western Australia, Carter settled at Broomehill in 1905 and from then on until 1922, when he finally returned to England, he carried out many collecting expeditions in all parts of the lower South-West, and to Peron Peninsula and Dirk Hartog Island. In the latter regions he re-discovered in 1916 two birds, the Black-and-white Wren and the Western Grass-Wren, specimens of which had not been obtained since they were first collected by the French naturalists Quoy and Gaimard in 1818. Carter published a number of papers on Western Australian ornithology in the journal of the British Ornithologists' Union, *The Ibis,* and in that of the Royal Australasian Ornithologists' Union, *The Emu.* Many skins collected by him are in the Western Australian Museum and others went to the collection of Gregory M. Mathews, these latter being, together with those that had previously gone to the "Tring Collection," now in the American Museum of Natural History, New York.

In the 'eighties Western Australia was visited by the well-known South Australian ornithologist, Captain **S. A. White** (1870-1954), son of the Samuel White who had sent James Cockerell to collect at King George's Sound in 1879. S. A. White collected at Perth and inland to New Norcia, but nothing was published by him on the birds he collected.

In 1889 **Archibald James Campbell** (1853-1929) of Melbourne, who was obtaining data for his book on the nests and eggs of Australian birds which was to be published in 1901, arrived at Albany, and travelled slowly north along the coastal roads to Augusta and Bunbury and Perth. From Perth he visited Geraldton and the Abrolhos Islands. On his return to Melbourne, Campbell published thirteen articles in the *Australasian* describing his trip and the scientific notes on birds and eggs collected appeared in the *Report of the Australasian Association for the Advancement of Science,* in the *Victorian Naturalist,* in *The Emu,* and in the *Proceedings of the Royal Physical Society of Edinburgh, Scotland.* The total natural history items collected amounted to over 1,000, and included 105 bird-skins and 600 eggs. His visit is particularly notable for the fact that he obtained at Torbay, in October 1889, a specimen of the Noisy Scrub-bird, and despite assiduous searching the species was not encountered again until December 1961. He was able to add a few field-notes to those furnished earlier on its habits by George Masters and William Webb. The visit generally did much to forward knowledge of the distribution and breeding habits of Western Australian birds.

In the 'nineties extensive ornithological work was done by two major exploring expeditions which examined the inland regions of South Australia and Western Australia. These were the Elder Expedition of 1891 and the Calvert Expedition of 1896. The Elder Expedition left Warrina on the South Australian Northern Railway, 633 miles from Adelaide, and reached Annean Station in the Murchison district in January, 1892. **Richard Helms** (1858-1914) who was later, from 1896 to 1899, biologist to the Department of Agriculture, Western Australia, was the naturalist with the expedition and collected a few bird-skins and eggs. These were reported on by E. C. Stirling and A. Zietz, director and assistant director of the South Australian Museum, in the *Transactions of the Royal Society of South Australia* in 1893. There are also notes on birds encountered during the expedition in the journal of the leader, David Lindsay, published separately at Adelaide by the Government Printer in 1893. It may here be mentioned that, in 1897, accompanied by O. H. Lipfert, the taxidermist to the Western Australian Museum, who will be referred to again later, Helms, while holding the appointment just mentioned, visited the Abrolhos Islands and in the following year published an account of the birds they met with, together with notes on their breeding habits. This paper is to be found in the *Producers' Gazette and Settlers' Record,* the forerunner of the present *Journal of the Department of Agriculture of Western Aus-*

tralia, in which it was reprinted in 1902. Helms also wrote a short account on the birds of the State in the *Illustrated Handbook of Western Australia,* issued in connection with the Paris International Exhibition of 1900.

The results of the Calvert Expedition of 1896 were much more important to the ornithology of Western Australia than were those of the Elder Expedition. **George Arthur Keartland** (1845-1926), a well-known Victorian ornithologist was the naturalist; he had previously, in 1894, been naturalist with the Horn Expedition to northern South Australia, which had resulted in the accumulation of much information on the birds of the inner regions of that State. Keartland, therefore, had the advantage of previous experience in collecting birds under difficult conditions. Unfortunately, owing to the tragic death of two members of the party some 300 bird-skins were abandoned at Joanna Spring, south of the Fitzroy River, but 167 specimens, belonging to 59 species, besides nests and eggs, eventually reached the South Australian Museum. These skins were reported on by A. J. North in 1898, in the *Transactions of the Royal Society of South Australia,* when he recorded also Keartland's field-notes on a further 122 species of which specimens were not brought back. Keartland's work contributed considerably to knowledge of the bird-life of the State in the arid region between the upper reaches of the Murchison River and the Fitzroy River.

In 1899 **Robert Hall** (1867-1949), another Victorian ornithologist, who was later curator of the Tasmanian Museum, Hobart, made a collecting trip to Western Australia. Arriving at Albany in September, in which locality he made initial collections, he then continued to Katanning, Geraldton and the Abrolhos Islands. Some 156 bird-skins were obtained altogether. The information obtained by Hall was included in a series of articles he wrote in the *Journal of the Department of Agriculture of Western Australia* between 1900 and 1903, and the bird-skins collected were scientifically treated in a paper in *The Ibis* in 1902. In the latter year he also gave a popular account of his visit in several issues of the *Victorian Naturalist.* Before this personal visit to Western Australia, one **Lindsay Cameron** had been sending Hall collections of bird-skins from Kalgoorlie. These were dealt with by Hall in two papers in the *Victorian Naturalist* in 1899, and in another in the *Transactions of the Royal Society of South Australia* in 1900.

Hermann Franz Otto Lipfert (1864-1942), usually known as O. H. Lipfert, arrived in the State in 1892 and, early in 1894, joined the Western Australian Museum as taxidermist and collector. During the earlier years of his service he spent much time collecting specimens for the Museum. He visited the Abrolhos Islands from October to December, 1894, again in September 1897, as already referred to with Richard Helms, and a third time in October 1898, this time with the late H. Murray Giles, who was for many years head-keeper of the Zoological Gardens, South Perth. In between these visits Lipfert made many other excursions in search of specimens to various parts of the State, one, particularly noteworthy, being to Bernier and Dorre Islands in 1910. The longest collecting trip of his career was, however, from April 1930 to August 1931 when he joined a party sent out to recondition the wells on the Canning Stock Route from Wiluna to Halls Creek, the 200 bird-skins brought back being divided between the Western Australian and the South Australian Museums.

For twelve years, between 1895 and 1906, **John Thomas Tunney** (1871-1929), of Gracefield, Kojonup, was employed by the Western Australian Museum in collecting natural history specimens and, as a result many bird-skins from all parts of the State were added to the collections of that institution. Neither Lipfert nor Tunney, however, published anything of importance regarding the bird-life they met with while on their collecting trips and the results of their work have to be looked for in systematic papers by other ornithologists, who dealt with some of the specimens they had collected, particularly A. W. Milligan, E. Hartert and G. M. Mathews.

In 1897 there had arrived in Perth one who was to do much to further knowledge of the bird-life of Western Australia. **Alexander William Milligan** (1858-1921) had gained considerable experience in ornithology in Victoria and had also studied birds in Queensland. For eleven years he resided in Perth and during most of that time acted as honorary ornithologist to the Museum. Not only did he do much field-work himself in the southern regions of Western Australia, but he wrote several papers in which he described collections by others. His most noteworthy personal achievement was, perhaps, the discovery of a form of the Rufous Bristle-bird in the lower South-West. Contemporary with Milligan a young cadet at the Museum, **C. P. Conigrave** (1882-1961), took particular interest in birds, and together the two, aided by the director, B. H. Woodward, considerably extended knowledge on the geographical distribution of birds in the State.

But the rate at which such knowledge, together with information on the breeding habits of the indigenous birds, was accumulated quickened considerably with the arrival in the State in 1901 of **Frederick Bulstrode Lawson Whitlock** (1860-1953). Whitlock had studied birds in England before coming to Australia and was already a competent ornithologist. To commence with, Whitlock collected in the Kalgoorlie district (1901), then in the East Murchison (1902-03) and after a visit to Mogumber, joined Milligan and Conigrave in a trip to the Wongan Hills (1903). A visit to Rottnest Island preceded one to Norseman in 1904, which, after a period at his home at Wilson's Inlet, was followed by a visit to the Irwin Valley (1907). At that time H. L. White of Belltrees, Scone, New South Wales, was engaged in amassing a collection of the eggs of the birds of Australia and was employing collectors in many parts. From 1908 until White died in 1927 most of Whitlock's time was spent in visiting various parts of Western Australia, in addition to an expedition into

Central Australia in a search for birds and their eggs. On these assignments Whitlock collected on the de Grey and Condon Rivers, and on the Abrolhos Islands (1908), in the Wiluna district (1909), at the Stirling Range, in the country in the vicinity of Wilson's Inlet, at the Nullagine River (1910 and 1911), on the Coongan River and on Barrow Island (1917), on Peron Peninsula and Barrow and Dirk Hartog Islands (1918), on the Nullarbor Plain (1921) and at the Fortescue River (1922). Visits to Central Australia and the Kimberley Division, with the birds of which we are not concerned in this *Handbook,* completed the most extensive range of collecting ever undertaken by any one ornithologist in Western Australia. Further, and it is a very important point, Whitlock had the gift of being able so to describe his field-work that the many articles he has published in *The Emu* and other journals are not only literary efforts of a high standard, but are also graphic word-pictures of the habits of the species with which he came in contact. No other ornithologist who has worked in Western Australia, and probably no other Australian ornithologist, has left for posterity such complete and readable accounts of the birds he encountered. Many bird-skins and eggs collected by him are in the Western Australian Museum, the National Museum, Melbourne ("H. L. White Collection"), the American Museum of Natural History, New York ("Mathews Collection"), and other institutions. Whitlock enjoys the distinction of being the discoverer of the last "new" bird to be found in Western Australia—that is, "new" in the sense of being considered a full species, previously unknown to scence. This was the Grey Honeyeater, which was described by A. J. North in 1910 from specimens collected the previous year near Wiluna.

Apart from resident ornithologists like Whitlock and Tunney, several visiting naturalists and collectors worked in the State for varying periods. **Edwin Ashby** (1861-1941) of South Australia paid three visits to Western Australia in 1901, 1921 and 1927, the last two coinciding with the holding in the State of annual congresses of the Royal Australasian Crnithologists' Union. From 1901 to 1911 **Charles George Gibson** (1878-1937), of the Geological Survey, travelled extensively in the State in the course of his official duties and, being interested in birds, studied the habits of those he met with. He has left valuable notes on their nesting, which are to be found in the pages of *The Emu, The Ibis* and *The Victorian Naturalist.* Many of his field-notes are also incorporated in A. J. North's four-volume work on the nests and eggs of birds found breeding in Australia.

In 1904 and 1905 the State was visited by a collector from the British Museum, **Guy Chesterton Shortridge** (1880-1949) (later director of the Kaffrarian Museum, King Williams Town, South Africa), who took back to that institution a large collection of mammals and birds. Shortridge collected at Albany, the Margaret River, Busselton and Bunbury; along the Great Southern Railway and 40 miles east; at Kalgoorlie, Southern Cross and Laverton; on Bernier Island, and in the Gascoyne district. A detailed account of the bird specimens appeared in *The Ibis* in 1909 and 1910 from the pen of W. R. Ogilvie-Grant, then in charge of the bird collections in the British Museum.

From 1912 to 1920 much work was done in ornithology in Western Australia by **W. B. Alexander** (1885-1965), the former director and later librarian of the Edward Grey Institute of Field Ornithology of Great Britain, who came from Cambridge University to take charge of the natural history collections of the Western Australian Museum. A paper he wrote in *The Emu* in 1921 on the birds of the Swan River district is outstanding as a stocktaking of the birds of the Perth area at that time. Also notable is his account of the birds he met with at the Abrolhos Islands whilst a member of the Percy Sladen Trust Expedition to those islands under the leadership of Professor W. J. Dakin. His very useful history of zoological discovery in Western Australia up to the year 1840 was published in the Royal Society's *Journal* (1914-1918) and has been partly drawn on for the preparation of this section, particularly for the summaries of the very early voyages. It may here be mentioned that Mr. Alexander was the author of the articles on birds contained in the two volumes of *The Illustrated Australian Encyclopaedia* published at Sydney in 1925-26, and several were incorporated, wholly or in part, in the 10 volumes of *The Australian Encyclopaedia* published in Sydney in 1958.

Another collector, this time working for the South Australian Museum, toured the State in 1912-13. He was **W. D. Dodd** (1891-), son of a well-known Queensland entomologist. Starting at Balbarrup on the Warren River in September 1912, he then collected at the Capel River in November. In December he visited the Abrolhos Islands and in January and February 1913 he collected at the Fortescue River and in the Hamersley Range. Dodd then went on to the Kimberley Division. As a result of this visit, some 200 bird-skins from Western Australia were added to the collections in the South Australian Museum.

In 1912 **Paul Denys Montague** (1890-1917) of Cambridge University, who later lost his life in World War I, visited the State under the auspices of the Royal Society of London and investigated the fauna of the Monte Bello Islands. In 1914 he published in the *Proceedings of the Zoological Society of London* an account of the visit, which included field-notes on the 25 species of birds he observed.

Two collecting trips to the southern portions of the State remain to be referred to. The first was by the well-known field-naturalist of New South Wales, **Sidney William Jackson** (1873-1946), who between October 1912 and February 1913 made an unsuccessful search for the Noisy Scrub-bird in the southern coastal districts between Denmark and Broke Inlet. The other was a visit in 1921 to the Archipelago of the Recherche by a team of collectors, which included **A. F. Basset Hull** (1862-1945) from the Australian Museum, Sydney. Published records of Jackson's visit were for long confined to popular articles in the *Sydney Mail,* but in 1952 extracts from his field notes were published in the *Western Australian Naturalist.* A detailed account of the birds met with at the Archipelago of the Recherche is to be found in *The Emu* of 1922.

That concludes, as far as it is proposed to detail in this account, the history of the discovery of birds in Western Australia and the work which has been done by ornithologists who had studied the habits of the birds occurring in the State. It is early yet to chronicle the work of those numerous naturalists who are still engaged in investigating the wild life of the western portion of Australia.

SECTION II—BIRD GEOGRAPHY

It will add much to the reader's appreciation of the geographical distribution of bird species if he has some comprehension of how the avifauna of Western Australia is related to that of the rest of the continent and whether any pattern can be recognised in the seemingly chaotic maze of species boundaries.

Even the briefest study of a series of maps illustrating bird distribution in Western Australia makes it apparent that the most strongly demarcated boundary is that between the Kimberley Division and the Pilbara district (Fig. 1). Many of the birds of the South-West are limited in their range northward by ecological and climatic barriers which may vary, such as moisture, and the threshold limits differ widely between species. It cannot be said that they combine to form any major breaks in distribution and it is only when we come to the Kimberley boundary that we find on the north of it such a distinctively different avifauna that one has the impression, as F. Lawson Whitlock has pointed out, that one is entering a new country.

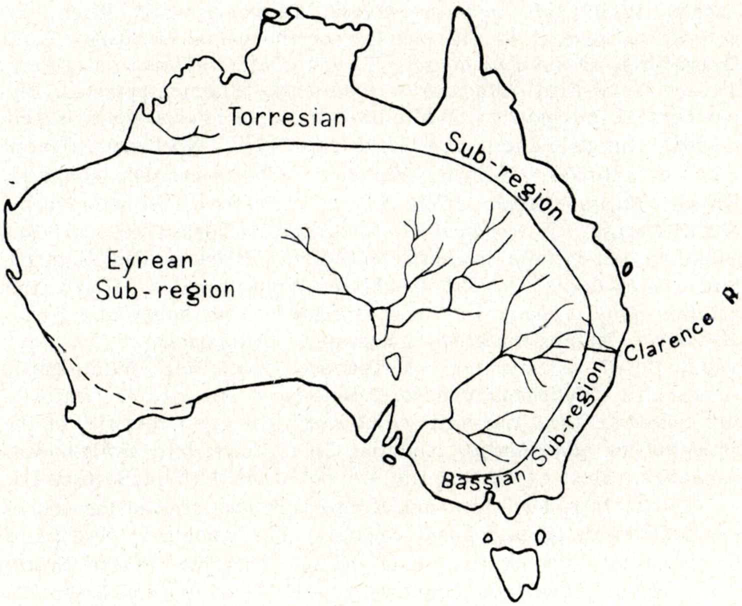

Fig. 1.—Main areas of distribution of Australian Faunas—modified from Baldwin Spencer. The approximate boundary of most of the Bassian elements in South-Western Australia is shown by the interrupted line.

The Kimberley birds not found further south include the Great-billed Heron *(Ardea sumatrana)*, Pied Heron *(Ardea picata)*, Burdekin Duck *(Tadorna radjah)*, Red Goshawk *(Erythrotriorchis radiatus)*, Crested Hawk *(Aviceda subcristata)*, Scrub-Fowl *(Megapodius freycinet)*, most of the Bustard Quails, Chestnut Rail *(Eulabeornis castaneoventris)*, White-browed Crake *(Porzana cinerea)*, Bush-Hen *(Amaurornis olivaceus)*, Jacana *(Jacana gallinacea)*, Red-crowned Pigeon *(Ptilinopus regina)*, Torres Strait Pigeon *(Ducula spilorrhoa)*, Green-winged Pigeon *(Chalcophaps indica)*, Rock Pigeon *(Petrophassa albipennis)*, Partridge Pigeon *(Geophaps smithi)*, Red-collared Lorikeet *(Trichoglossus haematodus)*, Scaly-breasted Lorikeet *(T. chlorolepidotus)*, Varied Lorikeet *(Psitteuteles versicolor)*, Sulphur-crested Cockatoo *(Cacatua galerita)*, Red-winged Parrot *(Aprosmictus erythropterus)*, Northern Rosella *(Platycercus venustus)*, Brush Cuckoo *(Cacomantis variolosus)*, Little Bronze Cuckoo *(Chrysococcyx minutillus)*, Koel *(Eudynamys scolopacea)*, Channel-bill Cuckoo *(Scythrops novaehollandiae)*, Rufous Owl *(Ninox rufa)*, White-tailed Nightjar *(Caprimulgus macrurus)*, Azure Kingfisher *(Ceyx azureus)*, Forest Kingfisher *(Halcyon macleayi)*, Roller *(Eurystomus orientalis)*, Rainbow Pitta *(Pitta iris)*, Papuan Cuckoo-Shrike *(Coracina papuensis)*, Cicada-bird *(Edoliisoma tenuirostre)*, Varied Triller *(Lalage leucomela)*, Lavender-flanked Wren *(Malurus dulcis)*, Red-backed Wren *(M. melanocephalus)*, Purple-crowned Wren *(M. coronatus)*, several of the warblers of the genus *Gerygone*, Black Grass-Wren *(Amytornis housei)*, Yellow Chat *(Ephthianura crocea)*, Tawny Grass-Bird *(Megalurus timoriensis)*, Lemon-breasted Flycatcher *(M. flavigaster)*, White-browed Robin *(Poecilodryas superciliosa)*, Rufous Fantail *(Rhipidura rufifrons)*, Northern Fantail *(R. rufiventris)*, Brown Whistler *(Pachycephala simplex)*, Brown Shrike-Thrush *(Colluricincla brunnea)*, Brown-breasted Shrike-Thrush *(C. woodwardi)*, Little Shrike-Thrush *(C. parvula)*, White-winged Sittella *(Neositta leucoptera)*, Black-headed Diamond-bird *(Pardalotus melanocephalus)*, White-throated Honeyeater *(Melithreptus albogularis)*, most species of the honeyeater genus *Myzomela*, Rufous-throated Honeyeater *(Conopophila rufogularis)*, Yellow-tinted Honeyeater *(Meliphaga flavescens)*, White-gaped Honeyeater *(Meliphaga unicolor)*, Blue-faced Honeyeater *(Entomyzon cyanotis)*, the Friar-birds *(Philemon)*, the great majority of the grass-finches *(Estrildidae)*, and the Great Bower-bird *(Chlamydera nuchalis)*. Most of these birds are not dealt with in Section III.

A few "intruders" from the Kimberleys have crossed the boundary southwards as permanent colonists, thus tending to obscure in the minds of some ornithologists the real character of the North-West region (namely the country between the Murchison River and the Pilbara district) but these northern tropical elements have no common southern boundary so we cannot recognise a Kimberley outlier in the North-West as can the botanists and students of some other animal groups. The numbers of such species diminish in frequency as we go south, but a few have managed to colonise the whole State. Such are the Black Bittern and the Brown Honeyeater, both of which in the east reach only as far as the Sydney district.

Eastwards the boundary of comparable significance to the Kimberley line, from the Australia-wide standpoint, is the Great Dividing Range (Fig. 1). Professor W. Baldwin Spencer in 1896, following on earlier work of Professor Ralph Tate and Charles Hedley, was the first effectively to establish the importance of both, and he bestowed the zoo-geographical names associated with these boundaries. The northern tropical area was called the Torresian sub-region, the eastern and south-eastern coastal strip, northwards to the Clarence River, the Bassian sub-region and the rest of Australia was the Eyrean sub-region—named after Lake Eyre in South Australia. The validity of the Great Dividing Range as a faunal boundary may be readily appreciated by naming some of the birds familiar in Western Australia which range across the continent to this boundary, or closely approach it, and are not found, except sporadically, across it in the east Australian coastal areas. Such are: the Major Mitchell Cockatoo, Mulga Parrot, Budgerygah, Black-eared Cuckoo, White-backed Swallow, Ground Cuckoo-Shrike, Western Warbler, White-face, Chestnut-rumped Thornbill, the western and inland relative of the Brown Thornbill, referred to in this book as *Acanthiza apicalis,* Redthroat, Blue-and-white Wren, Crimson Chat, Red-capped Robin, Crested Bell-bird, Black-capped Sittella, Red-tipped Diamond-bird, Black Honeyeater, White-fronted Honeyeater, Singing Honeyeater, Yellow-throated Miner, Spiny-cheeked Honeyeater, Black-faced Wood-Swallow, Grey Bell-Magpie or Squeaker, Spotted Bower-bird and Little Crow.

The Eyrean sub-region was the only one admitted by Baldwin Spencer when analysing the distribution of the Western and Central Australian fauna, and, according to the ideas prevailing at the time, the whole of that portion of Western Australia, which is treated in this book, would be occupied by one homogeneous bird fauna. However, though the fundamental importance of these primary divisions has remained undiminished, the data available to Baldwin Spencer were insufficient for a fair assessment of the situation existing in the South-West of Western Australa. Here there is a complexity of bird distribution which cannot be simply explained in terms of the old concept of faunal regions, and "instead of thinking of fixed regions it is necessary to think of fluid faunas" (Ernst Mayr).

However, though both the Eyrean and Bassian sub-regions were originally defined in terms of the old static zoo-geography their names can be used for the bird faunas which were evolved within their boundaries and which are still on the whole characteristic of them. The South-West is an area which at present cannot be classified as a simple zoo-geographical region, neither as a uniform portion of the Eyrean sub-region, as Baldwin Spencer originally thought, nor as a peculiar region on its own, as Tom Iredale suggested in 1928. It is a district occupied by an *intermingling of two faunas,* the Eyrean and the Bassian.

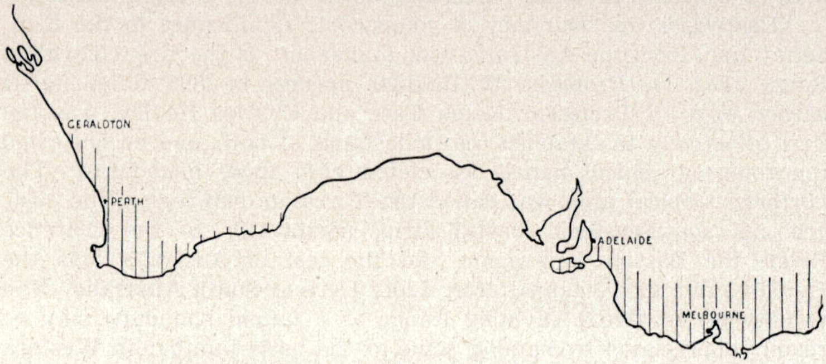

Fig. 2.—Distribution of the Western and Eastern Spinebills *(Acanthorhynchus superciliosus* and *A. tenuirostris).*

The birds of the South-West which are either specifically identical with those now occurring in southern Victoria, for instance, or but little differentiated, and which no longer have a continuous distribution through South Australia, include: the Painted Quail, Brush Bronzewing, White-tailed and Yellow-tailed Black Cockatoos *(Calyptorhynchus baudini* and *funereus),* Long-billed Corella *(Cacatua tenuirostris),* Western and Eastern Rosellas *(Platycercus icterotis* and *eximius),* Ground Parrot *(Pezoporus wallicus),* Fan-tailed Cuckoo *(Cacomantis pyrrhophanus),* the breeding range of the Golden Bronze Cuckoo *(Chrysococcyx plagosus),* Little Grass-bird *(Megalurus gramineus),* Scarlet Robin *(Petroica multicolor),* Southern Emu-Wren *(Stipiturus malachurus),* Western and Buff-tailed Thornbills *(Acanthiza inornata* and *reguloides),* Bristle-bird *(Dasyornis broadbenti),* the Yellow Robins *(Eopsaltria griseogularis* and *australis),* Golden Whistler *(Pachycephala pectoralis),* Shrike-Tit *(Falcunculus frontatus),* Spotted Diamond-bird *(Pardalotus punctatus),* Silvereye *(Zosterops gouldi* and *lateralis),* White-naped Honeyeater *(Melithreptus lunatus),* Western and Eastern Spinebills *(Acanthorhynchus superciliosus* and *tenuirostris),* Tawny-crowned Honeyeater *(Gliciphila melanops),* New Holland Honeyeater *(Phylidonyris novaehollandiae),* White-cheeked Honeyeater *(Phylidonyris niger),* Little Wattle-bird *(Anthochaera chrysoptera),* Red Wattle-bird *(Anthochaera carunculata),* Red-eared and Beautiful Firetails *(Zonaeginthus oculatus* and *bellus).*

In the case of some species, there still exists a narrow corridor of continuous distribution through the Nullarbor Plain or the coastal belt just south of it. Such species are the Spotted Scrub-Wren, perhaps the Dusky Wood-Swallow, and the Raven. The

White-fronted Chat has a somewhat wider connecting corridor. The case of the Yellow Robin is of interest in that the western form *(griseogularis)* also occurs on the eastern side of the Bight but is separated by Spencer Gulf and the dry country of South Australia from the eastern forms of the Yellow Robin (see Fig. 3).

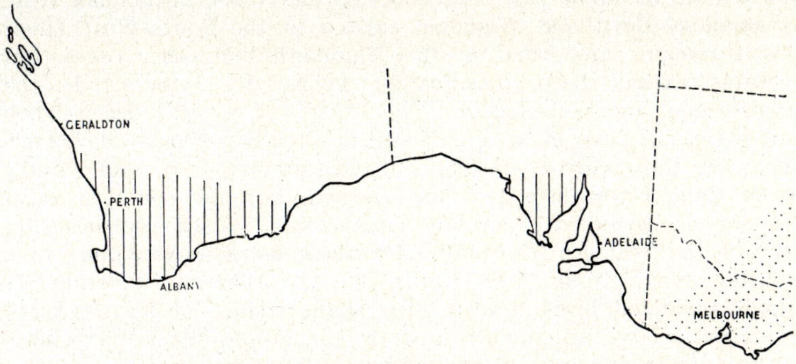

Fig. 3.—Distribution of the Yellow Robins—the Western (*Eopsaltria griseogularis*—vertical lines) and the Eastern (*E. australis*—dotted area).

Despite these elements of a common bird fauna in south-western and south-eastern Australia, the ornithological resemblances are consistently masked by the intermingling in the west of typical Eyrean forms. Thus occurring freely throughout the South-West are such Eyrean types as the Purple-crowned Lorikeet, Twenty-eight Parrot, Western Warbler, Banded Blue-Wren, Black-capped Sittella and Squeaker. The species of striated Diamond-bird which is abundant throughout the South-West is the Red-tipped *(Pardalotus substriatus)* and not the Bassian *P. striatus*. So also the representative of the Brown Thornbill group of species belongs to the Eyrean *apicalis* (=*albiventris*) series and not to the Bassian *pusilla*. Neither of these Bassian types occurs anywhere in Western Australia.

One may ask did they occur in former times, and were they replaced by their Eyrean relatives when the climate deteriorated during the Pleistocene and in sub-recent times? And associated with this question is another. What of the numerous Bassian species which are entirely absent from Western Australia?—birds such as the Little and Musk Lorikeets *(Glossopsitta pusilla* and *concinna)*, Crimson Rosella *(Platycerus elegans)*, Turquoise and Blue-winged Parrots *(Neophema pulchella* and *chrysostoma)*, Lyre-bird *(Menura novaehollandiae)*, Australian Ground-Thrush *(Oreocincla lunulata)*, Spotted Quail-Thrush *(Cinclosoma punctata)*, Superb Blue-Wren *(Malurus cyaneus)*, Striated and Little Thornbills *(Acanthiza lineata and nana)*, Speckled Warbler *(Chthonicola sagittata)*,

Pilot-bird *(Pycnoptilus floccosus)*, the Pink, Flame and Rose Robins *(Petroica rhodinogaster, phoenicea* and *rosea)*, Noisy Miner *(Manorhina melanocephala)* and Diamond Firetail *(Zonaeginthus guttatus)*.

It is conceivable that some or most of them may have once existed, as we know that creatures like the Koala, Tasmanian Wolf, Tasmanian Devil and Wombats existed in the South-West (their fossil remains are found in the Mammoth and other caves, but became extinct here, presumably, during dry intervals in the Pleistocene and Sub-Recent. The range of several species and sub-species of birds in the humid belt is already perilously restricted. Thus the Rufous Bristle-bird is confined to a strip of coastal country from Cape Naturaliste to Cape Leeuwin and the southern races of the Broad-tailed Thornbill *(Acanthiza apicalis leeuwinensis)* and of the Western Thornbill *(Acanthiza inornata mastersi)* to a coastal fringe from Cape Naturaliste to Albany. Unfortunately small woodland birds have a very slight chance of leaving fossil remains and we are unable to draw conclusions from that source as we are in the case of the mammals just enumerated.

At the present time the Bassian component of the bird-life of the South-West represents a selection only of that still existing in south-eastern Australia, and there has been comparatively little further evolution to create distinctive types in our area. The apparent exceptions will be referred to later. The South-West corner of Western Australia is an outlier of a formerly continuous Bassian sub-region in the sense originally employed by Baldwin Spencer, but with its fauna considerably reduced in number of species and partially replaced by an extensive invasion of Eyrean forms. As the climate deteriorated during arid intervals in the Pleistocene and later, and Bassian species died out in consequence, Eyrean ones moved in to occupy the vacant niches.

Apparently two separate factors contributed to the isolation of the Bassian bird-life in the South-West corner. These were the increasing desiccation of the interior, which caused a great shrinkage of the humid southern belt of country, and, secondly, the formation of the Great Australian Bight by down-faulting in comparatively late geological times—either late Pliocene or early Pleistocene (Dr R. W. Fairbridge).

The several dry intervals between the Pleistocene glaciations and the arid cycle following them, alternating with pluvial periods, would cause periodic "make and break" connections between the south-west and south-east around what is now the Great Australian Bight. This would provide opportunities for "double invasions" of the South-West both by Bassian and Eyrean species. As an example of the former, J. A. Keast and J. R. Ford have suggested the Yellow Robin group. The earliest Bassian representative evolved, during a period of isolation, into the White-breasted Robin *(Eopsaltria georgiana)*. When faunal traffic was resumed during a later, favourable, climatic interval a second invasion brought in the more typical present-day Yellow Robins, whose western form *(Eopsaltria griseogularis)* now has, through subsequent climatic deterioration, a broken distribution in the South-West and Eyre's Peninsula (Fig. 3). Eyrean invasions of this type have accounted for three forms of chestnut-shouldered Wrens *(Malurus elegans, pulcherrimus* and *lamberti)* coming to inhabit Western Australia and now behaving as good species towards each other. Another instance is that of the Ringneck or Twenty-eight Parrot *(Barnardius zonarius)*, in which the earliest population developed into the distinctive green-bellied form now confined to the South-West forests *(B. z. semitorquatus)*.

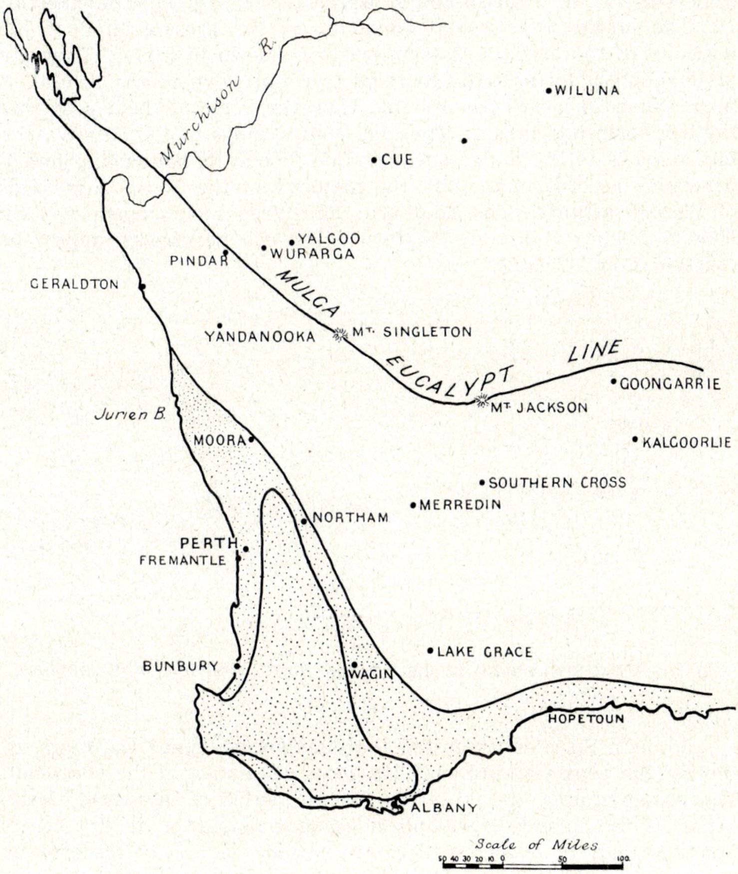

Fig. 4.—Distribution of majority of Bassian species in Western Australia (stippled)—the heavy jarrah forest belt is bounded by the continuous line within this area.

However, it failed to attain full specific separation before the arrival of the second invasion of yellow-bellied forms (typical *zonarius*) and there is a hybrid zone, or stepped cline, between them. A somewhat similar situation is provided by the Yellow-throated Miners *(Manorhina),* where the two forms merge gradually into each other —the darker *obscura* occurring in the South-West and the paler *flavigula* more inland and north. Other cases of Bassian and Eyrean double or "multiple invasions' have been proposed, and recent studies suggest that the White-tailed Black Cockatoo may be numbered as one of the Bassian examples. The phenomenon has been investigated in other animal groups, notably in the frogs by Dr. A. R. Main, in spiders by Dr. B. Y. Main and in March flies by Dr. I. M. Mackerras.

The northern or inland boundary at the present time of the majority of the birds of Bassian origin is shown in Fig. 4. They are predominantly forest birds confined to country where the rainfall is higher than 20 inches per annum. However, some of them penetrate further north and inland. Thus the winter range of the Red Wattlebird extends to the mulga-eucalypt line though its summer or breeding range accords more with the majority of the South-West birds of Bassian affinity. The Southern Emu-Wren occurs north to Dirk Hartog Island while the Western Silver-eye penetrates along the coast to Point Cloates.

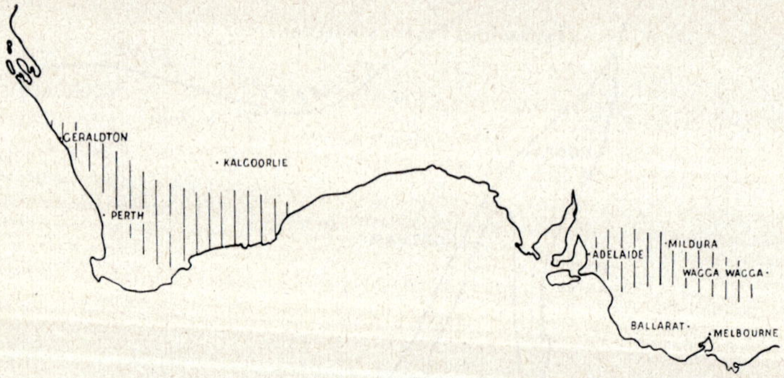

Fig. 5.—Main distribution of the Smoker Parrot *(Polytelis anthopeplus).*

Baldwin Spencer suggested that in Pliocene times the fauna of the Eyrean area was probably the most distinctive in the continent. The present fauna, developed during the period of increasing desiccation of the inland, represents a gradual adaptation to dry conditions. As may be expected there are varying degrees of response by birds of undoubted Eyrean origin to what might be termed the progressive advance or encroachment of the "dead heart" of Australia. There are those which have become most adapted to the arid conditions and can be considered as now typical of the desert or extreme arid conditions. Such species include the Alexandra Parrot, Night Parrot, Banded Whiteface, Slate-backed Thornbill, Rufous-crowned Emu-Wren, Wedgebill, Cinnamon Quail-Thrush, Grey Honeyeater and Grey-headed Honeyeater. It is not suggested that these species preferred the extreme arid environment, but they can be considered as the survivals which have been able to cope with the growing harshness of the conditions. The more favoured country would provide greater competition, and some of the dry country element may be considered as having found the desert an asylum or refuge from this more crowded habitat.

Another group of Eyrean birds forms an interesting assemblage associated with the dwarf eucalypts (mallees) of the southern part of Australia and the sand plain heaths which commonly go with them in Western Australia. The mallee belt is not continuous across Australia, being interrupted by a tongue of the central desert-steppe in the Nullarbor Plain. Therefore the Eyrean birds of the mallee group may show a discontinuous distribution, east and west, somewhat akin to that exhibited by the Bassian birds of the southern humid belt. The mallee and sandplain birds include: Mallee-Fowl, Purple-crowned Lorikeet, Smoker Parrot, Mulga Parrot, Blue Bonnet —reduced in Western Australia to a small remnant, Elegant Grass-Parrot, Southern Scrub-Robin, Chestnut Quail-Thrush, White-browed Babbler, Chestnut-rumped Thornbill, Gilbert's Whistler, Western Whipbird, Rufous Tree-creeper, White-fronted Honeyeater, Purple-gaped Honeyeater, Yellow-plumed Honeyeater and Squeaker.

It is an interesting fact that several of the Eyrean birds have differentiated into eastern and western forms, some so strongly marked that they now form good species. Thus we have the western and eastern representatives of the Ring-neck Parrots *(Barnardius zonarius* and *barnardi)*, the western and eastern forms of the Quail-Thrush of the arid interior *(Cinclosoma cinnamomeum castaneothorax* and *C. c. cinnamomeum)*, the western and eastern Whitefaces *(Aphelocephala leucopsis castaneiventris* and *A. l. leucopsis*, with, in the centre, an intermediate form, *A. l. whitei)*, the Banded and Black-backed Blue-Wrens *(Malurus splendens* and *M. callainus* and *melanotus)*, the Rufous and Brown Tree-creepers *(Climacteris rufa* and *C. picumnus)* and the Spotted Bower-birds *(Chlamydera maculata guttata* and *C. m. maculata)*. The Grey Shrike-Thrush which has almost a pan-Australian distribution may be considered here also. The eastern form, *Colluricincla harmonica*, though common in the Bassian area, ranges widely over the eastern portion of the Eyrean sub-region and is replaced in the western Eyrean, after the manner of the preceding birds, by *C. rufiventris*. What former barrier, in the middle of the Eyrean sub-region, which accounted for this evolution of western and eastern forms is now under discussion. The herpetologist, A. Kluge, has suggested that it was the great Pleistocene Lake Dieri (now represented by its saline remnants, Lakes Eyre, Frome and Torrens), as well as by the Flinders

Range and the sea bight to the south which existed at the heightened sea-levels during the inter-glacial maxima. At present the boundary between most of these species is a line passing northward from Spencer Gulf. This is the case, for instance, with the Ring-neck Parrots, the Tree-creepers and the Bower-birds, the central Australian birds being of the western forms.

The elements of the Eyrean fauna which in the South-West live side by side with Bassian birds comprise species of the mallee assemblage as well as some which are found over a wider range in the interior. Among these wide-ranging components are the Ring-neck or Twenty-eight Parrot, Western Warbler, Banded Blue-Wren (its Bassian relative of south-eastern Australia is *Malurus cyaneus,* not represented as such in Western Australia), and Red-tipped

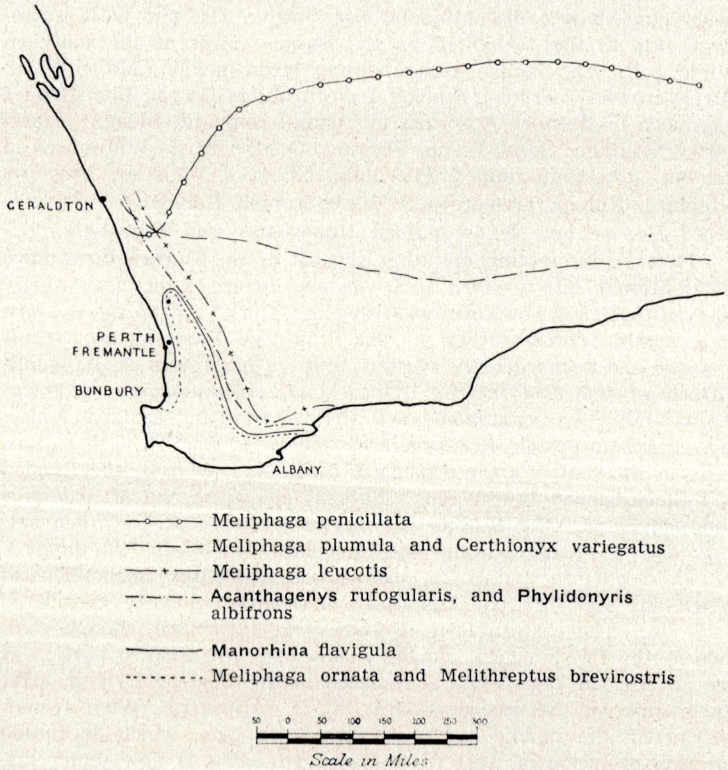

Fig. 6.—Southern and south-western boundaries of various Honeyeaters.

Diamond-bird. Those of the more limited mallee assemblage include the Purple-crowned Lorikeet, Rufous Tree-creeper and Squeaker.

The outstanding factor now delimiting the Bassian from the main Eyrean elements of the bird fauna in Western Australia is an ecological one: rainfall, with all the difference this makes in respect to vegetation and other dependent environment effects. It was plainly a shifting boundary, having been further inland in past times. As has already been stated, different species show a varying threshold limit in respect to their response to the environmental gradient. Fig. 6, giving the southern and south-western limits of occurrence of eight species of honeyeaters, illustrates the variation which exists. The boundary between the Bassian and Eyrean in the eastern states is a more positive one—the highlands of the Great Dividing Range which during the pluvial periods in past times may have been snow-capped and a more formidable barrier than now. But both east and west of this barrier there are rain and vegetational differences analogous to those found in South-West Australia. As in the west, so in the east there are many instances of Eyrean species crossing the boundary on to the Bassian region of the east coast. The Whiteface

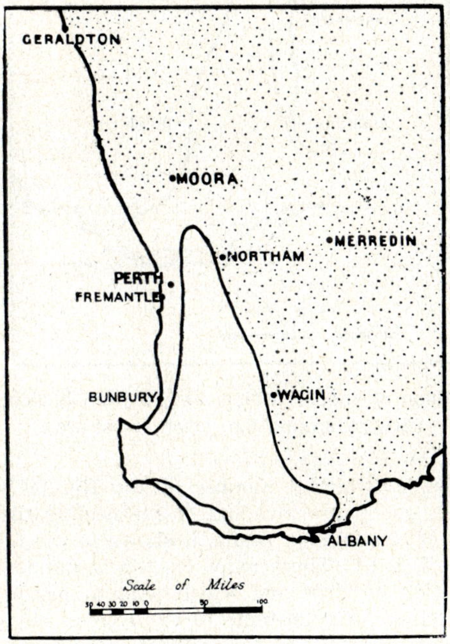

Fig. 7.—Distribution of species which avoid the South-West corner (stippled)—the heavy jarrah forest belt is bounded by the inner continuous line.

has done it near Sydney. The White-backed Swallow has penetrated the boundary at several places and is now an occasional breeding species near Sydney. *Strepera versicolor* also has crossed it at several places, as have the Striped Honeyeater *(Plectorhyncha lanceolata)*, the Red-tipped Diamond-bird *(Pardalotus substriatus)*, which has extensively inter-bred with the Bassian *P. striatus* to produce the variable hybrid population described as *P. ornatus* (=*assimilis*), the Crested Bell-Bird, Red-capped Robin, Red-backed Parrot *(Psephotus haematonotus)* and the Banded Stilt.

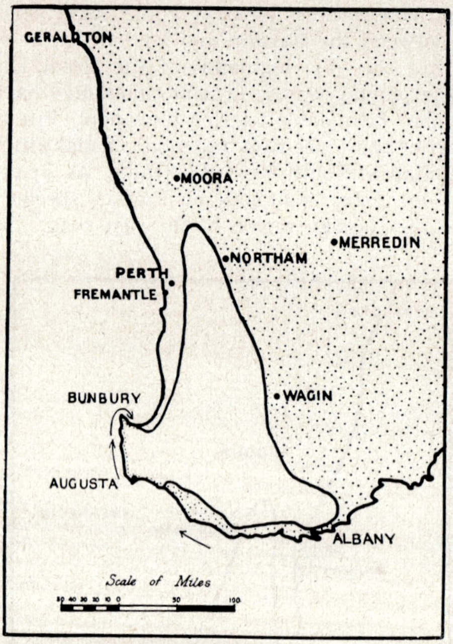

Fig. 8.—Distribution of species which are penetrating along the southern corridor around the jarrah forest area.

The manner in which a number of the Eyrean species of the mallee-sandplain group are invading or re-invading the South-West, the stronghold of the Bassian fauna, is shown in the following series of maps (Figs. 7, 8, 9). These maps explain a number of anomalies in the distribution of birds and which tend to puzzle beginners in ornithology. The greatest obstacle to the Eyrean birds is the thick continuous forest block of the South-West, comprising dense stands of jarrah *(Eucalyptus marginata)*, karri *(E. diversicolor)*, tuart *(E. gomphocephala)*, marri (red gum) *(E. calophylla)* and wandoo *(E. redunca* var. *elata)*. The Eyrean species which appear to be impeded in their distribution by the intervention of this heavy forest (Fig. 7) include the Southern Scrub-Robin, Chestnut Quail-Thrush, Redthroat, Crested Bell-bird, White-fronted Honeyeater, and the Zebra Finch. Others which do not approach so near to the forested area include the Australian Dotterel, Chestnut-rumped Thornbill and Spiny-cheeked Honeyeater.

However, the heavily forested area does not extend, as such, quite to the shoreline. Along the Swan coastal plain there is a tract of lighter country, including sandplain and open clay flats, while around the south coast there is a narrow strip, almost continuous, of open heath country (which contains elements of mallee flora), both providing avenues enabling Eyrean birds to outflank the forest. Along both corridors several Eyrean species have travelled varying distances.

Among the more interesting of these penetrations by the south corridor (Fig. 8) are those of the Mallee-Fowl which has reached the Busselton district, and the White-browed Babbler which has arrived at a similar point and, en route made limited invasions of the forest at several places, including Denmark, the lower reaches of the Warren River and Pemberton. The invaders from the north

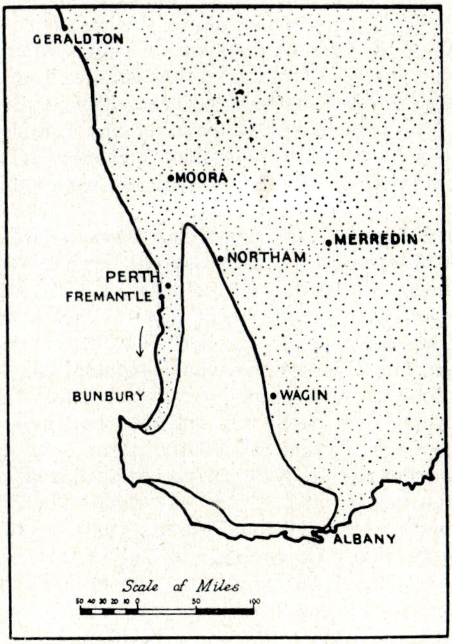

Fig. 9.—Distribution of species which are penetrating along the northern corridor into the South-West.

(Fig. 9), via the Swan coastal plain, include the Blue-and-white and Variegated Wrens, which have reached the Perth district, the Yellow-throated Miner, which has got down as far as Pinjarra; and the Brown-headed Honeyeater, Singing Honeyeater, Yellow-plumed Honeyeater and Black-faced Wood-Swallow which have reached to Busselton and beyond. The Galah and Smoker Parrot are recent invaders along this route and the White-backed Swallow and Black-throated Butcher-bird come down spasmodically. On the east of the forested area there has been some penetration of the jarrah and wandoo country in certain "salients." The most aggressive species appears to be the White-browed Babbler which, on the Toodyay Road, is within 43 miles of Perth. The Chestnut Quail-Thrush has also been recorded on the old York Road a similar distance from Perth.

It is rather an odd circumstance, and one difficult to explain adequately, that a few Bassian species show a distribution precisely similar to that of many of the above-mentioned Eyrean forms. They avoid the block of heavy-forested country in the South-West and have their headquarters in the savannah-mallee country or sandplain to the east and north. One which is entirely absent from the South-West corner (cf. distribution in Fig. 7) is the White-eared Honeyeater. The Brown Flycatcher is the analogue of the Eyrean group which has made a penetration along the southern corridor. The Shrike-Tit shows the type of distribution represented in Fig. 9. The Tawny-crowned Honeyeater has established itself on both corridors and completely encircled the forested area. The Southern Emu-Wren is another species which, whilst avoiding and completely encircling the jarrah forest block, has (as detailed in Section III) a distribution much more extensive in this State than elsewhere in its Australian range.

The distribution of these particular Bassian birds tends to suggest that the absence of so many Eyrean species from the South-West corner may not be due to historical but to ecological causes. The obstacle presented by the great block of heavy jarrah forest to the spread of several species—Bassian as well as Eyrean in origin—is clear enough, but it is not the whole explanation. Interesting differences in habitat preference have developed within the same species in east and west. One need only refer to two cases, those of the White-eared Honeyeater and the Southern Emu-Wren, already mentioned. The honeyeater in Western Australia has developed into an essentially dry-country bird, of the savannah-mallee woodland (see distribution limits in Figs. 6 and 7); and has none of the partiality of its eastern relatives (including the Yellow-throated Honeyeater, *Meliphaga flavicollis,* of Tasmania) for the humid coastal forests.

The encroachment and invasions just referred to are aided by clearing and deforestation following on settlement and it is evident that they will be accelerated in the future. The spread of the Western Magpie is a good illustration. Prior to settlement it was unknown in the heavily timbered country of the extreme South-West corner; in 1887 the ornithologist Tom Carter noted its southern limit as the Carbanup River near Busselton, but it now occurs in pockets of cleared land to Augusta. The process is now particularly evident in districts like Bridgetown, where the Singing Honeyeater and Hooded and Red-capped Robins were early arrivals since settlement. They have recently been followed by the Rufous Songlark, Shrike-Tit and Yellow-plumed and Brown-headed Honeyeaters. Since the 1950s the clearing of the forested area in the South-West has been greatly accelerated and this will aid the range extension of many Wheatbelt species through the jarrah forest block.

However, there is another factor operating which complicates the complete understanding of the distribution picture. During the past 40 years or so significant extensions of range have come under notice among several Eyrean species, which are steadily invading the outer portions of the South-West. Such species are the Crested Pigeon, Galah, Smoker, Budgerygah, Black-faced Wood-Swallow, Black-throated Butcher-bird and perhaps the Little Crow. During the same period other species have exhibited a notable increase in abundance in the South-West, including the Common Bronzewing and the Magpie Lark. It was formerly held that all these changes in status could also be attributed, as in the Western Magpie example just mentioned, to the effects of man-caused alterations in the environment. However, it now appears much more likely that these particular cases of range extension are due to a recent alteration in climate which is part of a world-wide phenomenon and accounting for considerable shifts in animal distribution, both on land and in the sea. The recent climatic change has led to an amelioration of the weather in high latitudes, in the northern hemisphere, with an increase of winter rainfall in the west wind zones and a decrease in the more arid areas. In Western Australia it has been shown that the annual rainfall of Perth has increased by about 6 in. and that of Albany by about 6½ in. (Dr. J. Gentilli). Outside of the wet South-West corner, however, the changes have been in an opposite direction and there is evidence of a slight deterioration of climate. Professor E. de C. Clarke and Mr. H. Tarlton Phillipps describe the drying up of the estuaries and the falling level of the water-table in the coastal strip between Albany and Hopetoun during the past 30-35 years. These changes can be expected to result in the dry country fauna closing in on the South-West, but the fauna in the South-West corner itself should hold its own, with the aquatic species there probably increasing in abundance.

These climatic fluctuations invite attention to the south-east coastal "bulge" between Esperance and Cape Arid. As first pointed out by J. P. Brooks in 1894, this section receives a slightly higher rainfall than the coastal strip immediately west of it, between Doubtful Island Bay and Fanny's Cove, and dropping off sharply inland from the coast. The Esperance-Cape Arid "bulge" is in some degree an outlier of the South-West region. In the case of some birds there is probably a broken distribution along the coastal strip eastwards of Albany, which must be accentuated by the recent slight deterioration in climate. The evidence so far, however, suggests that only the Red-eared Firetail is so affected. Its South-West distribution ends at the Hamersley estuary (west of Hopetoun) and the bird is not encountered again for about 135 miles until Lucky Bay (east of Esperance). It has been recorded at Mississippi and Duke of Orleans Bays and may occur even further east.

In passing it may be mentioned that some species, such as the Diamond Dove, Black-shouldered Kite, and Grass-Parrots (*Neophema* spp.) show periodic fluctuations in abundance which appear to be unrelated to the causes just discussed, though both, the results of settlement and climatic changes, may exercise important modifying effects. In addition the irruptions, due to chance vagaries of weather, referred to on p. 6, must be borne in mind when considering particular cases of range expansion.

The implication of this chapter is that the present-day bird fauna of this State is intimately related to the existing bird fauna of eastern Australia and there has been no marked evolution of new types or survival of distinctive ancient endemic species. There is one exception to this generalisation. The King Parrot (*Purpureicephalus spurius*) belongs to the only genus of Australian birds confined to the South-West. We believe it is the survivor of a form allied to the common ancestor of the ring-neck parrots (*Barnardius*) and the rosellas (*Platycercus*), and must have once had a wider distribution. Dr. A. J. Cain considers that its closest surviving relative is the Crested Parrakeet (*Eunymphicus cornutus*), a relict species in New Caledonia and the Loyalty Islands in the Pacific. Perhaps the disappearance of the King Parrot from most of Australia, other than the South-West, was due to the development of other *Platycercus* species, particularly the Crimson Rosella (*P. elegans*), which has either not reached Western Australia or not survived here. The White-breasted Robin (*Eopsaltria georgiana*) was once considered a puzzling endemic also, because of its deceiving superficial resemblance to the Mangrove Robin (*Eopsaltria pulverulenta*); its probable origin is discussed on p. 52. The Noisy Scrub-bird (*Atrichornis clamosus*), though distinctive because of its archaic character, is a close ally of the Rufous Scrub-bird (*A. rufescens*), now lingering as a relict form in some of the sub-tropical brushes of northern New South Wales and southern Queensland. These widely separated colonies are the final remnants of a formerly continuously distributed Bassian species now on the threshold of total extinction. Had the Noisy Scrub-bird vanished prior to the period of white settlement no ornithologist would have been daring enough to suggest its former occurrence in Western Australia on the basis of the present distribution of the genus in the Eastern States.

Before closing this section it may be desirable to set out in summary the principal physical and other environmental factors which play a part in controlling bird distribution within the main faunal sub-regions. That part of Western Australia dealt with in this book is portion of the great western plateau region of the continent and is topographically so featureless that it can be said that variations in altitude play no part whatsoever in affecting bird occurrances. Most of this area falls within the winter rainfall zone, except the Pilbara and adjoining districts, which lie, with the Kimberley Division, in the summer rainfall belt. In the interior rains are erratic and may fall at any time of the year.

The importance of the South-West heavy forest area, in respect to bird distribution, has been sufficiently stressed previously and its significance will be frequently referred to later in the accounts of particular species. Another important plant boundary-line, which will be often mentioned in the detailed accounts of bird distributions, is the mulga-eucalypt line (Fig. 4), which in the words of C. A. Gardner (former Government Botanist) is "remarkably well defined. One crosses it at Goongarrie to the north of Kalgoorlie; it runs exactly past Mt. Singleton, very close to Mt. Jackson, and it is intercepted again between Pindar and Wurarga. Exactly what happens to the east of Goongarrie is not definitely known, but the southern boundary [of the mulga] reaches the coast at Shark Bay." South of this line the woodland formations are dominated by eucalypts of various species and the ground water is salt. North of the line the woodland is predominantly mulga (mainly *Acacia aneura)* and in a very broad way, it can be said that the eucalypts, apart from a few desert species, are confined to the watercourses. Most importantly, the ground and surface waters in the mulga country are fresh, which has aided its pastoral development. The mulga line is the northern boundary, or probable northern boundary, of several birds, including the Purple-crowned Lorikeet, Brown Flycatcher, Rufous Tree-creeper, Yellow-plumed Honeyeater, White-eared Honeyeater, Red Wattle-bird, Dusky Wood-Swallow and the Squeaker. Others, such as the Wedgebill, White-browed Tree-creeper and Pied Honeyeater, do not cross it southwards, except in local penetrations here and there.

Another plant line of importance, though, in this case apparently only to a single species, the Spotted Bower-bird, is the "fig-line"—the southern boundary of one of the native figs *(Ficus platypoda)*.

The distributions of the various species in the following section have been given in ordinary geographical terms, as they convey to the reader a clearer picture than would be the case if the distributions were set out in the form of the various plant, soil or physiographical natural regions which have been proposed by Professors E. de C. Clarke and J. A. Prescott, Dr. L. J. H. Teakle, Mr. C. A. Gardner and others. The use of geographical terms also allows of greater precision in describing distribution of particular species as their ranges rarely coincide, even approximately, with such regions. Perhaps later, when fuller knowledge is gained both of bird distribution and the limits of natural regions, we may use the latter more freely for descriptive purposes.

SECTION III—THE BIRDS

EMUS and CASSOWARIES: Order Casuariiformes
EMUS: Family Dromaiidæ

Emu

Dromaius novæhollandiæ

Native Names: Kullya (De Grey River); Yalliberry (Shaw River); Galiberie (Nickol Bay); Jungoorna, Kalia (Coolawanyah); Yalla-biddee (North-West Cape); Yallo-biddy (Lyons River); Kalaya (Jigalong, Gibson Desert and Laverton); Jakeburi (Warburton Range); Yallebiri (Pidong tribe, Upper Murchison River); Yallabiri (Wallawe tribe, Mount Magnet); Kaliat (Shark Bay); Cullya (Murchison River, mouth); Kallilia (Northampton); Kulire (Geraldton); Ky-a, Wegie (New Norcia); Wadje (Perth); Widjie, Waitch South-West); Wait, Nurruk (Albany); Tula (Minniming tribe, Fraser Range).

Some vocabularies (e.g. Sir George Grey and George Fletcher Moore) give Noor-ruk or Nurrak as names of the Emu in the South-West. but according to J. A. Hassell this word means the egg of the Emu. Coodardy is the name for the Emu chick in the East Murchison country.

The largest bird in Australia and recognisable at sight without any description being needed. The other large non-flying bird, the Cassowary *(Casuarius casuarius)* does not occur in Western Australia. The sexes are similar and. despite bush lore on the subject, there is no universal rule on sex recognition. There is no difference in eye colour. which varies from yellow to dark brown. From studies in the East Murchison, S. J. J. F. Davies has found that the female is slightly larger than the male, but there is considerable overlap. In every mated pair, however, the female is the larger. During the four months before egg-laying the female is also darker than the male. She undergoes a body moult between February and May, whereas he moults after the eggs hatch in July and August, and, as a result, she is in new dark plumage at a time when he is still in old faded plumage. In addition, between February and May dark feathers grow on the female's head and upper neck, and the white bases of the neck feathers of the male become conspicuous, so that dark/light contrast between the sexes is exaggerated. In addition to weighing more than the male, the female member of a pair has a greater body length, longer tarsus, longer toe and longer beak. If mean measurements of a sample are used it is possible to show that males are larger or smaller than the female, depending on the way in which the sample has been taken, and it probably is less confusing to discuss differences between members of a pair.

Both sexes are able to give both a drumming call and a guttural grunt, but the male gives most grunts and the female does most drumming. Two types of drumming occur. One is given by birds that are slightly alarmed, and this is the gentle bubbling call for which captive emus are famous. The other is a much more resonant and louder booming call given by a female about the time she is laying and for some weeks afterwards. Pronounced expansion of the air sacs of the neck accompanies this booming.

The standing height (to the shoulder) is about 3 ft. 6 in. to 4 ft., and in the alert position the head may be 6 ft. from the ground. When feeding undisturbed the head and back are on a line parallel to and about 4 ft. above the ground. The body length of females averages 5 ft. and of males two inches less. The weight varies according to the season. In the autumn, before egg-laying and incubation. when the birds are in their best condition, females average 82 lb. and males 72 lb. The heaviest weight we have on record is a male shot at Beacon in August 1950 (one bird out of about a thousand) which weighed 113 lb. In drought years adult weights may drop to 52 lb. (data mostly from S. J. J. F. Davies). The average weight of 58 adult Emus (both sexes), weighed by J. L. Long, was 79 lb., with a range from 51 to 100 lb.

Distribution: In former times the Emu occurred all over the State and survives in varying abundance everywhere except in the closer settled areas. It still nests as near as 30 miles from Perth in the unoccupied country along the Toodyay Road, and also south of Mandurah. Wandering birds may appear as near as 11 miles of Perth (near Gnangara, B. Hutchison).

Nesting: The nest is almost always situated in open country and placed in the vicinity of a tree or large bush. In hilly or rough country a site on top of breakaways or hills is chosen. A flat bed of grasses, bark, leaves, etc., is usually made and trampled down, and on this the eggs are deposited. New laid eggs have a beautiful blue bloom which is rapidly rubbed off with the bird's body, probably even before it leaves the nest after laying the egg (S. J. J. F. Davies). The colour of the granulated shell is dark green and, as incubation proceeds, the eggs become polished by the wearing down of the granulations. They thus appear to become generally darker. The average size is 5¼ by 3½ inches, but occasionally abnormally large or abnormally small examples are found; among the largest recorded is an egg from the Pilbara district measuring 6¾ by almost 4 inches. and the smallest we have seen is one from near Yalgoo, 3 by 2¼ inches. The average weight of an egg. with its contents, is 20½ ounces (A. J. Campbell), ranging from 16 to 28 ounces (Dr. A. M. Morgan). The usual clutch size is 8, 9 or 10 eggs, varying up to 16.

The species is a regular winter breeder, egg-laying beginning at the end of April. It is the male bird alone which incubates the eggs and cares for the young. According to observations of birds in captivity by D. Fleay, in Victoria, the male Emu may brood over the eggs and newly-hatched young for 63 days, the hatching period of the individual eggs ranging from 58 to 61 days. The striped chicks are first seen in late July, and they lose their stripes by October. The birds take at least two years to reach maturity (J. L. Long). Captive birds may mature in eighteen months.

The birds feed on a wide variety of native fruits, vegetable matter and insects. In the North-West they eat fruits and flowers in spring and early summer, especially the green fruits of *Cassia*, emu berry *(Scaevola spinescens)*, turpentine or poverty bush *(Eremophila fraseri)*, sandalwood and quondong *(Santalum)*, the flowers of gidgee *(Acacia pruinocarpa)* and corkwood *(Hakea suberea)*, and the seeds of *Acacia*, particularly of *A. victoriae*. In late summer and autumn they eat grass and insects (caterpillars and beetles) that follow the summer rains, and in winter feed on the leaves and flowering heads of winter herbs, especially nulla nulla *(Ptilotis exaltatum)*, yellow everlastings *(Helichrysum)* and parakeelya *(Calandrinia)*. The last is an important food of very young chicks. In the South-West they eat the fleshy fruits of the emu bush *(Podocarpus)*, the red rind of zamia nuts and even sheoak *(Casuarina)* fruits. Plague grasshoppers are eaten in large quantities when available and also caterpillars. In the farming country they will eat cereals, introduced grasses and capeweed (preferably the flowering heads).

Motorists on country roads have often checked the running speed of the Emu. Cars can pace the bird up to 25-30 miles per hour and then overtake it.

Despite the fact that it has declined in numbers in the closely settled areas, the Emu in recent years has attracted considerable attention by developing into a major pest in the northern and north-eastern Wheatbelt. When the Vermin Act was originally passed by the State Parliament in 1918 the Emu was not important enough to figure in its provisions and the bird first found its way into official records as a serious pest in 1919 when the Upper Chapman Road Board, at Nanson, sought the removal of protection for Emus on the ground that they were a menace in the cropping season owing to large mobs trampling through the paddocks and feeding on the wheat. The agitation spread throughout the Northampton district and south to Mingenew. In 1923 protection for Emus under the Game Act was removed north of Latitude 30° S. (previously the northern limit had been the No. 3 Rabbit Proof Fence, passing east and west through Ajana) and district vermin boards were empowered to levy a bonus on Emu heads. The records are defective as to how many birds were destroyed in this period. Local boards paid about £180 as bonuses in 1926-27 and £136 in 1927-28, some boards paying 1/- per head, others 2/6. It was further reported that in the latter half of 1928 between 3,000 and 4,000 Emus were destroyed in the Ajana district alone, and one farmer stated that earlier in the year he had seen a heap of dead Emus on one holding 16 ft. high by 4 ft. square. In 1929 a particularly severe invasion of Emus in the Northampton district decided the Government to subsidise temporarily the local boards £ for £ in a uniform bonus of 2/6 per head. Payments were made on a total of 2,148 birds.

For some years trouble in the Northampton area waned, the rabbit proof fence checking the movement of the Emus southward from the Murchison country. In November 1930, the local vermin inspector reported that the No. 3 Rabbit Proof Fence blocked thousands of Emus going south — "They are lying dead all along the fence between Ajana and the junction of the No. 2 and 3 fences. . . . The farmers on the south side of the fence are jubilant."

In 1932 another focus of complaint against the Emu developed in the north-eastern Wheatbelt, in the Campion. Bullfinch and Walgoolan areas. Here an invasion of Emus took place for the first time on a scale large enough to make them a serious menace and the affected farmers persuaded the Commonwealth and State Governments to become involved in that bizarre interlude. the Emu War of November 1932. This was an attempt at the mass destruction of the birds with machine guns operated by a military unit. In October the Minister for Defence (Sir George Pearce) agreed to a request from the local settlers that a machine gun unit should be used to eradicate the large numbers of Emus which were then menacing the district. About 20,000 Emus were reported to be devastating crops in an area of about 40 square miles. The Premier (Sir James Mitchell) co-operated and the settlers agreed to meet the cost of the expedition. The offensive began on November 2, at Campion, when an armed party consisting of the Commanding Officer of the 7th Heavy Battery, Royal Australian Artillery (Major G. P. W. Meredith), with Sergeant S. McMurray and Gunner O'Halloran, took the field with two Lewis guns and 10,000 rounds of ammunition. About 50 settlers turned out to organise a 20-mile drive of Emus to a point on the No. 1 Rabbit Proof Fence where the fire would be concentrated. A special correspondent of the press was in attendance as well as a motion picture photographer from an international unit.

A preliminary skirmish on the first day accounted for about 12 birds out of a mob of 40 which approached Campion siding. However. the gunners' dreams of point-blank fire into masses of Emus soon vanished. On the second day it was realised that open warfare was useless as the birds had split up into innumerable small parties. Ambuscades were resorted to and the guns placed at dams where the birds came to drink. Before daylight on November 4 a mob of about 1,000 Emus approached such an ambush, the largest mob that had hitherto come within point-blank range. Gunner O'Halloran opened fire at a few hundred yards and about a dozen birds fell when the gun jammed. The biggest kill at an ambush took place two days later when about 50 birds were shot. By then about a quarter of the ammunition had been expended.

The disappointing results drew farcical references in the House of Representatives at Canberra and on November 8 Sir George Pearce ordered the withdrawal of the Lewis guns from the combat area. "No reason is advanced for this step," stated a Canberra press report, "but it is believed that the defence authorities do not desire to have a precedent set up in the destruction of noxious birds and animals in such a manner." The dismayed settlers, claiming that about 500 Emus had already been destroyed. urged the resumption of operations and their continuance at least until the allotted 10,000 rounds had been expended. Sir James Mitchell added his representations and the same military party again entrained for Campion on November 12. The campaign, however, had lost its glamour and the party was unaccompanied by the press or photographers. There were no spectacular results anyway. A day's casualties of 50 Emus was sufficiently important to be chronicled in an occasional news item. The Department of Agriculture looked coldly on the whole campaign. A spokesman told the press that the department had not been consulted when the armed party under Major Meredith commenced its offensive against the Emus and the whole affair had been arranged by the settlers themselves with the Defence Department.

In 1935 the Northampton and neighbouring districts suffered another severe Emu invasion and the Government agreed to pay a bonus of 1/- per head without insisting on corresponding subsidies from the local vermin boards. Altogether 57,034 birds were destroyed from August to January, 28,577 in the Northampton district alone, and the great majority of these were killed north of the No. 3 Rabbit Proof Fence. The trouble continued, but on a lesser scale, during the following year, and the bonus was reinstated from August to November 1936, 15,521 being destroyed. For years afterwards the heaps of Emu bones could be seen strewn alongside the Rabbit Proof Fence near Ajana.

The year 1936 also saw a revival of the Emu as a wheat pest in the north-eastern districts, but for some years the farmers' agitation waned both here and in the Northampton areas. In 1943, however, the Emus invaded the wheat country on both fronts. Instead of offering bonuses the Government on this occasion distributed free ammunition to local vermin boards for the organisation of Emu drives. A typical activity was that of the Walgoolan Emu Destruction Committee which conducted Sunday drives with motor vehicles. Up to July 1944 the committee accounted for over 2,000 Emus.

Until 1944 the Government had resisted the requests of farmers that permanent bonuses for the destruction of Emus be paid from the central vermin fund. Because this fund was contributed to by a State-wide tax it had been earmarked for dealing with pests declared as such throughout the State, and the Emu, despite its severity locally, was considered as a pest only in certain of the outer fringes of the Wheatbelt. However, the then Minister for Agriculture (Mr. F. J. S. Wise) yielded to the pressure, and in October, 1944, the Emu was declared "vermin" throughout Western Australia. Naturalists protested strongly at this step and, with the support of the majority of the district vermin boards themselves, succeeded in having the anomaly rectified in 1947, when the bird was restored to the protected list in the South-West corner. However, elsewhere bonus payments of 1/- per head continue to be paid from the central vermin fund and since they began to operate the payments have been made on the following numbers of Emus (to June each year): 1945, 16,278; 1946. 5,395; 1947, 8,836; 1948, 13,061; 1949, 27,244; 1950, 30,174; 1951, 22,106; 1952, 8,572; 1953, 18,436; 1954, 10,471; 1955, 12,905; 1956, 12,877; 1957, 25,890; 1958, 23,473; 1959, 9,639; 1960, 39,367 (this is the highest number on record since the uniform bonus has been paid); 1961, 10,732; 1962, 23,484; 1963, 12,477; 1964, 14,476; 1965, 19,381; 1966, 12,641; 1967, 11,032; 1968, 16,046; 1969, 8,462; 1970, 34,247; 1971, 12,233; 1972, 4,368.

Since February 1, 1948, the bonus on Emus has been increased temporarily, to 4/- per beak, provided that at least 1/- of this amount is paid from the district vermin boards' own funds. In May 1970 the bonus rate was reduced to 20c and the bonus was abolished altogether on June 30, 1972. A special temporary bonus of 6d. each for Emu eggs gathered in the agricultural areas only, was also authorised in February 1948, but no payments have ever been made under this heading.

In an effort to establish the South-West protected region (which in 1947 was defined as all that country to the south and west of the Perth-Kalgoorlie and Great Southern Railway lines), on a Vermin District basis (in this case also Shire District basis), the Agriculture Protection Board has recently arbitrarily included in the protected zone all those Shires where an appreciable portion of their districts came within this area. This had the effect of extending the area of protection a little further than was originally intended. Within the South-West area protection is occasionally lifted where Emus are claimed to be doing damage. The Shires in which an open season is declared at present are Greenbushes, Bridgetown, Collie, Nannup, Donnybrook and Gingin.

The success of the No. 3 Rabbit Proof Fence in arresting the seasonal movement of Emus southwards from the Murchison country encouraged the Department of Agriculture to extend the barrier system by building the Lake Moore Emu Fence between 1957 and 1959. It is approximately 120 miles long and runs in a slightly W.N.W. direction across Lakes Moore and Monger, joining the 80 mile peg of the northern section of the No. 1 Rabbit Proof Fence and the 138 mile peg of the northern section of the No. 2 Fence.

Recent banding studies organized by Dr. Davies have confirmed that Emus reaching the southern farming areas come from the North-West pastoral districts and also emphasise the effectiveness of the vermin-proof fences. Over 150 birds were banded outside these fences and none of the 35 recovered had got through them. There is a wide dispersal of marked birds from the banding sites. A bird banded on the No. 3 Fence (32 mile peg) on September 12, 1969, was recovered at Credo Station, near Kalgoorlie, 330 miles south-east, on August 15. 1970. Another banded at Mileura Station on September 10, 1969, was recovered at Warrora, north of Carnarvon, 310 miles north-west, on March 28. 1970.

PENGUINS: Order Sphenisciformes
PENGUINS: Family Spheniscidæ

Little Penguin (Fairy Penguin)

Eudyptula minor

Upper part a steely *bluish*-grey, *throat and entire under parts white*. There is *no crest*. Iris, blue-grey; beak, dark lead-grey with ridge black; legs, pinkish-white. Length, 15-17 in. Weight. 1 lb. 7 oz.-1 lb. 15 oz.; heavier during the moult.

Distribution: Occurs all along the south coast and on the west north to the Fremantle area, occasionally straying north as far as Pelsart Island (Abrolhos) and False Entrance, Shark Bay; its northernmost breeding island is Carnac. This island and Penguin, Shag, Seal and Bird Is.. near Safety Bay, appear to be the only breeding localities on the west coast. The species nests on several islands off the south coast, having been recorded on Seal and St. Alouarn Islands (C. Leeuwin), Breaksea and Rabbit Islands in King George's Sound, Eclipse Island, Bald I., Cheyne Island, the Doubtful Islands, and the islands of the Archipelago of the Recherche (Boxer. Rabbit, Charlie, Gull, Sandy Hook, Station, Mondrain, Termination, Round, Rob, Marts, Nares, North Twin Peaks, Wedge, Combe, Goose and Christmas (Daw) Islands).

Nesting: The Little Penguin usually nests under a rock or under a bush or a log, and a short burrow may or may not be excavated. The nest is made of leaves, seaweed, or sticks; one or two eggs form the clutch. and incubation takes about 36-42 days (L. E. Richdale). The eggs are oval, white, and 12 Western Australian eggs average 56 x 42 mm., varying from 52-60 x 37-44 mm. Weight 40 gm. As incubation proceeds the eggs become discoloured. The chicks when first hatched are covered with down, sooty-brown above and whitish below. They remain in the burrows 51-58 days (Richdale). Although Little Penguins may be found nesting singly, they usually breed in dispersed colonies.

This bird comes ashore during its nesting duties only at night, when the calls, at its island homes, may be heard continuously from dusk to dawn. It is very unobtrusive by day and even at sea is only occasionally seen, contrary to its habit in south-eastern Australia, where, as in Tasmania, one very commonly meets with small swimming parties by day. It is shy of human beings. It is, however, a very aggressive bird when disturbed in its burrow, biting with its formidable beak and beating an intruder vigorously with its flippers, each being used alternatively.

Rockhopper Penguin (Crested Penguin)

Eudyptes chrysocome (= cristatus)

Larger than the Little Penguin, with the upper parts dark bluish-grey, the *head and throat black* and the under parts white. Over each eye *plumes of yellow and some black feathers form a prominent crest.* The crest starts as a narrow superciliary line that does not reach the base of the beak. In the immature bird the crest is undeveloped, being represented by a yellowish eyebrow line and the throat is pale grey. The under flipper has a large black patch at the tip and a prominent black anterior margin. Iris, red; beak, reddish-brown; legs, pale flesh, soles black. Length, about 18-23 in. Weight 4¾-6 lb. — up to 10 lb. during moult.

Distribution: A straggler to the South-West coast from its breeding stations at St. Paul and New Amsterdam Islands in the southern Indian Ocean. Specimens have been observed north to Rottnest Island and Scarborough and east to Esperance (and further along the southern Australian coast eastwards to Victoria and Tasmania), but most of the records are from between Bunbury and Denmark.

There are a number of closely allied crested penguins which breed on the Sub-Antarctic Islands and range northward after the breeding season is over. The present species is the most frequent visitor to Western Australia and belongs to the sub-species *E. c. moseleyi* breeding at, among other localities, St. Paul and Amsterdam Islands.

It is often attempted to keep alive specimens of crested penguins which land on our shores. Two difficulties have usually foiled such endeavours. The first concerns the feeding of the birds; they refuse to feed themselves and have to be force-fed or hand-fed with chopped-up fish or meat. Otherwise the birds must be allowed access to a pool into which the food can be placed submerged, for penguins, accustomed to procuring their prey *under water,* seem to have little ability in adjusting themselves to taking their diet on land.

Another handicap in keeping penguins in captivity locally, even at the Zoological Gardens, South Perth, is their liability to develop a mould or fungal infection of the respiratory tracts. Dr. L. Le Souef, honorary prosector to the Gardens, has determined the mould to be *Aspergillus fumigatus. Aspergillus* infection is well known in captive birds, particularly in water birds. In America Aspergillosis has been as fatal to penguins in some zoos as it has been at South Perth, but in the San Diego Zoo experiments have suggested that predisposition to mould infection might be due to lack of iodine in the diet, and the trouble appears to be overcome by the daily administration of sodium iodide, approximately two grains per ten-pound bird, the dose being placed in one of the fish fed to it.

These penguins are not in the slightest degree shy in the presence of human beings and birds occasionally kept in captivity in Australia make engaging pets. The gait is a shambling walk, or if the bird wishes to move faster it hops with both legs jumping simultaneously. It swims with its flippers used as paddles. The voice is a loud hoarse trumpeting.

Thick-billed Penguin

Eudyptes pachyrhynchus

Similar to, but rather larger than the Rockhopper Penguin. Upper parts bluish black, the head and throat black and under parts white. The cheeks may show a streaked appearance when the white bases to the feathers are revealed. The yellow stripe or plume over the eye is more compact than in the previous species and not flared out at the tip; it originates in the angle formed by the culminicorn and latericorn and does not meet its fellow of the other side. The under flipper pattern shows much less black than in the other two crested penguins, being slight at the tip and the anterior margin. Iris, red brown; beak, reddish brown, the culmen when viewed from above is bowed at the sides; legs, pale flesh. Length, 20-28 in.

Distribution: A straggler to southern coasts from its breeding area on the west coast of the South Island of New Zealand and adjacent islands. Specimens are known from Scarborough, Leighton and Busselton. It has only recently been recorded from Western Australia (by the examination of museum specimens by K. G. Simpson, of Monash University). It is a fairly frequent visitor to Tasmania and south-eastern Australia.

Erect-crested Penguin

Eudyptes sclateri

The largest of the crested penguins known to visit Western Australia. Upper parts bluish black, the head and throat black; under parts white. The *yellow crest stands out very erect and brush-like,* not drooping, and does not meet in front. It originates at the base of the beak close to the gape and not at the angle between culminicorn and latericorn. The under surface of the flipper is *extensively black towards the tip* and along the anterior margin. Iris, brown; beak, brown, the culmen when viewed from above is parallel-sided; legs, pale flesh. Length, about 27 in.

Distribution: A straggler from the islands south of New Zealand (Antipodes, Campbell and Bounty Islands). It was not known to occur in this State until a live specimen was captured at Esperance in March 1972 and taken to the South Perth Zoological Gardens. It is an occasional visitor to Tasmania, Victoria and New South Wales.

Adelie Penguin

Pygoscelis adeliæ

A larger penguin than the preceding species. The upper parts bluish-black, head and throat black; head without crests or markings except for *conspicuous white eyelids;* under parts white. Iris, reddish-brown; beak, brick-red; legs, pale flesh-pink with black soles. Stands erect, to a height of about 2 ft. 4 in. Length, 27-31 in. Weight, 6-10 lb., heavier at beginning of moult.

Distribution: A straggler from Antarctic Seas. There are two records from Australia but neither are completely satisfactory. One bird was found near City Beach in March 1937, but as there was a possibility that it might have been a bird released from a Japanese whaling ship the record was omitted from earlier editions of this book. It has recently been learnt that in July 1933 a penguin was picked up at Portland, Victoria, and was identified as an Adelie by the late Professor F. Wood-Jones, an acknowledged authority on sea-birds. However, he never published the record in a scientific journal and the specimen is not preserved in any museum. No Adélie Penguins have ever been recorded in South Africa or New Zealand, which normally receive more Antarctic stragglers than Australia, and so the two supposed Australian occurrences should be held on the "suspense" list until further records are made.

The species breeds around the Antarctic continent.

TUBINARES: Order Procellariiformes
ALBATROSSES: Family Diomedeidæ

Wandering Albatross
Diomedea exulans

This is the commoner of the two largest albatross in local seas. The adult birds are pure white with only the ends of the wings black, both above and below. Young birds have the entire upper surface and under body dark, but the under wing (except for the tip) is always white—this character at once distinguishes the species from any other local albatross. Iris, brown, eye lid bluish; beak, flesh-white or cream; legs, pale bluish-grey. Length, 4 ft. 5 in.; wing span, 8 ft. 11 in. to 11 ft. 6 in. Weight, 13-25 lb. Culmen, 161-171 mm.; wing, 632-700 mm.

Distribution: The Wandering Albatross is a familiar attendant on ships crossing the Great Australian Bight and on the steamer route to Cape Town, but it is not often seen to the north of Fremantle. It approaches quite near Rottnest and the west side of Garden Island, but seldom ventures far into Cockburn Sound. It is very rarely found on the beaches after winter storms, only two specimens having been reported on the beaches near Perth in the past 40 years. In the Indian Ocean sector it nests at the Marion, Prince Edward, Crozet and Kerguelen Islands.

This grand albatross never fails to excite the admiration of all voyagers by its superb soaring ability as it glides for hours on end, with only a momentary wing-beat, behind and around a vessel at sea. To the traveller to Australia from the north the sight of the first albatross always evokes a thrill and the realisation that one is at last out of the tropics and in the southern seas.

An albatross of undetermined species figured, in 1887, in one of the most remarkable cases in bird-ringing annals. On September 18 of that year, near Trigg Island, some boys found a dead albatross with a rough tin collar around its neck with the following message punched in as if with a nail: "13 naufrages sont refugies sur les iles Crozet 4 Aout 1887" (i.e., thirteen shipwrecked sailors have taken refuge on the Crozet Islands, August 4, 1887). One of the then lads was Colonel W. O. Mansbridge, later resident magistrate at various centres, including Perth, and another Mr. A. A. Moffat, late of the Official Receiver's Office. They quickly brought the matter under official notice and the message was telegraphed to London immediately. For some reason the Foreign Office did not inform the French Government until October 6 and it was not until December 2 that a French warship, *La Meurthe,* arrived at the Crozet Islands from Madagascar to seek survivors. The Commander, Lt. Richard Foy, reported that in a hut at Hogs Island, one of the Crozet group, he found a letter from Captain P. Majou, of the French sailing ship *Tamaris,* which solved the mystery of the albatross's message, but unhappily, despite diligent search, there was no trace of the men. The captain's letter explained that the vessel, which had left Bordeaux for Noumea with a cargo of iron in December 1886, was wrecked on Penguin Island, in the Crozets, on March 9, 1887, and the ship's company made their way to Hogs Island in two boats, reaching it two days later. There they subsisted on food in a cache established by H.M.S. *Comus* in 1880, and after exhausting the provisions the men left for Possession Island on September 30. The unfortunate seamen, however, never arrived. It will be noted that the men released the albatross some five months after they were first marooned—possibly as a despairing "last straw" or maybe one of many in a systematic bird-marking scheme, but the captain's letter does not refer in any way to the experiment. At any rate, this particular bird was found in Western Australia twelve days before the fatal last voyage to Possession Island. Possibly the tin collar around the bird's neck contributed to its misadventure on a shore 3,027 miles from where it was marked, after 46 days, representing a mean daily travel of 66 miles.

It is traditionally held that the bird concerned was the Wandering Albatross, but in our first edition we suggested that it might have been a Yellow-nosed, which is the most common species visiting Western Australian seas. The Yellow-nosed Albatross, however, does not nest at the Crozet Islands. The only descriptions extant of the bird that carried the message are that it was "a white object" in the distance; that "its colouring was black and white. not distinct bands of these colours but more mottled," and that the wing span was about 6 ft. 6 in. The measurement, if remembered correctly (it was first mentioned by Colonel Mansbridge in 1922) would preclude the Wandering Albatross and would suggest one of the smaller species of mollyhawks. The only one which probably nests in the Crozets is the Black-browed Albatross, a rare visitor north of Cape Naturaliste. The late Gregory Mathews thought that the messenger was really a Giant Petrel in part-white plumage, but after inspecting a series of photographs we showed to him, Colonel Mansbridge was satisfied that the bird he found was actually an albatross.

Royal Albatross

Diomedea epomophora

Very similar in appearance to the adult Wandering Albatross, but the wings are more extensively black on the upper surfaces. The immatures resemble the adults. Iris, brown, *eye lid black;* beak, flesh pink, with *cutting edges of both mandibles blackish;* legs, bluish white. Length. about 4 ft. 6 in.; wing span, about 10 ft. Culmen 150-186 mm.; wing, 590-670 mm.

Distribution: This species breeds on the islands south of New Zealand and at Otago on the South Island. A few individuals have been observed in eastern Australia and the only record from this State is of a beach-washed specimen found at Ledge Point, 70 miles north of Fremantle in early March 1966. It carried a leg ring showing it was banded as a fledgling in the nest at Campbell Island in mid-September the previous year. This individual could have circumnavigated the globe via South America.

Black-browed Albatross (Mollymawk, Mollyhawk)

Diomedea melanophris

This and the following three species belong to a group of smaller Albatrosses (or Mollymawks) which are largely white but have the whole of the upper wings black, together with that portion of the back (the mantle) between the wings. Thus there is a *complete band of black from wing tip to wing tip,* while the head and neck in front, and the back behind, are white. Tail dark. The present species is distinguished from the others of this group by the *broad margin of black around the under wings* and by the *orange beak.* In the young bird, however, the beak is dark-brown but the black margin of the under wing is so wide in these immature birds that it

Fig. 10.—Black-browed Albatross—under wing pattern.

encloses only a very narrow central white area. *No other albatross has so much black on the under wing margin* and this feature alone practically diagnoses the species. There is a short black line in front of and a longer black line behind the eye. Iris, brown. Length,

33-34 in.; wing span 7 ft. 2 in.-7 ft. 8 in. Weight, 6¼-7lb. Culmen, 104-124 mm.; wing, 480-552 mm.

Distribution: This species is rarely seen north of Cape Naturaliste but beach-drifted specimens are occasionally found on metropolitan beaches after winter gales. It is common along the south coast and in the Great Australian Bight and is the most numerous albatross along the New South Wales coast as far as Sydney. Most of the birds seen on our south coast are adults with orange beaks. In the Indian Ocean sector it breeds at Kerguelen and Heard Islands and probably on the Crozets.

Yellow-nosed Albatross (Mollymawk, Mollyhawk)

Diomedea chlororhynchos

Native names: Wool-wool, Wulwul (South-West).

Like the preceding species in general colour pattern (white with black upper wings and mantle), but with *a narrower black edging to the under wings and sharply marked off from the white.* Tail dark. The *beak is black* in the immature birds and in the adult it has *a bright yellow ridge line* along the top of the beak, the terminal hook rose-pink. This is visible only when the bird is near at hand or a glass is used. Iris, brown; legs, milk white. Length, 2 ft. 4 in. - 2 ft. 8 in.; wing span, 5 ft. 10 in. - 6 ft. 9 in. Weight 4-5½ lb. Culmen, 107-120 mm.; wing, 445-520 mm.

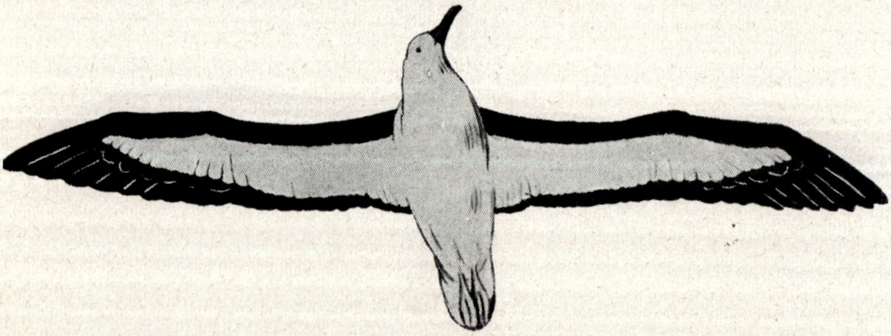

Fig. 11.—Yellow-nosed Albatross—under wing pattern.

Distribution: This is the most plentiful albatross on the west and south coasts and can be seen in numbers as far north as Point Cloates. It occurs in our waters between April and November. The extent of its northern wanderings is indicated by the finding of a dying bird on an atoll of the Cocos-Keeling Islands (Dr. C. A. Gibson-Hill, July 2, 1941). During an aerial survey from Shark Bay to Geraldton (205 miles) on July 25, 1946, with S. Fowler, one of us counted 175 of these albatrosses, 40 being in the largest single flock. Dr. Graeme Smith, of the C.S.I.R.O., saw about 300 birds settled on the water in groups up to 40 birds at the eastern end of King George's Sound on July 27, 1972. The species is a frequent attendant on the snapper fishing boats in the northern part of Shark Bay but is ordinarily a shyer bird than the Black-browed Albatross. On the west coast immature birds, with all-black beaks, appear to be in the majority, whereas on the south coast adult birds predominate.

In the Indian Ocean the species breeds at St. Paul Island.

Grey-headed Albatross

Diomedea chrysostoma

Like the preceding two species in the disposition of black and white in the plumage, but the *head and neck are slaty-grey* (though said to become white in old birds). The under wing shows the narrow black edging similar to that in the Yellow-nosed Albatross (but in the immatures the white central area may be reduced to virtual absence). The beak colouring is also like that of the Yellow-nosed, black with a yellow stripe along the ridge of the beak, but in addition along the lower edge of the lower mandible (ramicorn) there is another bright yellow stripe. Iris, brown; legs flesh colour. Length, 32-33 in.; wing span, 7 ft. Culmen, 101-118 mm.; wing, 482-530 mm.

Distribution: This distinctively-marked albatross may be seen along the steamer route to the Cape and in the Great Australian Bight. In the winter months it may range further north, up to, at times, the latitude of Shark Bay. Specimens are occasionally washed ashore on metropolitan beaches after winter storms. It breeds at Kerguelen Island.

White-capped Albatross (Shy Albatross)

Diomedea cauta

Like the preceding group of smaller albatrosses in general coloration (white with black on the upper wings and mantle) but with the *under wing pattern like the Wandering Albatross*—namely only the tips of the under wings are black. This is the only local Albatross with such a combination of characters. There is a dark line over the eye to the beak, which gives a "capped" appearance to the head. *The beak is light in colour,* pale yellow with a greenish tinge, but darker in the young. Iris, brown; legs, bluish-white. Length, 3 ft. 3 in.; wing span, 6½-8 ft. Weight, 7¼ lb. Culmen, 117-135 mm.; wing, 523-590 mm.

Distribution: Rarely reported in our seas but individuals have been recorded off Fremantle, Busselton and in the western portion of the Great Australian Bight. The nearest breeding grounds are Albatross Island (in western Bass Strait), and Pedra Branca and Mewstone (south of Tasmania).

Sooty Albatross

Phœbetria fusca

An *entirely dusky bird,* of a fairly uniform sooty-brown colour, with an incomplete white ring around the eye and, if the bird approaches near enough a *yellow stripe is noticeable along the lower mandible of the black beak.* The tail differs from other albatrosses in being wedge-shaped. The bird superficially resembles the Giant Petrel but the beak is black, not pale as in the petrel. Young birds are somewhat paler on the neck and body, and the coloured beak stripe is blue, not yellow. Length, 34 in.; wing span, 6 ft. 8in. Weight, 4¼-5½ lb. Culmen, 98-116 mm.; wing, 470-520 mm.

Distribution: The Sooty Albatross is observed in the Great Australian Bight and along the steamer route to the Cape, but is uncommon north of the Leeuwin. After severe winter gales specimens are occasionally washed up on the South-Western beaches. It breeds on Marion and St. Paul Islands.

Light-mantled Sooty Albatross

Phœbetria palpebrata

Differs from the preceding species in having the *back and under parts grey,* thus *contrasting with the dark greyish-brown head.* The *beak stripe is blue* or purplish throughout life. As immature birds of *P. fusca* are rather lighter in body colour than the adults and have a somewhat similar beak colouring to *P. palpebrata,* sight identifications of the present species are rendered difficult. The immature of *P. palpebrata* resembles the adult. Culmen, 98-112 mm.; wing 494-580 mm. Dr. R. C. Murphy states that the differential colour pattern of the head and body gives this bird "the fanciful appearance of a cowled monk."

Distribution: Breeds at the Marion, Crozets, Kerguelen and Heard Islands in the Indian Ocean sector, and has a more southerly distribution than the Sooty Albatross. It has only once with any degree of certainty been identified in the Great Australian Bight (in May 1917, by W. B. Alexander). No specimens have so far been collected in our waters, but they have been taken in Victoria.

PETRELS AND MUTTON BIRDS: Family Procellariidæ

Southern Giant Petrel

Macronectes giganteus

These are entirely dark petrels, approximating to an albatross in size. They can be distinguished by the *heavy pale-coloured beak* with a prominent nasal tube on the upper surface. Very rarely individuals belonging to a white phase occur, estimated at about 5% of the local population. Iris, brown; *beak cream, tinged green;* legs, dark iron-grey. Length, 34-39 in.; wing span, 6 ft. 1 in. - 6 ft. 9 in. Weight, 5½-7½ lb. Culmen, 78-105 mm.; wing, 430-550 mm. In the adult plumage, not seen in Australian seas, the head and neck are pale grey and the breast mottled brown and grey.

Until recent years it had been thought that only one species of Giant Petrel existed, and in our previous editions only one was recognised. In 1966 Dr. W. R. P. Bourne and Dr. John Warham demonstrated that two very similar, but reproductively isolated, species occurred. Individuals of both visit southern Australia but the present one, bearing the original species name of *giganteus*, appears to be the commoner, as immature birds are more given to wandering than in the relatively sedentary Northern Giant Petrel. The species are best separated by beak characters (greenish or whitish in the present species, brownish with dark markings in the Southern Giant Petrel).

Distribution. This petrel, ranking among the most common of the visiting species, is frequently washed up on the local beaches, and wanders as far north as Point Cloates. The nearest breeding localities to Western Australia are Marion, the Crozets, Kerguelen, Heard and Macquarie Islands.

The fact that some of the individuals visiting our shores originate from much further afield was first proved by the finding at Fremantle, on July 10, 1948, of a Giant Petrel which had been ringed by the Falkland Islands Dependencies Survey at Borge Bay, Signy Island, in the South Orkney Islands, off Graham Land, not quite four months earlier, on March 20. This bird had been ringed as a fledgling.

Since then the marking of Giant Petrels has been greatly extended by Antarctic research organizations. Besides Signy Island the birds have been ringed at Anvers Island (off Graham Land) and Macquarie and Heard Islands, with numerous subsequent recoveries in South Africa, southern Australia, New Zealand and South America. One of these marked at Signy Island was recovered at Mandurah on June 9, 1957, approximately 5 weeks after it had left its nest. Ringing and recovery points are about 7,500 miles apart and the flight represents a mean daily progress of 214 miles. Another fledgling marked at Anvers Island was recovered at Koks Island, Shark Bay (about 8,000 miles distant), on June 29, 1956, approximately 9 weeks after fledgling. All of these records are of first year and some second-year birds and it seems evident that it is only the immature birds which make the extensive migratory movements bringing them to Australian shores. The pattern of movement suggests that the young birds disperse from the nesting grounds so that the prevailing winds carry them in an easterly and north-easterly direction. When they reach a coast where food supply is abundant these scavenging birds tend to remain for part, at least, of the winter. The net result of the movements is that the birds probably circumnavigate the globe. Preliminary ringing results suggest that as they become mature the birds return to the places where they were hatched.

Northern Giant Petrel

Macronectes halli

Similar to the Northern Giant Petrel. Body plumage varying from dark sooty brown to dark grey, with light edgings to the feathers giving a freckled effect. No white-phase birds are known in this species. Iris, pale grey (dark brown in immature birds); beak, horn-coloured often with *a reddish tinge and with dark marks on the terminal nail;* legs, dark grey. Slightly smaller than the previous species. Culmen, 87-108 mm.; wing, 430-550 mm.

Distribution: The breeding islands are north of the Antarctic Convergence, and include the Chatham, Auckland, Campbell, Stewart, Antipodes, Gough, Marion, Crozet, Kerguelen and Macquarie Islands. Ringed individuals have been recovered in southern Australia: one ringed as an adult at Macquarie Island on April 10, 1967, was observed off Fremantle on July 1, 1968; two ringed as chicks at the Crozet Islands were recorded in this State.

Cape Petrel (Cape Pigeon, Pintado)

Daption capensis

A *dappled black and white petrel* which can be confused with no other sea-bird. Under parts entirely white, except the chin which may vary from white to speckled or black. Head and neck are black, back, wings and tail spotted black and white; two large white "windows" in each wing; tail with black bands. Iris, brown; beak, black; legs, black. Length, 15-16 in.; wing span, 32-36 in. Culmen, 29-34 mm.; wing, 250-270 mm. Weight, 11¾ oz.

Distribution: A winter visitor from the south, and seen here between June and September, breeding on Kerguelen and Heard Islands. On our coast it has been reported as far north as Exmouth Gulf and specimens are not infrequently washed up on the metropolitan beaches. In some winters it may be observed feeding outside Garden Island, near Rottnest, and in Cockburn Sound. Occasionally it may be seen, even from the mainland, battling against a gale and desperately attempting to get off from a lee shore.

Antarctic Fulmar (Silver-grey Petrel)

Fulmarus glacialoides

Light bluish-grey, paler on the neck and head, which is practically white; wings darker, slaty-black. Under parts and under wings white. Iris, brown; beak, pale flesh-colour, light blue nares and tipped black; legs, pale flesh-colour. Length, 16-20 in.; wing span, 3 ft. 9 in. Weight, 1 lb. 5 oz.-2 lb. Culmen, 44-48 mm.; wing 315-350 mm.

Distribution: This southern representative of the well-known Fulmar *(Fulmarus glacialis)* of the northern hemisphere is a winter visitor from the south, breeding in Antarctica. It is definitely rare in the seas immediately about southern Australia and it is only during the last 30 years that specimens had become beach-washed in Western Australia, the numbers increasing in recent years. The first was found at North Fremantle in June 1945 and the second came ashore at Rottnest Island during the winter of 1954. Thereafter the yearly records were as follows: 1968, Scarborough in September; 1970, Scarborough in August; 1971, City Beach in July; 1972, near the mouth of the Warren River some time during the winter; 1973, Cottesloe in July; 1974, Jurien Bay in July; City Beach in August and Augusta in September. A similar increase in beach finds after storms in recent years has been reported in south-eastern Australia.

Blue Petrel

Halobæna cœrulea

The upper parts are a *light blue-grey,* under parts white. The species resembles a prion closely in size and appearance but can be distinguished at once by the *white tip to the square tail.* The latter character is absolutely diagnostic of the species as it is the only petrel known which possesses a white-tipped tail. In general colour pattern this species is a miniature of the Soft-plumaged Petrel. Iris, brown; beak, black, pale-grey near the base; *legs, pale blue-grey.* Length, 11-12½ in.; wing span, 24-26 in. Culmen, 25-26 mm.; wing, 201-217 mm.

Distribution: A rare winter visitor from the south, breeding at Marion and Kerguelen Islands. It is sometimes washed up on South-Western beaches as far north as Quinn's Rocks.

PRIONS

Genus *Pachyptila*

There are five species of these birds ranging into local seas and which because of their exceedingly close similarity are preferably dealt with collectively. The colour pattern in each is essentially the same and most of the species can be told apart only by the character of the beak, of which exact measurements are essential, aided in some degree by general bodily dimensions.

The upper parts are a delicate dove-like *blue-grey,* the tips of the wing quills being black, the dark band so formed in the out-spread wing being linked across the back by black tips to the lower mantle feathers and these usually have narrow white edgings at the extreme ends. There is a *terminal black tail band.* The lores and eyebrows are white and there is a dark mark through the eye. The entire *under parts are white,* washed with blue-grey in the tail region.

The *beak varies greatly* in width, according to the species, but, unfortunately for convenience in identification, there is an overlap between the species, though the difference between the narrowest-beaked and the widest-beaked species is enormous. The inside of the upper mandible is fringed with a *row of comb-like lamellae* on each side, which in the two widest-beaked species are visible externally at the base when the beak is closed. The beak is a bluish iron-grey in colour, lighter at the tip. The *legs are blue,* the webs cream-coloured. The iris is brown.

The body length varies from 10 to 12 inches.

The characters used for distinguishing the species are as follows: 1, width of beak, taken with a pair of dividers at the base; 2, length of culmen or upper beak, taken with a pair of dividers by placing one of the points at the base of the beak where the feathering begins on the forehead and the other at the extreme tip of the hook of the bill—thus the chord of the beak is measured; 3, length of wing; 4, whether or not the lamellae are visible externally towards the base of the beak; 5, the relative size of the terminal nail or hook of the beak and the comparative length of the space between this hook and the nasal tube; 6, the width of the terminal tail band in relation to tail length. The accompanying key summarises these characters for the various species. The measurements are given in millimetres to obtain greater exactitude than is possible with English units.

Notes on the individual species follow:—

Fairy Prion

Pachyptila turtur

This appears to be the rarest of the Prions washed ashore on the local beaches, though it is the only Australian-nesting species. The nearest breeding grounds are on the islands of Bass Strait west to Lawrence Rocks, off Portland, and it is possible, though not proved, that our birds may originate from there. However it has been discovered that a breeding station may exist at Marion Island, in the south Indian Ocean and stragglers, aided by the prevailing westerlies, may occasionally reach our seas. The species lays in mid-October.

Fig. 12.—Fairy Prion; inset top view of beak.

This is the narrowest-billed species, which can be told apart from the Thin-billed Prion by the relatively heavier terminal nail or dertrum, with only a short space between it and the nasal tube.

	vittata	*salvini*	*desolata*	*belcheri*	*turtur*
Tail band:	Terminal tail band narrow, 20 to 25 mm. wide				Tail band **wide,** 35 to 40 mm., occupying about half the tail.
Dertrum:	Dertrum weak and separated from nasal tubes by an appreciable space.				Dertrum strong and with little space between it and nasal tubes.
Beak lamellae:	Beak lamellae well developed and **visible externally** when the beak is closed.		Lamellae **not** exposed externally when the beak is closed.		
Beak margins:	Beak margins with strong bowed outlines when viewed from above.	Beak margins very slightly bowed, almost straight.	Beak margins straight.		
Beak width:	19 to 22 mm. (from 16 mm. in immature birds)	13 to 17 mm.	11 to 14 mm.	9 to 11 mm.	8 to 11 mm.
Culmen length:	30 to 38 mm.	28 to 33 mm. (average 30)	26 to 30 mm.	23 to 27 mm.	22 to 23 mm.
Beak ratio:	Length $1\frac{2}{3}$ times the width (1.4 to 1.8)	Length twice the width (varies from 1.8 to 2.2 times)	Length twice width (varies from 1.8 to 2.3)	$2\frac{1}{2}$ times width (2.3 to 3.0)	—
Wing:	200 to 212 mm.	180 to 200 mm.	180 to 200 mm.	179 to 195 mm.	165 to 185 mm.

Also, one of the few plumage characters in the genus enables this species to be recognised at once from any other Prion—the black terminal tail band is very wide, being about *half the length of the tail,* or 35 to 40 mm. in width. In all the other species the band is a mere tip to the tail, being about 20 to 25 mm. in width.

Thin-billed Prion

Pachyptila belcheri

A not infrequent visitor, which nests on Kerguelen Island. Egg-laying takes place in mid-November.

This is the narrowest-billed of all those Prions with a weak terminal nail widely separated from the nasal tube. The bigger-billed

Fig. 13.—Thin-billed Prion.

birds of this species are difficult to separate from the Antarctic Prion and therefore the ratio of beak length/beak width must be taken into account. The length is 2½ times the width, the ratio varying from 2.3 to 3.0.

Antarctic Prion (Dove Prion)

Pachyptila desolata

This is the commonest species of Prion washed up on the South-west beaches; it breeds on Kerguelen and Heard Islands, laying at the end of December.

Distinction from the larger-billed species presents no difficulty, despite the fact that there is some overlap in bill dimensions, for in the Antarctic Prion the beak lamellae are hidden from external view in the closed beak, and the upper mandible has straight outlines when viewed from above. Difficulty is encountered in separating small-billed specimens of the Antarctic Prion from large-billed ones of the Thin-billed Prion, and recourse must be had to length-width ratios of the beak. The length averages twice the width, and ranges from 1.8 to 2.3 times the width.

Medium-billed Prion (Marion Island Prion)

Pachyptila salvini

A fairly frequent visitor to these shores, nesting on the Crozet and Marion Islands, the eggs being laid in mid-November. Unlike the other prions, this species has no breeding colony in the other oceans. The nesting islands have been only very rarely visited by

Fig. 14.—Medium-billed Prion.

naturalists and the majority of specimens in the world's museums originate from beach-drifted birds found on the shores of Australia and New Zealand.

This and the succeeding species are the widest-beaked prions and differ from all the others in having the comb-like lamellae clearly visible in the closed beak, towards its base; the effect is not unlike that of a snarling dog baring its back teeth, as the late G. M. Mathews first pointed out. The outlines of the beak, in plan view, are not quite straight but slightly bowed. Considerable difficulty is attached to separating the larger-billed individuals from the immature ones of the Broad-billed Prion. Distinction must be made on the balance of a combination of characters; thus the Medium-billed Prion has a wing length of usually under 200 mm. and a culmen length of under 32 mm., about twice the width.

Broad-billed Prion

Pachyptila vittata

This prion is cast up on our beaches only on very rare occasions. The species breeds at St. Paul Island, and an even larger form, which does not range in Western Australia, breeds in southern New Zealand. Egg-laying takes place in early September.

Pediunker (Grey Petrel)

Procellaria cinerea

Upper parts *grey* (darker on head, wings and tail), *under white,* the demarcation between the two shades being gradual and below eye level on the head. Iris, brown; beak, black, with part of the naricorn, latericorn and ramicorn greyish-green; legs, flesh colour. Length, 19½ in. Culmen, 45-50 mm.: wing, 340-360 mm.

Distribution: A winter visitor from the south; rarely washed up on the beaches and the only known record is from Busselton. It frequently follows ships on the Cape route, often following a boat for hours at a time, but it seems to have a decided aversion to approaching too near the land when away from its nesting grounds. In the Indian Ocean sector it breeds on Marion and Kerguelen Islands.

Shoemaker (White-chinned Petrel, Cape Hen)

Procellaria æquinoctialis

A large *brownish-black* petrel with a variable amount of *white on the chin.* Tail short and wedge-shaped. Iris, brown; beak, pale (greenish-white), of the long shearwater type but strongly built; Legs, black. Length 20-22 in.; wing span, 4 ft. 5 in.-4 ft. 10 in. Weight, 2 lb. 4 oz.-3 lb. 2 oz. Culmen, 48-56 mm.; wing, 340-400 mm.

Distribution: A common species in the southern oceans, following ships on the Cape route and across the Great Australian Bight, but has not been observed north of Cape Leeuwin. No storm-blown specimens were reported from anywhere in Australia until recent years. The first Australian find of a beach-washed bird was at Discovery Bay, Victoria, in March 1959, and the first from Western Australia, and the only one so far, was a dying bird at Scarborough found by Miss C. A. Nicholls on May 2, 1973 during a S.W. gale. The species breeds on Marion, the Crozet and Kerguelen Islands.

Little Shearwater

Puffinus assimilis

Black above and white below, *the dividing line on the side of the head passing through the eye.* Iris, dark brown; beak, greyish-black; *legs, light blue,* webbing blue. Length, 10½-12 in. Culmen, 22-25 mm.; wing, 170-184 mm.

Distribution: A non-migratory petrel which is known to nest at the following localities: Abrolhos Islands (in the Wallaby Group at West Wallaby, Beacon, Goss, Tattler, Pigeon, Dick's, Long and Seal Islets; in the Easter Group at Rat, Wooded, Suomi, Passage, Leo and an islet north of Passage; in the Pelsart Group at Pelsart and Murray Islands), islands between Dongara and Fremantle (Fisherman, Favourite, Escape, Cervantes, Green, Whittell and Buller Is.), Rottnest Island (on Parrakeet Island), Cosy Corner (on two small islets), St. Alouarn Island (off C. Leeuwin, reported by Dr. Mary Gillham, November 1959), Saddle Island (off Nornalup, reported by W. Nairne Clarke in May 1841), Eclipse Island and the Archipelago of the Recherche (recorded so far only on Boxer, Lion, Wedge and Combe Islands). Flocks are met with at sea, including for a day or so after leaving Fremantle on the Colombo steamer route, and birds are not uncommonly washed ashore on the beaches after winter storms.

Nesting: A winter breeder. On Eclipse Island where the life history of the species has been studied by Lighthouse-keeper A. V. Newman, the birds come ashore early in January and renovate their burrows or excavate new ones. The burrows may be up to 5 feet in length, terminating in a roomy nesting chamber, or may be openings under rocks or a hole under a tree. The nests may be merely de-

Fig. 15.—Little Shearwater.

pressions in the soil or lined with a few sticks or seaweed. One egg forms the clutch and the birds will not lay again if the first egg is taken or destroyed. It is pure white and is subject to much variation in size and shape. Ten eggs from Eclipse Island range from 49-56 x 34-37 mm., with an average of 52 x 36 mm. The laying season on Eclipse Island has been found to be protracted, but it usually takes place from the third week in June to mid-July (earliest record June 21 and a fresh egg was found in September). On Parrakeet Island, an islet connected by a reef with the main island of Rottnest, A. H. Robinson found four nests with young and three with eggs on September 8, and on September 10 two more nests with eggs. Both parents share in the incubation, changing over about every two days. On Eclipse Island hatching takes place about the end of August, after 52 to 58 days' incubation, and the young remain in the burrow until late October or early November—70 to 75 days. The chicks are fed usually every second night and at longer intervals later, up to five-nightly periods. The parents cease their visits to the burrows 8 to 11 days before the departure of the young. In September the birds as a whole begin to desert the burrows and all have vanished by December (A. V. Newman).

A characteristic habit of this small petrel is its fluttering flight. It progresses by very rapid wing-beats as compared with the flight of the larger species, but it shares that method of flight with the Fluttering Shearwater. Like all the petrels, it has the habit of skimming or shearing over the waves, a gliding flight with outstretched wings. The objectives of this tireless searching are small squid or cuttlefish, which form the major portion of the diet of all petrels, judging by the evidence of stomach remains. The jaws or "beaks" of these cephalopods are invariably found in them, with sometimes pelagic fish like sardines and anchovies.

Fluttering Shearwater (Hutton's Shearwater)
Puffinus gavia

Dark *brownish-black above* (but may show a blue bloom in the fresh plumage) and *white below.* The dividing line on the side of the head between the dark and white passes *well below eye level.* Iris, brown; beak, dark olive-grey; legs, *brown on the outer aspect of the tarsus and fleshy-white on the inner.* Length, 13¾-14½ in. Weight, 13-15 oz. Culmen, 36-38 mm.; wing, 215-229 mm.

There are two forms of this shearwater, which is confined to Australian and New Zealand waters, and there is warrant for regarding them as two separate species. The only one found in Western Australia is the larger (measurements given above), named *P. g. huttoni* (and called in the vernacular, by those who consider it as a full species, Hutton's Shearwater). The smaller race, *P. g. gavia,* has a culmen length of 30-35 mm., and a wing length of

180-221 mm. When specimens are able to be examined in the hand *gavia* can be readily told from *huttoni* by the rami of the lower mandibles meeting only at the tip whereas they are closely applied for about an inch towards the tip in *huttoni*. The plumage differences have been summarised by Dr. R. A. Falla as follows:

Character	*gavia*	*huttoni*
Under wing coverts	mainly pure white	variable smudgy with dark shafts
Long axillaries	greyish square-ended, white-tipped	brown, more pointed, rarely white-tipped
Under tail coverts	pure white	lateral coverts variably flecked with brown on outer webs

The species is a close relative of the Manx Shearwater *(Puffinus puffinus)* of the Northern Hemisphere.

Distribution: As far as is known this is a New Zealand breeding species, many individuals of which migrate across the Tasman Sea to southern Australian waters. The smaller race *(gavia)* has a more northerly distribution and visits New South Wales and east-

Fig. 16.—Fluttering Shearwater.

ern Victoria. The larger form *(huttoni)* visits mostly South Australian waters but occasionally wanders further westwards. Four records are known of its occurrence in Western Australia. One beach-drifted bird was found at Bunbury by F. Lawson Whitlock on April 23, 1937, the second at Swanbourne beach on June 13, 1956, recorded by J. A. L. Watson, the third near City Beach on November 3, 1965, found by Miss C. A. Nicholls, and the fourth at Augusta, in early December 1969.

Nesting: No Australian breeding grounds have been found, despite close searches. In New Zealand birds of the short-billed race arrive at the nesting islands by September, eggs are laid at the end of the month or in early October and the young leave the burrows towards the end of January (R. A. Falla). The nesting grounds of the long-billed race visiting South and Western Australian waters *(huttoni)* had long eluded the searches of the New Zealand ornithologists. However, in February 1965 the surprising discovery was made by G. Harrow, following clues received from local inhabitants, that this shearwater nested on tussock-covered ridges of 5,000 ft. high mountain tops, some 7 miles from the sea on the South Island of New Zealand. The locality was under Mt. Kaiterau in the Seaward Kaikoura Range, in the north-eastern part of the South Island.

At sea these shearwaters congregate in small flocks. They fly with a characteristic fluttering motion of the wings, similar to that of the Little Shearwater, from which the present species cannot safely be distinguished at a distance.

Wedge-tailed Shearwater (Mutton-bird)

Puffinus pacificus

Entire plumage dark brownish-black; *tail long and wedge-shaped.* Iris, brown; beak, *slender and leaden-grey; legs, white-flesh.* Length, 17½ in. Weight, 12½-13½ oz. Culmen, 35.7-39.0 mm. wing, 289-304 mm.

In the Shark Bay area two colour phases occur; the majority are all-dark birds just as described, but a small proportion (about 20-30%) have pure-white under parts.

Distribution: This petrel appears to be most numerous, in Western Australia, from North-West Cape to the region of Bedout Island, off Condon, and the bulk of the birds are to be seen along the outer rim of the arc of islands and reefs from North-West Cape to Legendre Island in the Dampier Archipelago. To the south, breeding colonies occur as far as Carnac Island, but individuals forage to Geographe Bay.

Nesting: The birds excavate long burrows in soft soil, preferring, in the North-West islands, reddish or brown loam areas and avoiding the white coral sands, though the latter only are available at the Abrolhos. On rocky islets they nest under crevices or beneath thick plant cover, such as the nitre-bush *(Nitraria schoberi)*. One pure-white egg is laid. Average size of 32 eggs, 62 x 41 mm., ranging from 58-69 x 39-44 mm. An abnormally small one from Carnac I. measures 54 x 37 mm. A full study of the nesting cycle has not yet been made, but laying takes place mostly during November. The earliest recorded date, October 29, is from Airlie Island, off Onslow; at the Abrolhos Islands the first eggs are reported in mid-November, at Lancelin I. on November 18, and at Rottnest Island the earliest known date is November 22. Thus as in the case of the Little Shearwater the egg-laying period is more prolonged than in the migratory species (Fleshy-footed and Short-tailed Shearwaters). The young begin to be hatched in early January, and the nestlings are covered with ashy-grey down, which is lighter on the throat and upper breast.

Known nesting islands are the following, many of the northern ones having been first located by S. Fowler while on aerial surveys for the C.S.I.R.O. since 1943; Carnac Island, Rottnest Island (Cape Vlaming and Green and Dyers Islands), islands between Fremantle and Dongara (Edward, Lancelin, Wedge, Buller, Whittell, Green Islets, Cervantes, Essex Rocks, Boullanger, Tern, Whitlock, Escape, Favourite, islets near North Head (formerly), Sandland, Fisherman, and Beagle Is.), Abrolhos group (Pelsart, Gun, Jubilee, Rat, Pigeon, Long and West Wallaby Islands), Shark Bay (Baudin, Double, Freycinet, Slope and Friday Islands), Onslow area (North and South Muiron, Long or Surrurier, Anchor, Solitary, Lockyer, Airlie, North Sandy, Fortescue, Round and North-east Regnard Islands), Double Island (off Barrow Island), Lowendall Islands, Montebello group (South-east Island and small islet north-west of Bluebell Island), and Forestier Islands (Sable Island—the furthest north-east nesting locality).

Fleshy-footed Shearwater (Mutton-bird)

Puffinus carneipes

Native Names: Por-rut (general); Kar-ryne-mun-dyne (Augusta).

Entire plumage dark brownish-black; tail, fan-shaped. Iris, brown; *beak, pale whitish-horn, with a darker tip,* of heavier build and longer than in the two preceding species; *legs, white-flesh.* Length, 16-18 in.; wing span, 39-42 in. Weight, 1 lb. 5 oz. - 1 lb. 11 oz. Culmen, 40.0-43.0 mm.; wing, 270-317 mm.

Distribution: The breeding area is confined to the south coast, from the vicinity of Cape Leeuwin eastwards to the Archipelago of the Recherche. Known nesting localities are: An islet off Cape Hamelin ("Cosy Corner"), Seal Island (off Cape Leeuwin), Sandy Island (near Pt. D'Entrecasteaux), Chatham Island, Saddle Island, Stanley Island, Mutton-bird Island (Torbay), Eclipse Island, Breaksea Island, Michaelmas Island, Doubtful Islands and the Archipelago of the Recherche (Figure of Eight, Boxer, Charley, Rabbit, Woody, Thomas, Gunton, Sandy Hook, Cliff, Remark, Long, Mondrain, Station, Owen, Stanley, George, North and South Twin Peaks, Douglas, Gulch, Corbett and Christmas (Daw) Islands).

Nesting: The one pure-white egg is laid at the end of a burrow, and on the average is a larger egg than that of the preceding species.

Average size of 46 eggs, 70 x 46 mm., ranging from 64-76 x 43-50 mm. Weight, 75 gm. (69-81 gm.). The bird is a summer breeder with probably a similar cycle to that of the preceding and following species, but full details are lacking. At Eclipse Island in 1943, A. V. Newman found that the bulk of egg-laying took place from December 4 to 6 and a few fledged young were still in the burrows by April 20 the following year. A. F. Basset Hull found eggs on November 29 at Charley Island, when he investigated the fauna of the Recherche Archipelago in 1921. The nestling is covered with brownish-grey down which is paler below than above.

The feeding range during the nesting season extends north on the west coast to Bunbury and eastwards across the Great Australian Bight into South Australian waters. There are no breeding islands in South Australia. Further west a breeding station was recently discovered at St. Paul Island.

This is a migratory species comparable in its movements with the Short-tailed Shearwater in south-eastern Australia. It is completely absent from our area in the winter months (May to September), moving into the north Indian Ocean. Observations on the migration are almost entirely lacking, but specimens have been recovered on the beaches near Fremantle in May and October, when the birds were presumably in transit. Beach-drifted and moulting specimens have also been taken in June and July in Ceylon. Dr. C. A. Gibson-Hill has suggested a wide migratory circuit of the Indian Ocean for these birds. The first arrivals are noted at Eclipse Island about the third week in September.

In the early days a small industry on the young birds was conducted by the sealers at the Archipelago of the Recherche and sporadic attempts have been made to salt birds on the islands near Albany. The French naturalists, Quoy and Gaimard, who visited King George's Sound with Dumont d'Urville in October 1826, reported that the aboriginal women living with the English sealers brought to the French ship large numbers of plucked and cleaned adult birds each day.

Short-tailed Shearwater (Tasmanian Mutton-bird)

Puffinus tenuirostris

Entire plumage dark brownish-black; *tail short and rounded.* Iris, brown; beak, slender, *leaden-grey; legs, blackish-grey, inner side tinged purple.* Length, 15½-17 in.; weight, 1 lb - 1 lb. 7 oz.; culmen, 29.3-33.9 mm.; wing, 262-290 mm.

Distribution: An occasional straggler from the eastern States into the western portion of the Great Australian Bight and adjoining waters, the westernmost known limit being Hopetoun. The bird nests on the islands around Tasmania, in Bass Strait, and off the Australian mainland from Broughton Island (N.S.W.) westward to St. Francis Island in the Nuyts Archipelago, South Australia.

Nesting: The birds arrive at their nesting islands, to renovate their burrows, from the last week in September to early October. Egg-laying begins in Bass Strait about November 21 and for the most part is completed by the end of the month. Incubation period, 52-55 days; male and female incubate the egg alternately, the male taking the first shift after the egg is laid and brooding for a period varying from 12 to 14 days. Hatching takes place between January 10 and 23. The young birds leave the burrows at the end of April and during the first few days in May. Clutch, one. Average size 71 x 47 mm., ranging from 63-78 x 43-50 mm.; average weight, 85 gm. (73-94 gm.).

Of all the petrels this is, perhaps, the best-known species to the general public, for it is the "mutton-bird" of commerce, an extensive trade being carried on in Tasmania and, in earlier times, in Victoria. The industry is now conducted mainly by the islanders of the Furneaux Group, and, as now stabilised, concerns itself with the young birds only, which are taken from March 27 to April 30 each year. The oil is expressed from the alimentary tract, the birds are plucked and the carcasses salted in barrels or sold fresh, being flown in freight planes to Launceston. Some trade is done also in feathers, oil and fat. Despite the fact that the industry has been carried on for well over a century, and in early days was exceedingly wasteful, a heavy toll being levied on both the adult as well as the young birds and eggs, the annual take is still very large. In a good season about three-quarters of a million young birds are captured.

In the autumn the birds leave the Australian area and migrate to the North Pacific. Specimens have been collected in the Japanese region, Bering Straits, Alaska, and even as far afield as the Pacific coast of North America south to the Los Coronados Islands in California. As there are no known breeding grounds other than in south-eastern Australia this indicates a very extensive wandering. Many non-breeders, probably mainly immature birds, are to be found at all times of the year in some part or other of the species' enormous range. From September onwards the huge flocks converge along the east Australian coast, migrating south to the breeding islands. At such times the birds may be subject to severe mortality, being washed up in considerable numbers on southern Queensland and New South Wales beaches. Flocks of feeding mutton-birds are frequently associated with schools of pelagic fish such as tuna, pilchards and anchovies. A basic food organism is a species of krill *(Nyctiphanes australis)* but the birds feed also on small school-fish like anchovies *(Engraulis australis)* and cephalopods.

Sooty Shearwater (King Mutton-bird)

Puffinus griseus

Very similar in appearance to the previous species, differing mainly in its *larger size.* Also most individuals have *whitish under wing coverts,* compared with dusky under wing coverts in the previous species, but some Short-tailed Shearwaters also have light-coloured coverts. Iris, beak and leg colours as in the preceding species. Length, 18½-21 in.; weight, 1 lb. 7 oz.-2 lb. 2 oz.; culmen 37-45 mm.; wing, 287-322 mm.

Distribution: A Pacific Ocean species, breeding extensively in New Zealand (where it is the local mutton-bird), and sparingly along the east Australian coast, where small nesting colonies occur. It forages to some extent in southern Australian seas, but up to the present the only records in this State are observations of small flocks between King George's Sound and Eclipse Island by Drs. P. J. Fullagar and G. F. Van Tets, of the CSIRO, August 2 and 9, 1973.

The migration pattern of this species, more complex than in the preceding, has not been fully worked out, and it is difficult to evaluate the significance of these early August records from Albany.

Nesting: Similar to the preceding. Eggs average 77-48 mm., ranging from 72-88 x 44-52 mm.

Great-winged Petrel (Grey-faced Petrel, Mutton-bird)

Pterodroma macroptera

Entire plumage dark brownish-black; tail fan-shaped. Iris brown; *beak and legs jet black.* The beak is short and deep, of the form characteristic of this genus (see Fig. 17). Length, 15½-17 in.; wing span, 3 ft. 4 in.; weight, 1 lb. 4 oz.; culmen, 32-37 mm.; wing, 274-315 mm.

Distribution: This is a non-migratory petrel, nesting on islands along the south coast between Albany and Cape Arid. The bird is commonly met with at sea off the south coast and in the Great Australian Bight. Storm-drifted birds have been found on the western beaches as far north as the Abrolhos (Pelsart Island), and, more than in any other petrel species, individuals are liable to be blown inland by winter gales. Such derelicts have been recorded at the following places: Chorkerup, Cherry Tree Pool (near Kojonup), Katanning, Woodanilling, Corrigin, Maddington, East Bullsbrook, Northam, Cunderdin, Wyalkatchem, Nungarin and Mukinbudin. The geographical position of the more northern of these inland localities suggests that a breeding area, hitherto undiscovered, may exist on the west coast—possibly at the Abrolhos Islands. The East Bullsbrook bird, collected on December 1, 1969, was a fledgling in good condition, with down adhering to the nape.

Nesting: Known nesting islands are Eclipse, Breaksea, Rabbit (or Mistaken), Gull and Coffin Islands — all in the vicinity of Albany; Bald Island; the Recherche Archipelago (Figure of Eight, Boxer, Termination, Bens (probably), Goose and Christmas (Daw) Islands), and probably Pasley Island, off Cape Pasley. The single white egg is laid sometimes in a short burrow (2 to 3 feet in length), but very frequently under boulders or slabs of rock or under thick plant cover, such as "pig-face" *(Carpobrotus)*. The material used in the nests consists of coarse dry grass and dry stems of "pig-face." Average size of 10 eggs, 66 x 47 mm., ranging from 64-68 x 46-49 mm. The species is a winter breeder and at Eclipse Island A. V. Newman first noted the birds renovating their burrows at the end of March; John Warham records egg-laying about May 8 and hatching at the end of June. Newman gives the start of laying as May 20 and hatching in mid-July, with full-grown young still in the burrows in the third week in October. Graeme Smith found young almost fledged on Bald Island on October 10. Warham reports the departure of young at the end of November. At the Recherche Archipelago V. N. Serventy found young in the burrows in the third week of November.

This bird is quieter and more docile when handled than the more aggressive species of the genus *Puffinus*. The cries are also more pleasing to the human ear, being a piping "ti-ti-ti" or "pi-pi-pi."

Kerguelen Petrel

Pterodroma brevirostris

A uniform dark grey. Iris, brown; bill black; legs grey-brown. Length, 13-14 in. Culmen, 25-26 mm.; wing, 250-260 mm.

Fig. 17.—Kerguelen Petrel.

Distribution: A winter visitor from the south, breeding at Marion and Kerguelen Islands.

This petrel is very rarely observed at sea and the only evidences of its local occurrence are beach-drifted specimens. The first record of its presence in Australia was a specimen collected at Mosman Park beach by F. L. Whitlock in June 1926 and since then it has not uncommonly been found as a victim of winter storms.

Soft-plumaged Petrel

Pterodroma mollis

Light blue-grey above, white below. A *grey collar across the upper breast.* The *under wings are dark grey* with a narrow white medial stripe. On the upper wing is a broad black stripe across the coverts. Iris, brown; beak, black; legs, white flesh colour, toes and webs, parti-coloured, white and blackish-brown. Length, 13½ in.; wing span, 33 in. Culmen, 26-30 mm.; wing, 240-260 mm.

In flight the dark collar on the foreneck is conspicuous and the grey under surface of the wings is easily seen (R. A. Falla).

Distribution: A winter visitor from the south; its remains are not uncommon on the beaches near Perth after storms. It nests at Marion, Kerguelen and St. Paul Islands.

White-headed Petrel

Pterodroma lessoni

Head and under parts white; back and tail grey; wings brownish-black; black patch around and in front of the eye. Iris, brown; beak, jet black; legs, white-flesh; toes, parti-coloured, white and black. Length, 16½-17½ in. Culmen, 37-38 mm.; wing, 290 mm.; wing span, 3 ft. 6 in.

Distribution: A winter visitor from the south and may be seen as far north as the Dampier Archipelago. It is occasionally washed up on the South-West beaches after storms. It breeds on Kerguelen Island.

STORM-PETRELS: Family Hydrobatidæ

Wilson's Storm-Petrel

Oceanites oceanicus

All *black,* except for the base of the tail which is *white* above and for the most part below. A *light wing stripe,* formed by pale edgings to the wing covets. Iris, brown; beak, black; legs, black with yellow webs, projecting beyond the tail when in flight. Length, 6-7 in.; wing span, 15-16 in. Culmen, 11-14 mm.; wing, 136-160 mm.

Distribution: A winter visitor from the south. Though it is rarely washed ashore on local beaches it is really not uncommon all around our coastline. The seas off north-western and northern Australia, including the Timor Sea, are part of the wintering area of the species. During the northern migration, between April and June, odd birds and pairs may be seen in Cockburn Sound, quite near to Fremantle, approaching very close to fishing boats. Birds have been observed in Shark Bay throughout the summer; these would be immature, non-breeding individuals not participating in the return migration south to the breeding grounds. It is of interest that the few specimens which have been washed up on local beaches have all been recovered after late April and May gales. The breeding areas for the Indian Ocean birds are Antarctica, Kerguelen and Heard Islands.

White-faced Storm-Petrel (Frigate Petrel)

Pelagodroma marina

Native name: Ne-geit (Augusta).

Upper parts *dark grey,* becoming brownish on the back and wings, with the wing quills and a broad terminal bar across the tail black. The base of the tail is light grey, sharply marked off from the tail band. *Entire under parts white,* except the flight quills and tail. There is a *broad white forehead* and this extends backwards as a prominent white eyebrow line. Iris, hazel; beak and legs, black, with the webs yellow, bordered black. Length, 8 in. Weight, 2 oz. Culmen, 17 mm.; wing, 156 mm.

In flight the legs project slightly beyond the square tail and there is no light wing stripe.

Distribution This is the only local breeding storm-petrel and nesting colonies have been reported from the Abrolhos Islands (on a small island three miles south of East Wallaby Island, on West Wallaby Island, Lagoon Island, Beacon Island, Wooded Island and Pelsart Island), islands between Dongara and Fremantle (East Beagle, Fisherman, Sandland, Favourite, Whitlock, Boullanger, Buller and Lancelin Is.), Bird and Seal Islands, in Safety Bay, an islet off Cape Hamelin (in "Cosy Corner"), St. Alouarn Island (off Cape Leeuwin), Rabbit and Michaelmas Islands (in King George's Sound), and the Recherche Archipelago (Marts, Rob, Woody, Lion, Douglas, Christmas (Daw) and Nares Islands). It has been observed at sea as far north as Point Cloates.

Nesting: The burrows differ from those of other local nesting petrels—apart from their much smaller diameter—in that instead of being straight simple tunnels, they may have tortuous twisting courses, possibly even altering in depth, and often with two entrances to the surface. Egg, one, sometimes pure white, at other times with the larger end thickly speckled with spots of rusty-red and again, sometimes, with some spots in addition over the general surface of the egg. Average size of 14 eggs, 36.3 x 26.5 mm., ranging from 34.3-38.6 x 25.1-28.1 mm. Weight, 13 gm. The eggs take almost seven weeks to hatch. The young are covered with soft grey down. The local nesting cycle has not been fully studied. but at the Abrolhos Islands chicks in down have been noted on December 15; in the Jurien Bay Islands fairly fresh eggs were being incubated from the first week in October; and the first chicks reported on November 18 (J. Ford); at Safety Bay chicks in down between January 1 and 18, and almost fully plumaged young on February 15. At St. Alouarn Island burrows were not occupied in the first week of November and John Gilbert found eggs at the beginning of December. At the Archipelago of the Recherche fresh and slightly incubated eggs were noted on November 26, and small chicks in down on January 15.

Black-bellied Storm-Petrel

Fregetta tropica

Sooty-black above, with a white rump. The feathers of the back are narrowly edged with white but this is not evident when the birds are seen at a distance. *Under parts* (including the middle of the under wings) *white,* except for the black throat and chest and tail, and a *longitudinal black line along the centre of the abdomen.* This last character is difficult to observe in life, as the bird glides and skips low over the water, rarely allowing a good view of the under parts. Though similar to the Wilson Storm-Petrel in the colouring of the upper parts it can at once be distinguished by the *absence of a light wing stripe.* Iris, brown; beak and legs, black. Length, 7½-8¼ in. Culmen, 15-16 mm.; wing, 160-167 mm.

Distribution: Despite its scientific name the Black-bellied Storm-Petrel is a colder-water representative of the White-bellied Storm-Petrel and nests on islands in the Antarctic region—Crozet and Kerguelen Islands being the only ones known in the Indian Ocean sector. In the winter months it disperses northwards and may be seen on the shipping routes between Australia and South Africa. It has not so far been recorded as washed ashore on our beaches after storms.

White-bellied Storm-Petrel

Fregetta grallaria

Very similar to the preceding species in colour pattern, *sooty black above with a white rump,* and differing from it in the absence of the longitudinal black line along the middle of the abdomen. Thus, except for the black throat and chest the *under parts are wholly white.* Iris, brown. Beak and legs, black. Length, 7½-8 in. Culmen, 13-16 mm.; wing, 156-170 mm. (R. C. Murphy).

Distribution: Until recently the occurrence of this species was not definitely proved for the Indian Ocean, though it was well known in the South Atlantic and South Pacific Oceans. In January 1952 Dr. J. Diallo captured near Amsterdam Island a specimen which is now in the National Museum of Natural History, Paris (culmen, 14 mm.; wing, 168 mm.). Dr. D. B. Peakall made a sight record of a bird north-west of Amsterdam Island in July 1959. No beach-drifted birds have so far been picked up in Western Australia.

This species nests on islands further north than does the previous species, islands situated near to the convergence of Sub-Antarctic and Sub-Tropical waters. It breeds on Lord Howe Island, in the Tasman Sea, various islands in the Pacific and at Tristan and Gough Islands in the Atlantic. Dr. C. Jouanin considers it may be the storm-petrel which formerly was known to breed on St. Paul Island in the southern Indian Ocean.

Sooty Storm-Petrel

Oceanodroma matsudeiræ

A wholly brown storm-petrel with *no white on the rump,* upper parts darker and the under parts lighter brown. Upper wing coverts lighter than the rest of the wing, the shafts of the primary flight quills being white. *Tail forked.* Iris, black. Beak short and thick, black. Legs, black. Length, 9-9¼ in.; wing span, 18-20½ in. Culmen, 16-19 mm.; wing, 178-194 mm.

This large species, with its all-dark plumage and forked tail, can be confused with no other local storm-petrel.

Distribution: This is another storm-petrel which was added to the Australian list only recently. It belongs to a species group breeding only in the Pacific Ocean (the species names of the others being *markhami, melania* and *tristrami)* and their inter-relationships have been a matter of dispute. Dr. Oliver L. Austin now considers them all separate species, *O. matsudeiræ* being a western Pacific form breeding on the Bonin Islands, south-east of Japan. Its wanderings are not certainly known but that it includes northern Australian seas, at least occasionally, is evidenced by the collection of two specimens by Roger Pocklington, of the Woods Hole Oceanographic Institution's research vessel, *Atlantis II,* in late July 1965. One was taken 240 miles from the Lacepede Is. (lat. 16° 06′ S., long. 118° 03′ E.) and the other west of Seringapatam Reef (lat. 13° 26′

S., long. 120° 19′ E.). On September 25, 1968, a bird landed on H.M.A.S. *Moresby,* when she was about 65 miles off the Monte Bello Islands.

Another species of fork-tailed storm-petrel is likely to be found in the southern parts of the State. This is Leach's Storm-Petrel, *Oceanodroma leucorhoa,* which nests in the North Atlantic and North Pacific Oceans and migrates southwards. It was first recorded in Australia, when a beach-drifted specimen was found at Tower Hill near Port Fairy, Victoria, on July 4, 1965. Probably it had reached Australian seas via the Cape of Good Hope. The bird is dark sooty-brown above and below, as the previous species, except for a white rump, grey in the middle. Length, 8-9 in. Culmen, 13-17 mm.; wing, 150-160 mm.

DIVING PETRELS: Family Pelecanoididæ

Kerguelen Diving-Petrel

Pelecanoides exsul

Glossy black with metallic glint. Chin white; *chest clouded with grey;* forming an *unbroken band;* abdomen white. Iris, brown; beak, black; legs, blue. Length, 8½ in.

Distribution: This Diving-Petrel breeds on Marion, Crozet, Kerguelen, Heard, Auckland and Antipodes Islands. The only specimen to be recorded from this State was found at Middleton Beach, Albany, on March 20, 1974 by Noel Whiteford and forwarded to the W.A. Museum by Harley Webster.

The Common Diving-Petrel of Australia, *Pelecanoides urinatrix,* breeding on islands in Bass Strait, does not range westwards of Robe and Brighton in South Australia; it has never been encountered in Western Australian seas. The Albany bird was examined by Dr. G. M. Storr and found to be *exsul,* not *urinatrix,* because of its bigger size (wing length, 125 mm.; tail, 45 mm.; and width of beak, 9.3 mm.). The corresponding measurements of Bass Strait *urinatrix* are, respectively, 116-125 mm., 37-38 mm., and 6.9-8.2 mm. In the broad beak the sides are moderately converging, whereas in *urinatrix* the narrow beak has almost parallel sides. There is no unbroken grey band across the throat and foreneck in *urinatrix.*

Mr. John B. Cox, a marine ornithologist at Adelaide, believes that a large diving-petrel found at Sellick's Beach on May 4, 1956, was probably *exsul.*

Some authors place *exsul* as a subspecies of *urinatrix.*

GANNETS, CORMORANTS AND THEIR ALLIES: Order Pelecaniformes

TROPIC-BIRDS: Family Phæthontidæ

Red-tailed Tropic-bird (Bosun-bird)

Phæthon rubricauda

Plumage a glistening white, with a rosy suffusion on the back; a black horseshoe-shaped mark from the base of the bill to and behind the eye; flanks and wing bar black. Tail with 16 quills, *central pair of feathers are produced into long narrow brilliant red filaments.* Iris, brown; *beak, red;* legs, faint blue, toes black. Length (excluding tail streamers), 18 in. Weight, 1 lb. 13 oz. - 1 lb. 14 oz.

Distribution: The only regularly occupied breeding station was at the Abrolhos Islands, but Sugarloaf Rock, Cape Naturaliste, has now been colonised. Birds travel north and south and have been seen in the Fremantle district, Busselton and Albany, and have wandered around the south coast as far as South Australia (three records are known; Pearson Island, Port Lincoln and Yorke Peninsula).

Nesting: The single egg is laid in a slight hollow in the ground, usually under a bush, no nest being made. Colour, reddish-white, marked usually, mostly at the larger end, with purplish-black streaks and spots and blotches of brownish-red. Mean size of 6 eggs from the Abrolhos Islands, 66 x 48 mm., ranging from 62-71 x 47-50 mm. The young when first hatched are covered with whitish or greyish down.

O. H. Lipfert found a pair nesting at Rat Island, Abrolhos, in November 1894, and nesting was reported about the same time on Pelsart Island. The birds have always been regarded as uncommon at the Abrolhos and only odd nesting pairs have been reported by observers. In latter years they seem to have increased somewhat at Pelsart Island where, though no breeding birds were noticed in the two seasons of 1940-41 and 1941-42, a few pairs were seen during each of the five seasons from 1943-44 to 1947-48. Three nests inspected in December 1945 were all in close proximity to the camp of the British Phosphate Commission (later used as tourist quarters and then abandoned). One bird laid an egg on December 11 on one of the roads frequently used by motor lorries; another nest, containing a large young, was seen under a nitre-bush right alongside the same busy track. Six nests were located in the season 1947-48.

On November 1, 1939, a bird was found brooding an egg on a much frequented beach area at Busselton—the first known instance of nesting on the Australian mainland.

In the autumn of 1957 three Tropic-birds appeared at Rottnest Island and in the following summer the first nesting was attempted, unsuccessfully. On October 7, 1959, a bird was found brooding an egg near the aerodrome. This hatched about a fortnight later. Another egg was laid on December 22 about 20 feet from the other nest site. Neither pair succeeded in raising their progeny (W. A. Farmer).

Since 1963 nesting has taken place at the Sugarloaf Rock, Cape Naturaliste, where in the 1969-70 season 34 pairs bred. Some nesting is also attempted on the mainland beach opposite the Rock (studies by Michael Tarburton).

Tropic-birds, as the foregoing account indicates, seem quite indifferent to the presence of human beings when nesting. Their ordinary call is a harsh cry and when feeding young they emit a clucking sound not unlike that of a domestic hen.

White-tailed Tropic-bird (Golden Bosun-bird)

Phœthon lepturus

Mainly white, with a *golden or apricot wash* on the head, back, rump, and under parts; a black horseshoe-shaped mark through the eye; black band on the wing. Tail with 12 quills, the *central pair greatly elongated and of apricot colour.* Iris, brown, lower eyelid blue; *beak, yellow;* legs, blue or greenish-white, toes black. Length (excluding tail streamers), 16 in.

Distribution: This species does not breed in Western Australian waters, and the distinctive race described above (the golden-toned *P. l. fulvus)* nests only on Christmas Island, between Cocos Island and Java, where the late C. A. Gibson-Hill estimated the population to number no more than 300 to 450 breeding pairs. It ranges as far north as Java Head and southwestwards occasionally to the Cocos-Keeling group, 530 miles away.

It was not reported in Western Australian seas until July 27, 1965, when Roger Pocklington, of the Woods Hole Oceanographic Institution's research vessel *Atlantis II,* collected a specimen 240 miles off the Lacepede Islands (Lat. 16° 06′ S., Long 118° 03′ E.).

The species as a whole has an extensive pan-tropical distribution, in the Atlantic, Pacific and Indian Oceans, but all these populations (including the one nesting on the Cocos-Keeling Islands) lack the decided golden hue of the Christmas Island birds.

FRIGATE BIRDS: Family Fregatidæ

Lesser Frigate-bird

Fregata ariel

Native name: Wannoo (North-West Cape).

General coloration black, with a grey-brown wing bar. The *male has a white patch on each side of the body* under the wing and a bright red distensible pouch under the beak. The wings are long and the tail long and deeply forked. The female is larger and has a black throat and foreneck, with the breast white. Immature birds have the head and neck russet. Iris, brown; beak, grey in male, pale or with bluish tinge in female; orbital skin black in male, red in female; legs, black or reddish-brown in male, red in female. Females average larger than males. Length, 2 ft. 8 in.; wing span, male, 5¾ ft., female, 6¼ ft. Culmen, 78-90 mm. Weight, male, 1½ lb., female, 2 lb.

Distribution: A northern Australian species whose southernmost nesting station is Bedout Island, near Port Hedland; nesting also takes place at the Lacepedes (West Island) and on Sunday and Swan Islands (King Sound), whilst further north, at Adele Island, there is the largest nesting colony in the State. Considerable nesting occurs also at the Ashmore Reef (East and Middle Islands). The birds forage south-westwards to the Dampier Archipelago, but during summer "willy willies" wander further, preceding the storm centre. At such times they visit Onslow and on several occasions have been reported as far south as Point Cloates and Exmouth Gulf, whilst reports of Frigate-birds at Shark Bay probably refer to this species. In January 1973, after a cyclone, John Darnell saw one at Shay Gap, 30 miles inland.

Nesting: The birds nest in small colonies, making a small flat nest of sticks on the top of a bush. One egg only is laid; pure white, thinly coated with lime. Average size of 10 eggs, 64 x 44 mm., varying from 62-65 x 42-46 mm. Eggs are to be found in March, April, May and June, the peak of breeding varying from year to year. Incubation takes about 41 days. The young bird is naked when hatched but soon acquires a covering of whitish down.

Like the skuas, these aerial hunters of superb powers of flight (they never alight on the sea) obtain a certain amount of their food by harrying other sea-birds and compelling them to disgorge.

Greater Frigate-bird

Fregata minor

The male is entirely dark in plumage, with *no white in the adult state* and with a red, distensible, throat pouch. The larger female resembles the preceding species in colour pattern but it has a greyish-white throat and foreneck and the lower neck, breast and sides of the body are white. The wings are long and the tail long and deeply-forked. Iris, brown; eye ring, blue or black; beak, black in male, light pale pink or blue in female; legs, black or brown or pink in female. Length, 2 ft. 10 in. - 3 ft. 4 in. Females are noticeably larger than males. Culmen, male, 95-98 mm., female, 108-122 mm. Weight, male, 2 lb., female, 3½ lb.

The young have the under parts white, tinged russet. Head and neck russet.

Distribution: An irregular visitor from the tropics and not breeding in Western Australia. It is apparently much rarer locally than the preceding species, but wanders southwards more frequently. A specimen captured alive on the Swan River estuary on May 4, 1917, after stormy weather, was of the race breeding on Christmas Island in the Indian Ocean. Another individual was seen at Fremantle on May 5, 1951. A skull was found in the coast dunes near Walpole in June, 1972.

Nesting: The species does not breed in Western Australian waters. The nest is similar to that of the previous species, but the eggs are larger. Average size of 7 eggs from the Coral Sea, 68 x 47 mm., varying from 64-71 x 46-48 mm.

PELICANS: Family Pelecanidæ

Australian Pelican

Pelecanus conspicillatus

Native names: Wallan (Godfreys Tank, Canning Stock Route); Ngurlawanda, Wirmalo (De Grey River); Thudaara (Shaw River); Ji-ro-na (Coolawanyah); Bimba (Lyons River); Tjootara (Jigalong); Churuna (Shark Bay), Jilakarby (Upper Sanford River), Marnbu (mouth of Murchison River); Arranger (Northampton); Knulumberry (Geraldton); Nirimba (New Norcia, Perth and South-West); Beo-dee-lung (Mandurah); Bootlung, Boodalan, Bo-dal-lang, Budtallang, Puttarlung (South-West); Peelabong (Pallinup River).

An unmistakable species mostly white in colour, with the wing and tail quills black and a light grey crest on the back of the head. Broad white wing bar; two broad black bands along sides of lower back; tail black. Iris, brown, the bare skin around eye, yellow; beak flesh white to pale pink, tipped yellow; pouch pale or buff-pink; legs, leaden blue. Length, 5 ft. - 5 ft. 8 in.; wing span, 7 ft. 10 in. - 8 ft. 5½ in. Recorded weights 9-17 lb., usually 10½-12 lb. The beak is 13½-18¾ in. long; the distensible pouch is capable of holding 2-3 gallons of water.

Distribution: Occurs all over the State in suitable situations, in estuaries, rivers, lakes, swamps, marine bays and sometimes in the open sea.

Nesting: The Pelican nests in colonies and the nests are usually slight hollows scraped in the ground and often without nesting material. The nest of each bird is sufficiently distant from its neighbours to avoid pecking. On occasions an attempt at nest construction is made by the bird, when dry grass or other herbage is added to the hollow. The clutch consists of 2 or 3 eggs, which are white with a thick coating of lime; elliptical in shape. When the limy coating is removed the egg is found to be glossy. Average size of 15 eggs (Shark Bay), 93 x 57, varying from 85-98 x 53-60 mm. The young are hatched naked, but the feather-tracts are discernible in the skin. The egg-laying in Western Australia, from Shark Bay northwards, takes place in the autumn. At Mandurah it occurs in the spring.

Only a limited number of nesting sites have been recorded within the State: Pelican Island in Shark Bay (where D. G. Bathgate estimated that about 60 pairs nested in 1960), an island in Exmouth Gulf, Little Rocky Islet (Mangrove Islands), North Turtle Island off Port Hedland, the Lacepede Islands (Middle Island), Sandy Island (Long Reef) and Pelican Island (east of Cambridge Gulf near the Northern Territory border). There exist reports that a small colony nested on an islet in the Wallaby Group at Abrolhos, but recent confirmation is lacking. In the season 1962-63 a colony started nesting at Peel Inlet near Mandurah (on Creery I. and Nirimba Cay), and the birds continued to occupy one or other site subsequently. The nesting strength was of the order of 50 pairs. In July 1974, nesting began on an artificially raised island off Yundurup.

GANNETS AND BOOBIES: Family Sulidæ

Australian Gannet

Morus serrator

Mostly white except for the wing quills and the middle tail feathers which are black. The *head is yellowish-buff,* but this is not evident unless the birds are very near. Young birds are dark grey spotted with white. Iris, silver-grey; beak, grey; legs, grey with a yellowish line along the crest of each toe, converging and uniting proximally on the tarsus. Length, 33-36 in.; wing span, 5 ft. 8 in. Weight, 5 lb. - 5 lb. 9 oz.

Distribution: Essentially marine and occurring around the south and west coasts, north to Point Cloates.

There are no breeding sites in Western Australia, the nearest being at the Lawrence Rocks, Portland, Victoria, and there are others in Bass Strait (Black Pyramid and Cat Island) and south of Tasmania (Pedra Branca and Eddystone rocks).

This gannet is only a winter visitor to the State, remaining usually from March to September, though some individuals arrive in January. Ringing operations have shown that New Zealand-bred birds forage freely in Australian seas and two such birds have been recovered in this State—one on the beach at Northampton in 1955 and the other at Rottnest Island in 1956. Both were first-year birds and had been ringed as fledglings at Hauraki Gulf, New Zealand. The birds are attracted to Western Australian seas by the large shoals of mackerel, pilchards and other pelagic fishes which school at the surface in that period. The birds, which associate in small and moderate flocks, dive from a considerable height on their prey, sending up great splashes as they plunge. A flock of feeding gannets, actively "dive-bombing" in this manner, is an impressive sight and may be witnessed from elevated parts of the coast such as the cliffs near Albany or from boats in Cockburn Sound and near Garden Island. Their presence is a useful indicator to fishermen of the whereabouts of pelagic fish schools.

The Australian Gannet has been considered a subspecies *(serrator)* of the European Gannet, *M. bassana* (as we treated it in our third edition). However, it is sufficiently distinct in having the four middle tail-feathers and the secondaries black, whereas in the European bird the tail and secondaries are wholly white. Another species *(capensis)* breeds on islands off the coast of South Africa.

Masked Gannet (Blue-faced Booby)

Sula dactylatra

General plumage pattern as in the preceding species, but *all the flight quills of the wing, and the tail, black.* Head, white. Iris, yellow; beak, greenish-yellow, bare skin at forehead and around eye, black; legs, grey. Length, 33 in. Weight, 4 lb. 6 oz. - 4 lb. 12 oz.

The sexes can be readily distinguished at their nesting grounds by the call. The male bird utters a piping whistle and the female has a harsh raucous cry.

Distribution: A North-Western species, breeding on Bedout Island, near Port Hedland.

The Masked Gannet appears to be rather a local bird in our waters; it is plentiful in the vicinity of Bedout Island and wanders occasionally south to Shark Bay. It occurs sparingly to the Dampier Archipelago. Further north, at Adele Island, there is a smaller nesting colony than at Bedout Island.

Nesting: The nest is merely a depression scratched in the ground The 2 eggs are bluish-white, with a dense covering of chalky lime. Size, 64-66 mm. (61-69 x 43-49 mm.). There are autumn-breeding and spring-breeding populations and eggs are to be found in April and October on Bedout Island. The young are hatched in a naked condition, but soon become covered with white down. Normally only one young is reared. With the Brown Gannet this species has established a nesting colony of considerable size on Bedout Island.

Another northern and rather similar species, the Red-footed Gannet *(Sula sula),* has not been reported so far in our area.

It occurs in north-eastern Australia and there are considerable nesting colonies at North Keeling and Christmas Islands, south-west of Java. It has *no black* on the tail and *only the primaries of the wing,* not the secondaries, are black. Birds from the Coral Sea, however, have black secondaries but white humerals. In this eastern area dark phase birds are common. The *legs are red.* The nest is a bulky platform, built in trees or shrubs. The eggs measure 56-66 x 38-43 mm.

Brown Gannet (Booby)

Sula leucogaster

All the upper parts and the head and breast are *dark brown,* the rest of the *under parts white.* Iris, silver-grey; beak, whitish; naked skin on face and base of bill rich blue in male, chrome yellow in female, with small dark patch in front of eye; legs, pale green in male, yellow in female. Length, 28-30 in. Weight, 2 lb. 3 oz.-2 lb. 12 oz.

Distribution: This northern species breeds on Bedout Island near Port Hedland (southern breeding limit) and also on the Lacepede Islands (West and Middle Islands), Adele Island and Ashmore Reef (East, West and Middle Islands). A large roosting island is White Island (Lat. 15° 05′ S., Long. 124° 20′ E.). At sea the Brown Gannet predominates overwhelmingly in numbers over the Masked Gannet. It is one of the commonest sea-birds around northern Australia but does not range south-westwards beyond the Dampier Archipelago.

Nesting: The nest is often merely a depression scratched in the ground with a few pieces of sponge strewn around. At other times the birds use a quantity of seaweed and build up a nest about 2 feet in diameter and 3 or 4 inches high. The 2 eggs (very rarely 3) are greenish-white, with a dense chalky covering. Average measurements of 10 eggs, 61 x 42 mm., varying from 55-61 x 37-44 mm. The young are hatched naked, with lead-grey skins, but soon become covered with white down. Only one is usually reared.

This bird was the main guano producer of the former extensive deposits in the Lacepede Islands, Adele Island and Bedout Island, where an important trade was carried on during the third quarter of the last century. The guano exploitation began without the cognisance of the administrators of the colony, the Government first becoming aware of the trade about 1876. In the following three years no less than 37,000 tons of guano were raised, but by 1879 the deposits were virtually exhausted. Other islands further to the north, such as Jones, Lesueur and Stewart, also were worked for guano in this early period, but it is a curious fact that at present they carry a very slight bird population—White Island being the northernmost island with any appreciable bird life (S. Fowler has seen about 5,000 Brown Gannets roosting on it)—and it appears that there has been a very considerable shift in bird populations since the guano deposits were originally produced. Browse Island, still further north in the Timor Sea and the site of a very important guano trade, producing 23,000 tons of guano between 1876 and 1888, now carries no breeding sea-birds; they may have been exterminated by the resident labourers in the "guano era." The small unvegetated sand cays at Scott Reef and Cartier Island are unsuitable for nesting but several species nest freely at the Ashmore Reef, including Brown Gannets, Lesser Frigate-birds and Common Noddy, Sooty and Crested Terns.

CORMORANTS: Family Phalacrocoracidæ

Black Cormorant (Shag)

Phalacrocorax carbo

Native names: Garbang-a (South-West), Cardacut (Pallinup River).

A large *entirely black bird.* In closer view the plumage has a greenish sheen, which is particularly evident on the wings and scapulars, which have bronze reflections. The *gular pouch and bare skin around the eye are bright orange-yellow* sometimes with a greenish tinge and behind this the feathers are whitish grey. In the breeding season there is a *patch of white on each thigh,* which is clearly visible in flight and tiny white feathers are scattered on the head and more thickly on the sides of the neck. Iris, green; beak, dark grey, lower mandible paler; legs, black. Length, male, 32 in.; wing span, 55 in. This is the largest Australian Cormorant, males weighing between 5 and 6½ lb., averaging 5 lb. 9 oz., and females from 3 lb. 6 oz. to 4 lb. 11 oz., averaging 4 lb. 4 oz.

The species is distinguished from the only other all-black cormorant, the Little Black Cormorant, by the *light-coloured patch at the side of the head.*

Distribution: Found mostly in the South-West corner from the mouth of the Murchison River around the coast eastwards to Israelite Bay, and inland to the Avon River. That it occurs sparingly in the north is indicated by the observation of the species in 1951 on the Fortescue River at Millstream.

Nesting: This Cormorant nests in colonies, often associated with the Little Black and the Little Pied; in trees growing in swamps. The nests are rough platform-like structures of sticks, leaves and various herbage found in the vicinity of water. The 3 to 5 eggs are of the usual cormorant type—coarse, elliptical and greenish-white, but soon become coated with lime. A clutch of 6 has been recorded at Coolup. Size 60 x 36 mm.

Cormorants, or shags as they are alternatively known, are expert fishers, and operate both on inland waters and the strictly littoral portions of the sea coast. No species of cormorant, however, ventures more than a few miles from land. The Black Cormorant, which is nearly world-wide in distribution, in our region is almost confined to inland waters, on estuaries, rivers and swamps, though occasionally it is seen on some sheltered bays. Though they depend so essentially on the water for their livelihood, cormorants as a group have not, in one important particular, adapted themselves to that environment so closely as have for instance some other aquatic forms like the ducks and grebes. They have failed to acquire a water-resisting property for their plumage. Thus after a period of immersion the feathers of a cormorant become so sodden that flight is difficult or even temporarily impossible. They are thus compelled to spend a considerable time in drying their feathers, which they do by standing on a perch with partly outspread wings—a habit confined to cormorants and the closely allied darters.

Because of their fish-eating habits, cormorants are blamed in some quarters for destroying fish of commercial importance and even causing depletion. Such critics are apt to overlook the fact that not *every* species of fish occurring in any water is of value for human consumption, and in fact the commercial kinds are far outnumbered by the non-commercial. Cormorants and other fish-eaters take as food the species of fish which can be caught with the least effort and which are present in greatest abundance. It so happens that in our waters, as almost everywhere else, the edible and sporting fishes are too agile in comparison with the other available food for the birds. In Western Australia the cobbler seems to be the principal food of the Black Cormorant but perch, herring, hardyheads, silver-fish, mullet, yellow-eyed mullet and yellowtail have also been found in their stomachs. Of all the local cormorants this species is the one most destructive of commercially important fishes, but the birds are not numerous and the quantities of fish taken are not great enough to occasion concern.

Little Black Cormorant

Phalacrocorax sulcirostris

Entirely black with *no white in the plumage,* except for minute white feathers scattered thinly about the head and neck in the breeding plumage. These, however, would not be noticeable except in birds examined close at hand. The bare skin on the face is purplish-grey. Iris, green; beak, black above, remainder lead-grey; legs, black. Length, 24½ in. In weight males vary from 1 lb. 12 oz. to 2 lb. 9 oz., averaging 2 lb. 2 oz., and females from 1 lb. 7 oz. to 2 lb. 6 oz., averaging 1 lb. 11 oz.

Distribution: Generally throughout the State.

Nesting: The Little Black Cormorant nests in colonies and is usually associated with the Little Pied Cormorant. Sometimes the Black Cormorant is also found nesting in the same colony. The nests are built in low paper-bark trees growing in swamps and are rough platforms of sticks, leaves, bark or aquatic herbage. The usual clutch consists of 4 eggs and they are similar in shape, colour and texture to those of the other cormorants, being coarse dirty white, but when the lime with which they are coated is removed they are found actually to be greenish-white. A series of 9 from Moora averages 48 x 32 mm., varying from 47-49 x 31-33 mm.

This is essentially a river and estuarine species rarely frequenting the open coast. On the Swan River estuary it assembles into large feeding flocks, which are one of the characteristic sights of the river. The big company of the birds may be seen flying very close to the surface, and when a shoal of fish is encountered the leading birds wheel round and settle on the water; the remainder continue coming up and the birds aggregate to a dense mass, harrying the fish. Contrary to popular belief the fish are not mullet or other commercial species, but hardyheads, bull-bait or smelt *(Pranesus endrachtensis)* and anchovies *(Engraulis australis)*. Other fish fed on are gobies and gobbleguts.

Pied Cormorant

Phalacrocorax varius

Native names: Toorna (Nickol Bay), Ma-dee, Medie, Mee-die (Perth).

Black above and white below, with black thighs. The naked skin of the face region is *orange in front of the eye and bright greenish-blue below.* When examined closely the apparently uniform black colour of the wings and mantle is seen to be really a dark bronze-green, with the individual feathers edged with black. Iris, brilliant green; beak, dark horn; legs, black. Length, male, 31-32 in., female, 27¼-29 in. Wing span, 3 ft. 6 in. - 4 ft. 2 in. Weight, male, 4 lb. - 5 lb. 7 oz., average, 4 lb. 14 oz.; female, 3 lb. - 4 lb. 9 oz., average, 3 lb. 13 oz.

In flight this species may be distinguished from other cormorants by the straight course kept by a flock and *very low over the water*.

Distribution: Along the coastline of the whole State, except the section near Esperance and the Archipelago of the Recherche inhabited by the Black-faced Cormorant. The Pied Cormorant also becomes rare south of Cape Naturaliste and is virtually absent on the south coast, east of Cape Leeuwin, but reappears in the Great

Australian Bight in the vicinity of Eucla. Apparently the boisterous steep-to shoreline of the south coast affords an unfavourable environment compared with the more or less reef-sheltered waters north of Busselton.

Nesting: The nests are strongly-built, substantial, bowl-shaped structures, about 14 in. in diameter, made of pliant stems such as nitre-bush branches or other material, including leaves of *Mesembryanthemum* and are built on the ground or on the tops of low bushes or on mangroves, in crowded colonies. Eggs, 2 or 3, sometimes only one or as many as 4, very occasionally 5; longish oval in shape, coarse shell, and greenish-white in colour. The eggs soon wear to a dirty white. Average of 23 eggs, 60 x 38 mm., varying from 55-67 x 34-40 mm.

The newly hatched young have naked black skins. The down when it appears is white below and black above. Large young, when just acquiring the contour feathers at which stage they are almost as big as adults and may weigh nearly 3 lb., have the iris grey, beak pale yellow, and the bare skin on the forehead and around the eye a flesh-grey.

The same nesting ground is not always used each year but shifts are frequently made. This appears to be due to the need for the natural vegetation to recover from the effects of the "live" guano, which would otherwise ruin the site. In the Safety Bay islands nesting starts at the beginning of April and continues at least till the end of July, with the peak in eggs in May (V. N. Serventy and S. R. White). This autumn and early winter breeding is characteristic of all the populations in Western Australia except the colonies which nest on the Abrolhos Islands—there the nesting period is in November and December. Recorded breeding localities are the following: Safety Bay islands (The Sisters, Bird Island, Shag Island), Carnac Island (Flat Rock), Rottnest Island (Dyers Island), islands between Fremantle and Dongara (Whittell, Buller, South Ronsard Rock, Cervantes, Essex Rock, Sandland, Milligan, Lipfert, Webb, Snag and East Beagle Islands), Abrolhos group (Pelsart, Square and Wooded Islands, Pelican and Lagoon Islands, and an islet south-east of Pigeon Island), Dirk Hartog Island (Quoin Bluff and Shag Mia), Shark Bay islands, Faure and Pelican Islands and the Freycinet Estuary islands—including Freycinet, Double, Mary Anne, Wilds and Leschenault Islands), Dorre Island, Frazer Island (off Point Cloates), sandy cay near N. Sandy Island, the Dampier Archipelago, North Turtle Island, Lacepede Islands (Middle Island), Adele Island, Low Rocks (Admiralty Gulf) and Long Reef (Sandy Island).

The Pied Cormorant enters the larger estuaries, such as the Swan River (to Guildford) and Peel Inlet. Such penetration, especially in the case of the Swan River, is made mainly by the females and the males rarely pass beyond Fremantle Harbour. Ringing has shown that young birds, in their first year, disperse randomly along the coast, for distances up to 200 miles from where they were bred. Older birds appear to move about less (J. Ford).

Perhaps nowhere in Australia is this bird more plentiful than in Shark Bay, where it is the most conspicuous sea-bird. Whilst on an aerial survey for the Commonwealth Scientific and Industrial Research Organisation on October 9, 1945, S. Fowler made an approximate census of between 15,000 and 20,000 cormorants along the east coast of Dirk Hartog Island, the bulk being at the roosting ground four miles north of Quoin Bluff. In 1920 F. Lawson Whitlock estimated the Quoin Bluff colony to contain over 2,500 nests.

The Pied Cormorant is the sole guano-producer in Shark Bay, where formerly large deposits were exploited. The first intelligence of the occurrence of payable guano was brought by the schooner *Pelsart* when she returned from a fishing trip to the Bay in March 1850. The sample of guano brought back was forwarded to Mauritius and received so good a report that by the end of the year no fewer than 10 vessels, aggregating 4,000 tons, were in the Bay loading guano for export. Egg or Tetrodon Island carried the main deposit, variously estimated at 3,000 to 6,000 tons, and Lt. B. J. Helpman, R.N., estimated there were 20,000 tons of guano on the islands in Freycinet Estuary at the south of the Bay. On Saturday Island, near Cape Heirisson, the mound of guano was 11 feet thick at the centre. To ensure the collection of royalties a military guard under Lt. Elliot, with a detachment of the 99th Regiment, was established in October 1850 at The Quoin Bluff, on Dirk Hartog Island, overlooking Egg Island. It was withdrawn a year later, by which time the larger part of the guano had been removed. However, the deposits continued to be worked until 1888, in the later years only for the local market. Early in 1858 one vessel, the *New Perseverance,* was able to gather guano from an island that had been well scraped some years before, reporting the fact as "a proof that the accumulation of guano is constant and rapid."

F. L. von Bibra, holder of the Shark Bay guano concession between 1880 and 1888, estimated that the cormorants renewed the guano at the rate of 80 tons per annum and pelicans added an additional 15 tons. At the end of 1938 the erection of navigation beacons was completed in Shark Bay and on the open perforated platform an accumulation of almost 3½ ft. of guano was observed in July 1946. In a survey by the C.S.I.R.O. in 1947 about 1,000 tons of guano was estimated to exist in the various islands in the Bay.

The legal protection of the guano birds exercised the attention of Government in 1887, and on the suggestion of Mr. von Bibra, the cormorant was added to the list of birds protected during the breeding season in a proclamation issued in February 1888, there being no power under the Game Act, 1874-1878, to impose complete protection as desired by the guano interests.

This protection was continued, north of the Moore River only, until 1917, when it was cancelled, since when these cormorants had no legal protection in the State until 1952 when the new Fauna Protection Act came into force. In 1897 a reward of 3d. per head was offered for the destruction of cormorants (any species) south of the Moore River, "to prevent the wholesale destruction of young fish by these birds," and increased to 6d. the following year. This lapsed in 1905 but renewed agitation against the cormorant led to a re-gazettal of the bonus of 6d. per head in January 1912, but it was discontinued in September 1914.

The staple food of this cormorant in the sea includes parrot fish, cobblers, rock whiting, leatherjackets and pipefish, and in the estuaries, gobbleguts (in one bird 139 were counted, crammed in the stomach), gobies, prawns and shrimps. Estimates of the amount of food taken per day are difficult to assess. At the National Zoological Park in Washington, D.C., the Double-crested Cormorant *(P. auritus)*, approximately equivalent to the Pied Cormorant in size, is fed at the rate of three-quarters to one pound of fish per day, given at one feeding for only six days in the week. On this ration the birds thrive and breed freely in the large flight aviaries.

Black-faced Cormorant (White-breasted Cormorant)

Phalacrocorax fuscescens

Black above and white below, with black thighs. The back of the neck is profusely adorned with small, narrow white feathers, present both in the breeding and non-breeding seasons, but not evident except at close quarters. The *naked skin* in front of and around the eye and the gular pouch is *black,* that in the loral region being dotted with tiny white feathers. In immature birds this skin is grey. Iris, brilliant green; beak, dark grey; legs, black. Length 24-27 in., wing span, 3 ft. 3 in. - 3 ft. 8 in. Weight, 3-4½ lb.

The bird is similar to the preceding species, but slightly smaller and is distinguishable by the absence of bright colours on the naked parts of the head.

Distribution: Predominantly an eastern States species which in Western Australia is found only on the islands of the Archipelago of the Recherche and the adjoining mainland, from Hopetoun eastwards to Israelite Bay.

Nesting: This bird nests in colonies on rocky ledges and terraces of islands, the nests being constructed of seaweed, such as *Fucus* and *Laminaria.* The clutch consists of 2-4 eggs, which are pale greenish with a limy coating. Average measurements of 15 eggs, 57 x 40 mm., varying from 49-61 x 33-38 mm. Weight, 56 gm. They are indistinguishable from those of the Pied Cormorant. Egg dates, August-January, but the earlier layings appear to be unsuccessful. The nestlings on hatching are naked, the skin being almost black. The down is black, with some whitish down on the under parts in older birds.

The Black-faced Cormorant is not represented in any great strength in this State and the nesting colonies on the Recherche are fairly small. Thus the one at Lion Island, in Esperance Bay, consisted of about 25 pairs when inspected by us in January 1944, and one hardly ever sees more than a score or so individuals away from the rookeries. There is another small nesting colony at Middle Island. There are no cormorants at the Eastern Group, or the outer islands. The sparseness of the birds at the Recherche Archipelago is due to lack of shallows, all of the shore-lines being steep-to. The bird is much more numerous in South Australia.

The birds feed on small rock fish and on the juvenile yellow leatherjackets *(Cantherides ayraudi)* which, to the annoyance of fishermen, seasonally attain almost plague proportions on the south coast.

Little Pied Cormorant

Phalacrocorax melanoleucos

Native names: Ka: ru (Warburton Range); Go-go-go (Perth).

Black above and white below; the lores and line above eye white. No black thighs. Some birds show a brown or rusty stain on the under parts, a discoloration due to iron compounds (ferric oxide) in some waters. Occasionally birds are met with aberrant plumages, with more or less black on the under surface. A specimen collected at Como in 1937 had a black breast band. Iris, brown; *beak, dusky yellow;* dark brown above; *gular pouch, yellow;* legs, black. Length, 23 in.; wing span, 33-36 in. Weight, males vary from 1 lb. 8 oz. to 2 lb. 2 oz., averaging 1 lb. 13 oz., and females from 1 lb. 4 oz. to 1 lb. 15 oz., averaging 1 lb. 10 oz. Immature birds lack the white line above the eye and have brownish upper parts.

Distribution: Generally distributed.

Nesting: As a nest, the Little Pied Cormorant constructs a platform of sticks in a tree (usually a paper-bark), growing in a swamp in water. The centre is lined with green fronds from a paper-bark tree and on this the bird deposits the 3 to 5 eggs which it normally lays. It usually nests in colonies, often in association with the Little Black Cormorant. The eggs are of the usual cormorant type—oval, and covered with a limy wash which increases as incubation proceeds. When freshly laid the eggs are a delicate pale blue; 46 eggs from Moora and Carnarvon average 46 x 31 mm., varying from 36-52 x 26-33 mm. Fledglings have black down.

This is the common cormorant of rivers and swamps and is less plentiful in purely marine localities. Like the other species of cormorants, it feeds on fish such as gobies and gobbleguts, and crustaceans like shrimps and jilgies. The last-named appears to be the prevalent diet of the cormorants in fresh water swamps of the South-West.

Australian Darter (Needle-beak Shag)

Anhinga rufa

Native name: Mimmal (South-West).

The fully plumaged male is *almost entirely black,* with buff centres to the wing covert feathers; head and neck brownish-black, with a whitish cheek stripe. Females and young males are dark grey above, with light lanceolate markings on the wing coverts and the *whole of the under parts white.* The head and neck are thin and snake-like in appearance. Occasionally some individuals have red in the plumage, being in extreme cases entirely rusty-red below with a rufous tone on the upper parts. As in the comparable case of the Little Pied Cormorant, this is an artificial discoloration due to an iron contamination in some types of natural waters. Iris, yellow; beak, yellow, olive above, slender and sharp-pointed; legs, flesh-colour. Length, 34 to 37 in. Weight, females, 3 lb. 11 oz.-4 lb. 8 oz., averaging 4 lb. 1 oz. One male examined weighed 3 lb. 4 oz.

Distribution: Found all over the State; rather rare in the South-West but local strongholds are the estuaries and rivers. In the Swan River it is more common from Perth Water upstream and at Mandurah it can always be seen perched on the breakwater near the bar and on the paper-bark trees lining the channel to Peel Inlet. It is also plentiful in the Leschenault Estuary, including that part within the Bunbury town limits, and the lower portions of the Collie and Preston Rivers. It is less common further south, but a few are to be seen at the Hardy Inlet (Augusta), and on the south coast it does not reappear in strength until the Pallinup Estuary. It is abundant at the Wellstead Estuary, Bremer Bay.

Nesting: Usually nesting in small colonies, the Darter's nest is just a platform of sticks in a tree standing in water in a swamp or river. A depression to hold the eggs is lined with a few leaves. The 3 or 4, occasionally 5, eggs are of the cormorant type and covered with a coating of lime. When fresh, however, the shell is greenish-white. The eggs are more pointed at one end than is the case with cormorants' eggs. Average measurements, 54-58 x 33-36 mm. Both parents incubate and brood the young (E. Lindgren). The nearest nesting place to Perth is Lake Coollelal, Wanneroo and a swamp south-west of Coolup.

The Darter appears to be a more expert fisher than the cormorants, being able to capture fish not ordinarily taken by the latter. The Darter catches its prey by first spearing it with its pointed beak, afterwards shaking it off and then swallowing it. Fish remains examined in stomach contents usually show these puncture marks. The diet includes yellow-eyed mullet, silver-fish, whiting, cobbler and rock whiting.

The bird possesses the ability to submerge without diving and when in the water may be seen to sink slowly like a submarine, with, perhaps, only the head remaining above the surface (C. F. H. Jenkins). The call of the bird may be likened to the clicking sound made by a broken clock spring when being wound.

GREBE-LIKE BIRDS: Order Podicipediformes
GREBES: Family Podicipedidæ

Crested Grebe

Podiceps cristatus

Native names: Ka-lee, Kali (Perth).

Upper brown, with a white band in the short wing; under parts a gleaming white. Around the head is a striking adornment, consisting of elongated feathers, chestnut at the base and black at the tip, forming *an erect ruff or "tippet"*. Iris, red; beak, dark horn colour; legs, dark olive-green, with pale yellow soles to the feet. Length, 20 in. Weight, 1 lb. 12 oz - 1 lb. 14 oz.

Fig. 18.—Crested Grebe.

Distribution: This grebe is confined to the South-West corner, from Dandaragan to Culham Inlet. It has been reported at Lake Matilda.

Nesting: A nest at Lake Jandakot in December 1941 (K. G. Buller) was approximately 50 yards from the edge of the lake, and was a platform of rushes and other aquatic plants interlaced one over the other. A quantity of mud assisted to stabilise the nest. At the back of the nest was a depression, intended, presumably, for one of the birds to rest on when not engaged in incubation. The nest was about 4 feet in circumference and was entirely independent of standing rushes, but pieces of broken rushes and other debris formed a link with the standing rushes and prevented its drifting away. In it were 5, nest-stained, incubated eggs. The eggs are actually a glossy greenish-white but that colour is obscured by a dull white limy coating which soon becomes stained and dark in colour as incubation proceeds. Size, 50-52 x 34-36 mm. As with other members of the genus, the eggs are elliptical in shape. Besides Jandakot, known breeding lakes are Bibra, Yealering and Toolibin.

The Crested Glebe is definitely an uncommon species, but individuals appear from time to time on reservoirs and larger bodies of water, including brackish estuaries. At Wilson's Inlet, Denmark, we have seen them more regularly than anywhere else. In diving, the bird remains submerged from 22 to 28 seconds (Cooloongup Lake, Rockingham).

Little Grebe (Dabchick)

Podiceps novæhollandiæ

Native name: Ngoo-nan (Avon River).

In the field the *dark head with the light* (actually yellow) *spot between the eye and the beak* distinguishes this species from the Hoary-headed Grebe. The upper parts are greyish-black, with a rufous stripe along the side of the neck behind the eye. The under parts are white, save for the grey lower neck. There is a white wing bar. In the immature and winter plumage the head is grey and the head markings are ill-defined. Iris, straw-yellow; beak, black; legs, olive-green. Length, 9¼-9¾ in. Weight, 5¼ oz.

Distribution: The species is distributed all over the State.

Nesting: As with other members of the genus the nest is a floating structure composed of rushes or other aquatic plants piled up to form a large floating mass often 2 ft. across, on which the eggs are deposited. Sometimes the eggs are so close to the water that they are wet. The sitting bird adds to the structure whilst engaged in incubation and it usually covers the eggs with material, such as is used in the construction of the nest, before leaving the eggs for feeding purposes. The temperature of the nest (72°-88° F.)

is higher than that of the water in which it lies (4-10° F. colder) according to observations by E. Lindgren. The eggs, 3 to 6, are of the usual elliptical type laid by the other members of the genus, and are bluish-white, but this is obscured by a thin limy coating. They darken in colour as incubation proceeds. Size, 31-38 x 24-26 mm.

This is essentially a fresh water grebe compared with the Hoary-headed species and is commonly found in the fresh water reaches of rivers, swamps and country dams. However, on the Swan coastal plain, particularly, the Hoary-headed Grebe may outnumber it on the larger swamps and it seems to prefer smaller bodies of water than that species.

Fig. 19.—Little Grebe

When diving the bird remains under water from 7 to 16 seconds, averaging 12 seconds; and the interval between dives is 7 or 8 seconds unless the bird is disturbed, when the interval is lengthened (data obtained from a bird feeding young with food obtained by diving at Hyde Park, Perth).

The call is a shrill, vigorous "chee-ee-ee-ee".

Hoary-headed Grebe (Dabchick)

Podiceps poliocephalus

Native names: Wy-oo-da, Wy-uda (Perth); Poorajean (Pallinup River).

Upper grey-brown, with a prominent white wing bar. Upper breast buff-brown, rest of under parts white. The head is black and adorned with *numerous fine white hair-like feathers*. Iris, buff; beak, dark grey, tipped lighter; legs, olive-green. Length, 11½ in. Wing span, 19 in. Weight, 9¼-10 oz.

Distribution: Generally distributed, but extending north usually only as far as Point Cloates and the Barlee Range. There is one breeding record from Derby (P. Slater).

Nesting: There is nothing either in the situation or in the construction of the nest of this grebe to differentiate it from that of the previous species, but the nest is usually much smaller and built in shallower water. The eggs, however, although similar in shape, are larger (and usually 3 and 4 in number but up to 6 known) and often have limy nodules on the outside of the shell; further the limy coating obscures a *greenish*-white instead of a *bluish*-white shell. A series of 40 eggs from Lake Grace and Carnamah averages 40 x 28 mm., ranging from 36-43 x 25-29 mm. Mean weight, 16 gm. This grebe is more gregarious in its nesting habits than the Little Grebe, up to six nests having been found together, about 3-4 feet apart.

Though common enough in fresh waters also, to which breeding is restricted, the Hoary-headed Grebe is partial to brackish and marine estuaries as feeding areas. These include the broad waters of the Swan River, Peel Inlet and the south coast inlets. Sometimes flocks of many hundreds are to be seen in such localities in the winter months. When disturbed by an approaching boat the birds will rise from the surface and fly vigorously for quite a distance,

Fig. 20.—Hoary-headed Grebe.

at a few feet above the water. This is different from the behaviour of the previous species, which can rarely be induced to rise at all from the water in daylight. Like the Musk Duck, all grebes seem to undertake their wanderings between different waters at nighttime. Besides extensive land flights, ocean crossings are sometimes made. Thus both of the smaller grebes have at times been seen on the salt lakes at Rottnest Island, which is 12 miles from the mainland, but neither species is a regular resident there.

The habit of frequenting salt estuaries, where commercial net-fishing is often carried on, causes a severe toll at times of this grebe. Many become entangled in the fishing nets at night in rough weather. At the Pallinup Estuary on one occasion (June 24, 1943) in experimental fishing tests, we found 11 dead grebes in a length of 120 yards of 2⅜ in. mesh net.

In contrast to the previous species this is a silent bird, its only call being a low squeaky note when attending its young.

HERON-LIKE BIRDS: Order Ciconiiformes

HERONS, EGRETS and BITTERNS: Family Ardeidæ

White Egret (White Crane)

Egretta alba

Pure white (with ornamental plumes on the back in the nuptial plumage). Iris, yellow; beak, yellow (largely black in the breeding season), long and sharp-pointed, bare skin at base pea-green; legs black, long and trailed behind body in flight. If birds are inspected close at hand it will be seen that the upper leg (tibia) is yellowish, the tarsus is black, but yellow on the inner part. Length, 30 to 35 in. Weight, 2½ lb.

Distribution: Occurs all over the State in suitable areas. Small numbers appear on the Swan River (Heirisson Islands) and neighbouring swamps in the winter months.

Nesting: The White Egret, like other members of the group, nests in colonies and is usually associated with Nankeen Night Herons, Little Black Cormorants and Little Pied Cormorants. Few nesting colonies of these egrets are, however, known to us in the area covered by this *Handbook.* A small colony was discovered by S. R. White on the coastal plain south of Bunbury where a few pairs nested in 1939. P. J. Fuller found a colony of about 50 pairs on the Collie River, north-east of Bunbury, in October 1960. T. Riggert inspected a nesting colony of about 50 pairs at Toolibin Lake in November 1965, unrecorded previously by naturalists but well known to duck shooters. The nests are platforms of sticks built in paperbark trees in a swamp. The 3 or 4 eggs are elliptical, light bluish-green, of coarse texture, and, when freshly laid, have a slight gloss. They soon, however, lose this gloss and become rough and nest-stained as incubation proceeds. Size, 55 x 39 mm.

Plumed Egret

Egretta intermedia

Intermediate in size between the White and Little Egrets. Pure white, ornamental breeding plumes on the back and breast but not on the nape, and extending about 6 in. beyond the tail. Iris, reddish orange; beak, orange-red, yellow at tip; legs, flesh-coloured above the "knee" and black below. In the non-breeding plumage there are no ornamental plumes and the facial skin, iris and bill turn yellow. Length, 24 in.

Distribution: A winter visitor from the Kimberley Division to the northern parts of the North-West. John Darnell found it a scarce autumn and winter visitor at the sewage ponds at Mt. Goldsworthy and on the De Grey River.

Nesting: No local breeding is known.

Little Egret

Egretta garzetta

Pure white, with a pair of ornamental plumes on the back in the nuptial plumage. Iris, yellow; bare skin of face, yellow; beak, black; base of lower mandible, yellow; legs, black. Length, 22 in. It is a long-necked and a more delicate and graceful-looking bird than the Cattle Egret.

Distribution: A Kimberley species which has been occasionally recorded at Port Hedland, Port Samson and the Lyndon River, south of Exmouth Gulf. In July 1959 Dr. G. F. Mees saw four birds at Carnarvon and in April 1965 J. R. Ford saw one at Lake Yangebup, south-east of Fremantle. R. H. Stranger saw one at Mandurah in April 1969. In September 1972 T. Spence saw about 50 at Carnarvon. In December 1973 E. H. Sedgwick saw one at Benger swamp.

Reef Heron

Egretta sacra

(formerly *Demigretta sacra*)

Native name: Kooroodor (North-West Cape)

This species exists in two distinct colour phases: (a) general body colour *blackish-grey*, with brown tone on the abdomen and a narrow white stripe down the centre of the throat; (b) *pure white*. Iris, yellow; beak, slaty-brown (yellow in the white phase); legs, yellowish-green. Length, 24-28 in. Weight, 10½-12 oz.

Piebald individuals occur. One, with head and neck slaty-grey and the wings and body white was seen at Port Hedland by John Darnell.

Distribution: On rocky coasts and reefs all around the State. Once recorded at Pelican Point, on the Swan River estuary, and in Peel Inlet at Yundurup.

Nesting: The nest is merely a pile of sticks placed on the ground, or on a ledge in a rock-face, with a few smaller sticks in the centre where the eggs are deposited. The 2 to 5 eggs are pale green. Size, 43-46 x 33-35 mm. Birds of the grey and white phases mate together and we have observed that when that occurs some of the progeny may be of the white phase and some of the grey.

The white phase of this species is more frequent in the north, occurring southwards in decreasing proportions. The exact ratios in various localities have not yet been determined. However, even in the extreme south occasional white individuals are reported. There are records of white birds from Albany, Cape Naturaliste and Rottnest Island. Care should be taken in field identifications of the grey phase to avoid mistaking it for the common White-faced Heron, which also may occur on marine beaches. Similarly the white phase may be confused with the Egret which in the north may frequent marine tidal areas; the Egret may be distinguished at once by its black legs, which are relatively very long, compared with the Reef Heron, when trailed in flight.

Cattle Egret

Bubulcus ibis

White. In the breeding plumage the crown and nape become *orange-buff* or deep biscuit coloured, with similar-coloured plumes on the throat and back. Iris, yellow; beak, *short, yellow;* legs, yellow turning to greenish-yellow in the non-breeding state, appearing dark. Length, 21 in.

This small egret resembles the Little Egret in size, but is a squat, short-necked bird, the yellow bill further distinguishing it from the Little Egret. It is the only member of the heron-egret group to associate with livestock and frequently perches on the backs of cattle and horses.

Distribution: The Cattle Egret is a newcomer to Australia. Apparently it was a natural immigrant from Indonesia into Arnhem Land where it became an attendant on the buffalo, which was originally introduced to Port Essington in the 1840s.

It is unknown when the Cattle Egret first appeared but it was encountered "in hundreds" by the National Geographic-Smithsonian Institution Arnhem Land Expedition in 1948 in the Oenpelli district. Subsequently observers have noted them "in thousands" throughout the Northern Territory, associated with grazing buffalo herds and breeding in great numbers at the Adelaide River (J. M. Hewitt). It is likely that the Cattle Egret had colonised Australia much earlier than 1949 and that the introduction of 12 birds from Calcutta to the Kimberley Division in 1933 (referred to in our earlier editions) had played no part at all in the species' spread.

The birds are now widespread in both eastern Australia (to Victoria) and in Western Australia. The first definite record in this State was a single bird seen in the horse paddock at Millstream Station, on the Fortescue River, in 1949. They were first observed in the South-West in 1952 at Three Springs, and in the following year at Norseman and Bunbury. The first birds were noticed in the Perth area in 1954. Since then they have been reported "from districts as far apart as Esperance, Ravensthorpe, Bunbury, Bremer Bay, Mt. Barker and Baandee, indicating quite an extensive invasion of the South-West" (C. F. H. Jenkins).

Nesting: There are no breeding records so far from Western Australia. Nests recorded by M. T. Goddard in New South Wales were smaller than those of the White Egret, in whose company the Cattle Egrets nested, were more concave and built of finer materials, 13 in. in diameter, and firmly supported by a flat fork or by short upright leafy branches. Clutch, 3 to 6 eggs, white with a delicate bluish tinge, 47-50 x 33-34 mm.

White-faced Heron (White-fronted Heron, Blue Crane)

Ardea novæhollandiæ

Native names: Wy-ang (North-West Cape); Korlolba (Lyons River); Tarkaroo (Jigalong); Bogga (Wooleen); Wy-an (Perth); Pullong (Pallinup River).

Slaty-grey, wing quills darker, with *white on the forehead, around eye and on chin.* Long lanceolate grey plumes on the back and chestnut feathers on the upper chest. Iris, yellow; naked skin around eye, greenish yellow; beak, black; legs, greenish-yellow. Length, 24 to 29 in.; wing span, 3 ft. 9 in.

Distribution: All over the State in aquatic habitats, even occasionally on the sea beaches.

Nesting: The nest is a platform of sticks built high up in a horizontal fork of a tall tree, not necessarily close to water. Several birds are often found nesting in the same vicinity, but the species does not appear to nest actually in colonies. The eggs, 3 to 5 in number, are plain blue without gloss and oval in shape. Average size of 17 eggs, 48 x 35 mm., varying from 45-52 x 33-36 mm. The incubation period is from 3½ to 4 weeks and the nestlings remain at the nest for a further month (W. Plumb).

This is pre-eminently the common heron-like bird in the State. Wherever there is any standing water, temporary or permanent, one or more of these birds may be met with, their colour impressing the observer with the effectiveness of field-grey for camouflage purposes. Their food includes frogs, fresh water cobblers, jilgies, various water insects and grasshoppers, and on occasions a bird might raid the backyard goldfish pond.

The White-faced Heron sometimes visits islands quite far from the shore; it has been observed at the Abrolhos Islands, at Rottnest and on Christmas Island in the Archipelago of the Recherche.

White-necked Heron (Pacific Heron)

Ardea pacifica

Native names: Jil-lee-mil-yun, Jili-mil-yan (Perth); Koon-jere (Avon).

Head, neck and breast white, shaded with grey; *rest of body dark grey.* There are a few reddish-chestnut feathers on the back and upper breast; abdomen, streaked white; there is white around the bend of the wing. Iris, pea-green; beak, black; legs, black. Length, 30-42 in. Wing span, 4 ft. 10 in. - 5 ft. Weight, 1 lb. 6 oz.

Distribution: All over the State.

Nesting: The White-necked Heron nests in tall trees, sometimes in small colonies, but often only in individual pairs. The nest is merely a platform of sticks. The 4 eggs, the usual number laid, are pure green. Size, 58 x 39 mm. The nestling is covered with pale dove-coloured down, and the iris is straw colour. There are very few reports of nesting in the southern parts of the State. Two pairs made an abortive attempt at nesting near Kellerberrin in September 1953 (P. S. Stone). P. J. Fuller found six pairs nesting in river gums at Yarlarweelor Station on the Murchison River, and seven pairs on Mt. Clere Station on the Gascoyne River, in April 1961. This nesting followed good rains consequent on a February cyclone. T. Riggert found four breeding pairs at Lake Toolibin, 24 miles south-east of Narrogin, in December 1964. Earlier (September 1957 and February 1964) P. Slater found evidence that the species was probably nesting near Katanning.

In the South-West this species is much rarer than the White-faced Heron, but periodical irruptions may occur; then large concentrations appear where there may have been very few, or none, for some years. Thus in the Perth area, hundreds of these herons were to be seen on Burswood Island between October and December 1923, and there was another invasion in November and December 1926. E. H. Sedgwick reports another in 1938. In late 1941 large numbers were seen at Lake Jandakot. There were widespread invasions during the winters of 1952 and 1957, when birds penetrated throughout the South-West, to Nornalup, and the Wheatbelt. Numbers were above average in 1958 and 1959, and there was a large invasion in 1975. A reference to an invasion in August 1865 is contained in the correspondence of G. E. Egerton-Warburton, of "St. Werburgh's," near Mt. Barker.

Nankeen Night-Heron

Nycticorax caledonicus

Native name: Gnal-gan-ning, Ngalganning (Perth).

Upper parts *bright brownish-chestnut,* under white. The crown of the head and nape are black and from the back of the head project, in the nesting season, three delicate white filamentous plumes. A prominent white eyebrow line begins at the base of the beak, and below this, between the beak and the eye, is a naked area of vivid yellow-green. The sharp, heavy beak is black; iris, yellow; legs, bright yellow. Length, 21½-25¼ in. Weight, 1 lb. 8 oz. - 1 lb. 13 oz. Immature birds are brown, *streaked white and pale buff,* and the adult plumage is not fully attained until the end of the third year.

Distribution: All over the State in suitable habitats.

Nesting: The Night-Heron makes a poorly constructed platform of sticks in a tree standing in water, or overhanging a stream. The 2 or 3, sometimes 4 and 5 eggs are elliptical in shape and light bluish-green in colour. Size, 55 x 37 mm. The bird often nests in colonies, but sometimes there are solitary pairs. In 1900 a colony of about 200 bred on Bird Island, just off the mainland in Safety Bay, nesting on the ground under plants with a mere ring of sticks to prevent the eggs from rolling away. That colony no longer exists, but from a group of seven birds introduced from there to the Zoological Gardens, South Perth, in 1914 a large nesting colony was built up, numbering several hundreds of birds. Between 1939 and 1941, on the ground that the birds were killing fish in the goldfish pond, the nests and young were destroyed and the colony dispersed. In 1947 a heronry of about 250 nests was started in flooded (blue) gums *(Eucalyptus rudis)* lining the Murray River at Blythewood, Pinjarra. There is another large rookery in a paper-bark swamp at Waterloo, near the Collie River. Birds were found nesting in the Wellering Swamp, Gingin, in 1953.

This is a heron of nocturnal habits; during the day it may be seen roosting in companies of varying strength in the trees lining river banks or in swamps, but sometimes even in suburban street trees, as at Osborne Road, East Fremantle, and parks (such as Queens Gardens, in Moreton Bay fig trees).

Mangrove Heron (Green-backed Bittern)

Butorides striatus

Upper parts varying from grey to rufous-red, crown black with a greenish tinge. A dark line below the eye extends backwards and the loral space is yellow and grey-black. Under parts grey with a line along the mid-throat which is white spotted brown. Edge of wing white. Iris, bright yellow; beak, black; lower mandible, greenish-yellow; legs, olive-green with the back of the tarsus and under toes yellow. Length, 18-20 in., male, larger; wing span, 28 in. Weight, 8¼-10¾ oz.

Distribution: Found in coastal mangroves along the North-West coast, as far south as Peron Peninsula, Shark Bay. The grey-backed form (the race *cinereus*) ranges from the Kimberley Division to the De Grey River mangroves; the red-backed bird (the race *rogersi*) occurs in the Onslow mangroves. The Shark Bay birds are grey.

Nesting: The Mangrove Heron builds a loosely-knit platform, about a foot in diameter, of stout sticks, in a horizontal fork of a mangrove. A slight depression in the centre holds the pale green eggs. They measure 40 x 31 mm. The clutch is 2 or 3; in the Sydney district, however, it is usually 3 and on occasions 4 (K. A. Hindwood). The young are covered with green-grey down and have olive-green legs and beak.

Black Bittern

Dupetor flavicollis

Sooty black; throat and breast streaked black and white, with prominent yellow feathers at the sides. Young birds have the plumage brown-tinged and the feathers of the upper parts edged buff. Iris, yellow-brown; beak, black above, pale below; legs, dark slate. Length, 25-26 in.

Distribution: Distributed all over the State.

Nesting: A platform of sticks is constructed on a limb of a tree overhanging water. The 3 to 5 eggs are white, with a greenish tinge inside. Size, 45 x 35 mm. Both birds assist in the construction of the nest.

In the South-West this bittern is pre-eminently an inhabitant of the wooded banks of rivers and near Perth is commonly seen along the Canning and Swan Rivers in the flooded (blue) gums, which fringe their upper courses. It used to nest in Herdsman's Lake before its drainage in 1925. It has become rarer with the years.

Brown Bittern

Botaurus poiciloptilus

Native name: Bur-den-etch, Bar-dan-itch, Bardanitch (Perth).

Upper parts brown mottled with golden-buff, particularly on the wing coverts. Under parts buff, strongly streaked with brown. Iris, yellow; bare skin in front of eye, pea-green; beak, brown above, yellow below; legs, green. Length, 26-30 in.

Distribution: Confined to the South-west corner, as far north as Moora. A specimen claimed to have been collected at Shark Bay in 1880 needs authentication as to locality. Recently the species was recorded in the Kimberley Division. It has been seen in the south-east at Shark Lake, Esperance.

Nesting: The Brown Bittern nests in the dense patches of rushes which fringe many of our swamps. The nest is merely a platform formed by bending over the rushes in a certain spot and on the flat place thus formed the eggs are deposited. The usual number is 4 or 5, occasionally 3 or 6. They are greenish-brown; size 52 x 38 mm.

This is the best-known of the bitterns in the South-West swamps and at night its loud braying-like call betrays its presence. Each bird seems to have its own particular feeding spot in the form of a thick and matted clump of rushes well trodden down. It is usually littered with the carapaces of jilgies *(Chaeraps quinquecarinatus)* and often, when the bird leaves in a hurry, with half-digested frogs. These feeding places should not be confused with nesting sites.

Little Bittern

Ixobrychus minutus

General appearance, dark reddish-brown and buff. *Crown of head, mantle and back, and wing quills, black.* Neck and wing coverts red-brown. Sides of and under neck and middle of wing are yellow-buff. Most of the under parts are buff streaked with white. Tail, black above, white below. Iris, golden yellow; beak, black above, sides and lower mandible dull greenish-yellow; the long legs are green, claws grey-black. Length, 10½-14 in.

Distribution: Found only in the South-West, as far north as the mouth of the Murchison River.

Nesting: The Little Bittern makes a flat nest, loosely put together, of rushes, usually on those growing in water, but also in low paper-barks. A site where the rushes are bent over forming a suitable platform is usually selected. The nest is about seven inches in diameter and a couple of inches deep. The 3 or 4 white, elliptical eggs, rather like those of a pigeon, are glossy. Average measurements of a set of four taken at Busselton, where the bird had been found nesting in October, 32 x 26.7 mm.

This is a secretive species, not often seen as it skulks quietly among the rushes. Specimens occasionally come under notice when they meet with some misadventure, such as through collisions when they fly about at night.

Yellow Bittern (Chinese Little Bittern)

Ixobrychus sinensis

Similar to the preceding species but the *back is yellowish-brown to reddish-brown* instead of black. Lacks the line of dark feathers between the chin and breast of the previous species. Iris, yellow; beak, yellow; legs, yellow. Length, 12 in.

Distribution: Breeds in the Northern Hemisphere, from northern China to India, migrating south as far as the Malay Peninsula, the Philippines and New Guinea. The only Australian record is of a live bird recovered in a weakened condition at Kalgoorlie in January 1967, after a tropical cyclone (reported by J. Ford).

STORKS: Family Ciconiidæ

Jabiru (Black-necked Stork)

Xenorhynchus asiaticus

Head and neck green with metallic reflections and remainder of *body white,* except for the bronze-brown wing coverts and inner wing quills, and the green tail. Iris, white; beak, black; legs, coral red. Length, 4½ ft., wing span, 6 ft. 8 in. Weight, 9 lb.

At a distance, in flight, a Jabiru appears as a white bird with a black head, and a black bar bordered all round with white, on the upper and lower wings. In flight the neck and legs are stretched out to their fullest extent. The birds mount in the air in a series of circles. The flight is slow; the bird gives about four to a dozen flaps of the wings and then soars, repeating the process.

Distribution: A northern species which at times intrudes into the northern fringe of our area from the Kimberley Division. The first record of its occurrence in the Pilbara district was made by the explorer F. T. Gregory in 1861, when he collected two specimens at Breaker Inlet, near Condon. K. G. Buller met with the species on a clay-pan near Well 49 on the Canning Stock Route (east of Godfrey's Tank) in 1943, but no further south. These two localities are in approximately the same latitude. In 1952 single birds were seen at Port Hedland and at Onslow. Odd birds have been seen in the Port Hedland marshes in 1954 and 1958. In 1972 a bird was seen at Roebourne.

Nesting: There are no records of nesting in our area. The nest is built in a tree, sometimes as low as 12 ft. but generally much higher, up to 80 ft. or so. In the Kimberley Division nests have been seen in the Leichhardt pine. W. T. White, from observations in Queensland, describes the nest as a pile of sticks 3 feet high and 4 feet broad, with a flattened top on which is a thin layer of grass or rushes. On these 2 to 4 whitish eggs are laid. Size, 70-76 x 51-56 mm. Both parents share in the incubation. Nestling period, 10 weeks (P. Slater).

IBISES and SPOONBILLS: Family Threskiornithidæ

Straw-necked Ibis

Threskiornis spinicollis

Upper parts black with purple, bronze and green metallic reflections when examined closely in certain lights. Hindneck, throat and under parts white, with a black band around the upper breast, surmounted by a patch of stiff straw-like feathers, yellow in colour. The *tail is white, contrasting conspicuously with the black body* when the bird takes wing. The naked skin of the head and face is black. Iris, dark brown; *long curved beak,* black; legs, red to just below the "knee" joint and black below this. Length, 27-30 in.; wing span, 3 ft. 9 in. - 4 ft. 5 in. Weight, 3¼-4 lb.

Distribution: Occurs all over the State where there are suitable habitats.

Nesting: As with other members of this family of birds, the Straw-necked Ibis nests in colonies among rushes or other vegetation in swamps. The nest is usually to be found just above water among rushes or in tea-trees, and consists of rushes broken down to form a platform, to which, to give more body to the nest, sticks and rushes from elsewhere are added. The clutch numbers 2 to 5 eggs which, dull white in colour, soon become nest-stained. They are of rough texture and the inside lining, when viewed through a hole in the shell, is greenish. Average measurements of 10 eggs from Gingin to Carnamah, 63 x 43 mm., varying from 60-70 x 41-44 mm. Average weight, 59 gm.

This is the common ibis of the State, but it is only locally distributed in the South-West. Thus in some districts, such as Pinjarra, considerable flocks have occurred regularly for many years in the open paddocks. On the coastal plain around Perth it is uncommon. W. B. Alexander considered that the species had extended its range from the north only in recent years, and support is lent to this view by the fact that John Gilbert did not find the bird in the South-West during his collecting period (1839-1843). T. Carter, who met with it south of the Minilya River in 1888, has recorded that he was informed that it first appeared at Albany in 1892. The late O. H. Lipfert informed one of us (D.L.S.) in 1923 that this ibis was unknown in the Perth district prior to 1892 (a drought year), when a large number of the birds arrived in Perth. Specimens were collected for the Western Australian Museum by J. T. Tunney at Busselton and Augusta in 1896. Apparently it had not increased to any marked degree during the next half-century, but in recent years, since the mid-1950s, large numbers are found near swamps at Bridgetown, Waterloo and Harvey.

This is one of the species which was seen in the State in greatly increased numbers during the "invasion year" of 1952 (see p. 6), and birds occurred everywhere from the North-West to the Nullarbor Plain.

The first observations on breeding in Western Australia were made by T. Carter in October 1900, when some hundreds of birds were nesting at Cardabia Creek, 60 miles south-east of Point Cloates. At a brackish lake at Carnamah I. C. Carnaby observed a nesting colony of about 85 pairs in November 1958. The nests were on an islet on the ground and on fallen *Casuarina* trees. Several hundred pairs bred at the Marrida and Bulrush Swamps, Moora, in 1958. The best-known nesting colony is the one near Gingin. For many years many hundreds of pairs bred in the paperbark trees at the Wellering Swamp, three miles south of Gingin, but no breeding took place here since 1956 due to low water levels and the fact that the vegetation had been burnt. In 1960 Ian Edgar, of Gingin, discovered a flourishing colony nesting in a paperbark swamp towards Muchea, about 6 miles to the south-east of Wellering Swamp. South of Perth, the only known rookery is in the Pinjarra district, in a paperbark swamp between Lake Mealup and the south shore of Peel Inlet.

Young birds ringed at the Gingin and Muchea rookeries were recovered during the following summer in the Gascoyne, North-West Cape and Ashburton areas. Evidences of an even more intensive migration were later indicated. A bird which was ringed as a nestling in November 1960 near Muchea was killed in May 1961 by Aborigines at Gudji billabong, near the Millingimbi Mission in Arnhem Land and another, ringed near Muchea in October 1961, was recovered injured in November 1962 at Clergate, N.S.W.; a third ringed as a nestling on October 24, 1953, was found dead on May 24, 1969, just south of Beaudesert, Queensland.

Ibis are well known for their partiality for grasshoppers and C. F. H. Jenkins has seen flocks feeding on these insects even in dry country, as at Mullewa.

White Ibis

Threskiornis molucca

White, with the exception of the tips of the wing quills, which are black. Beak and naked skin on head and neck, black. Iris, dark brown; legs, pink to upper joint, below purplish-brown. Length, 26-30 in.

Distribution: Common throughout northern Australia, and thence south along the eastern side of the continent. In Western Australia it penetrated south of the Kimberley Division only on rare occasions, but since the large irruption of 1952 (see p. 6) it appears to have become established in small numbers in the South-West.

Tom Carter, writing on the birdlife of the Point Cloates area at the close of the last century, stated that occasional individuals mingled with the Straw-necked Ibis. No authentic reports of later occurrences were forthcoming until 1952, when, among the flocks of Straw-necked Ibis in their great irruption into southern Western Australia, there were noticed odd individuals of White Ibis. The localities in which these were reported that year were Moogooree Station (180 miles north-east of Carnarvon), Coolup, Bunbury and Busselton.

Some of the White Ibis which reached the South-West in 1952 have evidently remained, for up to 5 individuals have been repeatedly seen in the Bunbury-Busselton area up to the present time. One or two birds have also been seen at Bibra Lake, near Fremantle, since January 1956, at Lake Jandakot, Gingin, and, in the Wheatbelt, at Lake Toolibin and near Lake Ewlyamartup. A party of 12 birds was seen by J. Long at Wattle Grove in January and February 1962.

Nesting: So far the White Ibis has not been observed nesting south of the Kimberley Division. Two birds were associated with a breeding group of Straw-necked Ibis at Wellering Swamp (Gingin) in September 1956 and may have been nesting. The nesting habits are the same as in the Straw-necked Ibis and the eggs are similar.

Glossy Ibis

Plegadis falcinellus

This ibis looks *quite black* in the distance but in nearer view the attractive details of bronze red-brown body with green-tinged wings can be perceived. The head, neck and under parts are dark reddish-brown and the back and wings very dark green with metallic iridescence. Iris, dark brown; the *long curved beak,* dark brown at the base and the rest light olive-brown; legs, black. Length, 19-22 in.

Distribution: Apparently likely to be met with anywhere where the environment is suitable, but records are very infrequent in the southern part of the State. It has been reported in the Perth area, Moora and Katanning.

Nesting: There is no record of the breeding of this species within the areas dealt with in this book. The bird consorts with other heron-like birds—Straw-necked Ibises, White Egrets, etc.—for nesting purposes. The colony of Glossy Ibises is, however, usually segregated from the other species. The nests are constructed of the leafy ends of branches of trees and are platforms in trees growing in water. A nest measures from 11 to 15 inches in diameter and the centre, or egg-cavity, is about 2 inches deep; 3 to 4 eggs form the clutch and occasionally 5 and 6. They are deep greenish-blue in colour, shell of slightly rough texture and lustreless; size 53 x 36 mm. (A. J. North).

Royal Spoonbill (Black-billed Spoonbill)
Platalea regia

An *entirely white,* long-necked and long-legged bird, with the *legs and beak black,* including the bare skin around the eye and base of the beak. The beak is large and expanded at the end into a wide spoon-shape. Iris, scarlet; brown in the young. Length, 29-31 in. Weight, 4 lb.

Nesting: There is no definite record of its breeding within the area covered by this book. T. Carter, however, has recorded that in October 1900 some spoonbills were taken by natives from a nest in a white gum tree standing in water near Winning Pool, and Carter, who saw the beaks and feet of the birds, believed them to be of this species. I. C. Carnaby found a nesting colony of 34 pairs at Langi Crossing, Fitzroy River, in the West Kimberley, in March 1960. The nests were built of broken green branches at the tops of trees. The clutch comprised 3 to 5 eggs.

These birds feed with a sideways swishing of the bill in shallow water.

Fig. 21.—Royal Spoonbill.

Yellow-billed Spoonbill

Platalea flavipes

White, like the preceding species in general form, but with the *beak and legs yellow.* The bare skin on the head is bluish with a pink tinge. Iris, white. Length, 35 in. Wing span, 4 ft. 10 in. Weight, 4 lb.

By comparison with the Royal Spoonbill, which is snow-white, the Yellow-billed Spoonbill looks "dirty white," giving the impression of soiled plumage (K. A. Hindwood).

Distribution: Distributed through the State generally, excepting (until recent years) the South-West, Moora being considered to be the southerly limit. However, in August 1954 one bird was seen on the Canning River at Cannington and in 1958 (in April and August) single birds were observed in the Bunbury district. A group of 5 were seen at Midland in July 1965. One bird, probably immature, was seen near Busselton in January, 1973.

Nesting: The only record of the nesting of this species within the area treated in this book is that of P. T. Sandland and C. L. E. Orton who found the bird nesting in the Moora swamps in October 1921. Three nests—platforms of sticks in paper-bark trees—were noted, containing 3, 3 and 2 eggs respectively. The eggs are white; size 68 x 45 mm.

DUCK-LIKE BIRDS: Order Anatiformes

SWANS, GEESE and DUCKS: Family Anatidæ
Subfamily: Anseranatini

PIED GEESE: Tribe Anseranatini

Magpie Goose (Pied Goose)

Anseranus semipalmata

A large distinctive, *black and white bird.* The head, neck, back, wings and tail are black; inner upper wing coverts white, linking up with a white saddle over the mantle; rump and under parts (except thighs) white. Iris, brown; beak, pale lead-grey, cere pink with dusky markings; legs, yellow. Length, male, 2 ft. 7 in. to 3 ft.; female, 2 ft. 4 in. to 2 ft. 8 in. Wing span, male, 4 ft. 5 in. to 5 ft. 6 in.; female, 4 ft. 4 in. to 5 ft. The weight varies with the seasons. "A male has been shot weighing more than 8 lb., but they average about 6 lb. and females about 5 lb." (S. J. Davies).

The sexes are similar in plumage but besides being smaller the females differ in certain other ways. Dr. Davies states: "The bone cells of the crown expand and vacuolate in the adults, forming a distinct knob on the head. This knob tends to be larger in males than in females. The cere is typically pale pink. It becomes redder seasonally with the onset of the breeding season, and, momentarily, with an increase in sex drive, it tends to be redder in males than in females. In males the trachea forms two external coils or loops, which lie on the breast, immediately under the skin. Such loops are rarely present in females and then only one, an inch or so long, is present. The greatly lengthened windpipe of the males gives their call a deeper and more resonant quality than that of the females."

Distribution: Essentially an inhabitant of the Kimberley Division, where it is very common. However, nomadic wanderings take place to the southern parts of the State and individuals have been met with at Broomehill (November 1905, T. Carter). In 1952, when many northern birds irrupted into southern Australia there was an extensive invasion of this species. Odd birds and small parties were seen between Geraldton and Mullewa, the Swan River estuary and at Madura.

Nesting: The Magpie Goose has not yet been found to nest in this State outside of the Kimberley Division. There it constructs a large open nest in a swamp. The nest is actually built right down to the mud below the water and a good deal of grass and water weeds are used to form the foundation which is about two feet in diameter. The eggs (5 to as many as 11 are laid) are glossy white and are pitted. Size 70-80 x 52-55 mm.

Fig. 22.—Magpie Goose.

The taxonomic position of this bird has been a matter of dispute. Ernst Mayr and Jean Delacour regarded it as an aberrant member of the perching ducks, and it was so placed in our earlier editions. On behavioural studies P. A. Johnsgard distinguishes it from all members of the family and suggests it may have affinity with the South American screamers (Anhimidae). On cranial characters G. E. Woolfenden would place it as a distinct family among the ducks. However, Drs. H. J. Frith and S. J. Davies feel that most authors exaggerate its differences from other members of the family and would not be prepared to separate it further from them than at subfamily level. Drs. Frith and Davies have established that, unlike other members of the duck order, this species is not flightless at the annual wing moult.

Subfamily: Anserinæ

SWANS and GEESE: Tribe Cygnini

Black Swan

Cygnus atratus

Native names: Kulla-eedoo (Coolawanyah); Kurillthu (Pidong Tribe, upper Murchison River); Kurulu (Wallawe tribe, Mount Magnet); Cooraloo (Wooleen); Gooroyl (north of Perth, Grey); Guroyl (north of Perth, Moore); Mal-lee (north of Perth, Gilbert); Gol-jak, Guijak, Kool-jak, Cool-echa (Perth); Marlee, Mele, Mal-yi (South-West); Marlie (Albany); Weeler, Marlee (Pallinup River).

Black; flight quills white. Iris, varying from white through pink to red; beak, red with narrow white sub-terminal bar across upper mandible; legs, dark grey. Length, 3' ft. Weight, 9-14 lb.

The cygnets are light greyish-brown, paler below.

Distribution: The Black Swan is distributed all over the State. In the Swan River estuary, where it was first discovered by the Dutch navigator Vlaming in January 1697 it was, apparently, never abundant in the broadwaters, but frequented the shallows at Lucky Bay and above the Causeway. The French explorers from the *Naturaliste* when they ascended the Swan River in June 1801 did not meet with Black Swans until they passed Heirisson Island. During the 1890s the authorities imported birds from elsewhere in the State and Victoria and set them free, pinioned, in Perth Water. The Black Swan is quite abundant in the South-West and large flocks are to be met with at Peel Inlet, Leschenault Estuary, Augusta, Wilson's Inlet, Pallinup Inlet, Bremer Bay estuary and other waters. It nests in small numbers on most of the swamps around Perth, including Butler's Swamp, Claremont and Shenton Park Lake, Subiaco.

Nesting: The nest is a bulky collection of sticks and rushes, placed on an island in a swamp. At times it is built in a clump of thick bottle-brush, or in open water, when the large collection of rushes, etc., themselves form a miniature island. The eggs are pale green and become paler as incubation proceeds; 4 to 9 form the clutch but I. C. Carnaby found a nest with 13 eggs at Nannine in 1963. Incubation takes 35-45 days. Mean size of 20 eggs, 102 x 67 mm., varying from 98-106 x 66-69 mm. Average weight, 247 gm. (8¾ oz.).

Fletcher Moore stated that the family ancestors of the Ballarokes, a section of the Bibbulman tribe of the South-West Aborigines, were reputed to be these birds changed into men.

Mute Swan (White Swan)

Cygnus olor

Plumage entirely white. Iris, brown; beak, deep orange, with black knob at base (smaller in the female); legs, dark grey. Length, about 5 ft.; wing span, 7 ft. 10 in. Weight, 24 lb.

Distribution: Introduced (original home: Europe and northern Asia). A few birds are to be seen on the pools at Queens Gardens and Hyde Park, Perth, and the species has been established on the Avon River at Northam, where it breeds. Nesting also has taken place at Coolbellup (or North Lake) (J. R. Ford) and Bro. Matthew has seen one bird on a swamp at Bindoon. Pairs have been liberated on some of the southern rivers, as at Bridgetown (Blackwood River) but the species has shown no inclination to extend its range or increase markedly in numbers. Undoubtedly some birds released have fallen prey to local gunners.

Nesting: The nest is a large structure of rushes or coarse herbage built close to or in water. Usually only 3 to 5 eggs are laid. They are dull greenish-white and measure 4½ x 3 in. Incubation takes 36 days. The young are cared for by the female.

White Swans feed on water plants and aquatic insects, but are vary partial to grain, bread-crusts, etc.

WHISTLING, or TREE DUCKS: Tribe Dendrocygnini

Whistling Tree-Duck

Dendrocygna arcuata

Whole of the upper surface, including the top of the head, *blackish-brown,* the feathers of the middle back edged with yellow-buff; wings similar but the *shoulder of the wing* rusty-chestnut,

Throat white, breast chestnut-brown spotted with dark brown, *lower breast and abdomen uniform rufous;* lower abdomen greyish-brown streaked dull white. The plume-like feathers of the flanks are striped black and white. *Under parts of wing dark brown and unbarred.* Sexes similar. Iris, red-brown; beak, black; *legs, black.* Length, 18½ in.

This species differs from the following in being *darker above* and more extensively rufous on the under surface, which *lacks cross barrings* here and on the under wings.

Distribution: Mainly a northern form but it has been recorded as far south as the Perth area (one specimen shot at Wanneroo in 1899).

Nesting: This duck, in the Kimberley Division, nests on the ground in herbage and merely deposits its eggs in a depression, no down being stripped by the bird. A large number of eggs is laid—up to 15—and they are slightly glossy and creamy-white, with both ends rather pointed. Size, 51 x 35 mm. It has not yet been recorded as nesting in Western Australia south of the Kimberley Division.

Plumed Tree-Duck

Dendrocygna eytoni

Crown of head buff-brown; back and wings, brown with an olive tone and narrow white tips to the feathers of the mantle; lower back, brown. Tail black, with a broad light buff band across the base. Throat, white; neck and upper breast buff; *lower breast chestnut, barred black; abdomen, white.* On the flanks are long plume-like feathers which are prominently striped buff and black. The *under wing and axillaries are barred brown and buff.* The sexes are similar in coloration though the drake has longer flank plumes. Iris, orange-red; beak, pink blotched black with slate tip; *legs, pink.* Length, 17 in.

These ducks have a goose-like carriage and the flank plumes are held upright, outside of the wings. When birds are roused the head feathers are erected to a crest.

Distribution: A northern species; the most southerly record is Carnamah.

Nesting: As with the previous species, there are no nesting records south of the Kimberley Division. Similarly, the nest is also in herbage on the ground and a large number—10 to 12—of eggs laid. Size, 47-49 x 36-39 mm. Both sexes share in the brooding of the eggs, the 24-hour shifts being changed in the evening. The eggs are also laid in the evening and the incubation period is about 27 or 28 days after the laying of the last egg.

CAPE BARREN GEESE: Tribe Cereopsini

Cape Barren Goose

Cereopsis novæhollandiæ

Native name: Ca-an-gan (King George's Sound).

Almost entirely *grey,* lighter on the forehead and crown, and with dark spots on the wing coverts, and lighter indistinct barrings on the mantle. *The tail and ends of wing quills are black.* A striking feature is the prominent *yellow-green cere* which overlies the greater portion of the black beak. Iris, light buff; legs, pink-red, webs and toes black. Length, 32-34 in.; wing span, 5 ft. 8 in. Weight, 8-13 lb. Sexes alike, female slightly smaller. The voice differs in the sexes; "it is loud and trumpeting in the male, low and grunting in the female" (Jean Delacour and Peter Scott).

Fig. 23.—Cape Barren Goose.

Distribution: This bird is confined to the islands of the south coast, from Albany eastwards. In recent times its range has become restricted and its principal stronghold is the Archipelago of the Recherche, but there are sporadic appearances west to Busselton. In 1921 L. Glauert saw two birds flying near the Mammoth Cave and in 1942 a bird was seen near Busselton. None were seen at Bald Island by G. M. Storr during his survey of the island in 1959. It is rarely seen on the mainland but in the Esperance area individuals may be seen occasionally on the bold granite headlands adjacent to the islands of the Recherche Archipelago. In October 1945 a pair alighted on the green at Hopetoun opposite the hotel, the first occasion the bird had been seen locally in the memory of the oldest inhabitant. The species is now much more plentiful on the Furneaux Group, in eastern Bass Strait—the easternmost end of its range—where we have seen companies of up to 500 birds.

Nesting: The nest, usually made on the ground, is saucer-shaped, constructed of grasses and lined with down. Occasionally it may be elevated a few feet in a thick bush. The eggs, 4 to 7, are white and glossy but covered with a thin limy coating which makes them coarse in texture. They are elliptical in shape with both ends rather pointed. Average size of 20 eggs from the Eastern States, 83 x 56 mm., varying from 79-92 x 53-58 mm. The bird is a winter breeder and egg-laying may begin in late May. G. A. Heumann found that incubation is carried out by the female and takes 40 days; according to Delacour and Scott, 35 days. The male assists in nest-building. Birds attain sexual maturity at 3 years.

Matthew Flinders, the navigator, replenished his larder with these birds when his ship the *Investigator* called at Middle Island in the Recherche Archipelago in January 1802. He obtained 65 geese at a bay he called Goose Island Bay at Middle Island, and he commented on how few birds, by comparison, he got on a second visit in May 1803, when only 12 geese could be procured. The geese have always been hunted for the pot and, until full protection was accorded the species in 1938, coastal vessels used to bring them in crates from the Recherche islands and from Bald Island for the Christmas and Easter trade.

Though essentially an island bird, it feeds entirely ashore, grazing on grasses and other herbage, over which it leaves an odour which is offensive to sheep, hence pastoralists dislike its presence. It can swim, but rarely does so, except before it is fully fledged and when wounded.

Its relationships are not certainly known and Ernst Mayr and Jean Delacour consider it to be a primitive member of the shelduck group, the colour pattern of the downy young being very similar. P. A. Johnsgard, from behavioural studies, admits that it probably constitutes a link between the true geese and the shelducks, but its behaviour has more in common with the former. G. E. Woolfenden, on cranial characters, places it as a tribe of the Anserinae.

FRECKLED DUCKS: Tribe Stictonettini

Freckled Duck

Stictonetta nœvosa

Upper surface *brownish-black,* each feather freckled with buff; neck and upper breast similar, lower breast and abdomen white, each feather barred grey. There is *no speculum* or coloured wing patch. Iris, brown; beak, shining leaden grey, at base, between forehead and nostrils pale reddish yellow, becoming bright red in breeding condition; spatulate and slightly upturned; legs, dark olive-grey. Length, 22-23 in.; wing span, 29-30½ in.; weight, 2 lb. 4 oz. (male). Female, smaller.

Distribution: Likely to be found anywhere in the State in fresh waters.

Nesting: The Freckled Duck often uses an old Coot's nest, which it hollows out, in which to lay its 5 to 7 eggs. If it builds its own nest it constructs one of fairly fine sticks on a leaning limb in a thick, bushy tea-tree *(Leptospermum),* just above water level, and lines it with finer twigs and tea-tree twigs. The bird uses its down in a manner different from other ducks by lining the cavity of the nest with a felting of down, instead of spreading it out all over and among the eggs, as is the practice of most other ducks. A series of 13 eggs averages 63 x 46 mm., varying from 60-66 x 43-48 mm. They are of cream colour, glossy and smooth, not greasy and the shell is softer than the eggs of true ducks.

Dr. H. J. Frith has demonstrated that instead of being an aberrant member of the Anatinae, as previously thought, its physical and behavioural characters indicate that it is Anserine, probably most closely related to the swans. It can be regarded as a short-necked swan.

This is one of our rarer waterfowl, but it is to be seen on occasions throughout the South-West, birds having been observed at Monger's, Herdsman's and Thompson's Lakes near Perth, as well as Busselton and Albany. At a shoot near Narrogin in January 1946, no fewer than 40 Freckled Ducks formed part of the total bag of 960 ducks of various species. At a shoot here in December 1965 out of a bag of 256 ducks there were 32 Freckled Ducks. Near Winchester in January 1966 there were 7 Freckled Ducks in a bag of 120. In August 1960 several were shot by sportsmen in the Kalgoorlie district (Rowles Lagoon and Gindalbie Station) and were regarded as strange birds to the district. Nesting may occur generally in the species' range and has been recorded in the South-West at the Benger swamps and Lake Toolibin (near Narrogin).

Subfamily: Anatinæ
SHELDRAKES: Tribe Tadornini

Mountain Duck (Chestnut-breasted Sheldrake)
Tadorna tadornoides

Native names: Goor-a-go, Goo-ra-ga, Guraga, Kooracha (Perth); Karbunga (South-West); N-yuneruk (Albany); Gnimeruk (Pallinup River).

In the *female* the head and upper neck are black, with *white around the eye* and the base of the beak and chin. Below the black neck a *white collar*. The *lower neck and chest are chestnut;* back and tail greyish-black, with greenish reflections on the tail. The abdomen is greyish-black, the feathers with narrow rufous barrings. In the wing the secondary coverts are white forming the *conspicuous white wing patch* so noticeable in flight. The primary wing quills are black and the outer secondaries green and the inner chestnut. The under wing coverts are white. *The male lacks any white around the eyes and beak* and is a buffy-yellow where the female is chestnut. Iris, dark brown; beak and legs, greyish-black. Length, male, 23-25 in.; female, 22 in. Wing span, 3 ft. 8 in. Weight, male, 2½-4½ lb.; female, 2¼-3½ lb.

Distribution: Mainly a southern duck usually north to Dirk Hartog Island, and the Ophthalmia Range, but occasionally the species ranges further north, to Mundabullangana Station (near Port Hedland), Jigalong, and even the Kimberley Division (T. Spence saw two pairs at Parrys Lagoon, Wyndham, in June 1970).

Nesting: The Mountain Duck lays 10 to 19 eggs in a hollow in a tall tree, often at a height of 30 feet from the ground, and not necessarily in the neighbourhood of water. It may nest in rabbit burrows and at Rottnest Island in deep crevices in the limestone. The record clutch of 25 eggs was found by T. E. Bush at Yerecoin in July 1967. More eggs are occasionally recorded which must be combination clutches, as was the one of 50 eggs found by I. C. Carnaby at Lake Grace in 1969. The eggs are creamy-white and glossy. Average size of 14 eggs, 69 x 48 mm., varying from 65-72 x 47-49 mm. The duck uses a lot of down to cover the eggs; it is grey with dark tips and white centres. The young in down, when about a month old, have the body a dusky white with three longitudinal stripes of dark brown on the back, the middle one darker than the others, running from the head to the tail. The beak is blue with burnt-sienna colour at the tip and the feet are pale greyish-blue (S. R. White).

During the annual moult period, after the breeding season, Mountain Ducks congregate on to large bodies of water where they are likely to be safe at this vulnerable flightless stage. In December 1964 the English racing driver Donald Campbell, when seeking a high speed record with his hydroplane "Bluebird" on Lake Dumbleyung, was seriously impeded in his efforts by the presence of hundreds of moulting Mountain Ducks which could not be scared away as they were still flightless. Subsequently similar moulting concentrations have been noted at Lake Preston and Peel Inlet (both in 1974).

This duck has a decided preference for estuaries and salt lakes, on which at times it may occur in very great numbers. It is distributed throughout the Wheatbelt and frequents farm dams and wet patches in paddocks. It is sometimes found along sea beaches and is a regular visitor to the salt lakes at Rottnest Island, where it nests. It is held in very low esteem as a table bird, the flavour being too rank.

RIVER DUCKS: Tribe Anatini

Black Duck (Grey Duck)

Anas superciliosa

Native names: Kigaribiloo (Godfrey's Tank, Canning Stock Route); Koon-dah-ree (Coolawanyah); Wongona (Lyons River); Ngarrawa (Pidong tribe, east Murchison River); Ngaruwa (Wallawe tribe, Mount Magnet); Gnow-era (Wooleen); Oonern, Oonanah, Ngwonana, Gwoom-na-na (Perth); Ya-et (Albany); Yate (Pallinup River).

Upper parts dark brown with the feathers narrowly edged with buff. Throat white, rest of under parts brown, with the feathers edged buff. The *white-striped head pattern* is absolutely diagnostic; there is a narrow white line above the eye and a broad white line from the base of the beak passing to behind the eye. Wing speculum, green; under wing, white. Iris, bright brown; beak, dull bluish-grey; legs, olive-yellow. Length, 19-22 in. Wing span, 32 in. Weight, 1 lb. 11 oz. - 2 lb. 12 oz.

The sexes are similar in colour.

Distribution: Found all over the State.

Nesting: For nesting purposes this duck often selects a hollow in a large tree. No nest is made except that, as with most ducks, as incubation proceeds, the duck plucks down from her breast in quantity sufficient to cover completely the large number of eggs, 8 to 12, which she lays. Occasionally the sloping top of a blackboy *(Xanthorrhoea)* is selected as a nesting site, and often the eggs are laid on grass on the ground, but, with the advent of the fox, birds given to this habit are being steadily eliminated. Although eggs laid by any one bird do not vary in colour, there is variation in coloration of sets laid by different birds. The surface is always glossy and greasy, but the colour may be light greenish-white, or light creamy-white with a faint greenish tinge. As incubation proceeds the eggs assume a slight brownish hue. They are elliptical in shape. A series of 24 eggs from Lake Grace averages 58 x 43 mm., varying from 56-60 x 40-44 mm.

This is the State's best-known duck, being in general the most common species and the principal component of the sportsmen's bags, though the Grey Teal must run it a very close second. All swamps and most watercourses and drains have a quota of these birds. As the summer draws on, great numbers gather in the Perth area and form remarkably tame companies at the artificial pools, at Hyde Park and Queen's Gardens, in the city. These disappear at the onset of the winter rains when the ducks take up breeding territories in the neighbouring swamps. The Black Duck is not such an extensive wanderer as the Grey Teal but ringing experiments have demonstrated that some individuals may make lengthy journeys. A bird ringed by the Fisheries and Fauna Department at Woodanilling in April 1958 was trapped on a dam at Menindie, on the Darling River, N.S.W., in November 1960.

Wild ducks are protected in the South-West from the beginning of May until an opening date of the shooting season proclaimed in the *Government Gazette* yearly; the nature of the season and its effect on the breeding of the birds are taken into considration. No person is permitted to snare or trap any duck, and no one person may kill by shooting more than 10 in any one day. An annual licence is required.

Chestnut Teal (Mountain Teal)

Anas castanea

Native names: Gnwool-ye-nug-ger-rang, Ngwol-yi-nag-girang (Perth); Calyung (Pallinup River).

The male is amongst the most brightly coloured of the local ducks. The *head and neck are blackish-brown with a brilliant greenish sheen* on the top and sides. The back is dark brown with rufous edgings to the feathers, under parts light rufous. Under tail coverts dark brown with a pronounced white patch on each side. Wing speculum white and green. Iris, red-brown; beak, bluish-grey; legs, grey-brown or dark grey. Length, 15-17¼ in. Weight, 1 lb. 1 oz-1 lb. 9 oz.; males average slightly heavier than females.

The female, as a rule, lacks the vivid colours of the male, and is mostly seen in the sombre dress resembling that of the next species, the Grey Teal, but is generally of a richer brown with a *pronounced rufous tone on the under parts.* The upper plumage is dark brown, the feathers edged with grey. The chest is grey-flecked brown and the feathers of the under surface are brown with buff fringes, giving a mottled effect. On occasions females are found with the fully developed plumage of the male.

Distribution: Mainly a South-Western duck, but there are records of its occurrence and breeding as far north as North-West Cape.

Nesting: The data given below for the next species, the Grey Teal, apply exactly to this bird, except that the nesting down is slightly darker. Average size of 10 eggs (Lake Balicup), 51 x 36 mm., varying from 49-54 x 35-37 mm.

This teal appears to be most plentiful in the south of the State, in the brackish estuaries and lakes eastward of Cape Leeuwin, including Lake Muir. We have found it commoner than the Grey Teal in the series of barred estuaries, and the brackish reaches in the rivers entering them, between Albany and Esperance. Flocks of thirty or more have been seen flying in the gorge of the Hamersley River where it cuts its way through the jagged ranges of Mid Mount Barren, and pairs and small parties can be flushed in pools in the Phillips, Young and other rivers in this region.

Grey Teal

Anas gibberifrons

Native names: Chin-bijee (Coolawanyah); Wang-anner (North-West Cape); Nimon (Godfrey's Tank), Canning Stock Route); Googagoora (Jigalong, all ducks); Calyung (Pallinup River).

Upper parts grey brown, with light edging to the feathers. *Throat white.* Under parts light grey, each feather prominently tipped with brown. Wing speculum white and green. No white on flanks. The sexes are similar in colour. Iris, bright reddish-brown; beak, above, leaden-blue, lower mandible, brown at base, rest dull yellow; legs, greyish-brown. Length, 14-17 in. Weight, 12 oz. - 1 lb. 6 oz.; males average slightly heavier than females.

Grey Teal can always be recognised from female Chestnut Teal by having a much more pronounced whitish throat and the under parts lacking any reddish tone, being shades of grey and dark brown. Both species of teal can be told at once from the Black Duck when in flight by the white feathers on the upper wing.

Distribution: Found all over the State, and rivalling the Black Duck from the standpoint of abundance. It is very numerous on the Swan River estuary (especially at Pelican Point and Como) and on other estuaries in the South-West. When rains fill the inland lakes and pans it is one of the first water-birds to put in an appearance.

Ringing experiments in recent years have demonstrated the remarkable wandering capacities of this species. Birds marked in the South-West have been shot in the eastern States, and vice versa. Many ringed at Humpty Doo in the Northern Territory have been recovered from many parts of the State, from the Kimberley Division to deep into the South-West. Other birds from the same source dispersed widely over south-eastern Australia. No other Australian duck travels so extensively. So far no clear pattern of movement has emerged, but it appears to be a combination of random scattering from areas where conditions become unfavourable to concentrations at areas made favourable by rains and floods (H. J. Frith). A particularly interesting movement record is that of a bird ringed at the Moulting Lagoon, Swansea, Tasmania, on July 22, 1958, and shot at Lake Austin, near Cue, on July 29, 1962.

Nesting: The nest may be just a slight depression in the ground in thick grass, or it may be in a hollow tree. Eggs are from 8 to 12 in number, but in some instances as many as 15 in one nest have been noted. Elliptical in shape; glossy and greasy, and pale, creamy-white. A series of 45 eggs from Lake Grace and Coolup averages 49 x 37 mm., varying from 47-53 x 34-39 mm. Down, greyish-brown with light-coloured tips and centres, and there is usually such a large quantity in the nest that the eggs are completely hidden in it. This teal is one of the most successful opportunistic breeders among Australian waterfowl and nesting will occur when rainfall and rising water levels make conditions suitable. The pre-egg-laying display period is considerably shortened compared with other ducks (10 days after stimulating rains R. A. Aitken found shelled eggs in the oviduct).

The call of the adult to the young is a muted "eep," which can be heard only when one is very close to the birds.

Blue-winged Shoveler (Spoonbill Duck)

Anas rhynchotis

Native names: Bar-doo-ngoo-ba, Bardunguba (Perth).

Upper parts dark brown with the body feathers edged buff. The head has no markings except a light buff line curving down from in front of the eye to the throat. Throat and breast grey, freckled with buff; abdomen reddish-brown spotted black. *Wing speculum blue, white and green;* under wing white. The female is brown above, under light brown, spotted darker, with a grey patch on the wing and no coloured speculum. Iris, yellow; beak, dark olive-brown, expanded at the tip into a broad spoon shape; *legs, orange-red or salmon* in the male, yellowish-brown in the female. Length, 18¾ in. Weight, male, 1 lb. 11 oz.; female, 1 lb. 8 oz.

This striking duck, with its spoon-like beak, blue in the wing speculum and salmon-coloured legs, can be mistaken for no other species. The female resembles the Black Duck in general body colour, but the beak characters are diagnostic—the profile is almost a straight line from the crown of the head to the tip of the beak, not being turned as in other ducks. The male has an eclipse plumage similar to that of the female.

Distribution: Throughout the area treated in this book, and extending into the West Kimberley Division (Derby, P. Slater). One of our less common ducks.

Nesting: The Shoveler nests on the ground in rank grass or bulrushes on the edge of swamps. The nest is just a slight hollow, lined with grass. Eggs, 8 to 12, pale cream with a green tinge. Size, 50 x 37 mm. Plenty of down is used to cover the eggs and it is brownish-grey, each down being lighter coloured in the centre and at the tips.

Garganey

Anas querquedula

The following field description of a male in full nuptial plumage was taken by Peter Congreve at Yanchep: "A broad white superciliary stripe ran from in front of the eye down the sides of the nape; the crown of the head was dark, the sides of the face and neck were rich chocolate chestnut merging on the chest into brown; the abdomen appeared to be white, sharply demarcated by the brown chest. With the bird floating on the water the wing coverts and flanks were seen to be grey, giving the mid section a light grey appearance to the water line. The scapulars, black with broadly white centres, fell like ornamental plumes diagonally across the flanks. In flight both birds exhibited a broad grey wing bar with a white stripe edging it, before the dark colour of the flight quills." In the eclipse plumage dark brown, with light borders to the feathers, white superciliary line and a thin, dark line from eye to nape, throat white, upper breast brownish mottled with rusty white, abdomen white. Female similar to, but duller, than eclipse male. Iris, brown; beak and legs, lead-grey. Length, 15 in.

Distribution: A Northern Hemisphere species, breeding in the temperate parts of Asia and Europe, "wintering" in Africa, southern Asia, the East Indies and, occasionally wandering to New Guinea and eastern Australia. The only Western Australian record is of two birds seen at Yanchep National Park on March 19, 1971, by Peter Congreve. The birds, a male and female, stayed several days.

Pink-eared Duck (Widgeon, Whistling Teal)

Malacorhynchus membranaceus

Native names: Wym-bin, Wimbin (Perth).

Head, wings and back brown; lower back black, followed by a narrow white bar across the upper tail coverts; tail black narrowly tipped white; *breast, nape and flanks finely barred white and black,* the barring extending to the side of the upper tail coverts; throat and middle of abdomen white; a dark patch around the eye and extending below it; a small inconspicuous patch of pink feathers behind the eye; narrowly white along trailing edge of wing and near shoulder. No wing speculum. Iris, brown; beak, lead-coloured, flat and greatly widened at the tip; legs, lead-coloured. Length, 15½-17 in.; wing span, 25-27 in. Weight, male, 14 oz.; female, 11-12 oz.

The dark eye patch surrounded by white is a good field-mark; the broad beak is like that of a shoveler but the head and beak profile are like those of an ordinary duck. In flight the white band across the upper tail is noticeable. The species has a whistling note both on the water and in flight.

Distribution: Likely to be found anywhere in the State, but until 1952 it was one of our rarer ducks in the South-West. Since then it has become quite plentiful.

Nesting: The duck generally selects an old Coot's nest, or an old nest of a Little Pied Cormorant in which to lay its eggs. In one instance an old nest of a White-faced Heron, 20 feet up in a tea-tree *(Leptospermum)*, was used (C. L. E. Orton). The eggs are creamy-white, glossy, oval, and from 6 to 9 in a clutch. Average size of 40 eggs from Carnamah and Moora, 50 x 36 mm., varying from 43-53 x 32-38 mm. Average weight, 33 gm. The down is grey with lighter centres and tips. Nesting was first recorded in the Perth area (at Butler's Swamp) in the summer of 1962-63.

The upper and lower mandibles are fringed at the sides with fine lamellae, protected by membranous flaps on the upper mandible. H. J. Frith has demonstrated that this is a filtering mechanism and the Pink-eared Duck feeds in shallow water near the shore filtering aquatic insects and small crustacea (Cladocera, Copepoda and Ostracoda) out of the water. They do not strain the bottom mud as do other ducks.

POCHARDS: Tribe Aythyini

White-eyed Duck (White-winged Duck)

Aythya australis

Native names: Er-roo-doo (Perth); Errudo, E-ro-to, Irodu (South-West).

Head, neck, back and breast a *deep brown;* wings with a bronze sheen and with a *wide white wing stripe* almost its full length; no wing speculum. Lower breast under wing and under tail, white; abdomen, brown. The male has a darker appearance about the head and neck than the female. Iris, *white in male,* brown in female; beak, blackish with white tip in male and slaty tip in female; legs, pale bluish-lead. Length, 20 in. Weight, male, 2 lb. 6 oz.; female, 1 lb. 14 oz.

The duckling is pale yellow below, with yellow throat and face. Upper parts brown. Iris, hazel; beak, brown, mauve-pink at tip; legs, grey marked yellow.

In the field there is a possibility of confusion with the males of the Blue-billed Duck, but they can be told apart by the colours of the beaks. The white in the wing is distinctive when the bird is in flight, hence the ducks are sometimes known as "white-wings" by shooters.

Distribution: All over the State.

Nesting: The nest of the White-eyed Duck is usually to be found in a thick, bushy, isolated tea-tree *(Leptospermum)* growing in bulrushes. It is constructed of trodden-down rushes for a foundation and on this a deep cup is made of bulrushes, often with a canopy of the same over the nest—possibly a precaution against being robbed by crows. It is also built in the butts of paper-barks and on the tops of bottle-brush clumps. The nest is deeper and neater than nests of either the Blue-billed or Musk Ducks, both of which build in similar situations. The eggs are creamy-white, are usually 8 to 12 in number but may be up to 18. A clutch of 12 eggs from Moora average 56 x 42 mm., varying from 54-58 x 41-43 mm. The nest-down is greyish-brown, each feather being lighter in the centre and at the tip.

According to Sir George Grey the tribal section of the Ngotaks of the South-West Aborigines (the Bibbulman people) are said to be these birds transformed into men.

PERCHING DUCKS: Tribe Cairinini

Maned Goose (Wood Duck, Blue Duck)

Chenonetta jubata

Native name: Mar-ang-an-ner (Perth).

Predominantly grey in colour—in the distance with a bluish tone. Head and throat brown with short mane-like black feathers on the back of the neck. Hindneck grey, mottled brown; mantle and wing coverts clear grey; lower back and tail black; wing quills black, with *green speculum between two broad white bands;* under wing coverts white. Breast mottled brown and white, flanks grey finely barred brown, middle of abdomen and under tail coverts brownish-black. The *female* differs in the head being brownish-grey, the wing quills brown instead of black and with a duller green speculum; the abdomen and under tail coverts are white instead of black and the flanks white, broadly barred with pale brown. Iris, dark brown; the short goose-like beak dark brown and the legs black. Length, 18-19½ in.; wing span, 34-35 in. Weight, 1½-2 lb.

Distribution: All over the State.

Nesting: Nest in a hollow spout of a tree standing in water. Eggs usually 6 to 10 in number, but clutches in excess of 10 eggs are not rare; T. E. Bush found a clutch of 18 eggs at Wickepin in August 1968 and I. C. Carnaby one of 20 eggs at Lake Grace also in 1968. These may be combination clutches. Creamy-white and glossy; covered with down, which is greyish-white, but almost white at the tip and in the centre. The eggs measure 51 x 35 mm. The female does the incubating.

This species has become a more or less regular visitor to the southern parts of the State, the numbers increasing in recent years. Birds are sometimes seen in the city parks and visit even the far south coast, as at Nornalup. Flocks of up to 100 have been observed at the Moora swamps. The bird is very partial to wet, grassy meadows and is thus often met with on irrigated flats such as at Waroona. It frequently perches on logs and trees.

Fig. 24.—Maned Goose, male.

Green Pygmy-Goose

Nettapus pulchellus

Upper parts *bright green,* top of head brown, a broad white band in the wing (separating the black primaries from the rest of the wing, above and below), *patch at side of head below eye white,* green collar, breast dappled green and brown, belly white. In the female there is no solid green collar, and the underparts are more extensively white, lacking any green on the throat. Iris, dark brown; beak, dark greenish-grey; legs, greenish-grey. Length, 12-14 in.; wing span, 19-23½ in. Weight, 11-15 oz.

The smallest of all ducks, swimming high in the water and rarely diving.

Distribution: Normally northern Australia, from the Kimberley Division to northern Queensland. Very occasionally individuals wander to the southern parts of the State.

The first specimen authentically collected south of the Kimberleys was a female taken at a duck shoot at Moora on January 8, 1966 (T. L. Riggert, Fisheries and Fauna Department). On March 4, 1967 B. D. Bond saw two at Lake Joondalup, Wanneroo. There have been previous reports of southern occurrences but in the absence of specimens these had been set aside as dubious. The earliest such report was made to John Gould by the explorer F. T. Gregory who, during his North-West expedition of 1861, said he met with the species only on the Sherlock River, "in flights of eight or ten together." In more recent years a pair was said to have been shot on a lake 60 miles north of Perth about 1902 (reported by E. A. Le Souef), some at Lake Muir (December 1911) and others at Lake Ewlyamartup, near Broomehill (both reported at second-hand by Tom Carter). In view of the Moora and Wanneroo records credence may now be given to these reports. Obviously, however, the species must be a rare visitor to the south.

Nesting: A. J. North describes nests as having been found in grass at the margins of swamps and also in hollow branches. There are few data on clutch size, sets of 8 have been found and up to 13 recorded. The eggs are blunt oval in shape, creamy white, with smooth, lustrous, very hard shells. Size, 44 x 32 mm., varying from 35-49 x 29-36 mm. (H. J. Frith, based on 106 eggs, from 12 clutches).

STIFF-TAILED DUCKS: Tribe Oxyurini

Blue-billed Duck

Oxyura australis

Native names: Boo-doo (Perth); Bo-ta, Buatu (South-West).

Male, *head and neck black;* upper parts and breast dark chestnut; breast and abdomen dark brown mottled buff. Iris, brown; *beak, a vivid sky-blue;* mandible, flesh colour; legs, grey. Female, upper parts dark brown inconspicuously freckled with buff; under parts dark brown mottled buff, throat and abdomen lighter. Beak, dark olive-grey. Length, 16 in. Weight, 1 lb. 5 oz - 1 lb. 14 oz., males averaging larger.

Distribution: Found in the South-West from the Thundelarra swamps (north end of Lake Monger), the Moora swamps to the Warriup Lake, east of Albany.

Nesting: Usually an old Coot's nest, or a nest of a Musk Duck, is used for the 5 to 8 eggs laid. Nests containing 9 and 10 eggs have been recorded from Butler's Swamp, Claremont, and may represent combination clutches. It is the experience of Messrs. Orton and Sandland in the Moora swamps that when the Blue-billed Duck makes its own nest, it usually treads down bulrushes growing in a tea-tree which has a thick bushy top. They have noted two instances when nests were made of sticks, and they have also observed that when a nest is built by the birds, it always has a dome of rushes. The eggs are pale greenish-white and of coarse texture and are large for the size of the bird—average measurements, 66-71 x 47-53 mm. The Blue-billed Duck uses less down in the nest than most ducks; sometimes there is barely any in the nest.

As in the case of the Musk Duck, the bird is hardly ever seen in flight and if disturbed will always endeavour to escape by diving. Apparently it makes all its journeys at night time. It is not an uncommon bird and may be seen in Shenton Park Lake, Subiaco (where it nests) in the heart of a built-up area, without the observer needing to alight from his car. It also occurs at Monger's Lake as well as on most of the larger swamps and lakes, including Jandakot Lake, Lake Cooloongup near Rockingham and Lake Mealup west of Pinjarra. It congregates in large numbers on these open waters after the breeding season in a similar manner to the Musk Duck.

John Gilbert referred to its voice as a peculiar inward tone which the natives described by saying, "it has no voice but makes a noise with its heart."

Musk Duck (Steamer, Diver)

Biziura lobata

Native names: Go-da-ra, Gaddara, Kattarr (South-West); Goojuk (South-West); Quodyuk (Pallinup River); Coatchuck (Albany).

Upper *black,* with faint lighter barrings not visible at a distance. These barrings are more pronounced on the under parts and the belly is mainly white with brown barrings. The short tail consists of 22 stiff narrow feathers. Both sexes have a *membranous lobe below the beak,* which in the male is large and very conspicuous. The beak and lobe are black and the feet leaden-grey. Iris brown. Males are 24-29 in. in length and weigh $3\frac{1}{4}$-$8\frac{1}{2}$ lb. Females are $20\frac{1}{2}$-24 in. in length and weigh $2\frac{1}{4}$-$3\frac{1}{4}$ lb.

Distribution: The main area of distribution is the South-West corner from Moora to Esperance. However, it may on occasion wander much further afield, even as far as the Point Cloates area (Cardabia Pool).

Nesting: The nest is made in a clump of rushes growing in water, or in a low branch of a tree growing in water. It is made of rushes or grasses and is well lined with grass and down. Usually, especially when built in rushes, a canopy is formed by bending over the rushes. The eggs are very large for the size of the bird, average of 12 eggs being 79 x 52 mm., varying from 69-84 x 50-55 mm. The clutch comprises 2 or 3 eggs; they are elliptical in shape, both ends slightly pointed, and rough in texture. This roughness is caused by the presence of a thin outer coating of whitish limey matter which, on being scraped away, discloses that the egg itself is greenish-white with a slight gloss. As incubation proceeds the eggs become paler.

The apparent flightlessness of this bird and its superb diving power lend to the Musk Duck considerable public interest. During the daytime the birds can hardly ever be induced to fly and if chased or hunted will either "steam" along, with their small wings fluttering rapidly and the powerful legs paddling, or take refuge in diving. Timed dives indicate that the birds remain submerged for normal feeding purposes between 9 and 16 seconds (usually 11 to 12 seconds) and the intervals between dives are from 5 to 6 seconds (Lake Cooloongup, Rockingham). How they arrive at isolated country dams and then as suddenly disappear has excited much speculation; they are met with at times even on islands out at sea and have been observed on the salt lakes at Rottnest. The solution is that they will fly *only at night,* although this has been only rarely observed. Occasionally hunters shoot them in flight at night in error for Black Ducks and they have been known to collide with lighthouse lanterns and to alight on galvanised iron house-roofs which, apparently, on moonlight nights resemble pools of water to a flying bird. However, even at night Musk Ducks prefer to walk from swamp to swamp if the distance is not too great, and at such times they may be run over by motor cars. A brief flightless phase is, however, passed through at the annual wing moult, when the primary and secondary wing quills are shed simultaneously, but the renewal is rapid. We have seen individuals in this flightless stage in February.

So fundamental a part of the bird's life is the diving habit that young ducklings react instinctively to it. An interested observer once placed a Musk Duck's egg under a broody domestic hen which hatched it out but was driven to distraction by the duckling, when it was disturbed or danger threatened, leaping to the hen's throat and clinging tightly. Apparently under natural conditions the duckling grasps the mother's neck at the instant she dives and clings to her until she surfaces again. The young duckling has been seen riding on its mother's partially-submerged back.

The male indulges in a striking attachment display towards his mate. The bird crouches very low in the water with the head back and the pendent lobe taut and touching the water, just cutting the surface. He swims steadily towards and around his mate, or in the near vicinity, emitting a loud metallic "plonk", the tail moving up and down. Often this call is terminated by a loud hissing whistle and the tail motion changes—it is held upright, fanned out and bent forward over the back, exposing the pin-cushion-like under tail-coverts. As an accompaniment to these sounds and tail movements the bird shoots out two powerful jets or splashes of water from the sides of the body. These are not produced by the feet, as commonly held, as there is no jerkiness in the bird's motion and it may even be quite stationary when splashing water. The jets are shot out too far forward to be caused by the bird's rather backwardly placed feet and sometimes, even, the jets are seen to be directed *forward.* They are produced by the wings opening out and shutting smartly against the body and on close observation one may catch a fleeting glimpse of the wing action when the bird turns end-on during its display. The Musk Duck may splash water with its feet at other times, when not displaying. On these occasions the splashes are more diffuse than the jets caused by the wings; there is no accompanying whistle and the lobe is flaccid, waggling about loosely as the head is turned.

The performances of the Musk Duck at breeding time provide one of the characteristic sounds in local swamps in winter and spring. Fletcher Moore, in his *Descriptive Vocabulary* of the Aboriginal language, lists the word "kopotjan," defined as: "to make the same noise as the Gaddara, or steamer-duck."

Musk ducks are not usually regarded as edible and in any case are too difficult to shoot owing to their rapid diving—which is grebe-like in its instantaneousness. However, they often fall victims to fishermen's set nets and this method of capture was used in a curious little industry which used to flourish in the Mandurah district.

Each winter large numbers used to be netted by the fishing community in the lake-like expansions of the Serpentine River. The birds were plucked by repeated immersions in boiling water, then dressed, split, dry-salted and smoked for use as food during the rest of the year. Lake Goegrup, on the Serpentine, in August was a well-known locality for this "fishery," which ceased on any large scale after 1898 when fishing was prohibited in the rivers and lakes, but it was still continued in a small way by some of the older inhabitants. One year a local fish-canning factory tinned a sample of the birds, but the experiment proved a failure.

Musk Ducks concentrate on our lakes and estuaries after the breeding season in far greater numbers than they do in the Eastern States and companies of many hundreds are frequently seen.

DIURNAL BIRDS OF PREY: Order Falconiformes

EAGLES, GOSHAWKS, Etc.: Family Accipitridæ

Black-shouldered Kite

Elanus notatus

The lightest coloured hawk in our area. The upper parts are a pale grey, darker grey on the flight quills with a *broad black patch at the shoulder* (upper wing) and a small black patch in front of the eye. The under parts are entirely white, except for the grey flight feathers and a *black patch at the base of the outer quills.* Iris, orange-red; cere, yellow; beak, black; legs, yellow; claws, black. Length, 13-14½ in.; wing span, 33 to 37 in. Weight, 10-11 oz. The juvenile is rusty-coloured on the head and mantle, and the eyes dark.

Fig. 25.—Black-shouldered Kite—under wing pattern.

Distribution: Occurs all over the State but until recent years was scarce in the South-West.

Nesting: To house the 3 eggs, which is the normal number laid, this kite usually builds its nest in the thick leaf canopy of a tall tree; in the case of a pair which nested at Bridgetown in 1943 the nest was in the top of a tall marri (red gum). The nest is a bulky structure, made of sticks, and lined with leaves. The oval, dull white eggs are blotched in varying proportions with chocolate-brown; sometimes they are blotched, more or less, all over; in other examples the colouring is limited to a cap covering one end. Average measurements, 40 x 31 mm. Nestling period, 5 weeks (P. Slater). Breeding is protracted but July to September are the main months.

In this State the Black-shouldered Kite is typically a dry country hawk, but appears to be subject to cycles of great fluctuation in numbers and in recent years has extended in some strength to the forested South-West corner. This cycle of abundance began about 1935, was greatly accelerated since 1943, and had waned by 1948. Another cycle of increase began in 1951 and birds became numerous throughout the South-West in 1952 and have remained so for several years.

The Black-shouldered Kite is a graceful species and hovers as skilfully as a kestrel, no other hawk sharing in the habit. Mice figure prominently in its diet.

Letter-winged Kite

Elanus scriptus

Similar to the previous species but differing in the underwing pattern. There are much more extensive black markings on the under wing—*these outline the whole of the wing bend* and the black pattern resembles the letter "M" when the bird is in flight. Iris, red-brown; beak, black; legs, pale yellow; claws, black. Length, 14½ in.; wing span, 33-35 in. Weight, 8½ oz.

Fig. 26.—Letter-winged Kite—under wing pattern.

Distribution: Normally this kite occurs in the inner arid regions of central and eastern Australia, with occasional irruptions into Western Australia. In 1943 I. C. Carnaby saw three pairs near Learmonth, Exmouth Gulf. In 1951-52 there was an extensive invasion into the State. Birds were seen at Adele Island, Abydos Station, Hooley Station, Coolawanyah Station, Roebourne, Dongara, Gutha, Pintharuka, Moora, Pithara, Miling, Booraan, Burracoppin, Kellerberrin, Cunderdin and Esperance.

Another invasion took place during 1970. In May of that year T. E. Bush saw four birds near Mullewa, and in June T. Spence saw birds at Derby and Wyndham, in the Kimberleys. Some birds stayed over until the following year when Graeme Chapman noted an attempt at nesting at Manmanning.

S. A. Parker has suggested that the irruptions of this species are closely associated with fluctuations in numbers of the native rat, *Rattus villosissimus,* both species having a common centre of distribution in the eastern Lake Eyre basin. When the rats reach plague proportions they range widely, even as far as the Western Australian border. When the declining population of rats fail to provide sufficient food for them the kites disperse at random.

Nesting: I. C. Carnaby observed nesting at Exmouth Gulf, the nests being at the ends of branches in high trees, overlooking dried water courses. S. W. Jackson described colonial nesting in the species on the Diamantina River in south-western Queensland in June-July 1918, during a plague of native rats. The nests were in coolibah trees, 11-35 ft. from the ground, situated near the ends of limbs leaning out over waterholes, built of twigs, and measured about 14 in. in diameter, some old ones 20 in. in diameter. The nesting cavity was 8 in. in diameter and 3 in. deep. The lining was of coolibah leaves and a mass of rat fur. The usual number in a clutch was 5 eggs, and varied from 4 to 6. The colour is similar to that of the previous species.

Black Kite (Fork-tailed Kite, Kite Hawk, Kimberley Hawk)

Milvus migrans

Native names: Kurr-gurr-ee (Coolawanyah); Pad-oon (North-West Cape); Tjoonin (Jigalong).

Upper parts blackish-brown, wing quills black. Under rufous-brown, streaked black. Tail, dark brown, barred black and slightly but distinctly *forked—no other local hawk has this feature.* Younger birds are lighter and are strongly streaked light buff and brown on the head, neck, and under parts. Iris, brown; beak, blackish-grey; cere, yellow; legs, yellow. Length, 19-22 in.; sexes similar; wing span, 40-50 in. Weight, 1½ lb.

Distribution: A northern and central Australian species, not regularly recorded in this State much further south than the Tropic of Capricorn.

Nesting: As with most hawks, for nest purposes the Black Kite uses an old nest of a crow, or of another hawk. Some sticks are, however, added and the centre is lined with bark, wool, or other soft material. At Derby, because of the striking similarity between them, P. Slater believed the local kites constructed their own nests. These were built in the first or second main fork of the river black-butt, or occasionally in boabs, about 2 ft. 6 in. in diameter and 2 ft. deep, of small sticks and lined with cow dung. I. C. Carnaby found similar nests in the West Kimberley, with lumps of horse manure, cow dung or mud in them.

The eggs, 2 or 3, are white or whitish and sparingly spotted and blotched with red. Average measurements, 51 x 49 mm., but there is much variation in size. When viewed through a hole in the side the inside of the shell is seen to be dark green. The nestling is clothed with rufous down; iris, light brown; beak, dark horn; legs, yellowish. Incubation period, 34 days; nestling period, 60 days (P. Slater).

This interesting and quite inoffensive bird ranges much further south in the inland parts of the eastern States than it does in Western Australia. The regular haunts of the Black Kite in this State are in the Kimberley Division. Occasional irruptions occur into the North-West pastoral country, where some individuals may stay to nest. Until 1952 the extreme southerly limit of such visitations was a line from about Carnarvon trending easterly to the Peak Hill district, Well 8 (150 miles north-east of Wiluna) on the Canning Stock Route, and the Warburton Range. In 1952 (a year in which many northern birds invaded the south, see p. 6) the Black Kites swarmed south on an unprecedented scale. They appeared throughout the North-West in thousands, the earliest being reported in February and before the middle of April had reached Perth. This invasion of the south was the first on record since the State was founded in 1829, and advanced the southern limit of the species' occurrence by some 500 miles. The southern limit was a line joining Perth, Narrogin and Pingrup.

At the northern stations the kites frequent the vicinity of slaughter houses and refuse dumps, and around homesteads will be often seen feeding with domestic poultry. Bush fires attract flocks of kites to the scene and they like to play over the vortex of a "willy-willy" as it draws debris aloft. The flight is characteristic, slow and buoyant, with the tail being frequently twisted and fluttered. The call is a peculiar curlew-like whine, or a shrill almost musical whistle (H. T. Condon).

The species is widely distributed in Europe, North Africa, and parts of Asia. Travelling Australians who may never have seen it in their own land are made familiar with it at Aden, where with extraordinary tameness it performs a very useful service as a street scavenger.

Square-tailed Kite

Lophoictinia isura

Native names: Ge-durn-mul-uk (York); Mar-arl (Toodyay).

The mantle and wing quills are black. The *head and under parts are bright rufous-brown, streaked with black;* whitish on chin and forehead. The *long tail is square cut at the end,* brownish-grey in colour with a dark cross-bar. The under wing is variegated; the shoulder angle and coverts are rufous-brown, the bases of the flight quills whitish (forming a circular patch resembling, but not so conspicuous as in the Black-breasted Buzzard) and the ends black barred white and grey. Immature birds are lighter, being dappled above chestnut-brown and black, with narrower streakings below. Iris, yellow, but brown in young; beak, slaty-black, darker at tip; cere, milky-blue; legs, flesh-coloured. Length, 22 in.

Distribution: Occurs all over the State.

Nesting: The nest, a large collection of sticks, is built in a tall tree; green leaves are used to line the egg-cavity. The 2 or 3 eggs measure 53 x 38 mm. and are dull white with spots and blotches of reddish-brown and lavender. Birds may breed before they develop the full adult plumage.

This is one of the rarer hawks and most of the records are from the southern part of the State, from the midlands to the south-east coast. It is partial to the sandplain scrub, over which it flies low like a harrier, the long wings giving it a very wide wingspread.

Black-breasted Buzzard

Hamirostra melanosternon

Native name: Goo-dap (Avon River).

In the adult plumage the head, back and wings are black, with a rufous-brown collar around the nape and rufous feathers interspersed on the shoulder and mantle. The throat and breast are black and the abdomen and thighs rufous-brown. There is a *circular white "bulls-eye"* in the wing, like a R.A.A.F. marking, which is distinctive in flight. The wing is carried in a high-angled dihedral (distinguishing the species from the Square-tailed Kite). The rather short, fan-shaped tail is grey. Iris, dark-brown; cere and beak, grey; legs, greyish-white. Length, female, 22 in.

In the immature plumage the under parts are a rich tawny brown.

Distribution: Found all over the State, except the forested areas of the South-West.

Nesting: I. C. Carnaby found this species nesting at Tunnel Creek, on the Ashburton River, in September 1938. The nest is a large structure of sticks about 4 or 5 feet across and 2 or 3 feet deep, and is built in tall trees. The egg cavity is shallow and is generously lined with fresh-picked green leaves. Clutch, 2 eggs, measureing 60-64 x 46-48 mm. They are similar to eggs of the Wedge-tailed Eagle, white with spots or blotches of dark chocolate and light purple. The young are covered with bluish-white down.

Rabbits are now the main food of this hawk; it does not molest sheep. The Drummonds told Gilbert of the Aborigines' story of the bird breaking emus' eggs by dropping stones on them from the air. The colonial diarist, George Fletcher Moore, in May 1837, reported a similar conversation with two natives "that a large hawk, when it discovers an emu's nest, takes a stone in its talons, hovers over the nest, and lets it drop among the eggs to break them." The same story was told by a native in more recent times to the late Mr. B. H. Sharpe, of Wooleen Station, on the Murchison. The call is a peculiar short cry, rather like that of the Whistling Eagle.

Whistling Eagle (Whistling Kite)

Haliastur sphenurus

Native names: Jan-doo, Jandu (Perth).

Head and under parts light, back and wings dark. The head is brown conspicuously streaked with light buff, which lightens the whole appearance of the bird. The under parts are light brown with similar conspicuous light buff streakings. Back and wing coverts dark brown, streaked and spotted buffy-white; wing quills black. Under wing buff, marked with brown, the pale band parallel with the wing margin and a hinder dark area almost rectangular in shape. The *long tail rounded at the end,* is brown. Iris, light brown; beak, dark brown; legs, white. Length, male, 20-21 in.; female, 21-23 in. Wing span, female, up to 4 ft. 5 in. Weight, male, 1 lb. 7 oz.; female, 1 lb. 14 oz.

This species may be confused with the Little Eagle and the distinguishing characters are discussed on p. 166.

Distribution: Found all over the State.

Nesting: The Whistling Eagle usually builds its own nest, but, on occasions will repair an old one of another species of hawk. The same nest is often used year after year. It is a bulky structure of sticks and the depression in the centre is lined with fresh leaves from the tree in which it is built, and in which it is placed high up in a vertical fork. The 2 or 3 eggs are bluish-white, rough and coarse

Fig. 27.—Whistling Eagle—under wing pattern.

in texture. Markings may be sparse but sometimes consist of large blotches of reddish-brown. Sometimes they are only sparsely spotted with reddish-brown and lavender. The inner lining of the shell is green. Average size of 10 eggs, 57 x 44 mm., varying from 54-61 x 43-46 mm. The nestling is covered with short white down, but that on the head is long and hair-like. Iris, light reddish-brown; beak and cere, light horn-colour; legs, bluish-white.

This is the commonest large hawk in Western Australia, and although it has a distinct partiality for frequenting the neighbourhood of water, both on the coast and inland, it is not strictly confined to such situations. It is a carrion eater and may often be seen feeding on poisoned and trapped rabbits, dead sheep and the carcasses of other animals. Up to 1938 the bird was practically never seen in the Bridgetown district. In that year rabbits increased greatly and this hawk from that year onwards has become one of the normal breeding birds of the district. A similar increase took place, we are informed by Mr. Orton, in the Moora district. Besides rabbits the bird feeds on grasshoppers, but does not harm lambs or poultry.

The loud whistling call, "chew-ew-ew," at once directs attention to its presence as it soars high overhead.

Red-backed Sea-Eagle

Haliastur indus

Native name: Indee-narrangee (North-West Cape).

Head, neck and breast *pure white;* back, abdomen, wings and tail, *reddish-brown,* the wing quills black and the tail tipped light brown. In immature birds the head and breast are dull white, streaked brown and the back and wings are brown without any red tone. In transition plumage, the reddish feathers are mottled with the brown. Iris, brown; beak, bluish-horn; cere, lead-colour; legs, greenish-yellow. Length, 18-19 in.; wing span, 47 in.

Distribution: A northern coastal species which extends south as far as Carnarvon.

Nesting: Nesting usually in mangroves along the coastline, it makes a nest of sticks in a mangrove tree and lays one or two bluish-white or dull pale green eggs which have a few marks of rusty-red. They measure 57 x 43 mm.

This graceful little sea-eagle is common around the northern coastline, but southern ornithologists who have travelled outside of Australia are, perhaps, more familiar with the birds in Colombo where they are numerous in the harbour and compete for foodscraps with the ubiquitous Ceylon Crows. They are essentially carrion feeders. The call-note is a curiously modulated "pee-ah-h-h". The bird is known in India and Ceylon as the "Brahminy Kite."

Australian Goshawk

Accipiter fasciatus

Native names: Charlalya (Lyons River); Good-jee-lum, Gudjelan (Perth); Kil-lin-gil-lee and Mat-wel-itch (Avon River); Carakine (Pallinup River).

The adult is slaty-green above. The *under surface is rufous-brown, finely marked with numerous narrow white bars.* The under wing coverts are similar and the under wing quills and tail are grey, barred brown. Tail long. In the immature plumage the upper parts are dark slaty-brown with each feather narrowly edged with rufous-brown; the under parts are coarsely barred brown and white, with the pattern altering to longitudinal streaks on the throat and chest. Iris, yellow; naked skin around eye, yellow; cere, greenish-yellow; beak, blackish-brown; legs, yellow. Length, female, 18-20½ in.; male, 16½-17½ in. Wing span, female, 3 ft.; male, 2 ft. 5 in. - 2 ft. 8 in. Weights, female, 1 lb. 2 oz. - 1 lb. 7 oz.; male, 9½-12 oz. In the field the small males are indistinguishable from the females of the Sparrow-hawk.

Distribution: Occurs all over the State.

Nesting: The Goshawk often uses an old stick-nest belonging to a crow, or another hawk—one that is high up in a horizontal fork of a tree—and "improves" it by lining it thickly with green leaves. However, a fresh one of sticks in a similar situation is usually built. P. T. Sandland, of Moora, estimated that the proportion of pairs that build their own nests to those that do not is 3 : 2. The clutch consists of 2 to 4 rounded-oval eggs and very rarely 1 or 5. They are white in ground colour and spotted, smeared and blotched in varying degree, but normally not heavily, with red. Occasionally eggs are without markings. When held to the light and viewed through a hole, the inside is seen to be greenish-blue. Mr. Sandland found that prior to the rabbit invasion the clutch was usually 2 and often 3 eggs, but is now usually 3 and often 4. Average measurements of 19 eggs, 47 x 37 mm., varying from 45-50 x 36-38 mm. Peter Slater has determined the incubation period to be 30 days and the nestling period, 30-31 days. The female alone visits the nest when incubation begins, she leaving the nest to be fed by the male. The young gain yellow eyes at three months.

This woodland hawk lives on live food, not carrion, and preys on rabbits, birds and grasshoppers, being *the* hawk harassing poultry runs. It is not protected in this State. The call notes may be rendered as "kikk-kikk-kikk" or "kek-kek-kek" (female) and a drawn-out "wit" or "weet' (male).

Collared Sparrow-Hawk

Accipiter cirrocephalus

Native names: Jil-lee-jil-lee, Jillijilli (Perth); Min-min (Avon River); Kil-lin-gil-lee (Toodyay).

In the mature plumage the upper parts are dark slaty-grey, with *a narrow reddish-brown collar behind the head. The under surface is rufous-brown with narrow white cross-bars.* The immature birds have the feathers of the upper parts edged with rufous-brown and below the white barrings are much coarser, and become arranged as longitudinal streaks on the throat. Iris, golden-yellow; cere, yellow; beak, black; legs, yellow. Length, female, 14-15 in.; male, 11¾-12½ in. Wing span, female, 2 ft. 5 in. Weight, female, 13¼ oz.

The colour pattern is very similar to that of the Australian Goshawk and in the field size-differences must be depended on for identification, but, as already pointed out, females of this species cannot readily be told apart from males of the other species. When handled the toes offer a quick identification—in the Sparrow-Hawk the toes are more slender than those of the Goshawk and the middle toe is more slender and longer than the other toes.

Distribution: Occurs all over the State.

Nesting: This hawk makes its own nest of small sticks. It is actually no more than a platform with a slight hollow in the centre which is sometimes lined with green leaves. A new nest may or may not be constructed each year. Sometimes an old crow's nest is used, which is lined with green gum leaves. The 2 to 4 eggs are globular in shape, white, and have a few blotches of light and dark brown. They are rather like the eggs of the Australian Goshawk but are, of course, much smaller. Average size of 23 eggs, 38 x 31 mm., and varying from 36-41 x 29-32 mm. Incubation takes about 19 days, and the young when first hatched are covered in white down.

Australian Little Eagle

Hieraœtus morphnoides

Native name: Wee-dun-ba (Pilbara district).

Upper parts brown, wing quills black. Head, neck and sides of breast chestnut-brown streaked black—very boldly on the crown and with *a black streak below the eye.* Tail grey-brown, barred dark brown; *very square cut.* Under parts variable; may be very light with chestnut on the breast, or the entire under parts may be dark chestnut-brown streaked black. The legs are feathered, white, light buff or chestnut, down to the toes. The under wing coverts are white in the light phase with chestnut near the root of the wing, the flight quills being black. In the darker-breasted birds the coverts are dark brown near the root of the wing, elsewhere grey with the flight quills black. Iris, brown; cere, grey; beak, dark brown or blue-grey; toes, pale greenish or blue-grey. Length, male, 18 in.; female, 20-21 in. Wing span, female, 4 ft. 1½ in. - 4 ft. 3 in. Weight, female, 2 lb. 2 oz. - 2 lb. 9 oz.

This eagle may be recognised when perched by the feathered tarsus, it and the Wedge-tailed Eagle being the only members of the hawk tribe with "trousered" legs. From the Whistling Eagle it can be told by the usually much lighter under parts, the shorter, square-cut tail, and the shorter and broader wings. Though the intensity of colouring of the under wing pattern is variable the Little Eagle always has an *oblique pale band,* and there is a *triangular dark area* at the hinder border of the wing. In the Whistling Eagle the pale band is parallel with the wing margin and the hinder dark area is almost rectangular (J. H. Calaby). Both have a whistling call which is distinctive. The Little Eagle tends to soar higher than the Whistling Eagle.

Distribution: Found all over the State.

Nesting: The Little Eagle uses an old nest of another hawk or a crow, repairing it if necessary and lining it with fresh green gum

Fig. 28.—Australian Little Eagle—under wing pattern; upper, light phase; lower, dark phase.

leaves. Sometimes it builds its own nest. The egg is white or bluish-white, coarse in texture, and smudged, streaked and blotched with reddish-brown in varying degree, sometimes almost without markings. Average size of 7 eggs, 55 x 45 mm., varying from 54-57 x 43-47 mm. Clutch, 1 or 2. When viewed through a hole, the inner lining of the shell is seen to be bright green. The nestling when first hatched is creamy-white; iris, deep orange; eyelids, black; beak and legs, bluish-grey. Peter Slater has found the nestling period to last 9½ weeks.

There has been an interesting change in the normal clutch size of this species since the early 'twenties, and P. T. Sandland, of Moora, considered this was due to the advent of the rabbit making a greater supply of food available. He stated that in the years up to about 1923 the ratio would have been 3 or 4 single eggs to each pair of eggs, but since then the ratio is 2 pairs to one single egg and in some seasons would be 3 : 1 in favour of a 2 egg clutch. The same observer had noted a similar increase in clutch size in the Wedge-tailed Eagle (2 egg clutches much more predominant than formerly) and in the Australian Goshawk (3 and 4 egg clutches commoner than they used to be).

It is more a denizen of the drier areas than of the forested South-West country, but specimens have been shot at localities such as Cannington and Pickering Brook and it has been seen at Monger's Lake (Leederville), Gooseberry Hill (frequently), in the rugged territory between Toodyay and Perth and at Yundurup.

Wedge-tailed Eagle (Eagle Hawk)

Aquila audax

Native names: Nelgie (Well 42, Canning Stock Route); Waladoo (Jigalong); Jabba-roongoo (Coolawanyah); Jabba-loong (Millstream); Warreedar (North-West Cape); Walada (Lyons River); Walawuru (Warburton Range); Warrida (Pidong tribe, upper Murchison River); Yarda (Fraser Range); Wold-dja, Waldja (Perth district and Avon River); Walgil (South-West); War-lock (Broomehill); Nailoit and Warlit (Albany); Warlick (Pallinup Estuary).

The fully adult bird is wholly *black,* except for the rufous, black-streaked nape. Younger birds are dark brown, wing coverts edged pale brown, wing quills and tail black; under wing dark; crown of head and nape chestnut-brown streaked black. *Tail, long and wedge-shaped. Legs feathered to the toes.* Iris, yellow; beak, dark brown; cere, whitish; toes, whitish. Length, male, 35-37 in.; female, 35-42 in. Wing span, usually from 6 ft. 3 in. - 7 ft. 3 in.; the record is held by a Tasmanian bird with a wing span of 9 ft. 4 in.

(D. Fleay). Usual weights, 7-8½ lb. T. Carter weighed four birds at Broomehill which averaged 12¾ lb., but these had gorged themselves on the carcasses of two poisoned ewes.

Distribution: All over the State.

Nesting: The Wedge-tailed Eagle makes a huge platform of sticks in a fork of a tall tree on which to lay its 1 or 2 eggs. The platform is flat and there is no nesting cavity, but fresh gum-leaves are placed in the centre and the weight of the laying bird so depresses this spot that it becomes sufficiently concave to prevent the eggs rolling out. The eggs are whitish and marked in various ways with spots, blotches and smears of various shades of reddish-brown—usually one egg of the set is less heavily marked than the other, often to the extent of having no markings at all. Average measurements of 16 eggs, 73 x 58 mm., varying from 68-78 x 56-62 mm. The nestling is creamy white. Usually only one young is reared in nests which contain 2 eggs. It is extremely difficult to approach an eagle at the nest. With its "eagle eye" it detects an intruder from afar off and the casual observer rarely has the chance of flushing an eagle at the nest.

On Dorre Island in Shark Bay where there are no trees, O. H. Lipfert in 1910 found the bird building on rocks. On a visit in November 1948, N. E. Stewart found three such nests at the north end of the island, all in use; all were close to the water's edge, one not more than two feet above highwater mark.

This magnificent eagle is still quite abundant despite the huge number shot and poisoned annually, on the ground that it is a pest in sheep-grazing country. From as far back as 1892, Government gave a reward of two shillings per head for its destruction and it was, in 1919, declared vermin under the Vermin Act. In the 41 years between 1928 and 1968 bonuses were paid on a total of 147,237 heads (the majority of which came from the Kimberley Division), or a yearly average of 3,591. The bonus was five shillings per head.

The bonus on eagle beaks ceased in November 1968. In January 1969 the State-wide declaration on this species was cancelled and it is now declared vermin only in the following districts—Ashburton, Broome, Carnarvon, Coolgardie, Cue, Kalgoorlie, Laverton, Leonora, Marble Bar, Meekatharra, Menzies, Murchison, Nullagine, Port Hedland, Roebourne, Sandstone, Tableland, Upper Gascoyne and Wiluna.

The Eagle is now rarely seen in the near vicinity of Perth, though it nests in the rugged country of the Swan valley in the Darling Range, 20 miles from the city. One or two are occasionally seen over the Helena valley between Gooseberry Hill and Boya. Although the bird has a reputation as a lamb killer, it feeds mainly on rabbits and carrion, including lambs that have died from mismothering or other causes. Observers state that it steals a good deal of its food from foxes, dingoes and the smaller hawks.

White-breasted Sea-Eagle

Haliæetus leucogaster

Native names: Tantagee (North-West Cape); Ngulor (South-West); Chillian (Pallinup River).

In the fully plumaged adult the head, neck and entire under parts, except the wing quills, are *pure white*. Back, wings and tail are *grey,* and the *wedge-shaped tail is broadly tipped white.* The immature birds are mottled brown, lighter on the head and with a largely white tail above—below it has a broad dark terminal band. The under parts are streaked brown, the under wing coverts mottled brown and white, with a white band across the bases of the flight quills. Iris, brown; cere, grey-brown; beak, slaty-grey; legs bare and unfeathered, yellow. Length, 28-31 in.; female, 32-35 in. Wing span, male, 5 ft. 11 in.; female, 7 ft. 2 in.

The adult bird is quite distinctive; in the young plumage it may be liable to confusion with the Wedge-tailed Eagle but the white tail is always diagnostic of the species. This will distinguish it also from the much smaller Osprey, which has a brown tail and the breast streaked.

Distribution: Occurs along the entire coastline.

Nesting: A huge nest of sticks is built either in a tree or on a rock along the coastline. An attempt at lining the egg-cavity is made by the addition of a few leaves. Some nests are more than three feet in depth. Two plain white eggs are laid, which, when held up to the light and viewed through a hole in the side, show that the inner lining is dark green. They measure 71 x 53 mm. Sometimes only one egg comprises the clutch. Nesting period (North-West Cape), June to August. The nestling is covered with white down.

This eagle, Australia's second largest, appears to be rather rare around the South-West corner, but is common further north. Many pairs have their headquarters on the Abrolhos. It may ascend considerable distances along some of the northern rivers and P. T. Sandland saw one at Mary's Crossing on the Fortescue, about 400 miles inland from the coast.

The White-breasted Sea-Eagle was declared "vermin" under the Vermin Act, 1918, in 1938 in the Roebourne and Tableland Vermin district, as the bird was stated to be preying on lambs. It was removed from the vermin list in 1970.

Spotted Harrier

Circus assimilis

Native name: Pillargo (N.W. Cape).

Head, grey streaked rufous-brown, *cheeks and facial disc rufous-brown.* Upper parts grey, dotted white on inner wing coverts and mantle, remaining wing coverts lightly barred dark brown, wing quills black. *Under parts rufous-brown,* darker on the breast and *spotted throughout with white.* Under wing coverts similar, the bases of the quills white and the tips greyish-black. Tail grey above, strongly barred blackish-brown and tipped white; under dark grey barred white. Iris, yellow; beak, black; legs, yellow. Length, male, 20-21 in.; female, 23½ in. Weight, female, 15 oz. - 1 lb. 1 oz.

The immature plumage resembles that of the Swamp Harrier, but the markings are less definite, a warm brown colour. The eyes are darker.

Distribution: Found over the State generally, though usually not in the South-West forested area south-west of Moora, Goomalling, Kellerberrin, Bruce Rock and Lake Grace. However, one individual was shot at Busselton in January 1934. In 1952 birds were seen at Coolup and Bridgetown.

Nesting: Unlike the Swamp Harrier, the Spotted Harrier builds a nest, a large flat one, in a tree. It is a substantial structure of large sticks, finished off in the upper portion with smaller sticks, and the egg-cavity, which is quite shallow, is lined with plenty of green leaves. The eggs, 2 to 4, are coarse-grained and often so nest-stained as to appear to be marked with rusty smears and blotches. They are, however, a white or bluish-white, without markings. When viewed through a hole in the side, the inner lining is seen to be dark green. Average size of 9 eggs from Lake Grace and the North-West, 53 x 41 mm., varying from 50-57 x 39-42 mm. Nesting dates, North-West Cape area, July to September (T. Carter).

This harrier is essentially an arid-country bird, its place in the wetter areas being taken by the Swamp Harrier. It may be seen soaring wide overhead, primaries outspread in the manner of a Wedge-tailed Eagle, and the light and dark bands across the wings and tail aiding identification. The name is derived from the bird's habit of beating low over the bushes and it often alights on the ground. It feeds on rabbits and grasshoppers.

Swamp Harrier (Swamp Hawk)

Circus approximans

Native name: Dil-yurdu (South-West).

Like many hawks this species is very variable in plumage. In the fully mature stage the upper parts are almost uniform brown, with buff streaks on the nape and on the rump, and a *conspicuous white band across the upper tail coverts.* Under parts light buff, strongly streaked dark brown; under wing coverts whitish-buff barred grey-brown, bases of flight quills almost white and the tips blackish-brown. Tail grey-brown barred black, and long. Sub-adult birds have a variegated appearance on the back due to light buff edging on each feather. Some birds, apparently young, are entirely dark, a uniform brown on the back and a dark brown, with a rufous tone, below. Iris, brown; beak, black; legs, greenish-yellow. Length, female, 21-23 in.; wing span, 50 in. Weight, 1½ lb.

Distribution: All over the State.

Nesting: This is the only inland hawk in Western Australia which makes a nest on the ground or in rushes growing in water. The large nest—often as much as 18 inches in diameter—is made of rushes and plants with the base often actually in water. Old nests are used year after year and are repaired by adding broken-up rushes or plants to the egg cavity, which is sometimes as much as 6 inches deep. The 3 to 6 eggs, like those of the Spotted Harrier, are coarse in texture and white or bluish-white in colour. They also tend to become nest-stained. When the inside is viewed through a hole, the inner lining is seen to be green. Average size, 53 x 40 mm. The nestlings are covered in creamy down; iris, brown; cere, yellowish; beak, blackish-blue; legs, yellowish.

As its name implies, this harrier frequents the neighbourhood of water, including estuaries, swamps and marshy places generally.

OSPREYS: Family Pandionidæ

Osprey

Pandion haliœtus

Native names: Weel-bra (North-West Cape); Yoon-door-doo (Perth).

Head and neck white, streaked brown, with a *prominent dark brown stripe behind the eye;* back and wings brown, each feather narrowly edged light buff; wing quills black; tail brown, barred darker above, white below, rather long, fan-shaped. Under parts white, streaked brown on the upper breast only; under wing white, barred light brown, flight quills grey. The head feathers can be erected into a low crest. Iris, orange; beak, black; legs, whitish. Length, 23-24 in. Wing span, 5 ft. Weight, 3 lb. 14 oz.

This and the White-breasted Sea-Eagle are the only two marine hawk-like birds found in the southern parts of the State. The Osprey is distinctly smaller and can always be recognised by the brown tail, which is always largely white in the Sea-Eagle.

Distribution: Along the sea-coast of the whole of the State.

Nesting: Unlike the habits of the species in Europe and America the Australian bird does not usually nest in a tree, but builds a nest on a rock on a promontory on the coastline, or on a rocky islet. It has even built on the gallery at Bathurst Pt. lighthouse, Rottnest. Composed of sticks and seaweed closely bound, the nest is added to year by year. A nest examined by us on the south coast had reached a height of 5½ feet and it could be described as a decapitated cone, the eggs being laid in a shallow depression on the top of the cone. These cone-shaped nests are to be seen on many rocky promontories on the coasts and islands of the State. On the more northern portions of the coastline the average number of eggs laid appears to be 2, but on the south coast 3 eggs are more often the rule. Rarely 4 may be laid. The shell is coarse in texture, and rough. The colour is dull white more or less marked with blotches of red, which sometimes form a cap, often at the more pointed end. When heavily marked, which is often the case, the eggs are beautiful objects. The shape is a blunt oval. Average size of 14 eggs, 61 x 45 mm., varying from 57-65 x 43-48 mm. In the more northern part of the coast eggs are usually laid from July to September, at Rottnest Island during September, while on the south coast, October and November are the usual months.

From studies at Rottnest Island between 1961 and 1963 W. N. Holsworth found that the three pairs nesting there remained at their nests between July (when fresh material is added to the nest structure) and January. Though three eggs are laid, usually only two young are reared; the incubation period is about 36-37 days, and the young remain in the nest about 50 days. The young (which were ringed by the investigator) dispersed widely from Rottnest during their first year. One was shot 70 miles to the north, at Lancelin, and two others on the south coast, one near Denmark and the other 50 miles north-east of Albany.

The Osprey not infrequently perches on the masts of fishing boats and other vessels. Rottnest Island is a regular haunt. It rarely ventures inland but is occasionally seen on the Swan River estuary, as far as the Causeway. In the Kimberley Division, however, it does ascend the rivers; F. Lawson Whitlock has seen it far up the Fitzroy River and C. F. H. Jenkins some 60 miles up the Ord River. The bird's call is a plaintive "pee-ee! pee-ee!"

FALCONS and KESTRELS: Family Falconidæ

Peregrine Falcon (Black-cheeked Falcon)

Falco peregrinus

Native names: Gwet-ul-bur, Gwetalbur (Perth, Avon); Moldern (Pallinup River).

In the fully adult plumage the upper parts are slate-blue, finely barred black. Tail grey barred black, with slight white tipping. Head, neck, cheeks and wing quills slaty-black. Breast clear chestnut-brown, abdomen chestnut spotted black, flanks, thighs and under wing coverts light chestnut narrowly barred black. In immaturity the feathers of the back are blackish-brown, each tipped rufous; the whole of the under parts are light rufous-buff, longitudinally streaked dark brown. The head and cheeks are blackish-brown. Iris, brown; beak, slaty-blue, black at tip; cere, greenish-yellow; legs, yellow. Length, male, 14¼-15¼ in.; wing span, 32 in.; female, length, 18½ in. Weight, male, 15 oz.; female, 2 lb. 2 oz.

Distribution: The species ranges throughout the State.

Nesting: No nest is constructed by this falcon, its 2 or 3 (rarely 4) eggs being laid either in a hollow in a tree or on a ledge of rock on a cliff-face. The eggs are buffy-white and are handsomely marked with blotches of reddish-brown, which tend to form a cap at one end, or even a band around the minor axis of the egg. Average measurements of 13 Western Australian eggs, 53 x 41 mm., varying from 51-55 x 40-42 mm.; weight of egg-shell 4.0 gm.

Though distributed over such an extensive area, the species occurs sparingly anywhere and is localised in the region of mountainous country and cliffs. It is pre-eminently the falcon of crags and precipices. It has been most frequently reported from the Stirling Range and also occurs on the high granite islands of the Archipelago of the Recherche. Occasionally it appears in the Perth area and other districts far from its typical habitat. T. E. Bush considers it is spreading throughout the Wheatbelt, using old nests of ravens for nesting purposes.

This falcon hunts rabbits, ducks and similar game. The species has a wide distribution elsewhere and in Europe is one of the premier species used in the sport of falconry.

Little Falcon

Falco longipennis

Native names: Perg-gee (North-West Cape); Wow-oo (Murray River).

A smaller edition of and generally similar in appearance to the Peregrine Falcon. The upper parts are dark slaty-grey with no barrings on the feathers but instead have the central shafts dark, giving an inconspicuously streaked appearance in very close view. The head, slaty-black; throat and forehead, white; under parts rufous-brown, with streakings on the breast. There is a whitish collar at the sides of the neck. No grey on the flanks and thighs. Under wing coverts rufous, spotted lighter and not barred, and the flight quills grey with rufous-buff oval spots or bars. Immature birds are dark brown above with buff edgings to the feathers, and the under parts are an almost uniform rufous-brown, in contrast to the Peregrine in this state of plumage. Iris, brown; beak, grey tipped black; cere, greenish-yellow; legs, yellow. Length, male, 12 in.; female, 13-14 in.

Distribution: Found all over the State.

Nesting: The Little Falcon, for nesting purposes, may adopt the old nest of another bird, or a new nest may be built. When a new nest is built, it is constructed of dry twigs, lined with leaves or bark and placed high up in a tree. The clutch consists of 2 or 3 eggs, although up to 4 are sometimes laid, and are, except for being much smaller in size (46 x 34 mm.), very like those of the Brown Hawk, being buff-white, spotted and blotched with varying shades of red. Sometimes the markings tend to form a cap over one end of the egg. When viewed through a hole in the side of the egg, the inside of the shell is seen to be buffish.

The female, during incubation of the eggs, leaves the nest to be fed by the male and takes food from him in flight.

This is a comparatively common species and may appear around poultry runs and pigeon lofts, being one of the hawks which harries the Zoological Gardens, South Perth. Some observers consider it to be the swiftest of the falcons when it makes its lightning dash, with half-closed wings, at its quarry. This may be a bird much larger than itself, such as a duck or a cockatoo, but it feeds considerably on insects, including grasshoppers. It can hover to some extent, but not so skilfully as the Kestrel or the Black-shouldered Kite.

Grey Falcon

Falco hypoleucus

Native name: Boorh-ga (Moore River).

Upper parts *blue-grey;* under surface light ash-grey; wing quills black. Owing to the short tail the bird appears thicker-set and smaller than the Brown Hawk. In the distance it may appear white, particularly from below when only the wing tips look black. Iris, brown; cere and around eye, pale orange; beak, yellow base and black tip; legs, orange. Length, female, 17-17¼ in.; male, 12-14 in. Wing span, 3 ft. 2 in. (female); 2 ft. 10 in. (male). Weight, 12 oz. (male).

Distribution: Of general occurrence, except in the South-West forested area.

Nesting: This species of hawk appropriates an old nest of another bird, often that of a crow, for nesting purposes, and the only alteration made to the nest is the lining of the egg cavity with soft material, such as hair, wool or shredded bark. The clutch consists of 3 eggs, sometimes only 2 and on occasions as many as 4, and they are, except as regards size, very similar to those of the Brown Hawk. They are blunt-oval in shape and almost completely covered with spots and blotches of reddish-brown, the ground colour, a pinkish-buff, being more or less obscured. Size, 49-53 x 36-38 mm. November appears to be the chief nesting month.

Apart from the two *Elanus* species this is the only grey or white-looking hawk in Western Australia—other than in the Kimberley Division where the dimorphic Grey (White) Goshawk

(Accipiter novæhollandiæ) is found. The species is an inoffensive one from the human point of view and feeds on rabbits, mice, reptiles, birds and insects. It does not molest domestic stock.

Black Falcon

Falco subniger

Entirely sooty black, both above and below, with a greyish-blue bloom to the feathers, their concealed bases being white. There is no barred or streaked pattern as is usual with hawks. Iris, brown; beak, dark bluish-grey, paler at the base; naked skin around the eye and cere, pale bluish-white; legs, pale blue-grey. Length, 21¼ in.; male is slightly smaller.

This species, crow-like in its uniform duskiness, can be recognised from even the darkest phase of the Brown Hawk by its larger size and more powerful beak and feet and by the absence of any trace of a pattern on the feathers, including the absence of the round white spots on the wing quills.

Distribution: One of the rarest of our hawks and the existing records of occurrences are insufficient to indicate the geographical range of species in this State. The only specimen ever collected in Western Australia was taken at Katanning in March 1946, but there are sight observations from Moora, Ninghan Station (seen by G. M. Storr in September 1959), Wembley, near Perth (seen by Dr. Storr in January 1967), Wiluna, Ethel Creek, and the Kimberley Division.

Nesting: The Black Falcon takes possession of the abandoned nest of another species of hawk or of a crow and uses it without making any alteration. The clutch varies from 2 to 4 eggs, 3 being the usual number. Ground-colour buff, well sprinkled with reddish-brown and purplish-brown spots. Size 54 x 40 mm. The nestling is covered in grey down which is hair-like on the head and neck. The bill is leaden in colour; the eyes hazel and the legs are yellowish-blue.

Brown Hawk

Falco berigora

Native names: Carra-cal-lu (Pilbara district); Kerra-jinga (North-West Cape); Don-don (South-West); Kargyne (Perth and Avon River); Worgener (Pallinup River).

Upper parts dark brown with the inner margins of the wing feathers broadly spotted and toothed with rufous. Tail barred with rufous. Throat and abdomen very light brown and the breast dark brown. There is a *black facial streak* from the base of the beak to below the eye. Under wings and tail barred light grey and light rufous. Iris, brown; beak, bluish-grey, naked skin around eye pale blue; legs, pale blue. Length, female, 18 in.; male, $15\frac{3}{4}$-$17\frac{1}{2}$ in.; wing span, female, 38 in. Weight, female, 1 lb. 4 oz. - 1 lb. 8 oz.; male, $13\frac{1}{2}$ oz. - 1 lb. 3 oz.

There is considerable variation in the colour of this hawk; very light birds may be entirely white below except along the flanks, while in the extreme dark phase the entire under parts may be dark brown or black, and there is every stage in between. These variations do not appear to be geographically correlated, but inland an attractive rufous form occurs in which a very ruddy tone is present on both the upper and lower plumage—this is the race *centralia.*

Distribution: The Brown Hawk is to be found all over the State and is perhaps our commonest bird of prey.

Nesting: This bird never builds a nest of its own, but uses old nests of magpies and crows, or of wattle-birds. Also it sometimes uses hollows in trees, in which however no nest is built. The eggs, 2 or 3 in number, are rather like those of the Peregrine Falcon, being buffy-white with blotches of reddish-brown. Average size of 20 eggs, 49 x 38 mm., varying from 47-52 x 37-40 mm.

In recent years the Brown Hawk decreased markedly in the area around Perth, but still remaining plentiful in the remainder of the State. In 1972 there was some evidence that it was increasing again. Though frequently shot when it appears around habitations, it is no menace to the poultry keeper. In its feeding habits it is a useful hawk, subsisting on grasshoppers, beetles and centipedes, varying its diet with snakes, lizards, rabbits and other small animals. It possesses some hovering ability, but this is much more limited than that of the Kestrel.

Kestrel

Falco cenchroides

Native names: Bin-birri (Pilbara district); Meginma (Lyons River); Windee windee (Jigalong); Migwin (Wooleen); Martiet (Pallinup River).

Above a beautiful *pale cinnamon-brown,* each feather marked with black in the centre. Wing quills dark brown. The tail is grey with a black subterminal band, the tip being white. In the female it is a reddish-brown with several narrow bands of black. The *under parts are white,* with the breast feathers having narrow brown shaft lines. There is a slight brown shading below the eye. Iris, brown; cere and skin around the eye, yellow; beak, bluish-grey; legs, orange-yellow. Length, $12\frac{1}{4}$-$13\frac{1}{2}$in.; wing span, 28 in. Weight, 5-6 oz.

This hawk may be recognised almost at once by its *hovering habit,* it having the ability to remain poised at a spot in mid-air for an appreciable period. The only other hawk of equal ability in this respect is the Black-shouldered Kite, which, being white below, may be at a distance difficult to tell apart from the Kestrel unless the chestnut back of the latter comes into view.

Distribution: Found all over the State and rivalling the Brown Hawk in general abundance.

Nesting: The Kestrel makes no nest but lays its 4 to 5 eggs usually in hollows or spouts of trees or in crevices in cliffs, and exceptionally in old nests such as those of crows, if hollows are not available. It occasionally even chooses a building; in 1926 a pair nested in the top of a drainpipe at Warwick House in St. George's Terrace, Perth, and about the same time, for several seasons, another pair nested in the tower of St. Oswald's Church, Swanbourne. In 1971 a pair nested in a turret of the Perth Town Hall, Barrack Street. In the Nullarbor Plain E. McCrum found Kestrels nesting underground inside the blowholes, one nest being 15 feet below the surface. The rather roundish eggs are white, blotched with brownish-red, and are sometimes so heavily marked that there is little of the white ground colour showing. Average size of 30 eggs, 38 x 31 mm., varying from 35-41 x 29-33 mm. The nestling is covered with white down. Iris, red; beak, whitish; legs, yellow.

The Kestrel is a denizen of open country and is particularly plentiful along the coastal dunes in the South-West. It is essentially an insect-feeding hawk and also follows mouse plagues. In the Nullarbor Plain McCrum found that the main diet comprised dragon lizards, dragonflies and butterflies. In November 1968 a bird was killed by collision with a plane at the Guildford airport at 200 ft. It weighed 154 gm. and was carrying in its talons a jew lizard $11\frac{1}{4}$ in. long and weighing 45 gm. In the bird's stomach was a small skink and several grasshoppers and tree-crickets.

FOWL-LIKE BIRDS: Order Galliformes
MOUND-BUILDERS: Family Megapodiidæ

Mallee-Fowl (Gnow)

Leipoa ocellata

Native names: Gnow, Ngow-o (general); Nganamara (Warburton Range).

Back brown with black-and-white cross-bars to the feathers, tail tipped white; crown of head black; neck slate-grey. Throat pale brown; under parts white, with a prominent black stripe down the centre of the upper breast. Iris, light hazel; beak, slate-black; legs, blue-grey. Length, 22-27 in. Weight, 4-$4\frac{1}{4}$ lb.

Distribution: Found in open country (mallee, sandplain) south and west of a line from about Cape Farquhar (north of Carnarvon) to Kathleen Valley, Carnegie and the Warburton Range. Within this area, however, it is absent from the South-West forested belt, except in the narrow coastal fringe around the south coast, having been recorded from the mouth of the Warren River, Capes Leeuwin, Mentelle and Naturaliste, and near Busselton (Fig. 8).

Nesting: Instead of incubating their eggs themselves, Mallee-Fowls excavate a hole on a sandy patch about 3 feet deep and 6 feet across; they then scrape a heap of sand and dry vegetable matter into a large mound, measuring up to 15 feet across and 2½ feet in height. The eggs are laid in the mound, which functions as a natural incubator. The eggs, when freshly laid, are a delicate pink, but after being in the mound for a few days assume a darker hue, becoming almost brown. Occasionally white eggs are due to the thin pigment layer scraping off, exposing the shell. They are thin in the shell and are large for the size of the bird. A series of 21 eggs from various localities between the Stirling Ranges and Hamelin Pool average 91 x 60 mm., varying from 88-102 x 56-63 mm.

The most comprehensive study of the extraordinary breeding behaviour of this species was made by H. J. Frith in the mallee at Griffith, N.S.W. The birds are engaged at the mound for the greater part of the year, the mound being the centre of the male's territorial activities and he never moves more than 100 yards from it. The female wanders over a slightly greater area. During the short period of mound abandonment, between breeding seasons in the autumn, adjacent pairs may briefly associate together at common food supplies. The birds begin constructing the mound early in the winter but egg-laying does not start until temperature maintenance (at about 93° F.) becomes possible. A constant temperature is maintained by the male by continuous activity, the nature of which alters with the progress of the seasons. The heat in the mound is due essentially to the fermentation of organic matter buried within the mound, and during the earlier part of the season the male's chief occupation is to keep the temperature low by heaping the mound high and preventing solar heat entering. By mid-summer solar heat is the chief source of heat but the amount still has to be controlled by mound manipulation. In late summer, with fermentation ceased, the birds remove soil to allow sufficient solar heat to penetrate and warm the eggs. As autumn advances, and this becomes increasingly difficult, the mound is abandoned. The male tests the temperature of the soil by means of his tongue or some other part of the mouth cavity.

The female lays the eggs at varying intervals, averaging 6 days between each, in niches that she digs in the wall of the central egg chamber after it has been opened out by the male. Once the egg is laid and covered up the parents ignore it entirely and fail even to recognise the resulting chick as theirs. Laying begins early in September and continues until the end of December, but in some seasons may extend until the end of February. The average number laid by a female in one season varies between 15 and 24, but may be as few as 5 or as many as 33. The earliest eggs will have hatched by the time the later ones are laid and there are never more than 14 in a mound at any one time. The average incubation period is 62 days, but varies greatly, from 50 to 90 days, depending on the temperature level maintained in the mounds. When the chick hatches it struggles through the covering soil, pauses for several minutes at the surface, then staggers down the side of the mound to the shelter of the nearest bush. When hatched it is clothed in down except for the wings, which are feathered. It leads an independent, solitary life henceforth and does not associate with its parents or fellow chicks.

The only comparable observations in Western Australia were those of Bruce W. Leake, of Kellerberrin. He noted that the birds began to attend their mounds at the start of the early winter rains and during the next few months scratched together rows of leaves and other plant debris. Near the end of August these were raked into the mound and covered with sand. As this material became heated egg-laying began towards the end of September. Mr. Leake also discovered that solar heat, later in the season, replaced the heat due to plant fermentation as the warming agent in these natural incubators.

In 1925 Mr. Leake conducted an interesting experiment by constructing an artificial nesting mound on his property at Wooloundra and, after netting it around, transferred to it eggs from a mound made by wild birds. These duly hatched and the young were kept under observation for some time. Mr. Leake found that the bird made three different calls. One, which was uttered by the young, was soft and low. The second was likened to "a human being belching in the throat backwards." The third was a boom like a tame pigeon, but much louder and deeper, and was uttered with the body expanded and the head bent right down near the ground and almost between the legs. This call was not made until the birds were nine months old. Frith describes a similar call and interprets it as the territorial call of the male, delivered in defence of his mound area.

As mentioned in the historical chapter (Section I) the nesting mound was described some time before the bird itself was obtained. In 1832 the Surveyor-General, John Septimus Roe, observed some nesting mounds during an expedition east of Northam. It was not until January 1837 that the bird was described by James Drummond in a letter to the short-lived newspaper *The Swan River Guardian* three years before John Gould gave it its scientific name. In January 1838 the visiting naturalist, James Backhouse, correctly noted that east of York the "Bush Turkey" as he called it, "hatches its eggs in hillocks of sand." In the *Perth Gazette* of April 10, 1857 reference was made to "the extraordinary numbers of the 'Tumulus' building bird, the Australian Pheasant, called by the natives 'Gnaw' which have appeared on the back sheep runs during summer."

During the last three decades there has been a significant increase in the numbers of Mallee-Fowl in the South-West of Western Australia and the birds have been reported from a wide area in the Wheatbelt whence they had hitherto been considered to have vanished or become very scarce. Such reports have been received from near the mouth of the Murchison River in the north to Esperance in the south-east and from York in the west to Norseman in the east. Active mounds at the Dryandra Forestry Station (90 miles south-east of Perth) are now regularly visited by naturalists. Mallee-Fowl are often seen at roadsides feeding on wheat spilled from cartage trucks. The natural food consists of ground-living insects and the seeds and fruits of shrubs, including *Acacia*.

No information is available on local population density. However, at Griffith, N.S.W., Frith has found a density as high as one pair to 45 acres in favoured country and the lowest density in his general study area was one pair to each 100 acres. Contrary to popular opinion he has found that the introduced fox is not the main cause of the decline in uncleared areas, and that it is more probable that sheep, and perhaps rabbits, enter into direct competition with the birds for food.

The Mallee-Fowl has been protected generally throughout the State since 1930.

PHEASANTS and QUAILS: Family Phasianidæ

Stubble Quail

Corturnix pectoralis

Native name: Barraberry (Perth).

Upper parts boldly marked black, brown and light buff, and *streaked with large, spear-shaped white streaks.* In the female the throat is almost white, the breast with *longitudinal dark brown and buff streaks* and the abdomen buff-white. In the male the throat is *light chestnut* with prominent *black blotches on the breast;* abdomen buff-white. Iris, reddish-brown; beak, blackish-horn; legs, pale flesh. Length, male, 7¾ in.; female, 7 - 7¼ in. Weight, male, 3½ oz.; female, 4 - 4½ oz.

Distribution: The regular distribution is mainly in the South-West, north to the mouth of the Murchison River, and east to Beacon, Merredin and Lake Grace. Recently there have been a number of records much further north, near Carnarvon (May 28), Shark Bay (end of August), Point Cloates (November 9) and La Grange (May 27), and also eastwards to Yalgoo (June 30). Perhaps a seasonal migration may be involved. The species is known to be subject to wandering movements and some years may pass between visits to particular localities.

Nesting: For its eggs the Stubble Quail works a hollow in the ground, which it lines with grass or other vegetation growing in the locality in which the nesting occurs. If the nest is in growing crop, such as oats or wheat, then the nest is lined with the straw of those plants; if in clover, then clover stems are used. The clutch comprises 8 to 14 eggs in the lower South-West; in the Lake Grace district the number is 5 to 8 (I. C. Carnaby). They are yellow, with many spots and blotches of dark brown. Sometimes spots predominate; at other times the markings are mainly blotches which are often large in size. Average dimensions of 26 eggs measured by us, 31 x 23 mm., the range being 28-32 x 21-24. In the Bridgetown district the species nests in November and December; in the Lake Grace district the period is September and October.

The call is a sharp double note. This is the common quail of the South-West, but under natural conditions quail are not abundant in the south owing to the scarcity of indigenous grasses. Cultivation and pasture improvement have extended their habitat.

The Stubble Quail is the only species of quail which may now be legally hunted. The open season is from December 1 to March 31 north of latitude 31° S. (passing approximately through Lancelin Island and New Norcia) and from March 1 to June 30 south of this line.

Brown Quail

Synoicus ypsilophorus

Native names: Moo-reete, Murit (South-West), Purinderoo (Well 45, Canning Stock Route).

The colour of the back is variable, from rich bright brown to dark brown, streaked with narrow white lines and black markings on the feathers. The under parts are buff to grey, with *broken black* cross-barrings, which are absent from the throat. Iris, red; beak, slaty-black; legs, olive-green. Length, 7 to $8\frac{1}{4}$ in.; wing span, 12 in. Weight, male, 3 to $3\frac{3}{4}$ oz.; the female is larger than the male.

This species approximates to the Stubble Quail in size and may be distinguished by the lack of coarse white streaks on the back and by the ventral pattern being transverse, not longitudinal as in the Stubble Quail. Specimens in the hand can be positively distinguished by the shape of the wing. In the Stubble Quail the *pointed wing tip* is formed by the first and second primaries, which are equal in length, and by the third which is only slightly less. The secondaries are short and reach to the level of the ninth primary only. In the Brown Quail the wing is *more rounded,* with longer secondaries reaching as far as the seventh primary and the *broad wing tip* is composed of the first four primaries, which are approximately equal, and the slightly shorter fifth. This and the

Stubble Quail are the only local "quail" in which the *hind toe is present.*

Distribution: The Brown Quail occurs generally throughout the State in suitable localities, but our knowledge of the distribution of most of the species of quail is defective. Thus in the South-West corner the Brown Quail, though apparently plentiful around the south coast, has never been satisfactorily identified in the Perth district. It has been reported at Moora, Northam, York, Harvey (nesting recorded by W. H. Mathews), Bridgetown and the south coast, including some of the islands, namely Breaksea Island in King George's Sound and the Archipelago of the Recherche. Carter recorded that it was fairly common at Point Cloates after rains. It occurs on some of the islands off the North-West coast.

Nesting: The Brown Quail, like the Stubble Quail, lays a considerable number of eggs—up to 11; on one occasion F. Lawson Whitlock found a nest with no fewer than 14 eggs, all of which hatched. Like the Stubble Quail, the Brown Quail makes a slight nest of grass and leaves in a depression in the ground which it probably scratches out itself. The eggs are pointed at one end and are dull white, more or less spotted finely with light brown. They measure 28-32 x 21-23 mm. C. L. E. Orton informed us that he once found a nest at Moora, in October, which contained 7 eggs of the Brown Quail and 4 of the Stubble Quail. Incubation period, 21 days. The young fly in 10 days after they are hatched.

The call has been syllabised as a whistling "ffweep," and a whistling "gopwaarr," rising towards the end (E. Lindgren).

CRANES, RAILS, Etc.: Order Gruiformes
BUSTARD QUAILS: Suborder Turnices
Family Turnicidæ

Painted Quail

Turnix varia

Native names: Moo-ro-lum, Murolang (coastal), Nani (Avon district).

Upper parts a beautiful chestnut and black marbling, with narrow white streaks. On the crown and neck are black and white spots and on the sides of the neck bright chestnut. The throat is white, breast grey with light buff spots, and the abdomen buff. Iris, red; beak, blackish; legs, yellow. Length, males, $6\frac{1}{4}$-$8\frac{1}{4}$ in.; female, $8\frac{1}{4}$ in. Weight, male, $2\frac{3}{4}$ - 4 oz.; female, $4\frac{1}{2}$ oz.

Bustard Quails have *no hind toe.*

Distribution: South and west of a line joining Shark Bay (Denham), Northampton, Coorow, Wongan Hills, Broomehill, Lake Grace, and Dempster Inlet. The species occurs also on the Abrolhos Islands (North Island and the West and East Wallaby Islands).

Nesting: The nesting habits of the Painted Quail are peculiar. The female does all the courting, indulges in display and constructs the nest, but beyond that her part in the rearing of the young is limited to the laying of the eggs; the male alone carries out the incubation and the caring for the young. During the breeding season the female utters, usually at dusk and during the night, a characteristic "booming" note, "oom! oom! oom!" which is rather like the call of the Common Bronzewing Pigeon and can be heard a considerable distance away. It is made, apparently, to attract the male, and when he appears the female engages in display, running round him in a circle, with tail erect and chest puffed out, often stopping and "booming" at him, while stamping her feet and scratching at the ground. The female makes a shallow saucer-shaped depression in the ground under the shelter of a small shrub or tuft of grass, by twisting round on her breast with her tail erect, at the same time pulling in a few pieces of grass to form a nest. During this time she makes a crooning sound and the male answers with a similar call. The clutch is of 4 eggs (3 only on the Abrolhos Islands). They are whitish-green, freckled all over with brown spots and purplish blotches; 17 eggs laid by one female in a single season in one of our aviaries average 28 x 21 mm. in size, the range being 26-30 x 20-22 mm. Incubation takes 14 days and once the eggs are laid the female takes no further interest in that particular nest; but, as she continues her "booming," she apparently looks for a second mate. The breeding season appears to be prolonged—July to October.

The Painted Quail is distinctly an insect-eater, some kept in captivity by us being so greedy for such food as to take moths, grasshoppers, beetles, etc., from our fingers. The species is probably of considerable value to the farmer, except that it does not usually frequent cultivated land, its normal habitat being the poorer type of bush country.

Little Quail

Turnix velox

Native names: Malpuntari, Punpukara (Warburton (Range); Kar-a-dong (Avon).

Upper parts like the Painted Quail, but brighter rufous. Throat white. *Upper breast rufous with faint cross bars of white.* Abdomen white. Iris, pearl-colour; beak, culmen and tip leaden-brown, remainder leaden-blue; legs, flesh-colour. Length, 5¾ in. The male is smaller and is more streaked above, including a median white line on the head.

As in the Painted Quail there is *no hind toe.*

Distribution: This small quail occurs all over the State.

Nesting: The nest is a slight depression, lined with grass, in the ground under the shelter of a tuft of grass. The 4 eggs are oval and yellowish-white, thickly spotted and blotched with varying shades of brown. A series of 20 eggs from the North-West averages 23 x 18 mm., varying from 22-24 x 16-19 mm. As this is a Bustard Quail, the nesting habits are similar to those of the preceding series. Incubation period 13 days (A. H. Lendon).

Little Quails are the common quails of the North-West and are especially numerous after good rains, in open grass and spinifex, where, as T. Carter expressed it, they may be seen running about like mice from clump to clump. They are rare in the South-West and only irregular visitors to the Perth area.

The call has been described as a loud "coo-oo" or a low, deep "oo-ah," uttered continuously and frequently at night.

CRANES and RAILS: Suborder Grues

CRANES: Superfamily Gruoidea

CRANES: Family Gruidæ

Brolga (Native Companion)

Grus rubicunda

General body colour silvery-grey; wing, dark brown; naked red skin at side of face and back of neck; crown of head, olive-green. Iris, yellow; beak, olive-green; legs, dark brown. Length, 3 ft. 4 in.-4 ft. 3 in.; wing span, 7 ft. Weight, 9½-14 lb.

Distribution: The Brolga is a bird of the Kimberley Division but it has now penetrated into the North-West and occurs regularly as far south as Onslow. It would appear that this is a more or less recent extension of range as the species was never recorded in earlier days in the North-West by resident and visiting ornithologists, such as Tom Carter, F. Lawson Whitlock and A. H. Robinson. Apparently the birds first appeared in small numbers in the 1920s (according to A. H. Clark, of Onslow). There was a very large invasion in 1952 (see p. 6) when some birds were seen on the Carnarvon Road near Hamelin Pool, and one bird even as far south as Beacon, in the north-eastern Wheatbelt.

Nesting: Two eggs comprise the clutch and they are laid on the bare ground, no nest being made. They are creamy-white, with a few spots of brown and lavender; the surface is covered with minute pits. Size, 84-91 x 60-62 mm. There is no record of breeding south of the Kimberley Division.

The birds associate in small companies of up to six individuals and are noted for their habit of indulging in graceful dancing ceremonies or "corroborees." These dances are characteristic of cranes generally and are regarded as an "expression of general liveliness or excitement" and not primarily concerned with sexual displays. The call is a hoarse croaking or trumpeting.

RAILS: Superfamily Ralloidea

RAILS, CRAKES and WATER-HENS: Family Rallidæ

Lewin Water Rail

Rallus pectoralis

Back streaked black and brown, head and nape streaked rufous and black, wings black barred narrowly with white. Chin pale grey, breast grey-bronze, abdomen like the wings narrowly barred with white. Iris, brown; beak, brownish-red; legs, flesh colour. Length, $8\frac{1}{4}$ in.

Distribution: A South-Western species whose more precise distribution cannot as yet be stated; it is known to occur in the Bridgetown district.

Nesting: The nest is a saucer of rushes and grasses built among rushes in a swamp, the surrounding rushes being drawn together over the nest to form a canopy, others being trampled down to form a pathway to the nest. The 4 to 6 eggs are pale stone colour marked with spots and blotches of red and lavender-grey. They measure 36 x 28 mm.

The call is a "cree-eek" repeated loudly two or three times.

Banded Landrail

Rallus philippensis

Native name: Kul-lee (Perth).

Back, wings and tail a *greenish-brown conspicuously spotted with white* and marked with black; grey stripe over the eye. Nape a rich chestnut and the head streaked brown and black. Chin white, throat grey, under parts finely barred black and white, except for a chestnut band about half-an-inch wide which passes across the breast. Iris, brown to red; beak, brown; legs, brown. Length, 11-$12\frac{1}{2}$ in. Wing span, $17\frac{3}{4}$ in. Weight, 5 oz.

Distribution: Generally distributed, but apparently no further north than North-West Cape and Port Hedland. Like many rails it has great wandering tendencies and individuals are not infrequently seen on islands off the coast. It has been noted at the Abrolhos Islands (Pelsart and Rat Islands), at Sandland Island (near Jurien Bay) and Lancelin Island, and at the Recherche Archipelago (Christmas Island).

Nesting: The nest is usually constructed in a hollow in the ground under the shelter of a tussock of grass. It is a solid shallow cup of grasses. The 5 to 9 eggs are pinkish-white with a few blotches of red and purple and some underlying ones of purplish-grey. Size, 32-36 x 24-27 mm. Incubation lasts 18 to 19 days.

Corncrake (Landrail)

Crex crex

Upper parts *yellowish-buff* streaked darker brown, *wings chestnut;* grey stripe over eye. Under parts paler, with the sides of the body chestnut barred whitish. Iris, brown; beak, pale brown, tipped darker; legs, pale flesh. Length, 10-11 in.

It can be distinguished from the preceding species by its predominantly buff colouring, absence of white spottings, and chestnut wings. It may be looked for in the same type of habitat, in low scrub, rough grass and the rushes at the edges of swamps.

Distribution: An occasional wanderer from the northern hemisphere. The Corncrake nests in Europe and western Asia as far east, probably, as the Lena River and the Altai Mountains. It migrates in the northern winter to Africa and south-western Asia, and has on very rare occasions been found in Australia and New Zealand.

The only record for Western Australia is of a bird which flew on to an American military transport on December 9, 1944, when the ship was off the coast about Latitude 30° S. (off about Jurien Bay). The only other specimen known for Australia was shot at Randwick, Sydney, in June 1893. One was taken near Nelson, New Zealand, about 1865.

There is one record of another northern hemisphere crake, the Malay Banded Crake *(Rallina fasciata),* at Broome, outside of the area dealt with in this book.

Australian Spotted Crake

Porzana fluminea

Back dark bronze-green longitudinally *streaked* with white; chest and breast dark slate-grey; *abdomen and flanks black cross-barred with white. Under tail white.* Iris, reddish-brown; beak, green, with the upper mandible red at the base; legs, green. Length, 6¾-8¼ in.

Distribution: A South-Western species occurring as far north as the bore overflow at Hamelin Pool. Another population occurs in the Kimberley Division (recorded at Derby by P. Slater).

Nesting: The nest is an open, flat platform and is usually constructed just above water in a clump of grasses or rushes in a swamp. The grasses or rushes are trampled down in one direction so that a track or platform is made by which the incubating bird can reach the nest. The 4 to 5 oval eggs are greyish or olive-brown, spotted with various shades of purplish-brown and red. Size, 30-32 x 23 mm. Incubation period, 17 days (P. Slater).

Marsh Crake

Porzana pusilla

Back dark olive *streaked with black and some white streaks.* Breast pale blue-grey; *abdomen and under tail black cross-barred with white.* Iris, red; bare skin around eye, dull orange; beak, dark brown, base, pea-green; legs, green or olive-brown. Length, 6½-7 in.

Distribution: Occurs all over the State in suitable localities.

Nesting: The nest is to be looked for in a clump of rushes growing in a swamp, and is a saucer-shaped structure of about four inches in diameter, built of rushes or other aquatic herbage. As with most other crakes, there is usually a staging trampled to the nest. The 4 to 6 eggs are dark olive, with spots of a still darker shade of the same colour. Size 25-28 x 19-20 mm. The nestling is covered with greenish-black down.

The call notes are a harsh "krek-krek."

Spotless Crake

Porzana tabuensis

Native names: War-ra-jah, Warraja (Perth).

An even-coloured bird with *no conspicuous barrings, stripes or special markings.* The under tail coverts are slightly barred and spotted with white. The upper parts and wings are dark bronze-brown, the head and under parts grey. Iris, red; beak, black; legs, long, dark brown or reddish. Length, 7¾-8¼ in.

Distribution: Fresh water swamps in the South-West corner, north to Hutt Lagoon (Utcha Swamp) and south-east to Bremer Bay. It may also occur on islands off the coast and has been recorded at the Abrolhos Islands, North Fisherman and Sandland Islands (near Jurien Bay) and Eclipse Island (near Albany).

Nesting: The nest is cup-shaped, made of grasses and placed in a clump of rushes growing on the edge of a swamp, and often in rushes growing actually in water. Some grasses or rushes are bent over to form a platform on which the nest is built. When built in rushes the nest is sometimes attached to them rather like the nest of the Reed-Warbler. The 4 to 5 eggs are greenish-grey, light stone, biscuit or straw-colour, spotted and blotched to a varying degree, but usually not at all heavily, with chestnut and some underlying ones of lavender. Average measurements, 30 x 22 mm. The nestling, like that of the Marsh Crake, is covered with greenish-black down.

This is the commonest crake in local swamps, but its shy, skulking habit prevents it being often seen. As often as not its local presence is first revealed by the domestic cat bringing a specimen home. The call is a "rhee-ee." The species has a strong wandering and colonising tendency and in the Pacific archipelagos it has come to inhabit a large number of islands in environments often strangely different from those of the mainland swamps. In this State it has settled on Pelsart Island in the Abrolhos, where it may not infrequently be seen foraging on the ground beneath the dense mangrove growth fringing the lagoons, and it also ranges into the dry, scrubby, knee-high salt-bush thickets miles from these situations. It has also been seen on a rocky islet near Rat Island, where there are neither mangroves nor water, and on Beacon Island in the Wallaby Group. At Eclipse Island, off Albany, it inhabits a sterile stony location having no standing water.

Black-tailed Native Hen (Gallinule)

Tribonyx ventralis

Native names: Chun-din-di (Coolawanyah); Chandilla (North-West Cape); Noondeegannan (Jigalong); Thannimidi (Pidong tribe, East Murchison River); Thullundee (Wooleen); Bel-gar-be-jal (Perth); Nol-yang (Avon River).

Upper parts dark bronze-brown; throat and breast bluish-grey; tail black; abdomen black, *spotted white on the flanks.* Iris, golden-yellow; beak, upper mandible and forehead shield, dark pea-green; lower mandible, orange-red, green towards tip; *legs, brick-red.* Length, 12-15 in.; wing span, 24-25 in. Weight, 13½ oz.

The tail is carried elevated.

Distribution: Generally distributed throughout the State, but unusual in the lower South-West.

Nesting: The nest is to be found in swampy situations. It is saucer-shaped and is usually well hidden in grass or among bushy herbage close to the ground or just above water. The 4 to 8 eggs are dark green and are marked, sometimes only very sparsely, with blotches and small spots of chestnut and purplish-brown. They measure 43-46 x 29-32 mm. The nestling is covered with greenish-black down and has the beak coloured as in the adult.

This native hen is a well-known nomad in the northern and inland parts of the State and numbers suddenly arrive in a district after rains and may disappear just as suddenly. In the South-West it is rarer than the other species but a few appear on some of the swamps each year and may nest. Sometimes, however, very large irruptions occur. A famous one in the Perth district happened shortly after the colony was founded. Governor John Hutt, reporting the event to John Gould, stated that "this bird invaded the settlers' fields and gardens in the month of May 1833 in amazing numbers; it had not been seen before, and has hardly been seen since." John Gilbert, who visited the colony a few years later added that the birds trod down and destroyed whole fields of corn in a single night and that the natives, "not having seen them before, attributed their appearance to the settlers, and for a long time termed them the 'white-men's birds'." Fletcher Moore recorded that thousands were shot and consumed for food. The *Perth Gazette* recorded another visitation in May 1835, stating that "their ravages are generally confined to the banks of the river, and we have seen some acres cleared by them as effectually as though a scythe had been passed through the ripening corn." In August 1842, Gilbert found the birds plentiful near Moora. No further invasions were reported until the winter of 1853, but details of its extent are lacking. In April 1858, the explorer F. T. Gregory "met with large flocks of the Gallinule, which have for so many years excited the curiosity of the colonists as to their habitat," on the Murchison River, near Mounts Murchison and Narryer, and later encountered them on the Gascoyne. Writing in August, 1865, G. E. Egerton-Warburton, of St. Werburgh's, near Mt. Barker, referred to a visitation of a strange bird, "like the English water hen" and "which I suppose comes from some other part; driven by drought."

In the winter of 1872 there were devastating floods in the Midlands causing the Moore River to rise higher than ever known before, New Norcia suffering considerably. Rica Erickson, in her book, *The Victoria Plains,* stated that "in the wake of the floods thousands of waterhens invaded the district, feeding voraciously on the young grass and such crops that had escaped the scouring of the waters. Men and children were employed all day in scaring them away from the fields." The birds were reported in the Williams district later the same year, in October. The next visitation of which we have record took place in 1886, when "vast numbers of these birds invaded the Geraldton district, doing great damage to the crops. In the same year, there were numbers to be found at South Perth and on the Swan River, just above the Bunbury Bridge" (A. W. Milligan). T. Carter recorded that in June 1897, when a flood broke up the great drought then prevailing, there were "countless thousands" of the birds in the immediate vicinity of Geraldton, doing great damage to the crops. A visitation in some strength occurred in the South-West, including the Perth area, in April and May 1919. The birds invaded Queen's Gardens and Hyde Park in the city. The irruption was noticed at Kellerberrin and Busselton, where residents informed Carter, who was there at the time, that it was 14 years since a similar irruption had taken place. L. Glauert saw birds at Milly Milly Station, on the Murchison, in

May 1922. In 1934 thousands appeared at the Mt. Margaret Mission near Laverton and it was the only time that A. G. Mathews (who was there between 1932 and 1936) ever saw them; it was a heavy rainfall year, with excessive winter rains and all the claypans had several feet of water in them.

There was a slight invasion into South-West swamps in 1948 and a larger and more extensive invasion everywhere in 1952. Thousands were observed in Carnarvon in the autumn and flocks of 400-500 at Lake Gundaring in December. Smaller parties were noted elsewhere but they were widely ranging and were reported from as far afield as Perth, Nornalup, Balladonia and the Mount Margaret Mission. A fairly considerable invasion took place during the winter of 1964 and birds were seen in the Wheatbelt (Perenjori, Buntine, Tammin) near Norseman and at the Warburton Range. A few reached the South-West (Coolup swamps). Large numbers invaded the Carnarvon district in the winter of 1972. A moderate invasion in the South-West took place in 1975.

Dusky Moorhen

Gallinula tenebrosa

Black with a brown tinge. The under tail coverts are white which in life show as *white patches at the sides of the tail.* Iris, brown; base of the beak and a plate on the forehead are red, the tip of the beak being yellow; legs, green with red at the knee joints. Length, 14½-16 in.

The newly-hatched chicks, to about three weeks of age, are covered in black down, with the crown almost bald revealing the deep blue supra-orbital areas. White hair-like feathers among the down form a sparse "beard". The base of the bill and forehead shield are bright red, separated from the yellow bill tip by a black band. This black band gradually extends towards the tip, obliterating the yellow with increasing age, while the red fades. Immature birds have the beak dark green at the base.

Distribution: Usually confined to the South-West corner, north to the Gingin Brook and south-east to Bremer Bay. In April 1975, a single bird appeared at Leonora, no doubt due to flooding of local lakes.

Nesting: The nest is in the form of a platform placed in the rushes or on the butt of a paper-bark tree *(Melaleuca rhaphiophylla).* It is constructed of rushes, paper-bark or other aquatic herbage and is about the same size as that of the Swamphen and might be mistaken for the nest of that species as the eggs also are similar. Although we have seen some clutches of eggs of the Swamphen which differ little in size from those of the Moorhen, the eggs of the latter are usually smaller in size. A series of 16 eggs from Butler's

Swamp and Coolup averages 48 x 35 mm., varying from 44-50 x 33-36 mm. The clutch varies ordinarily from 7 to 9, but up to 13 eggs have been found in a single nest. Two broods are raised in the season.

The Dusky Moorhen is a common inhabitant of swamps and rushy pools. It obtrudes itself less on the notice of observers than does the Coot, but it is not so shy as the Swamphen. Like the closely similar European Moorhen *(G. chloropus)* it is a frequent habitue of the ornamental ponds in city parks, nesting freely at Queen's Gardens and Hyde Park in Perth. Moorhens arrive on the southern swamps when the first rains fill them, earlier than the Coots, and, as already mentioned, raise two broods instead of the Coot's one and hold smaller territories than those of the Coot. As the water recedes in the drying swamps in the summer the birds build up new platforms in the water on which to roost. In the non-breeding phase the Moorhen does not gather into flocks as does the Coot. In life the tail is constantly jerked. The call is a harsh "Kerrk."

Swamphen (Blue Bald Coot)

Porphyrio porphyrio

Native names: Goo-le-ma, Gullima (Perth); Moo-lah (Gordon and Pallinup Rivers).

A large bird like a domestic hen in size, dark bronze-brown above, head black, and *rich azure-blue on the upper breast,* on the sides of the head, and edge and shoulder of wing. The remainder of the *under parts are purple,* except the under tail coverts which are white. The *massive red beak is continued as a shield* up the forehead to between the eyes. The long legs are usually brownish-red with dusky joints, but on occasion birds have been recorded with green legs. Iris, red. Length, 19-22½ in.; wing span, 35 in. Weight, 2 lb. 11 oz.

Distribution: In the southern part of the State the distribution of this species is restricted to the South-West corner, from Moora to Esperance—the area inhabited by the race *P. p. bellus,* which is described above. The Kimberley form, *P. p. melanotus,* lacking any azure on the breast and having the entire under parts purple, extends south to the Fortescue River. The intervening area appears to be uninhabited by Swamphens.

Nesting: When the Swamphen nests in rushes, its nest is made by bending over rushes towards the centre of the site selected to form a foundation and adding bitten-off rushes from elsewhere until a flat saucer-shaped nest is constructed. The nest is just above the water. This bird also has the habit of constructing, in a somewhat similar way, platforms which it uses only for feeding purposes; these may be recognised by there usually being pieces of the carapaces of jilgies *(Chaeraps quinquecarinatus)* lying on them which the bird has been eating. The eggs vary much in number and in size; 2 to 5, occasionally 6, form the clutch, and usually the smaller the number laid the larger the size. Measurements, 51-58 x 37-42 mm. They are light sandy in colour with spots and blotches of varying shades of brown and purple; they are very similar to those of the Dusky Moorhen but, in Western Australia, the latter lays a greater number of eggs and they average smaller than those of the Swamphen. When constructing a nest elsewhere than in rushes the Swamphen usually selects a horizontal branch of a tree leaning over water. The nest is then constructed of sticks, grasses, or rushes. In the lower South-West eggs may be found from July to November.

These shy, secretive birds usually allow the observer only a fleeting glimpse as they skulk among the rushes of swamps and alongside watercourses. Sometimes they will venture on damp grassland nearby but usually have none of the tameness which characterises the species in the eastern States. However, the birds at Shenton Lake, Subiaco, have become so accustomed to people that they wander freely on lawn verges.

Coot

Fulica atra

Native names: Mool-ya-wind-doo and Gid-jee-broon, Kijjibrun (South-West).

Black above, under dark grey. Iris, brick-red; *beak and forehead shield pale bluish-grey appearing white in distance;* legs, steel-grey. Length, 15 in. Weight, 1 lb. 3 oz. - 1 lb. 8 oz.

Young birds lack the white forehead shield and the beak is dusky. The newly-hatched young are covered in black down which has yellow hair-like tips.

Distribution: Occurs all over the State in suitable localities.

Nesting: The nest is usually built in rushes from which the bird bends over stalks to form a saucer which it lines with shredded rushes. It may also be built in the butt or stump of a paper-bark. The eggs are dull white, with spots of varying shades of brown; usually 5 to 7 in number, but up to 12. A series of 20 eggs from Carnamah averages 51 x 35 mm., varying from 47-55 x 34-36 mm. Mean weight, 30 gm.

The Coot favours fresh water swamps, lakes and brackish inlets and estuaries but tends to avoid those which have too marine a character. Thus in the Swan River estuary it is plentiful from the Causeway upstream, but it is only in winter that odd individuals may be met with as far downstream as Crawley. In the brackish Wilson's Inlet at Denmark the birds are very abundant, and this is the case with most of the barred estuaries of the south coast, such as Pallinup and Bremer Bay estuaries. Unlike any other member of this family of birds, the Coot congregates in large flocks on open waters. Timed dives show that the birds stay 5 to 6 seconds under water to bring up submerged vegetation (at Hyde Park, Perth).

BUSTARDS: Suborder Otides

BUSTARDS: Family Otididæ

Australian Bustard (Wild Turkey)

Eupodotis australis

Native names: Parr-oor-ra, Kullin-nunga (Coolawanyah); Paddera (North-West Cape); Parderua (Lyons River); Nalpintjee (Jigalong); Burridira (Wallawe tribe, Mount Magnet); Nganuti (Warburton Range); Barrul (Fraser Range); Be-bil-ya, Bee-bill-yer; Bibbilya, Burabur (South-West); Coolie (Pallinup River).

One of the largest of the terrestrial flying birds. Back, wings and tail dark brown, marked with light brown; on the wing coverts a patch of black patterned with white; crown of head and crest-like nape feathers black. Foreneck greyish-white, feathers elongated. Black chest band; abdomen white. Iris, white; beak, dull white to brown; legs, yellowish or cream-coloured. Length, male, 43-45 in.; female, 30-33 in. Wing span, male, 6 ft. 3 in. - 7 ft.; female, 5 ft. 5 in. - 5 ft. 9 in.

We have few first-hand data of the weights of local birds, but T. Carter refers to a fine specimen, presumably a male, weighing 16 lb. In South Australia E. F. Boehm has recorded that males range from 14 to 18 lb., being occasionally much heavier, up to 28 lb., the heaviest known being a Victorian bird weighing 32 lb. Females usually weigh between 10 and 14 lb.; one in poor condition, shot at Broomehill, and weighed by Carter, weighed 9 lb.; another, in good condition, weighed by us at Bridgetown in April 1949 scaled only 7½ lb.

Distribution: All over the State in open country.

Nesting: The clutch comprises 1 egg, sometimes 2, and they are laid on the bare ground without any attempt at nest-making, but usually close to some low bush or shrub. The egg measures 75-81 x 52-55 mm., and is light olive-brown to light olive-green, marked with irregular-shaped spots, blotches and streaks of olive-brown, evenly distributed over the surface of the egg, but sometimes more numerous at the larger end where they tend to form an indefinite cap. The young when first hatched are covered with mottled brown and black down and they leave the egg-site almost immediately after hatching. For their own protection they will "squat" when danger offers. The female incubates the eggs and on the approach of an intruder slips quietly away in a crouching attitude. The nesting season is from July to November.

The male Bustard is well known for its singular display. The foreneck is inflated and the elongated feathers form an apron, which is swayed from side to side. The tail is elevated and spread and reaches the head, which is thrown backwards. The wings are drooped and the bird struts about in a majestic manner, emitting a deep "who-o-o-o."

The food of the Bustard is mainly grass, seeds and the fruits of native plants. One of its favourite fruits in the North-West is that of the myandee or moonflower *(Capparis spinosa)* and so intent is it when feeding on these fruits that the Aborigines are said to catch numbers of the Bustards by digging narrow pits around the myandee bushes. The Bustard varies this diet with crickets, beetles, mice, and even young birds, and a particularly valuable factor in its economy is the partiality it has for grasshoppers. The crop contents of a bird examined at Yalgoo by R. P. McMillan comprised 150 large caterpillars, 50 ground weevils one inch in length, 10 ground weevils half-inch long and 35 millipedes.

Though the bird has been fully protected by law since 1935, a large amount of illicit shooting takes place each year. Notwithstanding, the bird is not really rare excepting in the well-settled districts. It visits the open coastal dune country near Perth occasionally and has been seen at the Gnangara pine plantation north of Perth. On the less frequented roads in the sandplains the traveller is sure to flush one or two birds during a car journey. In the back country the species still occurs in strength, flocks of up to hundreds of individuals being encountered.

WADERS, GULLS and TERNS: Order Charadriiformes

WADERS: Suborder Charadrii

The majority of the birds of this Suborder are only summer visitors to Australia, migrating south from their breeding grounds in northern Asia. The height of the exodus from the South-West appears to be in mid-April, but a number of individuals of many of the species may remain with us for the winter. The return in the spring becomes evident at the beginning of September, the birds returning in increasing numbers during that and the following month.

Most of the species have two more or less distinct seasonal plumages, a bright and conspicuous nuptial plumage, in which rufous-red and black predominate, and a quieter drab plumage in various tones of grey which is worn whilst they are in Australia. It is this non-breeding or "winter" plumage (namely, that of the northern hemisphere winter) which is described in the following pages, and references to the breeding plumage are given only in those species where traces of it are observable locally either at the beginning or close of their sojourn with us.

Field identification of these waders is attended by special difficulties owing to the general similarity of many of the species in their non-breeding plumages, their wariness and the lack of cover in their usual habitat, which prevents the observer from making too close an approach. However, as the birds often occur in some numbers at a time, and may be practically stationary or moving only slowly over a mud flat, they can be studied far more effectively with a good pair of field glasses or a telescope than can most of the bush-birds by such means.

Points useful for identification are:

1. An estimate of size, using familiar birds as standards, such as the Silver Gull or already known species of waders. On a bare mud or sand flat it is extremely difficult otherwise to estimate size with any reasonable degree of accuracy. This is particularly the case if the sun is low, or in front of the observer, and under such conditions birds like Curlew Sandpipers may appear huge, like Great Knots.

2. The nature of the beak, whether short or long, whether it is the dotterel-plover type of thick-set beak or the slender sandpiper type; whether it is straight, down-curved or up-curved. The colour is usually dark, but in the case of the Bar-tailed Godwit the particoloured beak is diagnostic.

3. The colour pattern of the rump should be noted as soon as an opportunity offers, usually when the bird rises into flight. Determine whether the rump is wholly white (of the Curlew Sandpiper type, Fig. 34), wholly dark like the back (Grey-tailed Tattler type), or dark only in the middle with the outer parts white (Sharp-tailed Sandpiper type, Fig. 35). This is a most useful character.

4. The colour of the legs often provides helpful information as species closely similar in appearance may have differently coloured legs. In sunlight with good glasses one may distinguish at once Curlew Sandpipers from Sharp-tailed Sandpipers by the black legs of the former and the olive-green legs of the latter. The yellow legs of the Pectoral Sandpiper distinguish it at once from the closely similar Sharp-tailed Sandpiper (green legs). This is the only field character, apparently, by which the Little Stint (black legs) may be distinguished from the Long-toed Stint (green legs).

5. Special features may at once identify particular species. Thus the black axillaries of the Grey Plover decisively mark off that species from the somewhat similar Golden Plover, apart from the decided difference in rump and tail colour patterns.

6. Relative proportions of the body are also helpful. Thus the Black-tailed Godwit can be distinguished from the Bar-tailed not only in the striking difference in the colour pattern of the rump and the slightly up-turned beak of the latter, but also in the relative lengths of the legs. This is well seen in flight, when the long legs of

the Black-tailed species can be clearly seen trailed beyond the tail, as with the Stilts. In the Bar-tailed Godwit the legs hardly project beyond the tail.

7. Special habits are very helpful in aiding determination. The stone-turning habit, for instance, is specific for the Turnstone. Many dotterels bob their heads, as if they were afflicted with persistent hiccoughs, and some, in addition, bob their tails. The only sandpipers which do this are the Common, Terek, Green and Wood Sandpipers and the Greenshank. The stiff, skimming flight of the Common Sandpiper, again, is a flight characteristic which is unmistakable.

8. Lastly, full use should be made of characteristic call notes. Thus no other character than the note is needed in distinguishing the Curlew from the Whimbrel.

JACANAS: Family Jacanidæ

Pheasant-tailed Jacana

Hydrophasianus chirurgus

A bird which differs strikingly in its two seasonal plumages. Birds visiting Australia will be in the less conspicuous eclipse plumage. The rather bizarre full breeding plumage, known only in the northern hemisphere, is a sharply contrasted dark brown (almost black) and white, with a golden nape, a long black tail, projecting up to 35 cm beyond the body; underparts dark; head and throat white; the black nape patch extends as a dark line along the sides of the neck; flight quills black, their inner vanes and the wing coverts a conspicuous white.

The most outstanding difference in the eclipse plumage is the reduction of the tail, which is now so short as to be barely apparent. The colours are lighter—grey-brown and white, instead of the sharply contrasting almost black and white. Underparts white. Nape brown slightly freckled with white; the golden area reduced to a narrow stripe on each side behind the eye; below this is a narrow dark brown line joining a wider dark brown breast band. Crown of head brown, sometimes reddish brown, forehead freckled. Wing quills black, with increasing amounts of white on the inner vanes, inwards to the body; upper wing coverts brown, freckled white. There is a sharp, inconspicuous spur on the shoulder of the wing. Long toes and heavy claws. Iris, dark brown; beak, pale blue; legs pale blue. Length, excluding tail, about 3 ft. Beak 25-29 mm. Middle toe and claw, 77-89 mm.; claw, 20-23 mm.

Distribution: A very rare migrant from the northern hemisphere. The only Australian record is of a bird closely observed by Mr. J. A. Smith on December 22, 1974 at Paraburdoo, in the NorthWest.

It was feeding at the edge of a sewage pond. The species ranges from Pakistan through southern China and Malaya to Formosa, the Philippines and Java. Like other Jacanas it favours waters with floating vegetation but is "more ready to frequent open water, and more accustomed to wander to flood water, streams, and similar spots free of weeds" (Hugh Whistler, describing its habits in India).

PAINTED SNIPES: Family Rostratulidæ

Painted Snipe

Rostratula benghalensis

Head, throat and chest blackish-brown, with a rufous band on the nape and a white ring round the large eye, the white extending backwards as a bar over the ear coverts. Along the middle of the head, from the base of the beak, occurs a light buff line. The back and wings are a dark olive-green, finely-barred black, and the *wings are ornamented with bright chestnut spots and black bars.* The short tail is grey, spotted with chestnut and narrowly waved with black. Lower breast and abdomen white. Iris, hazel; beak, long and straight (2 in.), yellow-brown; legs, green. Length, 9½ in.

The male is a smaller, less colourful bird, lacking the rufous on the hindneck.

Distribution: This bird has probably a widespread distribution throughout the State, but is rarely met with. The first record of its occurrence was a pair of specimen skins advertised for sale by Francis Armstrong in 1841 in the Perth newspaper, *The Inquirer.* It has been recorded from Herdsman's Lake, from near Armadale and Pinjarra, on the swamps at Moora, and at Bunketch. P. Slater recently recorded it at Derby.

Nesting: The Painted Snipe lines a depression in the ground close to water with grass or other herbage in which to lay its eggs. If the ground is dry, practically no nesting material is used and the eggs are laid on the bare ground. If, however, the nesting site is on land on which water is lying a solid nest of rushes and herbage is made, with a shallow portion in the centre to hold the eggs. The nest is then about seven inches in diameter and an inch deep. A canopy may, or may not, be added by interlacing the rushes, if any, over the nest. The eggs, 4 in the clutch, are rather peculiar in their appearance. They are "peg-top" in shape and creamy-white in colour, but this is obscured by a maze of irregular lines of black, twisting and turning over the surface of the shell. Sometimes these lines merge to form black blotches. Size, 30 x 27 mm. C. L. E. Orton has observed eggs in nests on the Moora swamps in October.

The nesting habits of this species are peculiar. As in the case of the Painted Quail, the female is polyandrous, but unlike the Painted Quail she does not even construct her own nest. This is done by the male, who also undertakes the duties of incubating the eggs and the raising of the young. Once the female has deposited a clutch of eggs in a nest prepared for her by a male, she takes no further interest in that nest, but mates again with another male who has arranged the necessary nest. How many clutches of eggs one female will normally lay in a season is not known. Incubation period, 15-16 days (P. Slater).

OYSTERCATCHERS: Family Hæmatopodidæ

Pied Oystercatcher

Hœmatopus longirostris
(formerly *ostralegus*)

Native names: Warrangarree, Widung-gurra (North-West Cape); Goran-goran (Pallinup River).

Head and breast, upper back, wings and tail *black; upper tail coverts and lower back and abdomen white.* Iris, red; beak, red; legs, red. Length, 19 in.

Distribution: Generally distributed around the coast, favouring sandy and muddy beaches, and non-migratory.

Nesting: The Pied Oystercatcher makes no nest but lays its 2 (rarely 3) eggs in a shallow scraping in the sand near the sea-shore. The eggs are grey-white (stone grey), with spots and blotches of dark brown, some of the markings appearing as if under the surface. Average size of 10 eggs, 59 x 41 mm., varying from 56-61 x 39-42 mm.

The birds are strictly territorial but the two species of oyster-catcher tolerate each other's presence when nesting in close proximity.

Oystercatchers, of either species, do not as a rule occur on estuarine beaches in this State, but are essentially denizens of the open sea coast where they occur in pairs and small parties. In parts of the eastern States, as in Tasmania, they may be seen, sometimes in hundreds, in estuaries and tidal creeks. In the Swan River estuary a pair frequented Pelican Point for a prolonged period in the summer of 1948-49, an altogether unusual occurrence. The penetrating "pee-pee-pee" calls are one of the familiar alarm-notes of the sea-beaches.

Sooty Oystercatcher

Hœmatopus fuliginosus

Native name: Wee-ardoo (North-West Cape).

Entirely black. Iris, scarlet; eyelid, orange-red; beak, crimson, shading to orange-red at the base; legs, pink; toes, orange-brown. Length, 19 in. Weight, 1 lb. 6 oz - 1 lb. 12 oz.

Distribution: Generally distributed along the coast, usually found on *rocky shores* and islands, in contrast to the foregoing species, though they not infrequently intermingle. Non-migratory.

Nesting: The Sooty Oystercatcher also makes no nest but lays its 2 eggs among seaweed, or on *Carpobrotus,* in the clefts of rocks along the sea-shore. A site well above high-water mark is selected. Average size of 18 eggs, 64 x 43 mm., varying from 61-68 x 40-45 mm. Besides being distinctly larger the eggs are darker than those of the Pied Oystercatcher. The ground colour is dark olive-grey (dark stone grey), strongly marked with brown spots and blotches and submerged lighter spots of dull purplish-brown. At the broad end the blotches may run together to form extensive dark splashes. Both sexes share in the incubation. The nestling is covered in sooty-brown down, and is black on the head and along two longitudinal lines on the back. The first plumage acquired by the young bird is entirely black, like that of the parent, but with the wing covert and mantle feathers narrowly edged and tipped buff-brown; iris, light brown; ring round eye, orange-brown; beak, orange-brown; legs, grey. The young is extraordinarily adept in hiding among rock crevices near the nest site until it attains the power of flight.

PLOVERS and DOTTERELS: Family Charadriidæ

Masked Plover

Vanellus miles

Upper parts light brownish-grey; crown of head black; wing quills dark brown; a broad white band across the upper tail, lower tail black; nuchal collar and *entire under parts white.* There is a sharp naked spur on the wing and a *prominent yellow facial wattle* projects out above the eye and lores. Iris, yellow; beak, yellow; legs, dark pink. Length, 14¾ in. Weight, 11 oz.

Distribution: Until very recently this north Australian species was known only from the Kimberley Division, though birds had occasionally been observed in the North-West as far south as Nickol Bay. However, in very recent years there has been a remarkable occurrence in the southern parts of the State, coinciding with an invasion of the related and closely similar Spur-winged Plover from the eastern States. The first Masked Plover was seen in the south by Eric Lindgren at Narrikup in June 1957. Subsequently the bird was reported at Moora and near Goomalling and a specimen was collected at Floreat Park by G. F. Mees and J. R. Ford in March 1959. All these records may refer to the same individual, as it frequently associated with Spur-winged Plovers. It may have accompanied them in a flight from eastern Australia, where the species normally occurs much further south than it does in Western Australia.

Nesting: The 4 eggs are laid in a hollow in the ground on the edge of a swamp; on occasions a little grass is added as nesting material. The pyriform eggs vary in colour according to the type of ground on which they are laid; they may be green or yellowish-olive. The markings consist of spots and blotches of olive and dark brown of varying shades. Size 41 x 33 mm. There is no record of the nesting of the species in Western Australia south of the Kimberley Division.

Spur-winged Plover

Vanellus novæhollandiæ

Very similar to the preceding species, but differing from it in possessing a *black nuchal collar which extends to the sides of the chest.* These black patches do not unite to form a complete breast band as in the Banded Plover. Iris, yellow; beak, facial wattle and wing spur, yellow; legs, dark pink. Length, 14½ in. Wing span, 2 ft. 10 in. Weight, 11½ oz.

Distribution: A species with an extensive range in south-eastern Australia and which has only recently colonised the South-West of this State. The first specimen was collected by W. H. Butler and V. N. Serventy at Lake Muir in November 1956 and the species has been subsequently observed at Narrikup, Bunbury, Floreat Park (near Perth) and near Goomalling, at times in the company of an individual of the previous species. In December 1965 a bird was reported at Denmark. In November 1973 one was seen just east of Eucla over the South Australian border.

The Spur-winged Plover has increased greatly in numbers in the eastern States in recent years and is evidently expanding its

Fig. 29.—Plumage patterns of the three land plovers: Spur-winged Plover, left; Masked Plover, middle; Banded Plover, right. Relative sizes not to scale.

range. Stragglers have been noted in New Zealand since the late 1880s and breeding colonies have been established there since the early 1930s.

Nesting: No nesting has so far been observed in Western Australia. In the east the nest is a depression in the ground, lined with dead sticks, etc. Clutch, 4 eggs (usually 3 to 5); ground colour stone, covered with reddish-purple and lavender spots—mean size of Tasmanian eggs 51 x 37, varying from 42-59 x 32-40 mm. (David Thomas).

Banded Plover

Vanellus tricolor

Native names: Kirrkee keen keen (Jigalong); Nidulyorong (South-West).

Upper parts brown; crown and sides of head black, continuous with *a broad black chest band;* white streak behind eye; white wing stripe; flight quills dark brown; broad white band across upper tail; lower tail black, narrowly tipped white. Throat and abdomen white. Iris, light yellow; eyelid yellow, red wattle between eye and beak; beak, yellow tipped darker; legs, red. Length, 10¼ in.

Distribution: Generally distributed in open country, north to the Fortescue River and Jigalong.

Nesting: The Banded Plover scrapes a circular depression in the ground in which to lay its 4, sometimes only 3, eggs. The eggs are of "peg-top" shape and are always found in the nest with the pointed ends inward. The colour varies with the type of ground on which they are laid; if the nest is in a green paddock, the eggs are olive-green in ground colour, but if on lighter coloured soil, then the ground colour is similarly light. The markings consist of brown spots and blotches and some appear as if under the surface of the shell. Average size of 16 eggs, 44 x 32 mm., varying from 41-48 x 31-33 mm.

On the approach of an intruder, the sitting bird slips quietly off the nest and joins forces with its mate and other birds in the immediate neighbourhood, to swoop at him with shrill cries.

Nesting commences in May and eggs may be found in nests as late as October. Both parents assist in the incubation of the eggs, which take from 23 to 30 days to hatch; all eggs hatch out together and the newly hatched young are mottled brown and black on the back with a white band on the neck and breast. The iris and beak are black and the legs dark green.

The Banded Plover has become common on pastures and in crop paddocks as the result of settlement. Birds are frequently heard flying at night, even over the city of Perth, and flocks of up to fifty or so may be seen on the eastern part of the Esplanade, where nesting also takes place. In 1965 a pair nested at Pelican Point. The birds are much given to sheltering in the shade of fence-posts, or in the shade thrown by young trees in orchards, in the South-West.

Grey Plover

Pluvialis squatarola

Upper parts brown, dappled with white; tail barred white and brown, with *a broad white bar across the upper tail,* the sparse brown markings on it being inconspicuous at a distance; entire under parts white, except the *black axillaries,* at the root of the wings, which flash out prominently as the bird is flying. At the beginning and end of summer some individuals may have varying amounts of black in the plumage. Iris, brown; beak, blackish; legs, dark olive-brown. Length, 11-12 in. Weight, 7¾ oz. Culmen, 28-31 mm.; wing, 189-201 mm.

Distribution: A summer migrant from the northern hemisphere. Small parties are not uncommon on the ocean beaches and also in estuaries and inlets, including the Swan River and around the South-West. They appear to be regular visitors to Princess Royal Harbour.

Eastern Golden Plover

Pluvialis dominica

Upper parts black, dappled with greenish-yellow, wings grey and blackish-brown. Pale buff eyebrow. Throat and breast marked grey and yellowish-buff, marked off from the white abdomen. *Tail and rump like the back, showing no white.* Axillaries grey. Eye large, iris brown; beak, short, rather heavily built, black; legs, slate-grey. Length, 9½-10 in. Wing span, 20 in. Weight, 4-5½ oz. Culmen, 22 mm.; wing, 164-175 mm.

Confusion is possible with the Grey Plover, despite the colour differences, but the Golden Plover can always be recognized by the *absence of white at the base of the tail* and by the *light* axillaries.

Distribution: A summer migrant from the northern hemisphere. Small parties of up to seven or eight individuals appear on local beaches and in the estuaries, but it is much less plentiful in South-Western Australia than the Grey Plover.

Oriental Dotterel

Charadrius veredus

Native name: Boolgarra-boolgarra (North-West Cape).

Upper parts entirely greyish-brown with a light rufous wash; wing quills dark; no white on tail; forehead and eyebrow line dull white. Throat whitish, breast sandy-buff, abdomen whitish. Under wing grey. Iris, brown; beak, black, lighter at base of lower mandible and fairly slender; legs, brown or dull yellow. Length, 9½-10½ in. Culmen, 22-25 mm.; wing, 166-175 mm.

In the field the upright carriage and longer legs—longer than those of other dotterels—are distinctive. The head is held high and "bobbed" frequently (A. R. McGill).

Distribution: A summer migrant from the northern hemisphere where it breeds in Mongolia and northern China.

It is plentiful along the coastal flats in the North-West, arriving very regularly about the second week in September and leaving about the middle of March. It occurs at inland waters and we have seen a specimen from Cosmo Newberry. It is rather rare in the south, but birds have been observed at the Jandakot Lake, south of Fremantle, and at the salt lakes north of the Stirling Range. T. Carter recorded that during the great heat-wave of 1895, on December 30 and 31, the beach at North-West Cape was for miles lined with Oriental Dotterels—probably driven to the water-line by the intense heat.

Fig. 30.—Oriental Dotterel, x 1.

Large Sand-Dotterel

Charadrius leschenaultii

Upper parts grey, wing quills darker. Forehead broadly white; lores grey. Lower parts white, with grey at the sides of the foreneck almost meeting below, at the chest. *Beak, rather long and robust,* greyish-black. Iris, brown; legs, olive-grey. Length, 8-9 in. Weight, 2½-3¼ oz. Culmen, 22-27 mm.; wing, 136-149 mm.

Distribution: A summer migrant from the northern hemisphere, occurring on the ocean and estuarine beaches. It is not so common as many of the other waders, but specimens have been collected, or observations made, as far south as the Swan River estuary, Augusta, Albany and Bremer Bay. It is more frequently met with at the Abrolhos and on the northern beaches.

Fig. 31.—Large Sand-Dotterel, x 1.

Mongolian Dotterel

Charadrius mongolus

Similar to the preceding, but slightly smaller and differing in having a *shorter and heavier beak,* with a strong dertrum. White at sides of tail. Birds in non-breeding plumage differ from similarly-garbed individuals of the Double-banded Dotterel in lacking any rufous tone on the feathers. On arrival and before departure for the north the nuptial plumage may be present in varying degree; reddish nape; rufous-red breast band lined with black above; lores and behind eye black; forehead and throat white. Iris, brown; beak, black; legs, grey. Length, 7½-8 in. Culmen, 17-20 mm.; wing, 125-135 mm.

Distribution: A summer migrant from the northern hemisphere and more plentiful in the northern part of the State. In Western Australia the species appears to be less plentiful than the Large Sand-Dotterel, the reverse being the case in the Eastern States.

Fig. 32.—Mongolian Dotterel, x 1.

Little Ringed Plover

Charadrius dubius

In summer plumage upper parts and crown of head, greyish-brown. Head pattern variegated; lores black, forehead band white, above which is a black band. Behind the head a white nuchal collar linking with the white throat. Between the nuchal collar and the mantle is a black band which encircles the body and ventrally becomes *a broad breast band.* Remainder of under white. Wing quills dark brown, with *no white bar.* Tail brown, tipped darker, outermost quills white. Birds in this bright plumage are unlikely to be met with in Australia, where they will be in the duller "winter" plumage: upper uniform drab greyish-brown, whitish forehead and no variegated black and white about the head. Breast band reduced to a diffuse brownish bar, often broken. Iris, brown, fleshy ring around eye yellow; beak, black, in summer yellow base to lower mandible; legs, long, pale yellow or flesh. Length 6-7 in.

Distribution: A northern hemisphere species, migrating to south-east Asia. Only one record from Western Australia. John Darnell saw two birds at the sewage ponds at Mt. Goldsworthy, in the Pilbara, on October 18, 1972, and they stayed several days. They were in winter plumage. The species is essentially a fresh water bird.

The long tail and wings give the body a slim appearance, particularly as the head is held well above the body. The call is distinctive, a low-pitched "tee-oo". The birds feed mainly in the water and not in the verge mud and have a habit of pattering with the feet to stir up small food organisms in the substrate.

Double-banded Dotterel

Charadrius bicinctus

Upper brown; forehead white; a black streak from the base of the beak backward to below and behind the eye. The *white under parts are crossed by two separated bands, a black upper and a lower chestnut band,* which is wider. The rump and tail pattern is like that of the Red-capped Dotterel, with white showing on either side of the central dark area. Iris, dark brown; beak, black; legs, green. Length, 7¼ in. Culmen, 16-19 mm.; wing, 122-130 mm.

Birds in the very distinctive nuptial plumage thus described are unmistakable, but diagnosis is difficult in the immature and non-breeding plumages when the two bands are missing or only partially developed. The black cheek stripe also becomes sandy-buff. In this plumage the larger size and more active habits will distinguish the species from incompletely plumaged Red-capped Dotterels; from the similar-sized Mongolian Dotterel the distinctions are discussed in the account of that species.

Distribution: An occasional migrant and wanderer to the south and south-west coasts from New Zealand.

Nesting: In New Zealand this dotterel breeds on sandy and shingly beaches along the coast, or inland on shingly river beds, on tilled ground, or in open fields. The nest is merely a hollow in the ground with or without a little nesting material. The 2 or 3 eggs are ovoid in shape, dark greenish-yellow, marked closely all over with irregular deep-brown spots and blotches. Size, 35 x 25 mm.

Formerly it was believed that this dotterel nested in Tasmania and south-eastern Australia, but it seems quite certain that all the Australian birds are migrants from New Zealand. A quota of non-breeding birds remain in Australia during the normal nesting season. The species is common in south-eastern Australia, westward to South Australia, but some individuals manage to penetrate into Western Australia and records are known for the south coast and on the west coast as far north as Leeman Salt Lake (48 miles south of Dongara), and at the Abrolhos Islands (West Wallaby). Such visitations appear to be, however, quite rare.

Red-capped Dotterel

Charadrius ruficapillus

Upper parts light greyish-brown; wing quills darker; outer tail coverts white, centre dark brown; *crown of head and nape rusty-red;* forehead broadly white, bordered above by a narrow black line; lores black; *under parts white;* a black and rufous patch at the sides of the neck, near the wings. Females lack the reddish head and black forehead line and have dark olive-grey legs. Iris, brown; beak, black; legs, black. Length, 6 in. Weight, 1¼ oz. Culmen, 14-15 mm.; wing, 99-108 mm.

Distribution: Generally distributed and resident along the sea coast, estuaries and on inland lakes. Rarely it may appear on some of the drying swamps on the coastal plain near Perth in the summer. Our birds are closely related to an almost cosmopolitan species, *Charadrius alexandrinus,* which in England is known as the Kentish Plover.

Nesting: No nest is made by this species, but the 2 eggs, which is the number invariably laid, are deposited in a shallow depression in the sand just above high-water mark on the sea-beach, or on the edges of inland lakes. The eggs are pale greenish, marked with dark brown to purplish-bluish spots; 12 from the South-West average 30 x 22 mm. Nesting regularly takes place behind Pelican Point.

Black-fronted Dotterel

Charadrius melanops

Native names: Limbee (Jigalong); Ne-dool-ye-rong, Nidul-yorong (South-West).

Sharply contrasted markings of black, white and brown. Upper parts brown with black forehead and a black line passing through the eyes to the black nuchal ring behind. Above, this is bordered with a white line passing over the eye and around the back of the head. Wing quills blackish, a white wing stripe and a chestnut patch on the shoulder. Upper tail coverts with brown central feathers and white at the sides. Under parts white, with *a black band across the chest.* Iris, brown, with a bright red orbital ring; beak, orange-yellow at base and black tipped; legs, flesh colour. Length, 5½-6½ in.

Distribution: This is one of the non-migratory dotterels, widely distributed and plentiful. A pair or two are likely to be found on any fresh water swamp, pool or dam anywhere in the State, but it is rarely seen on estuaries and never on the sea beaches.

Nesting: The eggs are laid in a slight hollow in a dry river bed, or on the edge of water. The 3 eggs are light stone in colour or with a yellowish tinge according to the type of sand on which they are laid, and are very closely marked all over with small specks and lines of umber and dark grey. The lines are sometimes so distributed as to look like hieroglyphics and even appear to form such letters as Z or K. A set of four from Geraldton measures 28 x 22 mm.

Hooded Dotterel

Charadrius cacullatus

Upper parts brown; head, throat and on upper back, wing quills and tail, *dark sooty-brown; a white collar behind the dark head;*

white wing stripe and outer tail coverts white. Under parts, except throat, white. Iris, brown; orbits bright scarlet; beak, orange at base, tip black; legs, flesh colour. Length, 7½ in.

Birds in immature plumage lack the sharply contrasted black and white head pattern. The head is grey with a narrow brown zone on each side of the white nuchal collar. The throat is white.

Distribution: Confined to the South-West, along the sea coast, estuaries and salt lakes, to the south of a line from Port Gregory, Wongan Hills, Lake Deborah (near Bullfinch), Lake Cowan to Balladonia. Its abundance on inland salt lakes is a great contrast to the distribution picture in the Eastern States, where the species is restricted to marine beaches.

Nesting: A hollow scratched in the sand well above high-water mark, constitutes the nest of the Hooded Dotterel and its 2 or 3 eggs have a sandy ground colour, spotted and blotched with blackish-brown markings. They are of the usual Limicoline pyriform type. Size, 36-39 x 26-27 mm.

Red-kneed Dotterel

Charadrius cinctus

Head, neck, and *a broad band across the chest black;* back, wings and middle of tail bronze-brown; flight quills black edged with broad white stripe; outer tail coverts white; throat and abdomen white; flanks streaked white and chestnut. Iris, brown; beak, reddish-brown, black tip, violet at base; legs, leaden; "knee" red or violet. Length, 7-7½ in. Weight, 2 oz.

Distribution: Distributed all over the State, but sparingly in the South-West. It occurs in the summer on drying-up swamps on the coastal plain and also, at times, on the samphire marshes of the estuaries like those of the Swan River by the Causeway and Pelican Point, but it does not visit marine environments.

Nesting: The nest is a slight depression under a salt-bush. Eggs 4, stone-colour and covered all over with wavy streaks and blotches of dark brown and black. They are pointed at one end. A series of 8 eggs from Lake Grace averages 31 x 22 mm., varying from 29-32 x 22-23 mm.

Australian Dotterel (Prairie Plover)

Peltohyas australis

Native name: Bindoon (Jigalong).

General colour, both above and below a *bright rusty-red,* the individual feathers of the back centred with brown. There is a black band across the top of the head, from eye to eye, and extending down the cheeks. A black nuchal band passes down the fore-neck and meets its fellow of the other side on the breast, thence extending as *a Y-shaped black mark towards the abdomen.* Lower abdomen dark chestnut. Under tail coverts whitish. No white on the upper tail. Wing quills dark brown. Iris, dark brown; beak, dark grey; legs, pale yellowish brown, toes tinged grey. Length, 7½-8¾ in.; wing span, 16-17½ in. Weight, 2½-3½ oz.

Distribution: This is a resident, Australian-breeding species which occurs over most of the State north to Onslow, the upper Ashburton and Jigalong, but not in the South-West corner (Moora, Cunderdin and Lake Grace are the known limits).

Nesting: The Australian Dotterel lays its eggs (3) on the bare ground—shaping a shallow hollow and using no nesting material. It has a habit of covering the eggs with pellets of sun-baked earth when it leaves the nest for feeding purposes. The pear-shaped eggs are buff-brown or light olive-brown, boldly blotched with dark brown, particularly at the broader end, and submerged dull violet spots. Size, 34-39 x 26-27 mm. The young in down are pale grey, slightly rufous down the back, striped and marked with dark brown and black in wavy patterns. The species may breed at any time following rains.

Like other dotterels this bird has the habit of bobbing its head. The call has been syllabised as "quoick."

CURLEWS, SANDPIPERS AND SNIPES
Family Scolopacidæ

Eastern Curlew

Numenius madagascariensis

Native name: Wid-joo-on-ong (Mandurah).

The largest of the family. Streaked brown and buff, both above and below, the under parts being lighter; ashy-white eyebrow line. Tail and coverts barred dark and light brown, *showing no white* above. Iris, brown; beak, black, long and deeply down-curved (chord, 5½ in.); legs, pale grey. Length, 21-24 in.; wing span, 42 in.

Distribution: A summer migrant from the northern hemisphere, occurring singly or in pairs on estuarine flats and salt marshes; it utters a mournful "carr-er! carr-er!" or a high-pitched "ker-lee, ker-lee" when flushed. It is not very common and is never seen locally in the large flocks which are known in parts of the Eastern States, where sometimes a thousand individuals may be feeding together over a mud flat.

Whimbrel

Numenius phaeopus

Upper parts variegated darker and lighter brownish-grey. Upper tail coverts barred light brown and white, giving the appearance of a *white rump* at a distance. Head brown with a *light grey median stripe* and *light grey eyebrow lines*. The under parts are strongly marked with light brown stripes on the throat and breast and bars on the abdomen, lower abdomen white. Iris, blackish-brown; beak, long, down-curved, black; legs, olive-grey. Length, 16 in.

The Whimbrel looks like a small curlew but a useful field-mark is the *whitish-looking rump,* and, on closer view, the striped crown. The *call is quite different,* being a whistling "ti-ti-ti," repeated several times.

Distribution: A summer migrant from the northern hemisphere. This is one of the rarer waders in the southern parts of the State, but it has been observed on occasions at Rottnest Island, in the Swan River estuary at Pelican Point and at Albany. It is more numerous further north and is plentiful in the tidal flats and mangroves at Carnarvon. However, as in the case of the preceding species, it would appear that the main body of migrating whimbrels passes down the eastern, rather than the western, shores of the Australian continent.

Little Whimbrel

Numenius minutus

Upper parts variegated dark and light brown, with a conspicuous light buff *eyebrow* and a *median stripe on the head.* Wing quills dark brown. Upper tail coverts brown with light grey crossbars. Tail grey-brown, barred dark brown. Throat and abdomen dull white, breast striped brown. Iris, brown; beak, comparatively short ($1\frac{3}{4}$ in.), strongly arched downwards; legs, grey. Length, 14 in.

Distribution: A summer migrant from the northern hemisphere. Until recently it had not been recorded south of North-West Cape. In the late summer of 1949 A. H. Robinson observed one bird on a millet crop growing on swamp-land at Coolup. It preferred grass land to actual swamp habitat. Two birds were seen in the same area in the late summer of 1953, in a dry pasture paddock in company with Banded Plovers. M. Tarburton saw one bird at the Swan River Causeway in December 1964.

Black-tailed Godwit

Limosa limosa

Upper parts greyish-brown, feathers centred darker; wing quills darker brown and a *broad white wing stripe; broad white band across upper tail coverts,* contrasting sharply with the black tail. Breast marked with grey; abdomen white. Iris, brown; beak dark, straight and long ($2\frac{7}{8}$ in.); legs, greenish-grey, which are trailed beyond the tail when in flight. Length, 14 in.

Distribution: A summer migrant from the northern hemisphere. It is more frequently met with in the northern parts of the State but small parties have been seen on the Swan River estuary at Pelican Point and Peel Inlet, Mandurah. It has also been observed at Kalgoorlie (on an artificial dam).

Bar-tailed Godwit

Limosa lapponica

Upper parts, mottled grey and brown. Light grey eyebrow line. Wing quills dark brown with narrow white wing stripe. Lower back, upper tail coverts and tail *barred white and brown.* Under parts dull white, clouded with grey on the breast and lightly streaked with brown. Iris, brown; beak, very long ($3\frac{1}{4}$ in.) and distinctly upcurved, basal half pale, distal half dark; legs, black and in flight do not extend much beyond the tail, as they do in the previous species. Length, $14\frac{3}{4}$ in.

Distribution: A summer migrant from the northern hemisphere. Like the other godwit it is more abundant on the northern beaches of the State, being numerous as far south as Carnarvon, but it appears to visit the southern areas more frequently than the other. It is to be seen at times on the Swan River (Pelican Point) and Peel Inlet.

Wood Sandpiper

Tringa glareola

Back and wings spotted, white eyebrow, even grey breast, and *wholly white rump.* Beak straight. Legs, olive-green, brighter than in the Sharp-tailed Sandpiper. Iris, brown; beak, straight, blackish grey. Length, $8\frac{3}{4}$-$9\frac{1}{4}$ in. Weight, 2 oz.

The birds bob their heads and flick their tails, with the body pulsating. The general upright body posture resembles that of the Greenshank, with the head high and the neck outstretched. The alarm note is a sharp three-noted whistle which can be variously rendered as "giff-giff-giff" or "chee-chee-chee."

Distribution: The Wood Sandpiper is a summer migrant from the northern hemisphere and though never very numerous it is a regular and apparently not infrequent visitor. Unlike most other waders it is not found in maritime situations but is a denizen of the grassy margins of drying fresh water swamps and the banks of river pools. Moreover, it has the distinctly un-wader habit of frequenting waters in woodland localities and even perches on trees, fences, etc. In the southern part of the State it has been observed at Guildford (Swan River), Kelmscott (Canning River), on the Serpentine River, Coolup (in fresh water swamps), Benger swamp, Lake Gundaring, Corycup Lake, Kalgoorlie and Esperance.

It is desirable to point out the differences between the Wood Sandpiper and another closely similar wader which, though it has not so far been recorded from Australia, may yet be found here. This is the Green Sandpiper *(Tringa ochropus)*, whose known migrations extend south to Formosa, Java, the Philippines and Borneo. The Green is darker above than the Wood, the plumage having an almost uniform appearance; there is more white on the tail, there being only three or four dark bars on the tail feathers compared with about eight in the Wood; the under side of the wing is blackish, not light greyish; the axillaries in the Green are sepia narrowly barred with white, and in the Wood mostly white, barred with sepia. The calls are also different, that of the Green being syllabised, as a "loud and sharp 'ku-wit, too-wit, kler-wheet' or 'gwi-i-wick'."

Common Sandpiper

Tringa hypoleucos

Upper parts grey-brown *with a slight greenish wash;* a pale eyebrow; under parts white, shaded grey at the side of the chest. The *outer tail feathers are toothed white and greyish-brown* and a series of white spots in the wing quills produce a conspicuous wing stripe, visible in flight. Iris, brown; beak, long and straight, dark grey; legs, pale olive-green. Length, 8½ in. Weight, 1½ oz.

It is distinguished from sandpipers of similar size by the *absence of white at the sides* or across the rump and its *peculiar undulating flight* with wings held out stiffly and low. It shares with the Wood and Terek Sandpipers and the Greenshank the habit of *bobbing its head and tail.*

Distribution: A summer visitor from the northern hemisphere.

Though never seen in large flocks like some of the other migratory sandpipers, this bird is among the regular and well dispersed of the waders. It particularly favours rocky beaches and in ones or twos may be seen on the shores of islands, ocean beaches, inlets (in the Swan River estuary even within the confines of the city) and on country dams and river banks. It sometimes alights on moored dinghies and is often seen perched on a rock—un-wader-like habits. The consistency of occurrence of this bird in Western Australia is in great contrast with the situation in the east, where it is an exceedingly rare visitor.

The word "Common" in the name has no relevance to the Australian situation; it is what the species is called in Great Britain.

Greenshank

Tringa nebularia

Upper parts grey, lightly freckled with white. Forehead and above eye white. *Lower back and tail white,* the tail feathers being crossed by light brown bars which are inconspicuous at a distance. Under parts white. Iris, brown; beak, black, long and very slightly up-turned; legs, green-grey and long. Length, 12-13¾ in.; weight, 6½ oz.

Distribution: A summer migrant from the northern hemisphere. though a few individuals may spend the winter here. Though not so commonly observed on the estuarine parts of the Swan River, it is more frequently noted up-river and around the margins of lakes. It is relatively plentiful on the estuaries of the south coast. Its loud and shrill warning call—"chewey-chewey"—often foils the endeavours of the sportsman creeping on to ducks or other game.

Marsh Sandpiper

Tringa stagnatilis

In plumage pattern this species looks very similar to a small Greenshank, but has longer legs, projecting beyond the tail in flight, and a finer beak. Iris, brown; beak, black; legs, olive-yellow. Length, 8¼ in.

Distribution: A summer visitor from the northern hemisphere. Until recently there was no record of the species in the southern parts of the State but in January 1950 a specimen was collected for the W.A. Museum at Lake Jandakot. In April 1956 J. R. Ford saw one bird among a flock of 16 Greenshanks at Lake Cooloongup and in November 1964 two birds at the bore overflow at Hamelin Pool. P. R. Colston and C. J. O. Harrison saw a bird at Albany in March 1966 and two at Roebourne in June. J. A. K. Lane saw two birds at Point Waylen (Swan River estuary) in March 1972, T. Spence saw 13 at Carnarvon in September 1972, Dr. A. A. Burbidge saw one bird at Carnarvon in December 1972, R. Garstone saw four birds at Wardering Lake and seven at Corycup Lake in February 1973. The Marsh Sandpiper frequents fresh water rather than marine environments.

Redshank

Tringa totanus

Upper parts greyish-brown, with *rump white.* Wing dark brown with *a broad white band along the hinder margin.* Throat and breast streaked brown, abdomen white. Iris, brown; beak, long and straight, pinkish; legs, orange-red. Length 11 in.

Distribution: A northern hemisphere bird which migrates to south-east Asia. It is a new additon to the avifauna of Western Australia and there are two recent records. On September 2, 1974 L. E. Sedgwick saw one bird at Coral Bay, north of Carnarvon, feeding at the muddy edge of a shallow salt lake. John Darnell saw single birds on two occasions at salt intake works, 14 miles east of Port Hedland in September and October, 1975.

The rather musical 3-syllabled call, "tyoo-wuh-wuh," is distinctive.

Grey-tailed Tattler

Tringa brevipes

Upper parts *almost uniform slaty-grey,* wing quills darker, *no white on the rump;* a conspicuous white eyeline. Under parts white. In the nuptial plumage the throat and abdomen remain white but the breast becomes finely barred with grey and white, and birds in this phase are sometimes seen in Australia just after or prior to migration. Iris, dark brown; beak, long and straight, dark olive; *legs yellowish.* Length, 10¼ in.

Distribution: A summer migrant from the northern hemisphere. Like many of these migratory waders it is more plentiful in the northern parts of the State, but some individuals have been seen on the Swan River estuary (they occasionally appear at Pelican Point and one was noted on the rocky beach near the Crawley Baths, Perth, in March 1931). The common call is "ter-wee," twice or thrice repeated in fluty tones; a second call note is "troo-eet."

There is a second species, the Wandering Tattler *(T. incana),* which migrates to eastern Australia and has not so far been recorded from Western Australia. It may, however, possibly occur here and hence some reference to it is desirable. It is practically impossible to distinguish the two from field observations as the main diagnostic character is the relative length of the narial groove in the upper beak. In *T. incana* the groove extends about two-thirds along the length of the beak, and in *T. brevipes* for only about one-half of the length.

Terek Sandpiper

Xenus cinereus

Upper parts grey, rump uniform with back, with slight amount of white on outer tail feathers; white tips to secondary wing feathers visible in flight. Under parts white. Iris, brown; *beak long and upturned,* black, *lighter at base; legs, orange.* Length, 9 in. Weight, 2 oz.

In habits the bird resembles the Common Sandpiper, being very active, and bobbing its tail like that species. It is slightly larger than the Curlew Sandpiper.

Distribution: A summer migrant from the northern hemisphere. It is more common in the northern parts of the State, but in recent years there have been several records from the South-West; Pelican Point, Swan River estuary (first recorded on September 9, 1936), the mouth of the Moore River, Mandurah and Albany.

Turnstone

Arenaria interpres

Upper parts dark brown, mottled black and chestnut; white wing stripe; *lower back white* with a *second white band across upper tail coverts.* Throat white, *a broad black chest band* with biconcave lower border; abdomen white. Iris, brown; beak, dark olive-green, straight and robust; *legs, deep orange-yellow.* Length, 9¾ in. Weight, 3 oz.

Distribution: A summer migrant from the northern hemisphere. It is found on the marine beaches, especially those in rocky areas. It has a characteristic habit of turning over small stones, sea-weed and debris with its beak. The Turnstone is particularly plentiful around the margins of the salt lakes on Rottnest Island, on the coral beaches of the Abrolhos Islands, and on the limestone islets of Shark Bay. Very occasionally it appears at Pelican Point, Swan River estuary. It is rather scarce on the southern mainland beaches, but becomes more plentiful north of the mouth of the Moore River. It is one of the commonest waders on the tidal reefs of the North-West islands.

Chinese Snipe (Pin-tailed Snipe)

Gallinago megala

Upper parts streaked dark brown and chestnut, central parts of each feather being darker, edged lighter (chestnut). But some individuals may be uniformly drab, lacking differentiation on the back. Crown dark brown with a pale medial stripe and a similar stripe above each eye; loral stripe brown. Abdomen off-white, barred brown on the flanks. Axillaries strongly barred grey and white. Throat off-white, chest mottled brown and chestnut. Iris, brown; beak, brown, long, about 2½ in. (58-70 mm.); legs grey. Length, 10-11 in. Wing, 135-149 mm.; tail, 53-60 mm.

The two species of snipe recorded in this State are so similar in plumage characters that they can be scarcely distinguished by this means. The Chinese Snipe is slightly smaller (note wing and tail measurements). The decisive distinguishing features are to be found in the tail. Both snipe have a fan-shaped tail; in the Chinese Snipe there are *20 tail quills,* the *outermost being extremely narrow,* not exceeding 3 mm. in width. A good field character, when the birds are standing, is to note the length of the primary wing quills in relation to the secondaries. In the Chinese Snipe the two are almost coincident; in the Japanese Snipe the primary quills protrude well beyond the secondaries in the closed wing. This character, however, fails should in the future other Asiatic snipes be found to occur here; it is serviceable only for distinguishing between the Chinese and Japanese Snipes.

Distribution: Breeds in eastern Siberia, migrating south to the Philippines, Borneo, New Guinea and northern Australia. Until recently it was unknown south of the Kimberley Division. Whilst he was resident at Mt. Goldsworthy, in the Pilbara, John Darnell found this species to be fairly plentiful and widespread. At Mt. Goldsworthy and Mt. Blaize he observed up to six individuals at a time between late September and the end of March each season from 1972 to 1975, as well as two records in flash flood pools at Shay Gap in November 1973. He has mist-netted both species.

Both the Chinese Snipe and the Japanese Snipe are restricted to fresh water habitats, not saline ones.

Japanese Snipe

Gallinago hardwickii

Plumage characters very similar to the preceding, but back always streaked. Slightly larger, length, 12½ in. Beak, 64-77 mm.; wing, 153-167 mm.; tail, 60-68 mm. The essential differences are in the shape of the outer tail quills. There are *18 tail quills* in the Japanese Snipe and the *outermost ones* though narrower than the central feathers, are *distinctly wider* than those in the Chinese Snipe, being 4-6 mm. in width. In the closed wing the *primary quills protrude noticeably beyond the tips of the secondaries.*

Distribution: Breeds in Japan, migrating south to northern and eastern Australia. It was unknown in Western Australia, south of the Kimberley Division, until John Darnell mist-netted specimens

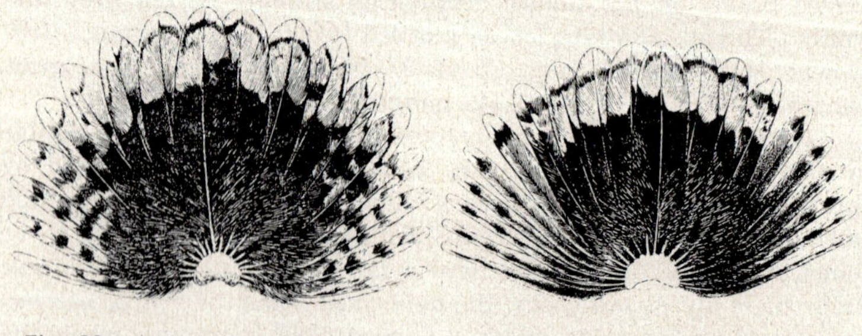

Fig. 33.—Left, tail quills of Japanese Snipe; right, of Chinese Snipe.
—After Lilian Medland

at Mt. Goldsworthy, in the Pilbara. He has noted single birds there in March, October and November, 1972; October, 1973, and February, 1975.

This snipe is a legal game-bird in eastern Australia.

Asiatic Dowitcher

Limnodromus semipalmatus

Back and wing coverts greyish brown, streaked paler. Upper tail barred brown and off-white, this pattern extending into the back. Under parts, chin and breast dappled brown and grey. Centre of abdomen white, flanks barred brown and white. Flight quills black. Pale whitish wing bar. Iris, dark brown; *beak, very long* (3¼ in.), straight and *slightly swollen at tip,* slaty black, pink tinge at base of lower mandible. Length, 13 in.

In the breeding plumage it is browner above and cinnamon-red below and some of this colouring is retained in new arrivals in Australia.

The Asiatic Dowitcher resembles the Bar-tailed Godwit, but is straight-billed and is slightly smaller. It is a bird of saline habitats and mud flats.

Distribution: Breeds in northern Asia, migrating to eastern India and China, the Philippines and Greater Sunda Islands. It has recently been found to be a fairly common migrant in the North-West. John Darnell found parties of up to six individuals feeding with godwits and other waders at the salt intake works 14 miles east of Port Hedland. On October 5, 1975, Ron Johnstone collected a specimen there for the WA Museum.

Great Knot (Japanese Knot)

Calidris tenuirostris

Upper parts greyish-brown, the feathers with dark brown centres; lighter on the forehead; wing quills dark brown; white wing stripe; a *complete white band across the upper tail coverts.* Under parts white, throat and upper breast spotted with dark brown. Iris, brown; beak, black, long and straight (1½-1¾ in.); legs, greenish. Length, 11-11½ in. Weight, 4¾ oz. Culmen, 41-47 mm.; wing, 175-184 mm. Size differences readily distinguish the two knots in the field, particularly if other waders are available for comparison. The Great Knot is conspicuous by its large size. It dwarfs the Curlew Sandpiper and the Sharp-tailed Sandpiper, and the Little Stint is a mere pygmy alongside it. It is larger than a Golden Plover and its body size is almost as big as that of the Grey Plover, but with its notably long beak taken into consideration it would be longer. Great Knots often wade into shallow water, sometimes up to their bellies.

Distribution: A summer migrant from the northern hemisphere. This is one of our rarer waders but in the South-West it has been observed at Garden Island, the Swan River estuary, the ocean beach at Mandurah, Bunbury and the south coast (Wellstead estuary, Bremer Bay). It does not frequent inland or fresh water localities.

Knot

Calidris canutus

Upper parts grey, with darker spots; wings darker, with a white wing stripe; forehead whitish with a white eyebrow line; upper tail coverts barred white and dark brown but *appearing wholly white* at a distance. Under parts white, breast marked with grey. Birds in assuming nuptial plumage have traces of rusty-red on the upper and under parts. Iris, brown; beak, dark olive, long and straight; *legs, olive-green.* Length, 9½ in. Culmen, 31-35 mm.; wing, 157 mm.

Knots are slightly larger in size than Curlew Sandpipers and Sharp-tailed Sandpipers, rather shorter-legged and of an upright carriage when on the alert. The rump is not so clearly white as in the Curlew Sandpiper.

Distribution: A summer migrant from the northern hemisphere. One of the rarer visitors but not infrequently seen on the Swan River estuary (Pelican Point).

Sanderling

Calidris alba

Upper parts *light grey,* the feathers centred brown; wing quills blackish-brown with *a broad white wing stripe;* forehead white; middle upper tail coverts brown, outer white. Under parts and sides of foreneck pure white. In the autumn some individuals assume traces of the nuptial plumage before migration—the back becomes blackish-brown with prominent white markings. Iris, brown; beak, black, straight; legs, black. Length, 7¼ in.

From other waders of its size group it can be readily distinguished by the light body plumage and the *very white appearance about the head.*

Distribution: A summer migrant from the northern hemisphere, the Sanderling is essentially a denizen of the ocean beaches. It rarely frequents the estuaries and never inland and fresh waters. It does not appear to be common south of the mouth of the Moore River.

Curlew Sandpiper

Calidris ferruginea

Upper parts greyish-brown; white eyebrow stripe; wing quills darker brown; white wing stripe; a *complete white band across the upper tail coverts.* Under parts white. In the spring and from February onwards traces of the rusty-red nuptial plumage appear, both above and below, and some individuals may become almost wholly reddish. Iris, brown; beak, black, long and *slightly down-curved; legs, black.* Length, 8-8½ in.; weight, 2 oz. Culmen, 33-42 mm.; wing 121-137 mm.

Distribution: A summer migrant from the northern hemisphere, but many individuals of grey plumage may winter here. It is one of the most plentiful of the visiting shore-birds and on the Swan River estuary its numbers are usually only exceeded by the Little Stint, though in some summers, as that of 1947-48, it may be comparatively scarce. It does not visit the fresh water swamps, but may occur on the inland brackish and salt lakes.

When the migrants begin to arrive in mid-September about 20% are in transitional summer-winter plumage, the majority being in winter grey. Body moult is general. Wing moult takes place after arrival and is completed by the end of December or early January. In the autumn the moult to the summer plumage begins and many birds before they leave in April are almost in full reddish summer plumage.

Fig. 34.—Curlew Sandpiper.

Red-necked Stint (Little Stint)

Calidris ruficollis

The smallest sandpiper. Upper parts greyish-brown, with darker centres to the feathers. Wing quills darker brown; wing stripe present. White eyebrow line. Middle feathers of the upper tail coverts dark brown, side feathers white. Under parts white. Traces of rusty-red feathers appear on the head, throat and back on birds getting ready to migrate. Iris, brown; beak, black, short and straight; *legs, black*. Length, 5½-6½ in. Weight, 1 oz. Culmen, 18-19 mm.; wing, 98-106 mm.

Distribution: A summer migrant from the northern hemisphere, and some numbers may winter here. This is the most numerous of the visiting waders and may occur in large and small companies on the sea beaches, estuaries and inland salt lakes.

The majority of the migrants when they begin to arrive in mid-September are in grey winter plumage but some individuals are in transitional summer-winter plumage. Wing moult begins immediately after arrival and the primary quills are completely renewed between mid-December and mid-January, when the secondary wing quills and tail feathers begin to moult. Birds in April show transitional winter-summer plumage and a few leave in full summer plumage.

Long-toed Stint

Calidris subminuta

Very similar to the previous species, and distinguishable in the field by its *yellow-green instead of black legs*. It is also rather darker in general appearance, with a less distinct wing bar. The wing is slightly shorter—91-97 mm.

Distribution: A rare summer migrant from the northern hemisphere. C. D. Fisher collected a specimen on Meeberrie Station, in the Murchison, in November 1966, and at Beverley in January 1968. Sight observations have been made on fresh water swamps near Perth.

Sharp-tailed Sandpiper

Calidris acuminata

Upper parts streaked grey, brown and black, *the crown being streaked rufous and black*. The *tail and rump are almost black in the middle, with white at the sides*. The breast is marked buff and brown, the throat and abdomen white, but the boundaries are diffuse and *not clear cut*. Iris, brown; beak, long and straight, black; legs, olive-green. The males are larger than the females. Length, male, 9 in.; female, 7 in. Weight, male averaging nearly 3 oz. (=74 gm., varying from 62 to 94 gm.); female, averaging 2 oz. (=56 gm., varying from 49 to 64 gm.). When the birds first arrive in the South-West from their northern migration they are much lighter in weight. Culmen, male, 24-28 mm.; female, 23-27 mm. Wing, male, 127-141 mm.; female, 124-134 mm.

The green legs and rump pattern at once distinguish the Sharp-tailed Sandpiper from the somewhat similar-sized Curlew Sandpiper, with which it frequently associates.

Distribution: One of the commonest of the migratory northern waders, found around the margins of fresh water swamps as well as on estuaries and inlets, and on the sea-shore. It occurs also on inland swamps and lakes and we have a record from Cosmo Newberry.

The winter exodus is complete and, unlike the situation in several other species of migratory waders (e.g. Bar-tailed Godwit, Little Stint and Curlew Sandpiper), no birds remain behind in southern

Fig. 35.—Sharp-tailed Sandpiper.

Australia during the winter months. An individual ringed at Pelican Point, Swan River, on January 5, 1961, was killed in Siberia, in the valley of the Yana River (lat. 67° 40′ N.) on May 28, 1961.

When the birds start to return about the end of September about half are in transitional summer-winter plumage and comparatively few in true winter "grey". Some arrive still in summer (rufous) plumage. Body moult is general, but individuals may retain some vestige of summer plumage throughout the season. By comparison with the Red-necked Stint and Curlew Sandpiper the all-grey period is very brief (end of December and early January). Primary wing moult begins in early October and is completed by mid-January, when the secondary wing quills and tail feathers are being replaced. Transitional winter-summer plumage begins to be apparent in February, when the numbers of birds start to show a decrease, but the wearing of full summer plumage before departure is rare.

Pectoral Sandpiper

Calidris melanotos

Very similar to the Sharp-tailed Sandpiper, upper parts brownish-grey, striped darker; a whitish eyebrow; tail pattern similar, but crown not so rufous. Throat white, breast streaked dark brown and white, *sharply marked off from the pure white abdomen.* Iris, brown; beak, blackish, basal third *yellow*; long and straight; legs, *yellow.* Length, 7-8 in. Weight, $1\frac{3}{4}$ oz. (female). Culmen, 28 mm.; wing, 130 mm. (female).

The abrupt demarcation between the heavily streaked breast and the white abdomen and the colour of the legs, appear to be the main characters by which the two species can be distinguished in the field. The paler base of the beak may be evident also and the darker head. If specimens can be examined in the hand they can be told apart by the form of the tail. In the Sharp-tailed Sandpiper the tail feathers are fan-shaped and regularly graduated to the long central feathers. In the Pectoral Sandpiper the terminal outline has a bi-concave appearance due to the relative great length of the central pair of feathers compared with the three outermost which are equal. The shaft of the first primary is entirely white in the Pectoral but has a brownish base in the Sharp-tailed Sandpiper.

Distribution: A summer migrant from the northern hemisphere. Its status is uncertain as so far only three specimens have been positively identified from Western Australia. The first was collected at Albany in March 1910, the second trapped at Pelican Point, on the Swan River, in November 1959 and the third trapped at a swamp at Moora in February 1973. Sight records have been made at Reid and near Roebourne.

In North America, where the species is common, it is stated that it favours wet, grassy meadows and mud flats, rather than sandy beaches. Its note is described as a hoarse, reedy chirp, recorded as "kerr" or "trrit-trrit," harsher than that of the Sharp-tailed Sandpiper.

Broad-billed Sandpiper

Limicola falcinellus

In general plumage resembles the Sharp-tailed Sandpiper and has the same type of tail pattern. It is, however, a smaller bird and approximates more to the Little Stint in size, but the *beak is longer and heavier* than that of the Sharp-tailed Sandpiper and is somewhat de-curved at the tip. A white eyebrow is present. Throat white. The breast pattern is not so well marked, being more sparsely speckled than in the Sharp-tailed Sandpiper. Iris, brown; beak, black; legs, greenish-brown. Length, about $6\frac{1}{2}$ in. Culmen, 29-33 mm.; wing, 105-110 mm.

Distribution: A summer migrant from the northern hemisphere. It has been observed at Broome and Onslow, and in November 1975, two birds were seen at Alfred Cove, in the Swan River estuary, by Gerry Nicholls.

Ruff (Reeve)

Philomachus pugnax

Upper parts greyish-brown, with darker centres to the feathers; narrow white wing stripe present. Wing quills black. Tail conspicuously white at sides. Breast grey, chin and abdomen whitish. Iris, brown; beak straight, blackish-brown; legs, grey. There is a slight web joining the bases of the middle and outer toes. The males are larger than the females, and in the summer plumage (most unlikely to be ever encountered in Australia) are boldly barred brown and white and with a conspicuous erectile neck frill of variable colour. Length, male, 11-12 in.; female, 8½-10 in.; young birds are smaller than the adults. Wing span, female, 16 in. Weight, female, 86 gm. Culmen, male, 34-38 mm.; female, 29-34 mm. Wing, male, 185-198 mm.; female, 150-168 mm.

The Ruff (Reeve) in winter plumage closely resembles the Sharp-tailed Sandpiper, but is larger, lacking the pale eye-brow stripe and rusty crown, and has a more erect stance.

Distribution: A very rare migrant from the northern hemisphere. It breeds from northern Europe to Asia and on its southern migration rarely reaches the Malay Archipelago and Borneo. Only three records are certainly known from Australia. The first was of a specimen collected at Buckland Park, Port Gawler, South Australia, on January 14, 1962. The second was of a bird recognised and collected on October 29, 1963, at the Lower King River, near Albany, W.A., by C. D. Fisher. Both birds were in the company of Sharp-tailed Sandpipers and were feeding in coastal salt marshes. The third occurrence was a sight observation by J. R. Ford in mid-November 1964 of one bird among Sharp-tailed Sandpipers at a fresh water lake, due to a bore overflow, at Hamelin Pool, Shark Bay.

These records, following so soon after each other, suggest that, now that the species is definitely known to occur in Australia, it may be a more regular visitor than first suspected. The two specimens collected were both females.

This species is one of the few kinds of birds in which there is no name for the species as a whole: the male is known as the Ruff and the female as the Reeve. The species is remarkable for the displays and behaviour of the Ruffs at nesting time. In their bizarre plumage they take up separate display grounds ("leks"), where elaborate posturings take place, and here they are visited by the females for mating. Ruffs are promiscuous and the Reeves alone incubate the eggs and care for the young.

STILTS and AVOCETS: Family Recurvirostridæ

White-headed Stilt (Black-winged Stilt)

Himantopus himantopus

Native name: Djan-jar-uk, Djanjarak (Perth).

White, except for the black wings (above and below); *black across the mantle* and a *black patch on the nape.* In the young the nape is grey and there is grey also on the crown of the head and behind the eye, the mantle and wings being tinged with brown. The *extremely long legs are flesh-pink* and are trailed far beyond the tail when the bird is in flight. There is no hind toe. The toes are webbed but the webs are small and do not reach to the end of the toes. The long delicate beak is black; iris, dark red. Length, 14-14¾ in.

Distribution: Occurs all over the State but is sparingly distributed in the South-West corner. In the coastal plain swamps, lakes and estuaries, however, it may be found as far south as Busselton. It has also been seen at the mouth of the Denmark River (C. F. H. Jenkins).

Nesting: In Australia this Stilt is inclined to nest in colonies. The nests may be fairly substantial with aquatic herbage, grass, rootlets, etc., and built up above the shallow water. At other times the Stilt merely lays its eggs in a depression in the ground just away from the water; a few pieces of grass are then used to line the nesting cavity. The eggs are olive, blotched and spotted with dark brown and are very like those of the Avocet, only smaller. Average size of 17 eggs from Lake Grace and Carnamah, 43 x 31 mm., varying from 40-46 x 30-32 mm. Average weight, 20 gm. They are 4 in number, pyriform, and are placed in the nest with the sharper ends inward. Nesting was recorded near Perth (at Herdsman's Lake) in September-October 1967 by Brian Hutchison.

Stilts have a distinctive call compared with other waders, a bark-like note, resembling the yapping of a small dog. Young birds, however, have a shrill sandpiper-like whistle. Family parties containing many young birds are seen in the Perth area and the coastal plain in the late summer. Flocks occur on the Swan River estuary over a large part of the year and odd birds appear from time to time on Rottnest Island.

Banded Stilt (Rottnest Snipe, Bishop Snipe)

Cladorhynchus leucocephalus

White, except the wings which are brownish-black, with portion of the trailing edge white. Adult birds have a *broad chestnut band across the breast* and a dark brown line along the abdomen. The *very long legs are flesh-pink;* the toes are fully webbed and there is no hind toe. Iris, brown; beak, long and slender, black. Length, 14¾-17¾ in. Weight, 7¼-9 oz.

The absence of black on the back (both the mantle and nape) distinguishes this species from the White-headed Stilt even in the immature stage, when the chestnut breast band is undeveloped. It is also shorter in the leg than that bird and has a slightly weaker, though essentially similar barking call. It does not bob its head, as does the other stilt. Its webbed feet allow it to paddle and swim, habits which the other Stilt does not share.

Distribution: The distribution of this stilt is correlated with that of the salt lakes and brackish estuaries. It has been recorded north to Point Cloates, the Pilbara and the Percival Lakes.

Nesting: The clutch consists of 3 or 4 eggs and they are quite unlike those of other stilts. There is much variation between different clutches and also between eggs in the same clutch. Some types approach in pattern the eggs of oyster catchers and the Crested Tern. The egg is chalky and without lustre; the ground colour varies from deep fawn to pure white. The markings are black and deep brown and usually form twisting and turning lines over the surface, though sometimes they are blotches of varying intensity. Average measurements of 30 eggs from Lake Grace, 54 x 38 mm., varying from 47-56 x 36-41 mm. The young when first hatched are covered with pure-white down—the only such instance in the order of wading birds.

Although the Banded Stilt was first scientifically named in 1816, its nesting habits eluded investigators until 1930, when the first known nesting colony was discovered at Lake Grace. In August of that year "thousands" of birds were reported nesting on a sandy spit on the lake by Mrs. B. E. Cannon, of Kukerin, and eggs and young birds were collected for the Western Australian Museum. By a remarkable coincidence, another nesting colony was located in the following December at Lake Callabonna in South Australia. It appears, however, that although not fully appreciated at the time, a colony nested at flooded clay-pans near Quinn's Find, about 40 miles south-west of Nannine, in June 1923. Since the report of the original discovery at Lake Grace evidence was gathered that they had nested at the same time at Lake King, 50 miles to the east. No further nesting was recorded until August 1945, when an abortive attempt was made again at Lake Grace, and another, also an abortive effort, in August 1946—on both occasions the nests were flooded out. In November 1960, I. C. Carnaby found preparations for nesting at Wagga Wagga Lake, in the Yalgoo district, but this failed owing to the lakes drying out. In August 1971, W. H. Butler encountered adults and young south of Durba Spring, and near Well No. 11, on the Canning Stock Route, and presumed nesting had taken place earlier in the year at Lake Disappointment. In July

1973, Police Sergeant A. Middleton reported nesting at Lake Ballard, 8 miles north-west of Menzies and sent newly hatched chicks to the South Perth Zoological Gardens. Nesting occurred here also the following July. In May 1975, following observations by Mr. and Mrs. A. G. Wells, of concentrations of about 3,000 adult birds at Lake Ballard, Mr. Nicholas Kolichis chartered a light plane, with Mr. J. Woods, and they made an aerial survey of the area. Lake Ballard was a vast sheet of water with numerous islets. On two of them Banded Stilts were nesting; on one there being hundreds of large flightless young. At Lake Marmion, 20 miles east of Menzies about 2500 nesting pairs were present (eggs and young on May 26). Nesting at some time in the recent past had also taken place at the Percival Lakes (lat. 21 deg. 22' S., long. 124 deg. 44' E.), near the northern section of the Canning Stock, for in September 1975 the W.A. Museum received fragments of eggs, stil pigmented, found by a Geological Survey party partly buried in clay.

The nesting of the Banded Stilt is evidently controlled by rainfall over the extensive salt lake system of inland Western Australia, and the birds gather to those lakes where optimum conditions exist—these must be related to the period of maxima attained by their main food items, the brine shrimps (*Artemia salina* and *Parartemia* spp.), which are probably controlled by water-levels and relative salinity.

The summer-time aggregations of these birds on the salt lakes at Rottnest Island have long been famous. Whilst the Governor's summer residence was maintained there they were preserved for his shooting parties, bush brakes being erected around the lakes for the purpose.

Avocet

Recurvirostra novæhollandiæ

Native name: Ya-jin-goo-rong (Perth).

General body plumage white, with *chestnut head* and a broad white band across the middle of the brownish-black wings. The mantle is white with two black stripes along it. The long, delicate black *beak is pronouncedly up-curved,* and the *very long legs are a light grey-blue.* There is no hind toe, and the toes are fully webbed. Iris, reddish-brown. Length, 18 in.

Distribution: Occurs all over the State but is rare or absent in the extreme South-West, but it visits the coastal plain estuaries in the summer months as far south at Busselton. Some birds are usually to be seen at most times of the year on the Swan River estuary (Pelican Point). They have also been seen at the mouth of the Denmark River (C. F. H. Jenkins).

Nesting: The Avocet lays its eggs in a hollow in a clump of salt-bush, or other vegetation, growing in or close to the water, on the margins of salt-lakes. A few rootlets are placed in the hollow. The clutch is of 4, sometimes 3 and rarely 5. The eggs are pyriform in shape and are always placed with the four pointed ends in the centre. They are olive in ground colour, marked with dark brown in the form of roundish blotches and large spots. They are very like those of the White-headed Stilt but are larger. Average size of 30 eggs from Lake Grace and Lake King, 51 x 35 mm., varying from 46-53 x 33-37 mm.

The flight is rather laboured, and when flying the Avocet carries its legs stretched out straight behind.

STONE-CURLEWS: Family Burhinidæ

Southern Stone-Curlew

Burhinus grallarius

Native names: Windoo (Godfrey's Tank, Upper Canning Stock Route); Weela-murra (Coolawanyah); Wee-lo (North-West Cape and South-West); Weelawodda (Lyons River); Wil-yu, Welojabbin (South-West); Wee-loo (Pallinup River).

A large brown bird with the upper parts *strongly streaked* with buff and white and the wings spotted. The under parts are white, boldly streaked with brown. Iris, bright yellow, and the naked skin in front of the eye black; beak, dark horn; legs, olive-green. Length, 22½ in.; wing span, 3 ft. 1 in.

Distribution: Resident all over the State.

Nesting: A real "ground-nester" in that it makes no nest at all, the Stone-Curlew merely lays its eggs on the ground. The eggs are 2 in the clutch and their colouring ordinarily assimilates well with the appearance of their surroundings. They can be described generally as stone colour blotched with dark brown, but the ground colour varies with the nature of the ground on which the eggs are laid. There is sometimes variation in size, not only between eggs of different pairs of birds, but between the two eggs in the clutch. Average measurements, 53 x 39 mm. For some days after the eggs are laid the sitting bird will slip away quietly if the site is approached. When hatching is near, however, the bird will not leave the eggs but will remain on them and stretch its head and neck flat on the ground. The bird has been known to move its eggs several yards away after they had been examined by a human being. Both sexes share in the incubation.

This is another of the wading birds which has forsaken watery habitats for the bushland, where it leads a nocturnal existence. The loud, penetrating call of "weeloo" is as characteristic of the night noises of the bush as the "morepork" call of the Boobook Owl and might be heard in the outer suburbs of Perth (e.g. Cannington, 1954). So secretive are the birds and so effectively concealing is the colour scheme that it is an extremely unusual experience to flush a Stone-Curlew in the daytime. So great is the bird's reliance on camouflage that it has been known to crouch on the ground with extended neck at our approach and allow itself to be picked up and again placed on the ground without altering its stiffened posture.

Beach Stone-Curlew

Esacus magnirostris

Native name: Wee-lo (North-West Cape).

Upper parts brown, *unstriped;* shoulder of wing brown, upper wing coverts light grey, white wing stripe. Around eye white, dark brown eyebrow stripe and a dark brown patch behind the eye. Chin white, throat and breast grey, abdomen white. Iris, straw-yellow; beak, large and heavy; legs, greenish-yellow. Length, 20-22 in.; wing span, 3 ft. 6 in. Weight, 2 lb. 3 oz.

In life the bird has the appearance of a miniature Bustard.

Distribution: A northern species which ranges down the North-west coast to Point Cloates.

Nesting: As with the previous species, no nest is made and the single egg is laid on the beach without any material being placed around it. It is creamy-white, heavily marked with blotches of olive-brown and, in some cases, hair-like lines of the same colour. Size, 63 x 45 mm.

PRATINCOLES and COURSERS: Family Glareolidæ

Australian Pratincole

Stiltia isabella

Upper parts rufous-brown, wings brown, with long black flight quills; tail black, with the outer tail feathers and the upper tail coverts white. Throat greyish-white; breast brown; abdomen rich dark brown; lower abdomen and under tail coverts white; under wings dark; with black axillaries. Iris, brown; beak curved, black, red at base; legs, brown. Length, 8¼-9¾ in.; wing span, 20-23 in.; weight, 2½ oz.

Differs from the Oriental Pratincole in being more rufous on the upper parts and in lacking the circular throat band. It also has *longer legs,* and the *tail is short* and not forked. The flight quills are finer and more attenuated.

Distribution: This pratincole ranges widely over inland eastern and northern Australia but it has only rarely been positively identified in the area treated in this book. In January 1959 J. R. Ford saw one individual at the Harding River near Roebourne and in August 1965 G. A. Lodge saw three at Yanrey Station, 70 miles south of Onslow. Michael Brooker saw 12 near Carnarvon in August, 1972. J. A. K. Lane saw one at Pelican Point in December 1972. That it may also penetrate into Western Australia from the south-east is suggested by an observation of three birds at the aerodrome at Koonalda in South Australia, 60 miles from the State boundary (V. N. Serventy, February 1957), and some were seen at Kanandah (near Naretha) in January 1971 and 50 miles N.E. of Rawlinna in September 1971 (M. G. Brooker).

Nesting: The nest may be a depression in the soil, or the eggs may be laid on the bare ground. The two eggs which form the clutch are blunt oval in shape, sandy brown in colour, irregularly spotted and blotched with brownish-black and with submerged grey markings. Size, 30-33 x 23-25 mm. Mr. Lodge found a nest at Yanrey, the first recorded breeding south of the Kimberley Divison.

Oriental Pratincole

Glareola maldivarum

Upper parts olive-brown, wings darker; *tail and coverts pure white,* tail tipped dark brown; a *black line passes* from the lores below the eye, meeting its fellow across the breast, enclosing the *buff throat;* rest of breast rufous-buff, abdomen white. Iris, dark brown; beak, black; legs, blackish-brown. Length, 9 in. In this species the *tail is long and deeply forked. The legs are short,* compared with the Australian Pratincole.

Distribution: A summer migrant from the northern hemisphere, occurring normally as far south as Point Cloates. However it has been seen at the bore overflow at Hamelin Pool, and, in January 1967, two birds were present at Pelican Point.

GULLS and TERNS: Suborder Lari

SKUAS: Family Stercorariidæ

Southern Skua (Great Skua)

Stercorarius skua

(formerly *Catharacta antarctica*)

Entirely dark brown in colour with a *broad white wing band* caused by the white bases of the flight quills. This shows conspicuously when the bird is viewed in the air from beneath. Iris, brown;

beak, dull black; feet, jet black. Length, 24½-25¾ in. Weight, 3½-4½ lb.

Distribution: A winter visitor from the south, between April and November, and on our coast recorded as far north as Geraldton, including the Abrolhos Islands. It is commonly seen, singly and in pairs, in Cockburn Sound and King George's Sound. It breeds on Marion, Crozet, Kerguelen, Heard and Macquarie Islands.

The bird is a scavenger on flotsam and jetsam at sea and is tamer than most sea birds in approaching fishing boats when the catch is being gutted—at times it will pick up a tit-bit from a person's hand. It is a common attendant on shipping in Fremantle Harbour, in some seasons as many as fifty birds being present at a time. All the southern ports are visited and at Albany the skuas may be seen from the town streets foraging with the gulls on the foreshore of the inner harbour.

Artic Skua (Richardson or Parasitic Skua or Jaeger)

Stercorarius parasiticus

This species exists in two colour phases: (a) Upper parts slaty-black; under white, with a mottled brown band across the upper chest and extending down along the flanks. A mottled brown and buff band passes behind the head around the nape. The under wings are dark. The *two middle tail feathers are produced nearly two inches beyond the rest of the tail, and taper to a point.* Phase (b). Similar in all respects but the under parts are wholly dark, uniform with the back, but sometimes possessing the light nuchal collar. Iris, brown; beak, strongly hooked, dark olive-grey, tip black; legs, jet black. Length, 18 in.; wing span, 40-45 in. Weight, 1 lb. 3 oz.

Distribution: A summer visitor from the northern hemisphere. The breeding range is circumpolar "throughout the barren arctic grounds" and the birds visiting Australia are no doubt those nesting in the tundra marshes of eastern Siberia.

Though common in eastern Australia during its southern migration and penetrating through Bass Strait into South Australia, this skua is less plentiful in Western Australia. However, individuals are generally to be seen in Cockburn Sound, between Fremantle and Rottnest Island, and around the south coast to the Archipelago of the Recherche. The birds are usually seen between August and April, but an occasional individual may winter in the south. One such was seen off Christmas Island (south-east of Israelite Bay) in July 1947.

Skuas harry terns and gulls into giving up any food they have collected, but feed quite considerably on their own account, being essentially scavengers. They even feed on land fauna blown to sea, such as grasshoppers. Like all skuas they show little fear of man and will approach very close to boats if food scraps are offering.

Pomarine Skua (Twist-tailed Skua)

Stercorarius pomarinus

This skua also has two colour phases, a light and dark. In the former the upper parts are slaty-black, with a light collar, suffused with yellow, around the nape. Under parts white but across the chest is a band of dark brown, mottled with white. Part of the inner vanes of the wing quills are white, showing up as a *white wing bar,* conspicuous in flight. The middle tail *feathers* are produced beyond the rest of the tail, but are *blunt-ended and twisted.*

Immature birds, even of the light phase, have the breast dark brown, narrowly barred buff, the under wing coverts mottled dark brown and white and the *upper tail coverts white or black barred white.* As such individuals are numerous the white in the wing and on the tail readily distinguishes them from the Arctic Skua, particularly as the species is in moult during its sojourn in Australia and the characteristic tail feathers may not be evident.

Iris, brown; beak, strongly hooked, olive-grey, tipped darker; legs, light blue tarsus and black toes in very young birds and wholly jet black in more mature individuals. Length, 19-21 in. Weight, 1 lb. 5 oz. - 1 lb. 10 oz.

Distribution: A summer visitor from the northern hemisphere. Nesting occurs mostly within the Arctic Circle and visitors to Australia probably originate from eastern Siberia.

Like the preceding species this skua is much more abundant in eastern Australia than in the west, where it has been reported between Fremantle and Rottnest Island and along the west coast. It is much rarer here than the Arctic Skua. The Pomarine Skua is the common summer skua along the New South Wales coast but it does not pass Bass Strait into South Australia in such numbers as its smaller ally, the Arctic Skua.

GULLS and TERNS: Family Laridæ

Silver Gull

Larus novæhollandiæ

Native names: Dje-je-nup, Djijnak (South-West); Jeranger (Pallinup River).

The common gull of our coasts and needing no detailed description for its recognition. Head, body and tail white, back grey. The outer wing quills black with white spots near the tips and the remaining wing quills with black sub-terminal bands. An orange to red eyelid; beak, from base to nares orange to bright red, and usually dark from nares to tip (rarely entirely red); legs, brownish-red, occasionally orange, bright red or dark wine red. The beak and legs are usually duller in colour during the summer and for a brief period in late winter. They appear to be at their brightest in late autumn and early winter and again in the spring (C. A. Nicholls). Length, 15½-17 in. Wing span, 3 ft. - 3 ft. 2 in. Weight, 11-12 oz. In young birds the back and wings are mottled with brown, the legs are brown and the beak and iris dark. These young birds have a weaker, shriller note than the adults, tern-like in quality.

Distribution: Occurs generally along the entire coastline, visiting the estuaries and coastal lakes, as well as Lake Muir, but is rarely reported inland.

Nesting: The Silver Gull lays its 1 to 3 eggs in a depression in the rocks or between rocks, or among short vegetation if nesting inland. A few pieces of grass are added to form a bed for the eggs which are subject to considerable variation in colour. Some sets have the ground colour pale olive-green and are thickly covered with various-shaped markings—blotches, spots or lines—of sepia or black. Other sets may have the ground colour blue, in others it may be white, and in others again pink. They vary also in shape and size, some being long and tapering and others short and broad. Average size of 78 eggs from Safety Bay and Rottnest Island, 53 x 38 mm., varying from 48-57 x 35-42 mm.

According to observations made by V. N. Serventy and S. R. White, the Silver Gull has two nesting seasons in the year on the islands of Safety Bay; the autumn nesting reaches its peak (for eggs) during May and the spring nesting in September, the earlier season having about twice as many breeding birds as the later. T. Carter published data which indicate a similar state of affairs at Lake Muir. J. R. Ford has established that a double breeding season occurs on the west coast islands north to Beagle Islands. A spring breeding, at least, occurs in the Abrolhos area, and both an autumn and spring one in Shark Bay. From counts at Safety Bay 70 per cent of nests contain two eggs and the remaining 30 per cent have one and three eggs in about equal proportions. Miss C. A. Nicholls proved, by ringing, that individual pairs produce two broods in the year, in autumn and spring. Nesting colonies have been reported at the Archipelago of the Recherche (Goose, Middle and Charlie Islands), Breaksea Island, Eclipse Island (at South-West Islet), Wilsons Inlet, on a small islet off the mouth of the Margaret River, Safety Bay (Shag Island, Seal Island, Bird Island and The Sisters), Lake Walyungup, Garden Island, Carnac Island, Rottnest Island (including Dyers, Parrakeet, Joan and Green Islands, Phillip, Clune and Armstrong Rocks), Swan River estuary (Crawley baths, formerly), islands between Fremantle and Dongara (Edward, Lancelin, Whittell, Buller, Cervantes, Essex Rocks, Tern, Osprey, Favourite, Sandland (or Green), Fisherman, Milligan, Lipfert, Webb, Drummond Rock, Snag and Beagle Is.), Abrolhos group (West Wallaby, Rat, Suomi, Passage, Wooded, Pelsart and Coronation Islands), Dirk Hartog Island (Quoin Bluff), Shark Bay (the islands at the head of Freycinet Estuary, Slope, Egg, Pelican, Dorre and Bernier Islands), and the Monte Bello Islands. Inland localities include a fresh-water swamp at Woodanilling (Woody Swamp) and Lake Muir.

The Silver Gull has been protected throughout the State since 1945.

Franklin's Gull

Larus pipixcan

One of the black-headed gulls and smaller than the Silver Gull. The most conspicuous difference from the Silver Gull is the dark head—black in the breeding plumage and a dull black half-hood in the non-breeding. There is a small circlet of white above and below the eye. Mantle and wings very dark grey and wings in flight showing a clear white trailing edge which leads into a white band extending across the primaries separating the black wing tips from the dark grey of the inner wing. Under parts and tail white, with pale grey central quills. Iris, brown; beak, reddish when breeding, blackish later; legs, reddish brown. Length, 13½-14 in.

Distribution: A North American gull breeding on lakes inland of the Rocky Mountains and migrating south to the Gulf of Mexico in the Atlantic and south to Valparaiso on the Pacific coast of South America. It is a nomad to the Galapagos and Hawaiian Islands.

On the above known distribution local ornithologists were incredulous when Timothy Dixon, an English ornithologist temporarily living in Bunbury, announced he had identified a Franklin's Gull in the Leschenault Estuary on August 10, 1957. Although his detailed description was confirmed as that of a Franklin's Gull by the late W. B. Alexander it was doubted whether this could have been a natural occurrence, and the species was omitted from our previous editions. Subsequently there were several sightings of the species. A black-headed gull was reported at Jurien Bav in August, 1974. Early in March, 1976 Gerry Nicholls photographed one at Geraldton, authenticating its identification as a Franklin's Gull. On April 24, 1976 L. E. Sedgwick saw another, or perhaps the same individual, at Geraldton among Silver Gulls and was able to make a detailed description. Therefore there can be no doubt that the south-western coastline of Western Australia must be included within the extreme migratory range of Franklin's Gull, even if only occasionally.

The mystery is how the bird reaches our shores. Though there are more observers in eastern Australia there are no records from there, nor from northern Australia. The most likely explanation, on the sketchy existing data, is that some individuals may wander further south along the Chilean coast and are caught up in the West Wind zone, being carried eastwards into the South Atlantic and Indian Oceans, ultimately fetching up in south-western Australia.

Pacific Gull (Mollyhawk)

Larus pacificus
(formerly *Gabianus pacificus*)

Native name: Nga-ga-la (South-West).

This very large gull occurs in *two distinct phases* which are frequently mistaken as separate species. The adult is *white with black back and wings* (latter with small white spots near tip); tail white with subterminal black band. The *massive yellow beak* is red at the tip. Iris, creamy-white; fleshy ring around eye, yellow; legs, yellow. Length, 23-26 in.; wing span, 4 ft. 6 in. - 5 ft. 2 in. Culmen, 56-63 mm. Wing, 429-480 mm. Weight, 2-2½ lb. Males are larger than the females.

Immature birds are similar in size but are a uniform mottled brown. Beak, grey-buff with black near tip. Iris, brown; legs, brownish-black.

Distribution: Occurs along the coast as far north as Point Cloates and has been observed at Long Island off Onslow (N. E. Stewart). It is essentially a coastal bird and rarely enters estuaries or inlets, but is found in Albany Harbour, the entrance to Oyster Harbour, and sometimes in Fremantle Harbour.

Nesting: The bird nests on islands off the coast, not on the mainland. The nest is compactly made in rocks or in herbage and is constructed of grass, flower-stalks, etc. The 2 or 3 eggs in the clutch are olive-brown, blotched with brown and grey. Average size, 76 x 51 mm., varying from 65-81 x 49-54 mm. Weight, 90 gm. Recorded nesting localities are: Archipelago of the Recherche (Christmas (Daw), Round, Salisbury, Middle, Goose, Cave, Combe, Mondrain, Nares, Thomas, Woody and Gull Islands); an island off the mouth of the Warren River; Penguin Island (Safety Bay); islands between Fremantle and Dongara (Edward, Lancelin, Wedge, Buller, Whittell, Green, Cervantes, Essex Rocks, Boullanger, Tern, Whitlock, Escape, Favourite, islets near North Head (formerly), Sandland, Fisherman and Beagle Is.), the Abrolhos group (Pelsart, Jubilee, Wooded, Suomi, Passage, Long and East Wallaby Islands); Shark Bay islands (Leschenault, North Guano, Charlie, Slope, Sunday, Pelican and Dorre and Bernier Islands). Eggs have been noted in early October and large young in November on the islands of the Abrolhos group.

The Pacific Gull appears to have only recently extended its range from Shark Bay to Point Cloates. T. Carter resided at Point

Cloates between 1889 and 1902, and never observed it during that period. He first saw it when on a later visit to Carnarvon in 1913, some birds being present also at Point Cloates. Carter considered that the operations of the Point Cloates Whaling Station (functioning intermittently between 1912 and 1928) were responsible for the birds' appearance.

A widespread habit of this gull is to pick up large *Turbo* shells from reefs and drop them from a height on to rocky platforms, then diving down and feeding on the shellfish exposed by the fractured shell. Large limpet shells and sea-urchins have been seen to be dealt with in the same manner and the birds apparently have special "dropping" platforms on which a litter of broken *Turbo* and other material accumulates. The Pacific Gull is a predator on other colonial-nesting sea-birds and regularly haunts areas where beach seining is carried on, scavenging on the offal left by the fishermen.

The call note is a sharply ejaculated "oh oh", somewhat resembling the call of the Northern Hemisphere species, the Greater Black-backed Gull.

Kelp Gull (Dominican Gull)

Larus dominicanus

A large gull very similar to the Pacific Gull but differing from it in two main characters (1) the *tail quills in the adult are entirely white,* lacking the subterminal black band of the Pacific Gull, and (2) the *beak is of the normal gull shape,* much less robust than that of the Pacific Gull. Iris, silver-grey; fleshy ring around eye, reddish; beak, yellow with large reddish spot at tip of lower beak; legs, greenish yellow. Length, 23 in. Males are larger than the females and have a heavier beak. Culmen, male, 47-58 mm.; female, 42-51 mm. Wing, male, 386-434 mm.; female, 350-400 mm. Weight, male, 2-2½ lb.

Immature birds, like the Pacific Gull, are similar in size but are a uniform mottled brown, difficult to identify at sight from the Pacific Gull on plumage characters.

Distribution: In Western Australia recorded along the south coast (Esperance and Albany) and on the west coast near Fremantle and Jurien Bay. It is a circum-polar sub-antarctic species which has only comparatively recently colonised Australia, presumably from New Zealand.

Nesting: Thus far the only nesting records are from New South Wales, Victoria and Tasmania. Eggs and nest similar to the Paciifc Gull. Clutch 2 or 3 eggs, slightly smaller than those of the Pacific Gull. Average size, 70 x 50 mm., varying from 66-73 x 48-51 mm. Weight, 80 gm.

The species was not known to occur in this State until at the camp-out of the Royal Australasian Ornithologists' Union at Albany in October 1963. Mr. J. D. Watson, of Melbourne, saw one at Middleton Beach, in almost mature plumage. Subsequently Mr. J. R. Ford collected an adult bird at Jurien Bay in February 1954 and Mr. L. J. McHugh reported one at Esperance in April 1964. That the species had occurred much earlier in Western Australia, but was unrecognised, was demonstrated by Mr. Ford finding an immature bird, in first year plumage, in the collections of the W.A. Museum. It had been collected at Claremont in August 1924 and was mislabelled as a Pacific Gull. A bird ringed as a nestling at Moon Island, N.S.W., on December 12, 1962, was recovered at Shoalwater Bay, Rockingham, on August 29, 1966.

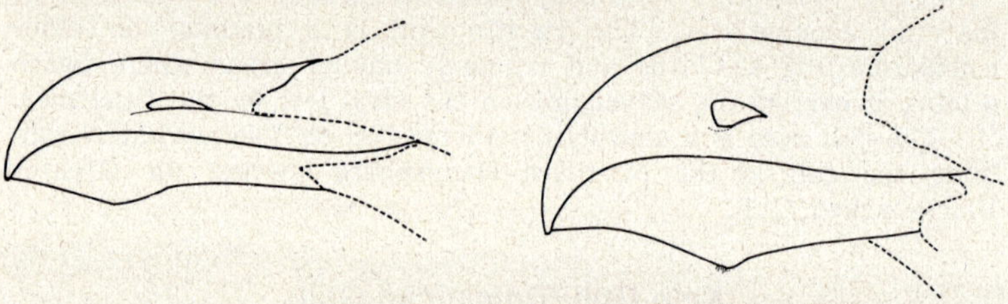

Fig. 36.—Left, beak of Kelp Gull; right, of Pacific Gull.

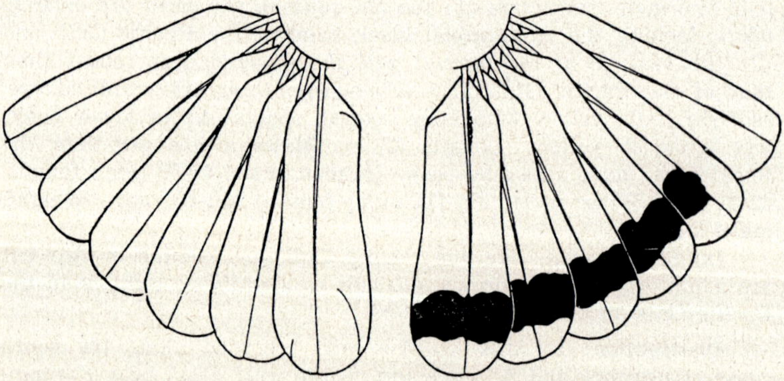

Fig. 37.—Left, tail quills of adult Kelp Gull; right, of adult Pacific Gull.
—After C. A. Nicholls

White-winged Black Tern (White-tailed Tern)
Chlidonias leucoptera

In the full breeding plumage, which is rare in Australia, the head and body are entirely black, wings grey (other than a small white patch on the shoulder) and the tail is a conspicuous white. Most of the birds likely to be seen in Western Australia have the head and body white with varying amounts of black mottling: the wings are grey, with white shoulder patches, there is a large black patch under the wing, and the tail is white. Young birds have brown mottlings in the plumage, under white and the tail grey. In the field the *pure white tail,* which is often spread out fan-wise, and the *contrasted speckled plumage,* are outstanding characteristics of the species (W. B. Alexander). Iris, dark brown; beak, vivid red; legs, orange-red, with the webs very much indented. Length, 8½-9½ in.

Distribution: This species breeds on the lakes and swamps of southern Europe, Algeria and central Asia, dispersing in the winter both north and south. It wanders to the Malay Peninsula and sometimes reaches the coasts of northern Australia.

In 1917 a remarkable invasion of thousands of these terns was witnessed in the South-West, the first occasion on which the species had ever been observed south of Cape York and the Northern Territory. The birds probably travelled from the north, for they were first reported at Doodlakine in the middle of March and they reached the coast, along a front extending at least from Wedge Island to Busselton, on April 8. Doubtless a strong easterly blow, which was then prevailing, directed their movements. Almost every swamp and estuary between Moora and Balingup seemed to have parties of the birds feeding over them and there were several hundreds on Herdsmans Lake. The birds remained about a month. Their chief food seemed to be a dragon-fly *(Hemianax papuensis)* which was present at the time in unusual abundance. The species was not reported again in the State until October 1945, when a flock was observed at sea near Broome.

In March 1956 there was another invasion in strength into the South-West, approaching that of 1917 in magnitude. They were first seen in the Perth area on March 4 and hundreds were seen on the coastal lakes to Mandurah. The last stragglers were noticed in the first week of April. A small irruption also occurred in March 1960. In February 1970 another small irruption reached the Perth area, birds being seen between Yanchep and Rockingham. All of these visits to the south were associated with tropical cyclones.

Marsh Tern (Whiskered Tern)

Chlidonias hybrida

This tern presents important differences between the nuptial and non-breeding plumages, but birds in the various phases are often seen together. The upper parts, including the wings, are grey, and the under parts white. In the full breeding plumage the crown of the head, to eye level, is entirely black, showing *no white on the forehead,* the throat is white and the rest of the *under parts slaty-black.* In the immature and non-breeding phases the under parts are wholly white and the black of the head is freely speckled with white, particularly on the forehead. Iris, brown; *beak and legs, wine-red.* Length, 10¼-10½ in.

The call is harsh and reedy, not so penetrating as that of other terns.

Distribution: A non-marine species, likely to be found anywhere in the State.

Nesting: There are few actual observations of the nesting of this bird in Western Australia. On September 28, 1917, C. L. E. Orton found a colony of 12 pairs nesting in a swamp 2 miles west of Moora. The nests are slight structures of a few fine rushes placed on top of samphires. In August 1948 P. J. Fuller noted a colony of about 7 pairs nesting on samphires on an islet in a salt lake about 17 miles east of Dowerin. In April 1960 Mr. Fuller found a colony of about 30 nests in a small islet in a claypan north-east of Nannine. Clutch 2 or 3 eggs, occasionally 4, measuring 35-38 x 26-28 mm. in size; colour greyish-olive to brownish-olive, sometimes dark stone, blotched and spotted with dark brown, often concentrated at the larger end. There are usually a few underlying spots of grey.

This tern is most frequently met in inland fresh water situations, though it does venture on salt water estuaries and occasionally in more marine habitats. On the Swan River it is a fairly frequent visitor, birds in the nuptial plumage being not uncommon. The metropolitan fresh water lakes and swamps are also its haunts, including, occasionally, such urbanised ones as Shenton Park Lake in Subiaco. Generally it is an irregular vagrant and may appear in distant parts of the State including the Nullarbor Plain, only after heavy rains and flood conditions.

Gull-billed Tern

Gelochelidon nilotica

Upper parts pale grey with black on crown of head and nape, and *no white forehead;* tail and under white. In the non-breeding plumage the head is streaked white and black. Iris, brown; beak *short and robust, black;* legs, *black.* Length, 14½-17 in. Weight, 10¼ oz.

Distribution: This is one of the inland terns, but may occur also on estuaries and sea-coasts. Until the 1950s it had been very infrequently reported by ornithologists and never in the South-West. However since then small parties have been observed at Morawa (1953, following flooding of the salt lakes), near Moora (1955), Lake Cooloongup and Lake Grace (1956), Hutt Lagoon (a flock of 40 seen by B. V. Teague and J. R. Ford, 1957), Murray River near Peel Inlet (1957), and Wanneroo and Jandakot Lake (1959).

Nesting: Until recently the only record of the nesting of the Gull-billed Tern in Western Australia was that of T. Carter, who, in May 1900, found a small colony of about five pairs commencing to nest on a swamp near Point Cloates; two nests were found, each of which contained a single egg. In September 1960 P. J. Fuller discovered a colony of about 60 pairs with eggs and empty nests on a samphire island in a brackish lake about 30 miles south-east of Yalgoo. The nests, 3 to 5 ft. apart, were depressions in the sand, some 1½-2 in. deep and completely lined with pieces of dead samphire. He also noted he found the birds nesting at Lake Nabberu in April 1965. Tom Spence noted nesting at Wooleen Station on the Murchison in October 1968.

The full clutch in the Yalgoo colony usually comprises 2 eggs, but I. C. Carnaby has found clutches of 4 and 5 eggs. The eggs vary in ground colours from buffy-white to various shades of light brown and from greenish-grey to light olive-green. The markings consist of irregular spots and blotches of blackish-brown, with underlying blotches of various shades of grey. Usually oval they are subject to some variation in shape, some eggs, even within the set, being oval as against eggs with one end rather pointed. Average size of 18 eggs from the Yalgoo colony, 52 x 37 mm., varying from 47-57 x 35-39 mm.

The call-notes have been recorded as "kuh-wuk, kuh-wuk," repeated several times, and a "che-ah" (H. T. Condon).

Caspian Tern

Hydroprogne caspia

The largest local species of tern, grey above with a black cap and nape, white below. There is *no white on the forehead.* Iris, dark brown; *beak, massive, bright red; legs, black.* Length, 23½ in.; wing span, 4 ft. 6 in. Weight, 1½ lb.

Distribution: Generally distributed all around the coast, also visiting the estuaries and some of the coastal lakes. It is frequently seen at Pelican Point, on the Swan River Estuary.

Nesting: The Caspian Tern nests in odd pairs or small groups, the largest colony reported being about one hundred pairs. The nest is merely a slight hollow in the ground—often in pig-face *(Carpobrotus).* The eggs are oval and of coarse shell; they are stone-grey or very pale brown and are sparingly marked with spots and blotches of greyish-black and umber. Average size of 19 eggs, 64 x 45 mm., varying from 61-68 x 43-46 mm. Weight, 60 gm. In about 60 per cent of nests, one egg only forms the set, in the remaining 40 per cent 2 eggs occur. Nesting localities recorded are: Israelite Bay, Archipelago of the Recherche (Goose, Douglas and Canning Islands), Green Island (Oyster Harbour), Porpoise Rocks and

Pelican Island (Wilson's Inlet), Nornalup Inlet (on an islet), Seal Island (C. Leeuwin), at Safety Bay (Penguin and Shag Islands), Rottnest Island (Green, Dyers, Clune, Parrakeet Islands and Armstrong Rocks), islands between Fremantle and Dongara (Edward, Lancelin, Green, Cervantes, North Essex, Boullanger, Tern, Favourite, Sandland, Milligan, Lipfert, Webb, Drummond Rock and East Beagle Islands), the Abrolhos group (West Wallaby, Pigeon, Long, Wooded and Pelsart Islands), Shark Bay (Pelican, Faure, Slope, Friday, Egg, Five and Sunday Islands), Frazer Island and on the mainland at Point Cloates, on many of the islands of the Dampier Archipelago off Onslow (reported on Enderby and a small islet off Double Island, Barrow Island), and the Lacepede Islands.

Roseate Tern

Sterna dougallii

Like the preceding species this tern has *no white on the forehead* in the nuptial plumage and the black of the head extends to below eye level. Back and wings grey, under white with a delicate rose flush on the breast. The *tail is white, strongly forked and the outer feathers projecting like streamers.* Iris, grey; *beak, red at the base and black towards the tip; legs, red.* Length, 13½-15¾ in.; wing span, 24 in. Weight, 3¾-4½ oz. Culmen, 34-42 mm.; wing, 192-210 mm.; tarsus, 19-20 mm.

Closely similar species are the migratory Arctic and European Common Terns and the distinguishing characters are described under those species.

Distribution: Predominantly a northern tern, until recently the furthest south nesting station being the Abrolhos Islands. In April 1960 a nesting colony was discovered at the Green Islets (100 miles north of Fremantle, between the mouths of the Hill and Moore Rivers). Apparently the species is spreading south, for until 1958 there was only one definite record of a Roseate Tern being seen in the Fremantle area (a storm-bound bird at Scarborough in July 1939). Most of the records of individuals further south and in South Australia (cited in our earlier editions) have since proved to be wrong identifications of the Arctic Tern. During 1958 and subsequently there have been several sight observations of Roseate Terns in Cockburn Sound, Rottnest and Carnac Islands.

Nesting: The nest is a flattened-out circular area on stunted saltbush or on the ground among coral fragments. Eggs one, very rarely 2, oval in shape, colour stone of various shades and marked with large blotches and spots of umber, grey and sepia. Average size, 41 x 29 mm. An autumn-breeding population was found by G. M. Storr on small cays in the Wallaby Group, Abrolhos, in June

1959, and others by Storr and J. R. Ford in April 1960 on Green Islets (south of Jurien Bay). Spring-nesting populations at the Abrolhos appear to breed later than the other terns. Egg-laying was recorded there in November 1926, and eggs were noticed starting to hatch on December 12, 1945. Heavily incubated eggs were recorded at Point Cloates in mid-November 1893, but at Bedout Island, off Port Hedland, J. T. Tunney found the species nesting in May.

Recorded nesting localities are as follows: Green Islets, Whittell Island, Buller Island, Cervantes Island and Beagle Island. At the Abrolhos nesting groups of nearly one thousand birds were observed at the north end of Pelsart Island (V. N. Serventy, 1940-41) and other groups have been observed nesting on Wooded Island (about 60 birds) and West Wallaby Island (about 150 birds). In December 1945 a colony of about 250 pairs nested at Wreck Point, Pelsart Island, and in January 1947, a colony comprising 2,582 nests was carefully enumerated by S. R. White and V. N. Serventy—the largest nesting aggregation ever recorded for this species. In 1948 the colony was of much smaller dimensions. It also nests on Beacon Island. T. Carter recorded a small nesting colony one year at Frazer Island, off Point Cloates, and J. T. Tunney found it breeding, but not in numbers, at Bedout Island.

The call is a harsh rasping cry.

Arctic Tern

Sterna macrura (= paradisæa)

Similar in size and plumage pattern to the Roseate and European Common Terns but the *under parts and sides of neck are usually pearl-grey* instead of pure white, leaving a well-defined white streak under the cap. In the nuptial plumage there is no white on the forehead, the black cap extending to the base of the beak. Upper parts slate-grey, rump and upper tail white; *long outer tail streamers.* Iris, brown; beak, *wholly red,* with no dark tips; legs, red. Length, 12-15 in. Culmen, 28-32 mm.; wing, 250-280 mm.; *tarsus, very short,* 15-16 mm. Wing span, 29 in.

In the eclipse, or non-breeding plumage, the forehead and lores become white and the beak and legs black.

The Roseate Tern can be distinguished from this species by its very much longer tail streamers which are wholly white and not dark grey on the outer vane as in the Arctic and European Common Terns. From the European Common Tern the Arctic may be recognised by the greyer under parts and much shorter legs. If specimens can be examined in the hand the pigment pattern of the outer primaries is a useful character in helping to separate the three species (Fig. 38).

At its wintering grounds the Arctic Tern utters a quiet call note not unlike the Australian Zebra Finch.

Distribution: A breeding species of the Arctic and Temperate regions of northern Europe, Asia and North America. In the northern winter it migrates across the Equator to Antarctic seas, among

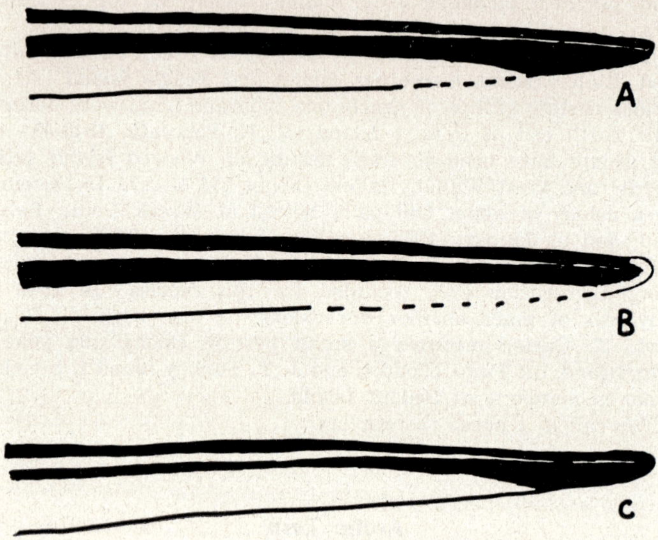

Fig. 38.—Pigment patterns (semi-diagrammatic) on the outer primaries of (A) European Common Tern from Coogee beach; (B) Roseate Tern from Bedout Island; and (C) Arctic Tern from Swan River. At a distance of 75 mm. from the tip of the feather the widths of the black band on the inner webs are 3.4, 3.6 and 2.4 mm. respectively. In the Arctic Tern the band on the inner web is lighter grey than on the outer, while in the other species both are a similar colour. In unabraded wings of the Roseate Tern the white extends round the tip of the feather as indicated, but in abraded feathers this character cannot be seen.

—After G. M. Dunnet

the pack-ice. It is a passage migrant in South Africa on its way south along the east Atlantic fly-way. G. M. Storr has explained the occurrence of individuals in southern Australia between October and December as birds which had rounded the Cape of Good Hope, from the northern Atlantic, and being "incapable of penetrating the zone of the 'roaring forties' . . . are carried for various distances eastwards before they break through to Antarctic seas."

There are eight positive records of Arctic Terns in southern Australia, five from this State and one each from South Australia,

Victoria and Tasmania. The occurrence of the Arctic Tern in the Australian region was recognised only in 1952, when W. B. Hitchcock identified a specimen in the National Museum of Victoria, which had been found washed up on the beach at Bunbury, Western Australia, on October 26, 1926. It was an adult female in nuptial plumage. Other Western Australian specimens have been found at Pelican Point, Swan River estuary (October 10, 1955), Naval Base, 8 miles south of Fremantle (May 16, 1956), C. Leeuwin (about October 8, 1956), and Madora Beach, north of Mandurah (June 10, 1963).

Two of these birds had heightened interest as they carried leg rings of northern European origin. The bird from Naval Base had been marked as a juvenile on July 5, 1955, in the Kandalaksch sanctuary on the White Sea, about 125 miles south of Murmansk. The Mandurah bird had been ringed as a nestling on June 27, 1962, on an islet in the Archipelago of Stockholm, Sweden. Apparently both the Russian and Swedish ringed birds had perished on their way north after their first season in the Antarctic.

Common Tern

Sterna hirundo

Very similar in appearance to the Arctic and Roseate Terns and with a similar seasonal change in plumage. From both of these terns it can be distinguished by the *shorter tail streamers, which do not project beyond the closed wing tips*. From the Arctic it differs in the *purer white under parts* and *longer legs,* and (in the nuptial plumage) by the *black tip to the red bill.* Identification of specimens in the hand can be aided by wing quill characters illustrated on p. 242. Iris, dark brown; beak, orange-red with black tip (in non-breeding state largely black); legs, red (reddish black in non-breeding state). Length, 12-15 in. Culmen, 33-39 mm.; wing, 261-267 mm.; tarsus, 19-21 mm.

The above is the description of the European Common Tern *(S. h. hirundo)*. The Eastern Common Tern *(S. h. longipennis)* differs mainly in the colour of the beak, which is wholly black, minutely tipped with horn. As in the case of the Common Sandpiper the word "Common" in the name has no significance in the Australian context; it is the name of the species in Great Britain.

Distribution: Breeds in the temperate regions of northern Europe, Asia and North America, south of the breeding area of the Arctic Tern. It migrates in the non-breeding season to South America South Africa and Australia.

Until recent years this tern was not considered to migrate as far as Australia, but we now know that individuals from both extremes of the breeding range, in western Europe and eastern Siberia, migrate to southern Australia. Only one western specimen has ever been recorded from this State and that, by a remarkable coincidence, was a ringed bird. It was found in an exhausted condition on Coogee beach, four miles south of Fremantle on January 7, 1956. It carried a Swedish leg-ring which showed that it had been banded as a nestling on July 9, 1955, at Marum, near Norrtalje in the Swedish province of Uppland. "This specimen had therefore travelled 13,000 miles (around the Cape of Good Hope) in six months" (G. M. Dunnet).

Birds of the eastern subspecies (*S. h. longipennis*), breeding in north-eastern Asia from east of Lake Baikal to Kamchatka and the Commander Islands, migrate to coastal parts of northern and eastern Australia, from Port Hedland to Victoria, and there is one record from South Australia. That they did so was not recognised, however, until 1941 (Ernst Mayr and K. A. Hindwood). Between Port Hedland and C. Keraudren John Darnell has found it a plentiful and regular summer visitor from early September to the end of April. Pre-departure flocks may swell to 800-1,000 individuals.

Crested Tern (Diver, Black-headed Gull)
Sterna bergii

Native names: Kal-jeer-gang, Kaljirgang (South-West).

The second largest of our terns; grey above and white below. The crown of the head and nape black, and *forehead white.* The black crown feathers are capable of being erected as a shaggy crest for display purposes in the breeding season. Iris, brown; *beak, yellow with a greenish tinge;* legs, brownish-black. Length, 17-19¾ in.; wing span, 39 in. Weight, 12½-13½ oz.

Distribution: Generally distributed along the whole coastline, and entering the inlets and larger estuaries.

Nesting: The nest is merely a depression in the ground, among sticks and shell fragments. One egg in the clutch, but occasionally 2 eggs are found—probably the product of two different pairs. The eggs are very variable in ground colour and markings. The former may be stone-grey or brown. The markings consist of streaks, smudges and blotches of different colours—usually black—and the streaks often have a form like hieroglyphics. No two eggs are exactly alike and in this, and in their markings, they are analogous to the eggs of the Banded Stilt. Average size of 26 eggs from Rottnest and Abrolhos Islands, 61 x 41 mm., varying from 56-68 x 40-44 mm. Recorded nesting localities are: Rocky Island (east of Hopetoun),

Glasse Island (Bremer Bay), Gull Rock (Nornalup), Seal Island off Cape Leeuwin, Hamelin Island and the old jetty at Hamelin Bay, Safety Bay, Carnac Island, Dyers Island and Green Island off Rottnest Island, islands between Fremantle and Dongara (Edward, Lancelin, Cervantes, South Essex, Tern and Sandland Islands), the Abrolhos group (West Wallaby, Beacon, Dick's, Goss, Wooded and Pelsart Islands), Shark Bay (Sunday, Meade, Mary Anne, Dorre and Koks Islands), Frazer Island off Point Cloates, Anchor Island, Bedout Island, Monte Bello group (islet north-west of Bluebell Island), Low Rocks (Admiralty Gulf), Long Reef (Sand Island), and Ashmore Reef (Middle Island). On Seal Island the birds lay late in November; at the Abrolhos they lay between July and November, but on Frazer Island T. Carter found eggs at the end of April, while on Bedout Island J. T. Tunney found eggs in May. Autumn nesting has also been observed on the south coast at Rocky Island.

This is the commonest tern on our South-Western coast and on the Swan River it is always to be seen, upstream as far as Maylands and on the Canning River to Riverton. It does not visit coastal swamps and lakes but a few do appear on Lake Muir, an extensive brackish lake 30 miles north of Nornalup. It is particularly numerous on the south coast and the Archipelago of the Recherche, but seems to be less plentiful in the north. On the Abrolhos, for instance, it must be ranked as one of the less abundant species.

The Crested Tern obtains its food by diving on shoals of pelagic fish and at the Abrolhos has been observed feeding its young on large sea pilchards *(Sardinops neopilchardus)*. On the south coast it ranks with the Gannet as an important indicator species of the presence of pilchards.

Lesser Crested Tern

Sterna bengalensis

Similar to the Crested Tern but smaller and with the *beak orange in colour instead of the lemon-yellow* of the other bird. In the nuptial plumage the black area of the hindneck and crown extends practically to the base of the beak, and in the field the forehead, however, is extensively white. Like the Crested Tern the black crown feathers can be erected into a crest for sexual display. Iris, brown; legs, black. Length, 14½-16½ in.; wing span, 35 in. Weight, 8½ oz.

Distribution: Occurs along the north-west coast as far south as Peron Peninsula, Shark Bay. At sea this species is appreciably less plentiful than the Crested Tern.

Nesting: The Lesser Crested Tern has been recorded nesting at Adele Island (young in early May), and at Bedout Island (eggs in early May). The egg is different from that of the larger Crested Tern and is also smaller, averaging 55 x 36 mm. It is light stone in colour, blotched and spotted with very dark purple and with lavender marks appearing as if below the surface of the shell. It is laid on the sand without any nest being made.

Fairy Tern

Sterna nereis

The *smallest tern,* very pale grey above, white below, with the crown of the head and nape black. *Forehead white.* In the distance the birds look wholly white with the top of the head black. Iris, brown; *beak, orange-yellow* (occasionally with a minute black tip); legs, yellow or orange. Length, 8½-10½ in.; wing span, 20 in. Weight, 2 oz.

Distribution: The Fairy Tern is generally distributed along the coast, north to King Sound.

Nesting: In the South-West this is a comparatively late nesting tern, eggs being found from early November to early January. The eggs, 1 or 2 in a clutch, sometimes 3, are laid on the ground among shell fragments and beach flotsam. They are yellowish-grey, blotched with brown and grey. Average size of 9 eggs from Rottnest Island, 35 x 25 mm.; varying from 33-36 x 24-26 mm.

Small colonies of these terns nest not only on islands but on sandy beaches on the mainland. Recorded localities are Seal Island (off C. Leeuwin), Safety Bay, Lake Walyungup, Garden Island, Rottnest Island (Lake Bagdad, Lake Herschell and Government House Lake), the Abrolhos group (Pelsart, Wooded, East and West Wallaby Islands), Dirk Hartog Island and in Shark Bay (near Denham on Peron Peninsula, Koks and Dorre Islands), Blowholes north of Carnarvon, Dampier Archipelago (Egret Island), and the Lacepede Is.

These dainty but noisy little terns loudly protest against trespass in their haunts and will follow one around even when no sign of nesting is apparent. In the summer months parties ascend the Swan River and visit some of the more open coastal lakes. Nowhere, however, are they more plentiful than along the North-West coast, and huge flocks have been seen from the air by S. Fowler (C.S.I.R.O. aerial fisheries observer) between Point Cloates and Cossack, with lesser numbers northwards. Individual flocks may contain up to two or three thousand birds and even 15,000 were computed in one area, in the vicinity of Sholl and Stewart Islands. An important roosting location is Frazer Island near Point Cloates. The birds forage over the sea to the western side of the chain of islands along this coast rather than on the sea-ways between the islands and the mainland. Their rapid flight with sudden instantaneous changes of direction remind one of sandpipers rather than terns.

Little Tern

Sterna albifrons

Similar in appearance and size to the Fairy Tern, and differing mainly in the beak coloration and head pattern. The forehead is white, as in the Fairy Tern, but the white extends further back, as *a white line above and beyond the eye.* The *beak is black-tipped,* instead of being entirely orange-yellow. The black tipping is a quarter to one-third the length of the beak. In the non-breeding plumage the beak may be wholly dark and the cap much more extensively white. Length, 8-11 in. Weight, 2 oz.

Distribution: This is the common small tern of eastern and northern Australia and formerly it was not considered as occurring in our area, the southernmost limit being regarded as Broome. Latterly, however, it has been recorded much further south, in the North-West, and overlapping the range of the Fairy Tern. John Darnell has observed the species at Cape Keraudren and at the salt intake works 14 miles east of Port Hedland, the birds being both in eclipse and full plumage. In October 1975 Ron Johnstone collected three specimens, in non-breeding plumage, at the salt works for the W.A. Museum.

The species, in Australia, does not nest west of Cape York, in Queensland. Birds in non-breeding plumage have been observed in the Port Hedland area between the end of September and early April, which is out of phase with the breeding regime in eastern Australia. Thus it is likely that the birds visiting the North-West are migrants from some breeding station north of Australia.

Sooty Tern (Wideawake)

Sterna fuscata

This is one of our two terns which are *entirely blackish above and white below*—the other being the Bridled Tern. These two species are very similar in appearance and the distinguishing characters are detailed under the description of the Bridled Tern. The Sooty Tern has the whole of the upper parts brownish-black, looking jet-black at a distance. The under parts, including the under wings are white, though the under surfaces of the flight quills are dark grey with white shafts. There is a black line from the eye to the base of the beak, and a broad white forehead which does not extend behind the eye. The tail is deeply forked, the *long outer tail feathers having the appearance of streamers.* Iris, brown; beak and legs, black. Length, 18½ in. Weight, 5¼ oz.

Distribution: A northern tern, the Abrolhos Islands being the furthest south nesting area. It is an oceanic species which may be met with far from its breeding grounds.

Nesting: One egg forms the clutch. It is white, sometimes with a pinkish tinge, and is marked with spots and blotches of reddish-brown, chestnut and underlying purplish-brown; often the markings are congregated at the larger end and in some instances are altogether absent. The inside lining of the shell is yellowish-white. Average size, 53 x 37 mm., varying from 49-58 x 35-38 mm. The Sooty Tern usually nests in close association with the Common Noddy, both being in vast communities. The Noddies build their nests on tops of bushes while the Sooties lay their single egg in a scrape on the bare ground beneath. At the Abrolhos the largest colony, numbering many thousands of birds, is at the south end of Pelsart Island, and another, much smaller, at Wooded Island in the Easter Group. Formerly a large colony, exceeding that of Pelsart Island

Fig. 39.—Sooty Tern.

in strength, nested on Rat Island in the Easter Group, but the last time this was seen by an ornithologist was in 1913 (W. B. Alexander), though there is evidence that it lingered on until the middle 30s. Three former colonies in the Wallaby Group (on Pelican, Pigeon and Long Islands) have also disappeared since Alexander's visit. It appears that the egging activities of crayfishermen and other human depredations have been responsible for this local extermination of the species. Further north the species has been reported nesting only on Bedout Island, off Port Hedland, and, outside our area, it nests at the Ashmore Reef (East and Middle Islands).

The scene on the islands during the nesting period is remarkable. The crowds of birds barely take notice of the human visitor, who can readily catch the adults at the nest. The birds also ignore industrial activity and during guano operations, when motor trucks and tractors were employed, they were not concerned in the slightest. When ashore during nesting duties they are in a constant state of activity and even at night, when their neighbours the Noddies are quiet, the animated calling of the Sooties continues. Their cries, which may be syllabised by their well-known sailormen's name of

"Wide-awake," are not harsh and fall pleasanter on the ear than do the majority of tern calls.

The migrations of the Sooty Tern are very imperfectly known, and as Dr. R. C. Murphy, of the American Museum of Natural History, has said, they are one of the great unsolved mysteries of the sea. At the Abrolhos the birds arrive at the end of July or the beginning of August. Nesting is in full operation in September and the peak of the hatching appears to be towards the end of November. About the end of February or the beginning of March all the birds vacate the islands entirely and disappear from the neighbouring seas. The birds nesting at Bedout Island, in the North-West, have a different breeding rhythm, egg-laying taking place in May. They are obviously a different population from the Abrolhos birds.

Where the birds go to after nesting is unknown, but probably they disperse far and wide over the Indian and Southern Oceans. Even during the nesting period the Abrolhos birds feed to the westward of the islands and are never seen in the 35-45 mile-wide Geelvink Channel between the Abrolhos and the mainland. Some hint of where part of the wandering population goes may be gathered from the storm-bound individuals reported away from the nesting islands. One was captured on the Swan River estuary in December 1917, and in March 1937 an immature bird was picked up at Bunbury beach after a blow. The most numerous visitation took place following a very severe and unseasonable gale on November 10-13, 1946, when large numbers of Sooty Terns were reported throughout the South-West, many in the Perth district, and as far south as Manjimup and inland to Moulyinning. Further afield are a few sight records from South Australia and Victoria, which from geographical considerations are probably members of the western stock. In November 1906 an adult bird was seen near Robe, South Australia, another in early April 1937 at Kangaroo Island, and a third captured alive at Brighton, South Australia, on March 28 in the same year. In January 1937 an adult was seen at Port Fairy, Victoria. Though the majority of these records were made during the nesting season it is known that sea-birds of different species may miss a breeding season at intervals and remain in their distant feeding grounds. Possibly the odd individuals seen at various points of southern Australia may be outliers of the main oceanic body of birds.

The diet of the local birds has not been closely studied, but small cuttle-fish have been noticed fed to their young by the parents at the Abrolhos nesting grounds.

Bridled Tern (Brown-winged Tern, Panayan Tern)
Sterna anœthetus

So alike is this species to the Sooty Tern that an observer who is not familiar with either finds it a matter of some difficulty at first to discriminate between them. There is the same general plumage pattern, but the *back and wings are rather more brownish,* not so dark as the Sooty, but both appear quite black when at a distance. The under parts have not the pure whiteness of the Sooty, but show a grey clouding—also a character not readily appreciable in the field. The chief distinguishing feature is the head pattern; in the Bridled Tern there is *a narrower white forehead band* than in the Sooty and this extends much further back *as an eyebrow line some distance behind the eye.* The size is smaller, length, 16½ in.; weight, 4¼ oz., but this again is one of the differences not easily perceptible in the field.

The Bridled Tern can also be readily identified by its call-note. Instead of the liquid, even melodious, "Wide-awake" note of the Sooty, the Bridled utters a *sharp bark,* like the yapping of a small dog or the barking notes of a Stilt.

Fig. 40.—Bridled Tern.

Distribution: Occurs along the coast from the Kimberley Division as far south as Cape Leeuwin.

Nesting: The single egg is light stone in colour and is marked all over with spots of rich chestnut and grey—the grey ones being small. Average size of 14 eggs from Rottnest and Carnac Islands, 46 x 32 mm.. varying from 43-49 x 31-34 mm. This tern does not form such huge nesting aggregations as the Sooty, but nests in small, more scattered groups and usually in altogether different situations. The single egg is often laid under rock ledges on small islets or stacks, overlooking the water, but sometimes the bird will nest like the Sooty under bushes. At the southern nesting islands the birds nest under ledges almost exclusively, but at the Abrolhos, though some individuals nest under rock ledges, the majority lay on the ground under saltbush or nitre-bush shrubs. The Bridled Terns do not intermingle with the Sooties, though, as at Wooded Island, they may nest in close proximity to them. They do not sit as closely as the Sooties do, but dash out from their nesting bushes when one makes too near an approach; the Sooties will usually continue to sit in such circumstances. The nesting season is appreciably later than that of the Sooty, commencing in the latter part of November, the eggs hatching after mid-January.

Recorded nesting islands are as follows: Seal Island (off Cape Leeuwin), some small limestone islets and stacks north of Cape Leeuwin, including an islet off Cape Hamelin (in "Cosy Corner"), Hamelin Island and Peak Island; stacks off C. Bouvard; in the Safety Bay area, Bird, Seal and Shag Islands and rocks off Point Peron; the small islets and rocks around Rottnest Island (Dyers, Phillip, Duck and Clune Rocks); islands between Fremantle and Dongara (Little Island off Whitford's Beach, Edward, Lancelin, Green, Whittell, Buller, Cervantes, North Essex, Whitlock, Fisherman and Beagle Islands); Abrolhos Islands (Pelsart, Wooded, Rat, Pelican and Pigeon Islands); Shark Bay (Dirk Hartog, Slope, Friday, Freycinet, Double, White, Sunday and Egg Islands), Monte Bello group (islets south of Hermite Island), Bonaparte Archipelago (Warn Island), King Sound (Sunday Island), Montalivet Islands (Warn Rock), and Low Rocks (Admiralty Gulf). At the Abrolhos the species appears to have increased considerably in population strength this century. W. B. Alexander, during his visit in March 1913, found it the rarest of the terns as he observed only one pair, on an islet near Rat Island. On two visits in the summers of 1940-42 V. N. Serventy located nesting groups of several hundred birds at Pelsart, Wooded and Pigeon Islands. In December 1945 we saw nesting birds in considerable numbers on Wooded and Pelsart Islands; the Pigeon Islands were not visited.

The species is apparently extending its breeding range southwards along the coast. John Gilbert did not report it from the Cape Leeuwin islands when he was there in December 1842 and W. B. Alexander gave the Safety Bay islands as the southern limit in 1920.

Though the Bridled Tern nests on islands near the mainland and haunts rocks and stacks at which it can be seen from the mainland shore, it nevertheless shuns the coast. The bird is a migrant, leaving its southern nesting areas in late February and returning in mid-October. Further north (as at Green Islets and the Abrolhos) it does not leave until April (G. M. Storr). At Shark Bay it arrives in late October, lays in mid-November and leaves in late February (R. G. Mitchell). One ringed bird has been recovered outside Australia. This had been ringed as a fledgling by Ron Johnstone at North Fisherman Island, 135 miles north of Perth, on January 7, 1973; it was recovered in coastal north-east Borneo on December 28, 1973.

Common Noddy

Anous stolidus

This species and the Lesser Noddy, which are with difficulty distinguishable from each other in the field, are the *only local terns which are entirely dusky in colour,* above and below, except for the paler shade of the crown of the head. The general colour is a dark greyish-brown, with the wing quills black. The crown of the head is the palest grey, becoming *almost white on the forehead and sharply defined from the black lores.* Iris, brown; beak, black; legs, brownish-black. Length, 16 in. Weight, 6-8 oz.

Immature birds do not have a light cap and are a uniform brownish-grey, except for the black lores.

Distribution: A northern tern whose most southerly nesting grounds are the Abrolhos Islands. Only two instances of wandering further south are known (Leighton Beach, November 1946 and Geographe Bay, June-July, 1948).

Nesting: The two species of Noddies are our only marine terns which build nests of any pretensions. This species builds its nest on the tops of the maritime saltbush *(Atriplex cinerea)* and to

Fig. 41.—Common Noddy.

some extent in the taller nitre-bush *(Nitraria schoberi),* being composed of trampled-down portions of the bush reinforced with a little Fucoid seaweed. This is not present in every case and seems to be a kind of token decoration for it does not appear to be an essential part of the structure of the nest. Whilst they are incubating their eggs the Noddies continually come down to the beach to pick up portions of *Fucus* lying dead on the strand-line but the practice ceases after the eggs hatch. They may nest also on the ground among samphire or even lay their eggs on the bare ground without any nest. The single egg is dull white and of rough texture, marked with blotches of rufous-brown and purplish. When held to the light and viewed through a hole, the egg is seen to have a green lining. Average size, 53 x 37 mm., varying from 50-58 x 35-41 mm.

A nesting colony of Common Noddies is an even more impressive sight than that of the Sooties for the birds nesting in the open are more conspicuous. There is an extraordinary indifference to the presence of man; the birds sit close on their nests and can be fairly easily caught by hand. There is a continuous clamour among them all day but unlike that of the Sooties it subsides during the night.

A large colony, numbering many thousands of birds, is located at the south end of Pelsart Island, partly overlapping the Sooty Tern colony, and this is the only surviving one on the Abrolhos. Formerly there was an even larger nesting colony on Rat Island and a smaller one on Wooded Island which, like those of the Sooty Terns, have now disappeared, probably from the same cause. In December 1889, when A. J. Campbell visited Rat Island he and G. H. Beddoes (manager of the guano establishment) estimated that about 300 acres were occupied by birds and averaging at least one bird per square yard calculated there were 1,452,000 birds on the small island. Now none remains and the island is the most desolate and least interesting of any of the Abrolhos group.

The only other island in our waters where this species has been observed breeding is Bedout Island, off Port Hedland, where nesting takes place at the opposite time of the year, in May. Further north however, nesting takes place at the Lacepede Islands (East Island, where a spring-breeding population also occurs; eggs in October) and at the Ashmore Reef (East Island, autumn-breeding population only).

The migrations of the Common Noddy are as little known as those of the Sooty Tern. On the Abrolhos the birds first appear suddenly about the middle of August. The breeding season parallels that of the Sooties fairly closely and they disappear from the island about March or April. We have very few records of Common Noddies from southern Australia, even as storm-blown birds, as we have for the Sooty. One bird was found at Leighton in November 1946, following a very severe gale, but the species was, at the time, in residence at the Abrolhos. Of much more interest, therefore, was the occurrence of some immature individuals among the large number of Lesser Noddies which were driven ashore at Dunsborough, near Cape Naturaliste, by the great storm of June 13 and 14, 1948. We found a specimen also the following month (on July 20) at Busselton. This was the first evidence that some of the Common Noddies spend the winter in the seas off our South-West coast. Another observation bearing on the possible movements of the species was provided by T. Carter, who reported that sometimes about the middle of May immense flocks are seen over the low sand-banks adjoining Frazer Island, near Point Cloates. These, however, might be part of the population nesting on Bedout Island.

The guano accumulations of the Abrolhos Islands appear to have been mostly due to the Noddy and Sooty Terns. The deposits were discovered during the survey by H.M.S. *Beagle* in 1840 and exploitation began in 1844; then the cutter *Waterwitch,* on a fishing and sealing trip brought a load of Pelsart Island guano to Fremantle. Fishing schooners continued to bring small supplies for the local market, but it was not until after John Forrest made a survey of the group in 1879 for its guano resources that steps were taken for regular production. This was begun in 1885 by the firm of Broadhurst and McNeil, then by J. & W. Bateman, and later by F. C. Broadhurst alone. He ceased operations in 1904 when the State prohibited the export of guano, but the lease was taken over by Fallowfield & Co. who continued until 1915. During this period a total of 56,000 tons of guano was known to have been produced from the Abrolhos—the islands concerned being Pelsart, Gun and seven other islets in the Southern Group; Rat, Wooded, Third and Beacon Islands in the Easter Group; and West Wallaby, Pelican, and North and South Pigeon Islands in the Wallaby Group. In 1943 the industry was revived by the British Phosphate Commissioners on Pelsart Island, because of urgent war needs, and 10,700 tons were raised when operations ceased at the end of 1945. The extent of the guano deposits suggest that colonial sea-birds once occupied a greater number of islands in the Abrolhos group than they do now.

Lesser Noddy

Anous tenuirostris

This bird is a slightly smaller replica of the preceding but the size-differences are barely appreciable in the field unless the two are seen close together. The Lesser Noddy is darker on the back, wings and under parts than the Common Noddy and the whitish-grey cap extends further backwards over the nape. The chief diagnostic character is the lores—the space between the beak and the eye—which are grey and there is *no sharp demarcation here between the grey and the dark brown* such as is found in the Common Noddy. Occasionally some birds are met with having darker lores, approaching the condition in the other species. Though the bird itself is smaller than the Common Noddy, its beak is very nearly as long, or longer, than in that species but more slender. The wing is 8-8½ in. in length as compared with 10-11½ in. in the Common Noddy. Length, 12-14 in.; wing span, 23-25 in. Weight, 3½ oz.

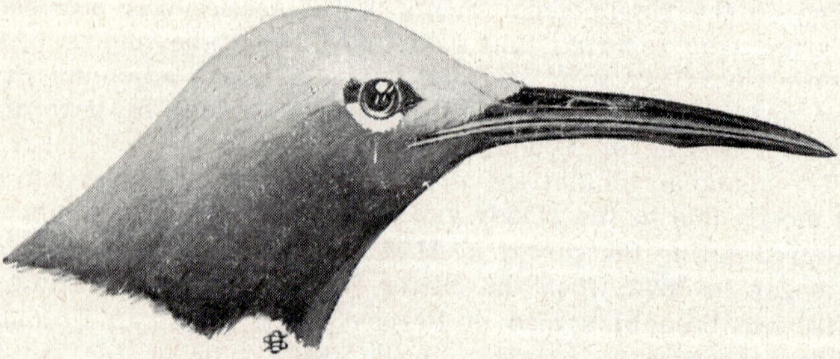

Fig. 42.—Lesser Noddy.

Immature birds possess a lighter cap.

Distribution: The Abrolhos Islands are the only localities where this bird breeds in Western Australia. An allied species, the White-capped Noddy *(A. minutus)*, nests along the Great Barrier Reefs, in Queensland, and in New Guinea. Though similar in size and habits, its head pattern is like that of the Common Noddy.

Nesting: The nests are built on branches of mangroves and are substantial platforms, with a very shallow egg cavity, constructed of a Fucoid seaweed, whitened and hardened with the bird's excreta. Only one egg is laid. It is smaller than that of the Common Noddy and very similar to it, but when held to the light the lining is of a lighter green. It is like the Common Noddy egg in being rough in texture, dull white and with the rufous-brown and purplish blotches. Average size of 16 eggs, 45 x 31 mm., varying from 43-48 x 31-32 mm. General hatching appears to take place by the middle of November, but eggs may be found later in the year. The birds are very docile, sit very closely on the eggs and can be easily handled, being far less aggressive than the Common Noddy.

Large nesting colonies exist on the mangroves surrounding lagoons on Wooded Island and on Pelsart Island (2½ miles north of Wreck Point). There has been a great variation in abundance of this species on the Abrolhos and at one time it was feared that the birds would eventually die out. The great nesting colony on Pelsart Island, where the vast flights of these noddies excited the admiration of the early observers (Lort Stokes in 1840 and John Gilbert in 1842) suddenly disappeared about the turn of the century, for reasons which are obscure. A. J. Campbell in 1889, O. H. Lipfert in 1894 and 1897 and Robert Hall in 1899 found the colony still of great size. But when the island was next visited by an ornithologist, C. G. Gibson, in 1907, he reported that the colony "has been abandoned for several years past," the Wooded Island colony being the only one then surviving. This was the state of affairs also when W. B. Alexander visited the Abrolhos in 1913, and he found evidence that even the Wooded Island colony itself appeared to be declining. It was not until 1936 that the ornithologist, P. T. Sandland, visited the islands—Pelsart Island only—and he found the colony 2½ miles from Wreck Point re-established and flourishing; there was also evidence that nesting had taken place in a former season in the mangroves at the Big Lagoon, four miles from Wreck Point. Later visitors, between 1940 and 1949, have found the 2½ mile colony continuing to prosper, a census in January 1947 giving an approximate count of 27,000 nests; there has, however, been no sign of renewed nesting in the mangroves at the Big Lagoon. Likewise, the Wooded Island colony is still in a flourishing condition and in 1940-41 V. N. Serventy found another colony in mangroves on an island just north of Wooded Island. This did not exist in 1945.

Unlike the Common Noddy and the Sooty Tern, the Lesser Noddy appears to be non-migratory and resident at the Abrolhos, but the evidence is conflicting. No continuous observations have been made by ornithologists, but from reports by the guano operatives it appears that the birds are to be seen in the vicinity of the nesting site throughout the year, with perhaps temporary absences. In the winter months, after north-westerly gales, individuals are occasionally washed up on the mainland beaches south to Busselton. In the winter of 1948 there was a phenomenal irruption of these birds on the beach at Dunsborough, near Cape Naturaliste. Following a terrific gale on June 13 and 14, thousands of the terns settled on the beach and on the small bushes and trees in the vicinity. When the storm abated on June 15 most of them disappeared but more than 200 dead birds remained. A later gale, about a month afterwards, caused another, or possibly part of the same flock, estimated at a hundred birds, to take shelter on the jetty at Busselton on July 19. On the next day we found several lying dead on the beach. On July 24, 1959, after a gale, birds were forced ashore at Fremantle and Bunbury.

PIGEONS and DOVES: Order Columbiformes
Family Columbidæ

Domestic Pigeon (Rock Dove)
Columba livia

Plumage variable. Typically blue-grey with a green-purple iridescent sheen on the sides of the neck; two black bands across secondary wing quills and often with other black wing markings; tail tipped black. Sometimes the under parts are light grey or whitish and some birds may be tinged with brown on the breast and back. Iris, orange-red; orbital ring and cere, whitish, sometimes grey; beak, black or dark grey; legs, bright coral red to dark purplish red. Length, 13½ in. Weight, 13¾-14 oz.

Distribution: Introduced from Europe. Pigeons, of various domestic breeds, are commonly kept in lofts in the metropolitan area, country towns and farms. Many have escaped from captivity and live as wild birds. It is mainly the homing pigeons (which most closely conform to the parental Rock Dove in appearance) that tend to become feral. At Fremantle harbour large flocks of 60-100 individuals congregate in the vicinity of the wheat silos. Along railway lines and the nearby ocean beaches parties of 10-30 birds are more usual. In native bush the birds are usually seen singly, in pairs or in small flocks of 4-5 birds. The species is feral on Garden Island.

Nesting: Most nests are found in artificial surroundings such as stone ledges, eaves of buildings and skillion roofs. Birds which have taken to the bush have been seen nesting, in the metropolitan area, in hollow limbs in tuarts *(Eucalyptus gomphocephala)*. At Nangeenan E. H. Sedgwick found a number of nests in hollows fairly high up in salmon gums *(E. salmonophloia)*; a Pied Butcher-bird was once seen to remove a Pigeon's egg from a hollow tree and Ravens were seen searching the spouts for eggs or young. I. C. Carnaby has seen pigeons entering salmon gum hollows near Moora. We have seen Pigeons using hollows in salmon gums near Trayning and in gimlets *(E. salubris)* near Yorkrakine Rock. A. G. Mathews states that for many years Pigeons have nested in the open cuts at Mt. Charlotte, Kalgoorlie. The nest is a simple platform of sticks, including eucalypt twigs 4-18 in. long, and dried grass. Eggs, 2 in clutch, white with smooth texture. A clutch from a feral pair at Pelican Point measured 39 x 29-30 mm. Reference books on Pigeons state that the eggs vary between 32-40 x 25-32 mm. in size.

Feral Pigeons were declared vermin under the Vermin Act in the municipal district of Toodyay in 1950.

Spotted Turtledove (Necklace Dove, Indian Dove)

Streptopelia chinensis

Upper parts brown, the individual feathers marked with darker brown; a *conspicuous black patch on the nape spotted with white;* wing quills darker brown; outer tail feathers black, broadly tipped white. Under parts cinnamon-brown; under tail coverts grey. Iris, salmon-pink; naked skin around eye, grey; beak, black; legs, pink. Length, 12½ in. Weight, 4¾-5 oz.

Distribution: Introduced from the South Perth Zoological Gardents since 1898 (original home: south-east Asia). It is well established in the city and suburban areas and has reached some outlying centres, including Rottnest and Garden Islands, at various centres along the eastern railway to Wooroloo; sporadically southwards as far as Quindalup, with isolated occurrences elsewhere such as Dongara, Kalgoorlie and Katanning (E. H. Sedgwick).

Derek Goodwin, of the British Museum, after examining some Perth specimens, found they tallied with the nominate race *chinensis* from southern China but others seen showed resemblances to the race *tigrina* from the Indo-Malayan region.

Nesting: This dove has a protracted nesting season during the spring and summer, though some nesting takes place throughout the year. The nest is an untidy cup-shaped structure of stems and rootlets, which, as brooding proceeds, become cemented together by excreta, 5-6 in. in diameter; 2 white eggs, size 27-31 x 22-23 mm.

Like the following species it habitually frequents city and suburban backyards, feeding with the fowls on fallen grain, etc. In such circumstances it is liable to become infected with the stickfast flea *(Echidnophaga gallinaceae)* and is a carrier of that pest.

The call of the male has been syllabised as "coocoo, croo(oo)," "coo-coo, krroo-oo," or "coo-coo, krroo, koo."

Senegal Turtledove (Laughing Dove)

Streptopelia senegalensis

Upper parts brown, head and throat with a lilac tone, mantle with a chestnut shade. *Wing coverts, bluish-grey,* quills brown. Outer tail feathers dark grey, broadly tipped white. Under parts delicate lilac-brown (vinaceous) with a chestnut breast under which the black feather bases are sometimes visible, particularly when the chest is puffed out during display. Under tail coverts white. Iris, brown; naked skin around eye, pink; beak, dark brown; legs pink. Length, 9¾-10 in. Weight, 3-4 oz.

Smaller than the Spotted Turtledove, and distinguished by the bluish wing and the absence of the black and white spotted neck patch. The feathers of the chestnut band across the upper breast have black bases which are sometimes revealed as black spots when the bird turns its head. The other species has no spots on the undersurface.

Distribution: Introduced from the South Perth Zoological Gardens since 1898 (original home: Africa, Arabia, Palestine, Persia, India). It has spread further out in the bush from the Perth metropolitan area than the Spotted Turtledove and has colonised more of the country towns than that species. It is found on Rottnest and Garden Islands and numerous centres in the Wheat-belt between Geraldton and Tambellup and east to Beacon and Merredin. There are isolated colonies at Kalgoorlie and Esperance and it has been seen at the Hamelin Pool turn-off on the Geraldton-Carnarvon road (E. H. Sedgwick).

Derek Goodwin found that Perth birds agreed in general appearance with African specimens south of the Sahara (the nominate race), and could be matched with specimens from East Africa.

Nesting: The nest is a crude structure similar to that of the previous species, with excreta accumulating as brooding proceeds; rather smaller—about 5 in. in diameter—and of frailer appearance, with light showing through the fabric. The 2 white eggs are slightly smaller than those of the preceding species; a series of 11 eggs from Perth averages 26 x 21 mm., varying from 25-27 x 20-21 mm. There is a 35-hour interval between the laying of the eggs, the first being laid late in the evening and the second in the morning. Incubation period, 13-14 days. Nestling period, 21 days. Birds raised in captivity have begun egg-laying at 7½ months (Miss C. A. Nicholls).

The call has been syllabised as "co-oo, coocoo," and the display call as "coo-oo-oo-coo-coo."

Peaceful Dove

Geopelia striata

Back grey, with a brown wash, *cross-barred with black, with whitish cross-bars on the nape and sides of the neck, and extending on to the chest.* The crown is a plain unmarked bluish-grey, the tail slaty-grey. The throat is greyish-white, the abdomen pale greyish-white and the under tail feathers have whitish tips. Iris, white, *naked skin around eye, pale green;* beak, milk-blue; legs, pink. Length, 7-8¼ in. Weight, 1½ oz.

Distribution: A northern species extending south as far as the Murchison River. It is to be seen commonly in the street trees at Carnarvon.

Nesting: The nest is of the usual type constructed by members of the family—a frail, flat platform of a few twigs and built usually on a horizontal limb of a small tree. The 2 eggs are white and elliptical in shape, and, as with the eggs of related species, are glossy. They measure 21-22 x 15-16 mm. The birds will attempt to decoy an observer away from the nesting site. In display the male runs some steps towards the female, bows, and, spreading his tail, fans it out and displays the white ornamentation. Preening of the partner is also part of the display.

Fig. 43.—Peaceful Dove.

The call has been syllabised by different naturalists as "zo-coco," "oodle-oo" and "coo-oo-oo."'

This dove favours wooded areas alongside creeks. It has a characteristic undulating flight, unlike the Diamond Dove. In sunbathing one wing is lifted up.

Diamond Dove

Geopelia cuneata

Native names: Calli-gora (Pilbara district); Calig-yoo-roo (Coolawanyah); Cula-cuga (Godfrey's Tank, Canning Stock Route); Goolagoogoo (Jigalong); Man-na-brun-ka (South-West).

Back and wings brown, the *wings marked with small white round spots;* crown of head and nape bluish-grey; tail grey, with outer tail feathers largely tipped white. Breast bluish-grey, paling towards abdomen. The female differs in having the sides of the neck, fore-neck and chest, pale brown. *Iris and naked skin around eye, red;* beak, olive-brown or black; legs, flesh pink. Length, 7¼ to 8½ in. Weight, 1 to 1½ oz.

Distinguished from the preceding species, not by size differences which are inappreciable, but by the *absence of transverse barring* and the *presence of white spots on the wings* and the *red*

Fig. 44.—Diamond Dove.

instead of the *pale green* area around the eye. Hybrids between the two species are known.

Distribution: Found throughout the North-West and interior, the normal southern limit being the Murchison country, but sometimes visiting the South-West.

Nesting: The nest is a frail platform of dry grasses or twigs and is usually placed in the fork of a small tree or a low bush, within a few feet of the ground. The 2 white eggs are roundish. Size, 6 eggs from Exmouth Gulf average 19.8 x 15.7 mm., varying from 18.6-20.9 x 15.1-16.5 mm. Incubation takes about 12 days; nestling period, 14 days. The call is a "whoo-whoo-whoo."

Though not a regular inhabitant of the South-West its occasional irruptions south, known since the earliest days of settlement, may indicate long-term cycles of abundance in the species. John Gilbert reported that "during my first visit to this part of the country [1839] only two situations were known as places of resort for this species, and I did not meet with more than five or six examples; since that period it has become extremely abundant, and now [i.e. in 1842-43] a pair or two may occasionally be seen about most of the settlers' houses on the Avon, becoming apparently very tame and familiarised to man." G. C. Shortridge collected the species at Boyadine, near Beverley, in January, 1906. In 1912 a pair nested close to the homestead at Petworth Park, Moora, but C. L. E. Orton had not seen the species in the district since. J. B. Higham of Narrogin, reported numbers in the Williams district in 1917. In the North-West this dove occurs in drier country than the Peaceful Dove.

According to Gilbert, the name "man-na-brun-ka" was applied to the bird by the Aborigines from a tradition that the bird originally introduced the "man-na," a gum which exudes from a species of *Acacia* and which was one of the favourite articles of native food.

Bar-shouldered Dove

Geopelia humeralis

Nape rufous; back and wings greyish-brown—all cross-barred with black. The wing quills in flight show an orange-brown band. Central tail feathers grey-brown, outer feathers red-brown and tipped with white. Head bluish-grey, back of head with black-cross-bars. Throat and chest bluish-grey, abdomen white with mauve tint. Under wing light chestnut. Under tail chestnut-brown with prominent white tipping. Iris, yellow; naked skin around eye greyish-pink; beak, light blue-grey; legs, bright pink. Length, $10\frac{3}{4}$-11 in. Weight, 4 oz. In flight it appears as a grey bird with rufous on the hind-neck and with rufous under wings.

Distribution: A northern species which extends into the Pilbara district as far south as the Hamersley Range, but along the coastal mangroves it penetrates as far south as Onslow. It is also found on the islands of the Dampier Archipelago, Lewis, Barrow, Lowendall, Monte Bello and North Sandy Islands.

Nesting: The nest of the Bar-shouldered Dove follows the usual type constructed by most members of the family in being a frail, loosely made platform of light sticks or twigs about 4 in. in diameter and one inch deep. It is usually situated in a bush or a low tree at no great height (up to 10 ft.) from the ground. The 2 eggs are roundish and, as is usual with the family, are pure white. They are smooth and glossy. They measure 26-28 x 20-21 mm.

The powerful, penetrating call has been noted by F. Lawson Whitlock as resembling the syllables "ollywattle." The bird is very vocal just before sundown.

Common Bronzewing

Phaps chalcoptera

Native names: Man-goota (Well 42, Canning Stock Route); Mow-gra (Jigalong); Manpi (Warburton Range); Woodah, Wodta, Oo-da (South-West); Moorhait (Albany); Gnampion (Pallinup River).

Back brown, each feather tipped with buff, outer tail feathers greyish with black sub-terminal band, head with purplish tinge. A *broad white forehead,* white line below the eye, extending backward, *chin white,* sides of neck bluish-grey. Breast mauve, becoming paler towards the abdomen. On wing coverts a brilliant metallic bronze and green speculum. Under wing chestnut. The female has a grey forehead, no grey on the sides of the neck and paler under parts. Iris, reddish; beak, black; legs, coral-pink. Length, 11-13½ in.; wing span, 20½-22½ in. Weight, 11 oz.

Distribution: General throughout the State.

Nesting: The nest is a flimsy, flat structure or platform of twigs, so scanty that the 2 eggs can often be seen from below. Usually it is placed in a horizontal fork of a tree, or on a tree-stump. On occasions, however, the bird has been known to construct quite a compact saucer-shaped nest of dry grasses. The eggs are pure white, elliptical and glossy. Mean size of 8 eggs 34 x 25 mm.; varying from 32-36 x 24-26 mm. The species is double-brooded and it may nest at any time of the year.

The flight is swift, strong and low, with continual wing beats.

This species may exhibit long-term cycles of fluctuations in numbers as do some other pigeons. It was very abundant in early colonial times; Lieut. H. W. Bunbury wrote of it as occurring in thousands in the summer of 1836. However it steadily declined in numbers in the southern portions of the State during the first and second decades of the present century but about 1934 it began to increase and the cycle of plenty still continues. At Bridgetown, for instance, flocks of up to 70 birds have been seen, and in some wheat-belt towns birds may be seen in the streets and feeding on spilled wheat at the railway sidings. In the outer parts of the Perth metropolitan area, where the bird had practically ceased to exist, it re-appeared in numbers and is now commonly flushed from the sides of the roads. The increase has been general, in closely and sparsely settled areas alike, in improved and virgin country, and it has been remarked upon in centres as far apart as the extreme South-West corner and Hamelin Pool, near Shark Bay. In the North-West the species has apparently maintained an even population, being locally present in the vicinity of springs in dry country.

Bronzewing Pigeons, both this species and the next, are given to feeding on the seeds of the box-poison plant, *Gastrolobium bilobum,* and as a result their entrails and bones, but not the flesh, are poisonous to dogs and cats. After eating such pigeons, dogs and cats are apt to have fits, become mad, bite at anyone within reach, and finally die in convulsions.

Brush Bronzewing (Box-poison Pigeon)

Phaps chalcoptera

Native name: Oo-da (South-West).

Back bronze-brown, *nape and sides of neck rich rufous,* crown bluish-grey, *forehead pale brown. Chin chestnut* with whitish below beak. *Remainder of under parts blue-grey,* this colour extending up to the side of the head. Lores black: *rufous line behind the eye.* On the wing coverts a metallic bronze and green speculum. Under wing, chestnut. The female differs in having the crown grey. Iris, brown; beak, black; legs, pink. Length, 11-12½ in. Weight, 7¼ oz.

This species can be readily distinguished from the Common Bronzewing by the absence of the white forehead and chin; by the presence of a rufous nape and chin; a reddish instead of a white line passing behind the eye; and the bluish instead of lilac under parts. It also has a shorter tail.

Distribution: Confined to the South-West corner, north to Freshwater Point (about 25 miles south of Dongara) and inland to Wongan Hills and Lake Grace; south-east to the Esperance region. It is also found on the Abrolhos Islands (West and East Wallaby Islands; formerly it was also on North and Pigeon Islands). At the Recherche Archipelago the Brush Bronzewing has been reported on Mondrain, Woody, Sandy Hook, Middle, Round and New Islands.

Nesting: A nest similar to that of the Common Bronzewing is made, but it is placed in situations much nearer the ground, sometimes even on the ground itself. The 2 eggs are indistinguishable from those of the Common Bronzewing. Size, 31 x 24 mm. Incubation takes 15 to 18 days.

This beautiful bird has seriously declined in numbers during the present century and it is now scarcely ever seen except along the south coastal district. It is still plentiful at the Abrolhos Islands and the Recherche Archipelago. The fact that it nests usually on or close to the ground has probably been the reason for its diminution in numbers, due to its offering itself as an easy target for the fox and the domestic cat.

Crested Pigeon (Topknot Pigeon)

Ocyphaps lophotes

Native names: Cal-cal-goa (Pilbara); Ca-ga-looloo (Coolawanyah); Bawkadeerna (Jigalong); Yaralapulpa (Warburton Range); Waltharrie (Wooleen).

Back and rump bronze-grey, head blue-grey carrying *a long dark plume;* tail dark metallic purple-brown, tipped white; wings pale brown barred with black and with a green and purple metallic speculum, the feathers of the wing coverts edged white; flight quills brown. The under parts are a delicate blue-grey. Iris, reddish-orange; eyelid, bright pink; beak, dark-grey; legs, bright pink. Length, 12¼-13¾ in. Wing span, 17 in. Weight, 6½-8 oz.

Distribution: Over most of Australia. In this State it was originally not found much further south than the Geraldton and Mullewa districts, but is now extending its range. Mrs. Minilya Sharpe states that it was not even present at Wooleen in the early days and first appeared in 1915, later becoming very abundant. In 1946-47 it made a temporary appearance at Kellerberrin. Later it was recorded as far south as Coorow, Dalwallinu, New Norcia, Pithara and Shackleton. By 1956 it was an occasional visitor to the Perth area. In 1967 it was seen at Lake Grace and has increased in numbers since. In 1970 a bird was seen at Woodanilling. Breeding has been observed east of Wickepin.

Nesting: The Crested Pigeon makes the usual pigeon type of nest—a flat, frail structure of a few twigs. It is usually placed in a bush or small tree and the 2 eggs are pure white, glossy, and blunted oval in shape. A set of 10 eggs from the North-West averages 29.5 x 22.2 mm., varying from 27.8-31.5 x 21.2-22.9 mm. Incubation takes 19 days.

In flight the wings are flapped for a period and then held out stiffly in a glide. During the flapping a characteristic whistling sound is produced through the emargination of the primary wing quills, particularly the third primary whose vane is very narrow. This provides opportunity for a signalling system among the birds (K. Immelmann). The tail is lifted on alighting. In the courtship display the tail is spread simultaneously with the wings, the special markings of the latter becoming conspicuous.

Red-plumed Pigeon

Lophophaps ferruginea

Native names: Too-low in, Jew-la-oola (Coolawanyah); Mah-boo-goo (Mulga Downs); Kool-brit (Tallangee tribe, North-West Cape), Mawgootagoota (Jigalong).

The general body colour is a *rich rufous brown, both above and below,* with black and pale grey barrings on the back and shoulders; forehead bluish-grey. The *crown bears a plume of rufous feathers.* The eye is encircled by a black line; chin black; upper throat white, followed by a black band. A narrow slaty-grey and black band encircles the lower breast. Metallic iridescence in the wing. Iris, orange-red; skin around the eye, red; beak, slaty-grey; legs, slaty-grey. Length, 7¾-8¼in. Weight, 3-3½ oz.

Distribution: This species occurs only in North-Western Australia, from the Pilbara district south to the Wooramel River (on the Byro-Gascoyne Junction road) and inland to Meekatharra. In northern and central Australia it is represented by a closely similar species, the Plumed Pigeon *(L. plumifera)* which differs from it in having the lower breast and abdomen white.

Nesting: The 2 eggs are laid on the ground in quite exposed positions, there being practically no nest made. They are the usual pigeon-type eggs and are creamy-white. Size 23-25 x 19-20 mm. The nesting season apparently depends on the rainfall as eggs have been found in May, September and October. More than one brood appears to be raised in the year.

The species appears to be sedentary and does not move long distances, the habitat being rocky ground in the vicinity of water. After the nesting season the birds congregate in small parties of up to six in number. They rise from the ground with a sudden loud whirr. The call has been syllabised by F. Lawson Whitlock as a deep guttural "coo-r-r-r".

Plumed Pigeon

Lophophaps plumifera

Like the preceding species but with the *lower breast and abdomen white.* There is a white line between the rufous breast and the black breast bar. It is also grey-brown in general colour instead of rufous and is slightly larger in size. The plume is longer (2-2½ in. as compared with 1¼-1¾ in.). Iris, golden yellow; skin around the eye, red; beak, dull black; legs, greyish black tinged purple. Length, 8¼-9¼ in.; wing span, 13½ in. Weight, 4 oz.

Distribution: In northern and central Australia, including the Kimberley Division. It qualifies for inclusion in this book by its occurrence in the Rawlinson Range, as the westernmost outpost of the central Australian population. There is a case for placing all forms of the Plumed Pigeons, including the preceding, as geographical representatives of one variable species. However, pending further study of the situation it seems preferable to retain the distinctive Red Plumed Pigeon as a separate species.

Nesting: Similar to the preceding species.

The habits of the two species are virtually identical. In a study of the Plumed Pigeon in the Rawlinson Range, Derek Goodwin, of the British Museum, has noted that in flight the bird "bursts" upwards with a rapid clatter of wings, then gliding on and down, sometimes with one quick wing beat in the middle of the glide. Long distance flights resemble those of the Crested Pigeon. He comments that this form of flight is normally found in ground birds and does not occur in most pigeons. It may have originated as an adaptation to terrestrial life.

Flock Pigeon (Plain Pigeon)

Histriophaps histrionica

A strikingly marked pigeon, with *black and white about the head, contrasting sharply with the rufous back and grey under parts.* The upper parts are light reddish-brown; wing quills bluish grey and their coverts edged white; leading edge of wing white. Metallic purple speculum on inner wing quills. The outer feathers of the short tail are bluish-grey spotted with white. Head and throat black, but the forehead and a patch below the lores, a line behind the eye and another below it at the side of the head, are white. Lower throat white and the remainder of the under parts bluish-grey. Under tail buff-brown. The female is duller in coloration, lacking the white forehead and the black is replaced by brown. Iris, brown; beak, black; legs, pink-grey. Length, 11¾-12 in. Weight, 8 oz.

Distribution: A northern and inland species occurring through the North-West as far south as the Minilya and Upper Murchison Rivers.

Nesting: This pigeon lays its 2 eggs on the ground in grass or low scrub and makes no nest. The eggs are of the usual elliptical pigeon type and are white with a creamy tinge. Size, 32-34 x 24 mm.

A nomad of the northern grasslands, this pigeon associates in large flocks in certain localities, according to whether there is a plentiful supply of grass seeds. In mid-west Australia it was first recorded in June 1858 when Surveyor F. T. Gregory noted in his journal that in the neighbourhood of Mount Augustus the birds "congregate in flights sometimes of a thousand together." T. Carter noted some on the Minilya River plains in 1887 and a few on the coast early in 1890 and again in 1893. He did not see any more until

Fg. 45.—Flock Pigeon, male.

1900, which was a record wet year, and the birds were present in enormous numbers on the lower Lyndon River. "The pigeons were watering in countless myriads" he wrote, "flock succeeding flock, and the roar of their wings was like the noise of a heavy surf on a reef." When feeding on the open plain, in country which had been recently burnt, "the flocks were greedily feeding in such numbers as to hide the ground." More recent observations have been very sparse. F. Lawson Whitlock met with a few on the De Grey River in 1908 and A. H. Robinson saw a flock of 16 at Chitawa claypan near Onslow in September 1927.

Since then there has been a remarkable dearth of information on the species until the autumn of 1957, when H. Hall, of Wallareenya Station, south of Port Hedland, reported its presence as a pigeon hitherto unknown in the district. This heralded a substantial invasion of the North-West during the following year. Flocks of 400-500 birds were seen at "Old" Onslow (14 miles south-west of the present site of the town) in April 1958 and several parties of about 30 birds were encountered by a W.A. Museum party at Millstream Station in July. In late September the same year, during a field trip with A. J. Marshall and I. C. Carnaby, we saw about a thousand birds coming to drink at Mardie Pool, near the mouth of the Fortescue River, in the hour before sundown. The Flock Pigeons came in flock after flock, each of 10-50 birds, circled over the water and landed. They dipped their heads and drank, remaining only a few seconds and were off. A few birds were seen also at the Yanyare River. Apparently the species had become conspicuous in the Northern Territory and the Kimberleys about four years previously. P. Slater observed it at the Kimberley Research Station (70 miles south-east of Wyndham) in 1955-56 and K. Immelmann saw up to 30 birds there in 1959-60.

PARROTS and COCKATOOS: Order Psittaciformes
Family Psittacidæ
LORIKEETS: Subfamily Loriinæ

Purple-crowned Lorikeet (Parrakeet, Zit Parrot)
Glossopsitta porphyrocephala

Native names: Cowara, Cower, Kowar (South-West).

Our smallest member of the parrot tribe. The upper parts are bright green, with a bronze shade on the nape, and the under parts are a light greenish-blue. On the forehead is a prominent orange-red band and behind this a plum-purple patch. Behind the eye is an orange-red patch. There is some blue on the shoulder of the wing while the under wing and the sides of the body are a vivid crimson. The purple patch is absent from the forehead in immature birds. Iris, brown; beak, black; legs, grey. Length, 5½-6½ in. Weight, 2 oz.

Distribution: In the south and south-west, south of the mulga-eucalypt line (see Fig. 4), the Great Victoria Desert and along the coastal mallee fringe to Eucla when the eucalypts are in flower. The birds follow their flowering and between such periods may absent themselves from a district for several months.

Nesting: No nest is made, but the 3 to 6 white eggs are laid in a hollow spout, or in a hole in a tree. No nesting material except wood-dust, is used. The eggs are without gloss; and roundish in shape; size, 19-20 x 17 mm. Incubation takes about 18 days, and is undertaken by the female. Morning and evening, when she comes off the eggs, she is escorted to and from her feeding area by the male bird.

These are tree-top birds and difficult to observe ordinarily, but they reveal themselves by their shrill cries as they fly rapidly overhead in small flocks. On account of their call they are known to children as "zit" parrots. Occasionally individuals may be picked up which have collided with telegraph lines or other obstructions.

The traditional belief that the brush-tipped tongue of lorikeets was a device for nectar-feeding has been dispelled by an authoritative paper published in 1970 by Dr. D. M. Churchill, a former Western Australian botanist and now Government Botanist of Victoria, and Mr. P. Christensen of the Western Australian Forests Department. They prove that the staple diet of these birds is pollen and the structure of the tongue enables the pollen to be harvested almost pure without detaching most of the anthers which carry it. Pollen is the birds' source of nitrogen for protein synthesis. Nectar is eaten also, but is at most a dietary supplement to the pollen.

Rainbow Lorikeet (Blue Mountain Parrot)

Trichoglossus moluccanus

Back, including mantle, wings and tail, green. Head, including cheeks and chin, streaked purplish blue. Band around hind-neck yellow-green. Breast a fiery orange-red; lower breast dark purplish-blue, red on sides; abdomen green; under tail coverts yellow. Iris, orange-red; beak, red; legs, grey. Length, 10-12 in. Sexes similar. The young have dark brown beaks.

Distribution: Eastern and south-eastern Australia. Early in 1968 an invasion of these birds was noticed in the Perth metropolitan area, and they are still present.

The occurrence has been documented by Dr. G. M. Storr, of the Western Australian Museum, who has records of the birds from Wembley, Daglish, Shenton Park, Nedlands, Crawley, and also at Gooseberry Hill. They are particularly evident in the University grounds, Crawley, feeding on the flowers of the lemon-scented gum, marri, red-flowering gum and coral tree. Dr. Storr notes that "their plump bodies, rapid wing-beats, direct unwavering flight, and occasional screeching call-note at once distinguish them from all local parrots of similar size."

The nearest population of Rainbow Lorikeets is that of Eyre Peninsula, 1,200 miles east of Perth. Dr. Storr suggests that the invasion may have been stimulated by the failure of most eucalypts to flower in south-eastern Australia during the great drought of 1967. Much of the intervening country along the coast is suited to the species, except for a belt of 1-200 miles around the Head of the Bight, but such a crossing would not be difficult for a strong-flying bird like this lorikeet.

Nesting: Has been known to take place, at least, in the University grounds, Crawley. The eggs, 2, sometimes 3, are laid in the hollow of a tree, averaging 23 x 27 mm. in size.

COCKATOOS: Subfamily Cacatuinæ

White-tailed Black Cockatoo

Calyptorhynchus baudini

Native names: Oo-lack (Perth); Ngo-lak, Ngol-ye-nuk (Avon River); Woo-lock (Albany); Gnular (Pallinup River).

Brownish-black; feathers narrowly edged light buff and on the under surface more conspicuously with buff-yellow. There is a *broad sub-terminal white band across the tail,* excepting the central pair of feathers, and a *large white patch on the cheek.* Iris, dark brown; beak, heavy, light horn or blackish; legs, brown. Length, 21-23 in. Wing span, 3 ft. 3 in. - 3 ft. 8 in. Weight, 1 lb. 4 oz. - 1 lb. 11 oz.

The sexes can be distinguished as follows. The ring of bare skin surrounding the eye is pink in the male and dark grey in the female. The beak is light horn in the female and black in the male. The cheek patch is a clear white in the female but a dull or dirty white in the male. The breast feathers in the female have relatively broad buff edgings; in the male these are narrower and less noticeable (Dr. A. H. Lendon).

Distribution: Confined to the South-West corner, from the lower Murchison River to the eastward of Esperance (recorded to Mississippi Bay). It extends inland as far as Wongan Hills, Kellerberrin and Norseman.

Two well-marked sub-species occur. In the forested South-West corner is a form with a long narrow bill *(baudini)* and inland

and to the east of Albany, in the sandplain and mallee areas, are the headquarters of a heavier-billed race with the mandible shortened *(latirostris,* I. C. Carnaby).

This bird is represented in the eastern States by the closely similar Yellow-tailed Black Cockatoo *(C. funereus),* the white on the head and tail of our bird being replaced by yellow.

Nesting: The salmon gum is the preferred nesting tree, with second preference to the wandoo, and occasionally the York gum is used. Observations are lacking on nesting in the extreme South-West. The 1 or 2 white eggs are laid in a cavity high up, though at Cockleshell Gully F. Grigson once found a nest in a wandoo where the entrance spout was about 7 ft. above the ground and the 2 eggs were about ground level. Often, in a clutch of 2, one may be infertile. If both eggs of a clutch of 2 hatch one of the nestlings is always smaller and disappears after a few days (C. L. E. Orton). The egg is oval and smooth in texture. A series of 8 eggs from Moora and Jurien Bay averages 47 x 35 mm., varying from 45-52 x 32-37 mm. A series of 6 from Lake Grace is larger, averaging 52 x 36 mm., and varying from 49-55 x 35-38 mm. The nesting season lasts from August to November. Only the female incubates and broods the nestling for about three weeks. Both parents feed the nestling which remains in the nest for 10-11 weeks. The young are fed by the parents for some time after they leave the nest.

A considerable amount of local wandering is indulged in by this species, which in the Perth area and the coastal plain becomes a regular passage movement. Flocks pass over the district travelling to the north in the autumn and early winter, from March to June, and to the southward from August to November. The appearance of the travelling birds near Perth in the autumn has earned for them a reputation as harbingers of rain. Their presence on these movements is heralded by their strange wailing cries, from which the Aborigines gave them the name "Oo-lack".

Their food, when the gum-trees are in blossom, consists of the nectar from the flowers and large quantities are consumed, so much so that if a bird is shot when feeding on the blossoms the crop will be found to be crammed with fluid which will pour out of the beak when the head is held down. After the blossom is over the cockatoos feed on the "gum-nuts", which they tear open to get at the seeds. At Boyup Brook G. A. Lodge reports that numbers arrive in September and October to feed on the native Geranium, older birds feeding large young. They are also very fond of the grubs which are to be found under the bark of dead trees—usually jarrah, which they strip very thoroughly in their search. In the drier inland areas they feed on the hard fruits of *Hakea crassifolia.* Black Cockatoos at times occasion great loss to apple-orchards; the trees not only receive considerable damage through the heavy birds alighting on the weak branches, but the fruit is stripped and the crop ruined. In consequence the species was "proclaimed" under the Vermin Act in the Plantagenet district in 1926, in the Denmark district in 1933, in the Preston, Bridgetown, Balingup and Collie districts in 1942, West Arthur in 1949, Greenbushes and Manjimup in 1951, and Albany in 1959. In 1971 the birds were declared vermin only in the Plantagenet, Denmark and Donnybrook districts.

Like other cockatoos they are occasionally eaten. In her private diary Mrs. Currie, wife of Admiral Mark Currie, recorded that she tasted black cockatoo when she dined with the Governor in July 1829. Flocks of the birds still fly over and feed in Government House grounds.

D. A. Saunders, now engaged in an ecological study of this cockatoo for the C.S.I.R.O. Division of Wildlife Research, has recently revealed an extraordinarily interesting story on the two forms of the species. There are no apparent physical differences between them other than the proportions of the beak. The long-billed birds are found in the heavily forested South-West corner, in the triangle bounded by Perth, Albany and the coast. They appear to be resident in this area, or at most indulge only in local wanderings. They eat mainly marri seeds and insect larvae found under the bark, and it is this long-billed form which has become a pest in apple and pear orchards, raiding the orchards when the marri fruit crop is poor. The short-billed form behaves and feeds very differently. Birds of this type may be found anywhere in the South-West corner, but the breeding range is suspected to be in the lower rainfall areas. After the breeding season they move out in December-January and amalgamate into bigger flocks near the coast, returning in June-July. It is this bird, and not the long-billed form, which has developed the habit of feeding on pine seeds, though their food is normally seed from proteaceous shrubs which have small, hard seed pods. Some of their flocks in the non-breeding season become very large and several thousands of them may congregate on the Forests Department pine plantations, staying around until most of the available seed has been eaten.

Red-tailed Black Cockatoo

Calyptorhynchus magnificus

Native names: Bildera (Lyons River); Biddiarra (Pidong tribe, upper Murchison River); Keer-jan-dee (Victoria Plains); Kar-rak, Koorark and Kor-rar-ra (South-West); Korridg-e-cup (Harvey); Curraak (Albany); Carrer (Pallinup River).

The fully adult male plumage is *a uniform black,* with *a broad fiery-red sub-terminal tail band,* excepting across the central pair of feathers. Iris, brown; the heavy beak is black; legs, grey-black. Length, 21-24 in.; wing span, 40 in. Weight, 1 lb. 6 oz.

Records of birds kept in captivity indicate that the males do not reach maturity for four years. The following details are based on observations by J. Grogan of Bassendean, who has been outstandingly successful in rearing the two species of black cockatoos:—

1st year: The head and wing coverts are spotted with yellow and the breast feathers are barred with yellow or buff; the tail feathers, instead of the entire broad red band, are crossed by yellow or orange bars, a quarter to half an inch wide. Birds in this plumage are often considered a separate species by country people.

2nd year: Following the annual moult the yellow specks become fewer in number; the breast barrings are less conspicuous; the tail bars become broader and more reddish and the beak is darker and harder.

3rd year: Very few yellow specks remain and many of the narrow black bars subdividing the orange-red bars on the tail disappear.

4th year (mature plumage): The whole plumage is now shining black, the tail band an unbroken uniform red and the beak black.

The female never alters materially from its fledgling plumage and corresponds with that of the first-year male just described. The beak is light horn.

Distribution: Two widely separated populations exist — in the Kimberley Division and in the South-West. The latter is now becoming scarce, its headquarters being the Midlands and Victoria Plains districts. Birds are seen north to the Murchison River (with scattered records to the Upper Ashburton) and east to Cue, Thundelarra and Merredin.

Nesting: The single egg, which is all that is laid by this bird, is laid in a hollow high up in a tall tree. The egg follows the usual parrot form and is pure white. Size, 47-55 x 33-36 mm.

The Red-tailed Black Cockatoo in the South-West consorts in smaller flocks than the white-tailed species. The call is very distinct; it is a loud grating "kree" or "krurr," the native name in the South-West districts being evidently onomatopoeic. Like the Long-billed Corella the bird has declined very seriously in numbers in the southern part of the State since settlement began, but near Perth pairs or small parties may very occasionally be still met with in the Darling Range.

It never associates with the preceding species, each of the black cockatoos keeping to themselves. The Red-tailed Black Cockatoo is not recorded as doing any damage in apple-orchards or pine plantations. Among its food in the South-West is the seed in matured marri capsules.

Major Mitchell (Pink Cockatoo, Chock-a-lok, Chockalott)

Cacatua leadbeateri

Native names: Labya (Well 42, Canning Stock Route); Jockalock (Wooleen); Jak-kul-yak-kul, Jackal-yackal (Avon River); Jak-kul-ya-ra (Wongan Hills).

A handsome white and pink bird. Body generally white with a pink flush. Head and breast salmon pink with *a large erectile crest.* The crest feathers are light pink at the base and white-tipped, the main intermediate portion being orange-red with a yellow or orange band in the centre in the male. Iris, orange-red in the female and dark brown in the male, the bare skin around the eye white; beak, light grey; legs, grey. Length, 15-16 in.; wing span, 32-33 in. Weight, 11 oz.

It can be readily distinguished from the Corellas when in flight by the under wings showing *pink* instead of yellow.

Distribution: Generally distributed, but avoiding the South-West corner south to about Jurien Bay and thence the boundary line passes through Toodyay and south-easterly to Esperance. In July-August R. P. McMillan observed a pair in the vicinity of the Guildford Grammar School, and there was an abortive attempt at nesting.

Nesting: The Major Mitchell lays its eggs in a hole in a tree or in a hollow spout of a tree, such hollows often being two or three feet deep. The same nesting hollow is often used in successive seasons by the same pair of birds. The 3 to 4 eggs which form the clutch are roundish, with one end slightly more pointed and are pure white and slightly glossy. Size, 36-37 x 27-29 mm. When disturbed the female, which undertakes the incubation of the eggs, flies screaming from the nesting hollow.

Little Corella

Cacatua sanguinea

Native names: Birr-ear-ra (Coolawanyah); Kogga-je (North-West Cape); Keenirri (Jigalong); Kakalyalya (Warburton Range); Boolie (Wooleen); Bulli (Pidong tribe, upper Murchison River).

Plumage white, with sulphur-yellow on the under tail quills and in the middle of the under wing. The concealed bases of the head and neck feathers are salmon-red and there are varying amounts of red on the feathers between the beak and the eye. An area of naked skin around the eye is blue. Iris, brown in both sexes; beak, bluish-white; legs, bluish-grey. Length, 18 in. Weight, 1 lb. 2 oz. - 1 lb. 5 oz.

Distribution: A northern species formerly ranging south to Shark Bay, Wooleen, Meekatharra and Wiluna. However, it appears to be one of the species which in recent years has been spreading southwards (cf. p. 61). Like the Galah it may have been restricted to the river gum habitats of the North-Western rivers and creeks. Flocks have now been reported south at the mouth of the Murchison River and we have collected the species at Mullewa. In January 1961 J. R. Ford observed it near Northern Gully, about 20 miles east of Geraldton. Nesting was recorded 10 miles south of Three Springs in 1975 (N. Kolichis). Since the early 1960s Little Corellas have frequented the metropolitan area west of Perth (seen at Nedlands, Claremont, Butler's Swamp and Bicton).

Nesting: The nest is usually in a hole in a tall tree, but the bird also nests in cavities in cliffs, and in treeless districts it sometimes nests in the tops of large ant-hills. The clutch consists of 3 (sometimes 4) white eggs. Average size of 9 eggs from Ullawarra and Landor Stations, 41 x 29 mm., varying from 39-44 x 29-30 mm. The breeding season ranges from August to October.

Fig. 46.—Little Corella.

Fig. 47.—Long-billed Corella.

This is the common white cockatoo of the North-West. It occurs all along the rivers, inhabiting the river gums, and where these are scarce or absent—as in parts of the North-West coast—it makes its headquarters in the cliffs. It forages over the neighbouring flats and has the reputation of being a good check on the "double-gee" *(Emex australis)* pest. For this reason, since 1911, it was protected under the old Game Act in areas at the mouths of the Gascoyne and Minilya Rivers.

Sometimes enormous flocks are met with, one observed at Wyndham being estimated at between 60,000 and 70,000 birds. They are common habitues of North-West towns as far south as Carnarvon and forage in the streets in the early mornings. They frequent garden trees and roofs with the same aplomb as pigeons and doves do in the south.

Long-billed Corella

Cacatua tenuirostris

Native names: Mannich, Mennich, Mannite or Manyt (South-West); Mannitch (Pallinup River); Munnit, Manup (Albany).

Similar in most respects to the preceding species, except for the bill, which has a relatively *enormously lengthened upper mandible*. The degree of development of the hook of the bill varies greatly. Length, 18-18½ in. Weight, 1 lb. 10 oz.

There are no other field differences between the two Corellas, in calls or otherwise. Previously there was a wide geographical gap between their ranges but the spread southwards of the Little Corella has almost extinguished this, and the species are now coming into contact with each other.

Distribution: The Long-billed Corella is confined to the South-West from about Geraldton to the south coast.

Nesting: The 2 or 3 eggs are laid in a hollow high up in a tall tree. The species nests at the Perup River, south-west of Manjimup, where it has been noted as using hollows in wandoos *(Eucalyptus redunca* var. *elata)*. C. L. E. Orton informs us that west of Moora, where the bird nests, it also prefers wandoos to marris for nesting purposes. The egg is dull white and oval, with one end rather pointed. Average size of 13 eggs from Jurien Bay, 44 x 31 mm., varying from 40-48 x 30-35 mm.

This bird is no longer found near Perth, where it occurred abundantly, as elsewhere in the South-West, when the colony was first settled. The botanist, Charles Fraser, who accompanied Captain James Stirling on a survey of the Swan River in March 1827, observed "large flocks" of this cockatoo at the eastern part of what is now the City of Perth near the Swan River, and noted that the birds "feed on the roots of orchideous plants, for which they scratch to a considerable depth." George Fletcher Moore, the Advocate-General, wrote in his diary on June 23, 1835, that at Guildford the "white cockatoos are becoming very troublesome upon the wheat, as well as the crows. One is obliged to keep a boy to drive them away, or to make some contrivance to frighten them. . . . It is singular to see a field spotted black and white with these depredators 'piebalded'." They have not been seen in such numbers near Perth within living memory, but in other parts of the South-West they remained in strength until comparatively recent times. Tom Carter wrote in 1924: "Some of the original settlers around Broome Hill told me how, when boys, they had been sent out by their fathers to frighten the flocks away from the corn crops." Small colonies survive in isolated parts of the Darling Range country and further south, particularly around Lake Muir, and they were known until lately at Augusta. In 1944 a small party appeared in the neighbourhood of Mandurah. The species' headquarters are now between Moora and Geraldton and east to about the Wongan Hills-Mullewa railway line and Kalannie. Within this region and westerly to the coast they are comparatively plentiful. A somewhat similar eclipse has been suffered by the species in the Eastern States.

Among the new foods which this species exploits is the introduced Guildford "grass" *(Romulea rosea)* whose corms it digs up with its long beak. This digging for food is a characteristic habit of the species as compared with the Little Corella. In consequence the plumage of Long-billed Corellas is often soiled reddish because of contact with the ground, whereas Little Corellas remain an immaculate white.

Galah

Cacatua roseicapilla

Native names: Pit-ahn-di (Coolawanvah); Bilya-coo (Mulga Downs); Culerrit (North-West Cape); Bilyendi (Lyons River); Cogalala (Well 38, Canning Stock Route); Bee-arkoo (Jigalong); Biarka (Laverton); Jindurka (Warburton Range); Biagu (Pidong tribe, upper Murchison River); Be-ahlee (Wooleen); Bee-ogu (mouth of Murchison River).

Pink and grey. The back, wings and tail are grey, lighter on the rump and darker at the wing and tail tips. The head is rose-pink, the under parts deep pink, this colour extending also under the wing along the leading edge. The rest of the under wing and the under tail coverts and tail are grey. Young birds have a greyish suffusion through the pink of the breast. Iris, pink in the female and very dark brown in the male; fleshy eye-ring, white or greyish-white; beak, horn-white; legs, light slate. Length, male, 14-14½ in.; female, 13-14½ in. Wing span, 30 in. (male). Weight, male, 12 oz.-1 lb.; female, 10 oz.-1 lb.

Distribution: This cockatoo is now fairly generally distributed in the State and in recent years has spread rapidly into the southern parts, particularly into the Wheatbelt. It would appear that prior to white settlement it did not occur south of the mulga-eucalypt line (see p. 53) and was restricted to the river gum habitats of the North-Western river systems, south to the Murchison. It was not mentioned by John Gilbert in his records or in the exploring journals of the Gregory brothers. The early ornithologists of this century

(A. W. Milligan and F. Lawson Whitlock) did not encounter it when surveying the bird fauna of the Midlands and Wongan Hills districts (1903) and Yandanooka (1904). One of the earliest southern records was made by W. B. Alexander who saw a pair at Waddouring Hill (south of Bencubbin) in October 1915—the first time they had been seen in the locality. In 1928 Galahs were very abundant at Mingenew. By the early 1920s they had reached the outskirts of the north-eastern Wheatbelt and odd birds appeared sporadically near Kellerberrin (1919, Bruce W. Leake) and even as far as Williams (1918, J. B. Higham). However, these intruders did not persist at the time. Apparently it was not until the 1930s and early 1940s that the birds became really plentiful in the northern Wheatbelt. By the close of the 1940 decade they had penetrated in strength to a line approximately from the Hill River to Bolgart, Goomalling, Wickepin and Lake King. Since 1957 they have become fairly plentiful in the Perth district, along the coast, and, avoiding the heavy jarrah forest country, have extended deep into the Wheatbelt. Breeding takes place as far south as Nyabing (1956) and Katanning (1958) and a few birds have appeared at Broomehill. Eastwards they have not yet reached Esperance but have been recorded at Fraser Range and Eucla.

A similar southerly extension of range has taken place in South Australia and in the eastern states. It is customary to explain this spread as the effect of an improvement in the bird's habitat requirements (building of dams, extension of cereal growing, etc.). However, though these factors have undoubtedly accelerated the process it seems that fundamentally the extension of range was initiated by a slight deterioration of the inland climate during this century (see p. 61).

Nesting: The Galah nests in a hollow in a eucalypt or mulga and lines the cavity at the bottom with a plentiful supply of fresh gum leaves. The usual number of eggs in a clutch is 5 and they are laid over a period of a fortnight. Up to 7 eggs have been recorded (I. C. Carnaby). The nesting hollow is usually well-bitten round the entrance. The eggs are pure white, average size of 19 eggs, 35 x 29 mm., varying from 23-38 x 25-29 mm. Two broods are raised in favourable seasons. The young assume the adult plumage with their first feathering.

The Galah usually feeds on the ground—on seeds and on bulbous plants which it digs out of the ground. The bird does considerable damage in the Wheatbelt, not only to standing crops but to bagged wheat. It has, therefore, been "declared" under the Vermin Act in most of the principal wheat-growing areas: in 1940 it was "declared" in the Mukinbudin district, in 1943 in the Koorda district, in 1945 in the Victoria Plains, Morawa and Dalwallinu districts, in 1946 in the Upper Chapman, Goomalling and Ballidu districts, in 1947 in the Moora, Quairading and Mount Marshall districts, in 1948 in the

Gingin and Westonia districts, in 1952 in the Corrigin and Wickepin districts, in 1953 in the Kununoppin-Trayning district, in 1958 in the Bruce Rock district, and in 1964 in the Coorow and Mullewa districts. In 1971 the list of districts in which it was declared vermin was reduced to Mukinbudin, Morawa, Moora, Westonia, Coorow and Mullewa. In 1942-43 the Mukinbudin Vermin Board paid bonuses on 3,529 Galahs destroyed and in the following year on 4,311 individuals. The bonus paid was one halfpenny per head.

PARROTS: Subfamily Psittacinæ

Cockatiel (Cockatoo-Parrot, Quarrion, Weero)

Leptolophus hollandicus

Native names: Bula-doota (Godfrey's Tank, Canning Stock Route); Wee-arra (North-West Cape); Toorir (Jigalong); Wamba (Gascoyne River); Weeroo (Lyons River); Wauraling, Wuraling, Woo-ra-ling, Woorerling (South-West).

Dark grey above and below, the long central tail feathers lighter; head and throat yellow, with an orange cheek patch and a permanent *yellow and grey crest; broad white wing bar* with yellow tinge. The female has the face and crest dull olive-yellow and the tail barred brown. Young birds have barred tails with a yellow tinge below. Iris, brown; beak, greyish-black; legs, light grey. Length, 11½-12¾ in.; wing span, 12 in. Weight, 3-4 oz.

Distribution: All over the State except the south-west jarrah forest area. Occasionally wandering birds penetrate as far as Bridgetown. In 1974 there was a general invasion of the species in areas they had rarely been seen previously.

Nesting: Nests in a hole in a tree, or in a hollow limb, or in a hole in a stump; usually in dead trees. No nesting material is used. According to the season the clutch varies from 4 to 7 eggs; glossy-white and oval rather than round in shape. Size 26-28 x 19-21 mm. Incubation takes 21 days. The male bird broods during the daytime (Dr. A. Lendon).

The normal song, uttered in flight, sounds like the words "queel-queel."

Red-winged Parrot

Aprosmictus erythropterus

Head and under parts bright green. Mantle blue-black. *Conspicuous bright red shoulder patch.* Lower back bright blue. Upper tail coverts pea green. Wing and tail quills green. Iris, orange-red; beak, orange, red at base; legs, leaden brown. Length, 12-13 in.

Distribution: A common bird in the Kimberley Division and it barely extends into the extreme north of the area covered by this book. It occurs in the pindan country, which forms a narrow belt along the south coast of La Grange into the northern part of the Pilbara. W. H. Butler saw a party of six of these parrots 21 miles south of the Anna Plains turnoff. There are numerous observations of the birds at the turnoffs to Frazier Downs, Nita Downs and Anna Plains.

In this area the birds were feeding on the seeds of a broad-leafed Acacia.

Smoker Parrot (Regent Parrot, Black-tailed Parrot, Rock Pebbler, Wockengull, Walkingale, Plaide-wing Parrot)

Polytelis anthopeplus

Native names: Wouk-un-ga, Waukanga, Walkinger (Avon River).

General appearance yellowish-green. *Upper parts olive-green,* rump lighter and the tail dark bluish-green. The wing quills are blackish with blue on the outer vanes, the wing coverts are bright yellow-green and dark red. The *under parts are a smoky yellow-green,* brighter on the throat and abdomen. The under tail feathers are black. Iris, red; beak, bright red; legs, iron-grey. Length, 15-16½ in. Weight, 4 oz.

The sexes have generally a similar colour pattern but the male is much yellower and the female relatively much greener. In the female, also, the under tail feathers are bordered on the inner vanes with salmon-pink.

This is a narrow- and long-winged swiftly flying parrot whose almost uniform smoky yellow-green coloration and red beak distinguishes it from all the other large local parrots.

Distribution: Found in the South-West portion of the State, north to Ajana, east to the Kalgoorlie area, the Dundas district, Esperance and Israelite Bay. The furthest east record is from Karonie. It does not usually occur in the jarrah forest area and the extreme South-West corner (Fig. 5).

Nesting: The Smoker nests in a hollow in the main trunk of a tree, and often the eggs are laid at the bottom of a hollow of considerable depth — sometimes as much as 15 feet down. The eggs, 3 to 6 in number, are pure white and roundish. A series of 14 eggs from Moora and Lake Grace averages 30 x 25 mm., varying from 29-32 x 24-26 mm. Incubation takes 21 days.

This is one of the species which has notably increased in abundance and range of distribution since the settlement of the Wheatbelt, in sharp contrast to its response to these factors in the eastern States. In South Australia, Victoria and New South Wales the Smoker is a bird of the mallee areas, but its numbers have dwindled with settlement. Here the reverse has happened and so numerous have the birds become that they have been "declared" under the Vermin Act in the following districts: Victoria Plains (1931), Wickepin and Narrogin (1936), Gingin (1941), Morawa and Broomehill (1945), Katanning and Goomalling (1946). Morawa was removed from the list in 1968. The species has now been removed from the vermin list in the last two districts.

The species is now spreading from the Wheatbelt into the forested South-West corner, where it had been unknown up to the first quarter of the present century. Stray individuals have appeared in the area since the early 'twenties and the records have become more numerous in later years, when the visitations have become more prolonged. The centres visited include Wooroloo, Chittering, Middle Swan, the Perth area, Pinjarra, Collie, Boyup Brook, Bridgetown and Albany. The bird's distribution, originally that represented in Fig. 5, is thus tending to become of the type illustrated in Fig. 9.

However, in more recent years the Smoker appears to be declining generally in the Wheatbelt.

Ringing experiments by Ray Aitken have proved that the Smoker is a greater wanderer than the other common parrot of the Wheatbelt, the Twenty-eight. A Smoker ringed at Nyabing in August 1954 was recovered two months later 60 miles to the north. Another ringed at Dumbleyung in July 1955 was found at Highbury, 33 miles north-west, in the following month. A third bird, ringed at Dumbleyung in September 1955 was shot at Grass Patch, nearly 200 miles east, in February 1958.

Alexandra Parrot

Polytelis alexandræ

Native name: Billbul (Jigalong).

Back light olive-green; head and neck, light blue; rump, violet; long tail green, margined with rose pink. Yellow-green on wing coverts. *Chin, side of face and throat rose pink;* yellow-green patch below eye; rest of under parts olive-green becoming very grey on abdomen. Flanks deep blue and lilac; dull red on thighs. The female is duller. Iris, orange; beak, coral red; legs, dark grey. Length, male, 17-19 in.; female, 15¾-16¼ in.

Distribution: An inhabitant of the arid interior, occurring north almost to the Fitzroy River and south to Wiluna, Sandstone and occasionally Menzies. It occurs eastwards through the Great Victoria and Gibson Deserts to the central Australian range country.

Plate 2.—Races of the Western Rosella

From left: male, female and immature of *Platycercus icterotis icterotis*; male and female of *Platycercus icterotis xanthogenys*.

Nesting: The clutch of 4 to 6 white eggs (28 x 33 mm.) is laid in a hollow limb or hole in a tree.

The food consists chiefly of the seeds of spinifex *(Triodia irritans)* and *Portulaca,* and the bird is almost restricted to spinifex country.

Western Rosella (Stanley Rosella, Yellow-cheeked Parrot)

Platycercus icterotis

(Plate 2)

Native names: Goota-un-goota-un, Guddang-uddan, Guldanguldan (Perth); Moy-a-dung, Moyadung (Avon River).

Red the prevailing colour. The *head and the entire under parts are red,* with yellow on the cheeks. The feathers of the back are black with green edgings, the rump is green, the tail blue and green and the wing blue and black, with green coverts. In the inland areas red tends to replace the green on the back — this is the race *xanthogenys*. Iris, dark brown; beak, bluish-horn; legs, slate-grey. Length, 10-11¾ in.

Females are slightly smaller and duller and greener. The breast is a dull red and more or less tinged green, the red colour on the head and hind neck is lacking or very poorly developed, and the cheek patch is a paler yellow. In the race *xanthogenys* the back is dark grey with the feathers edged buff.

Immature birds have white spots on the inner webs on the under side of the secondary feathers of the wing. These form a "wing bar," and are a feature of most of this group of broad-tailed parrots. In successive moults the spots become smaller and fainter and eventually, when the birds are fully mature, disappear altogether. The birds are greener than the adult female and lack the yellow cheek patch.

Distribution: Confined to the South-West corner, north to Moora and east to Wongan Hills, Moorine Rock (near Southern Cross) and the Dundas district. The male bird of the race *P. i. icterotis* illustrated on Plate 2 is from a specimen collected at Marrinup and the female from the Kalgan River, Albany. The specimens used for portraying the race *P. i. xanthogenys* were collected by F. Lawson Whitlock at Lake Dundas (male) and Wongan Hills (female).

Nesting: The Western Rosella uses a small hollow spout or a hole in a eucalypt tree as a nesting site and no nesting material is added. The clutch ranges from 3 to 7. A series of 17 eggs from Lake Grace averages 25.9 x 21.6 mm., varying from 24.6-27.7 x 20.3-22.5 mm. The nesting season is from August to December, September being the main month in which eggs may be found.

The call-note may be rendered as "chink-chink."

The Rosella was declared "vermin" throughout the State in 1921, on account of the damage it does to orchards, although in this respect it is not as destructive as are the Twenty-eight and King Parrots. However, it was removed from the State-wide list in 1963 and the declaration restricted to the Armadale-Kelmscott and Bridgetown-Greenbushes districts. The Rosella lives chiefly on seeds and is usually flushed from the ground, whence it will fly up and perch in a neighbouring tree before departing elsewhere.

King Parrot (Red-capped Parrot, Hookbill)

Purpureicephalus spurius

Native names: Djar-rail-bur-tong, Djarrylbardang (Perth); Djalyup (South-West); Chelyup (Stirling Range); Jul-u-up (King George's Sound).

The most gaudily-plumaged of our larger parrots. The male in mature plumage is bright deep green on the back and wings, with the flight quills dark blue and black. The *rump is bright greenish-yellow.* The tail is deep green with blue on the outer feathers. The *top of the head is a deep crimson.* The throat and sides of the head are a bright pea-green, contrasting brilliantly with the rest of the under-parts which are a deep *purple* and *red near the legs and the under tail coverts.* The adult female is almost as highly coloured as the male, but the cap is never quite as intensively red and has a slight green admixture. The breast is also slightly more greyish-violet than purple. Immature birds have a row of white spots on the under side of the flight quills, and during the first year are nearly uniform green, with the characteristic colouring of the adults only just apparent; there is a small red frontal band; the breast and abdomen are a pale greyish-violet and the under tail-coverts are red margined with green. Iris, dark brown; beak, grey-horn with a bluish tinge, and the upper mandible greatly elongated; legs light brown. Length, 14-14½ in.; wing span, 19 in. Weight, 4½ oz.

The King Parrot can be easily distinguished when in flight from the Twenty-eight by its *light-coloured rump.*

Distribution: Found in the South-West corner north to Dandaragan, Gingin and Mooliabeenee, and generally west of the Great Southern Railway, but in the south extending eastwards to Lake Grace (rare), and along the coast to Esperance.

Nesting: The nesting site is a hollow high up in a tree, the eggs being laid at the bottom of the hollow, without any nesting material. They are roundish in form and pure white. Normally 5 eggs form the set laid, occasionally 6, but up to 9 have been found. Measurements, 28 x 22 mm.

Under natural conditions these birds feed on the seeds in the capsules of various eucalypts, including the marri *(Eucalyptus calophylla)*, cones of sheoaks and other seeds. They are often seen feeding on the ground. They bite off the capsules and hold them in one foot while extracting the seeds with their long beaks. In parts of the South-West the Red-capped Parrot exceeds even the Twenty-eight in abundance and is frequently a pest in orchards. The species was "proclaimed" under the Vermin Act in the Collie and West Arthur districts in 1943.

Twenty-eight Parrot (Ring-neck, Port Lincoln, Yellow-collared Parrot)

Barnardius zonarius

Native names: Pun-bunba (Pilbara district); Pun-bun (Coolawanyah); Mullincoorin or Mullin-goora (North-West Cape); Mabenga (Lyons River); Doondroon marda (Jigalong); Patilpa (Warburton Range); Peel-baal (Yandanooka); Dum-ul-uk, Dammalak (Avon); Daran, Dow-arn, Tow-arn (Perth); Towerrin (Broomehill); Toonat (Katanning; Towern (Albany); Toornart (Pallinup River).

The upper parts are a *brilliant green,* with the wing quills blue and black and the outer tail feathers blue. The head is black with a varying amount of brick-red on the forehead above the beak, purple-blue on the cheeks and a yellow collar behind the head. The *under parts are either entirely green* or *with a broad yellow band on the lower breast,* separating the deep green of the chest from the bright pea-green of the abdomen. The under tail is pale blue. Iris, dark brown; beak, white horn; legs, grey. Length, 13-16¾ in. Wing span, 20 in. Weight, 5-7 oz.

Immature birds have white spots on the underside of the secondary wing feathers forming a "wing-bar."

Distribution: Generally distributed north to the Pilbara district and eastwards into Central Australia and South Australia.

Nesting: The normal number of eggs in the clutch appears to be 5, but up to 6 are known; they are laid at the bottom of a hollow spout of a tree, or at the bottom of a hole in the tree itself. They are white and are inclined to be more pointed at the sharper end than is usual with parrots. A series of 27 eggs, from localities between the Barlee Range and Lake Grace, averages 29.6 x 23.9 mm., varying from 27.3-32.0 x 22.3-25.0 mm. An abnormally small egg from Jurien Bay measures 20.0 x 17.5 mm.

This is Western Australia's commonest and most widespread and consequently its best-known parrot, both in the wild and as a cage-bird. It is plentiful both in the heavily-timbered South-West forest and in the more open areas inland and further north. It feeds on the pulp of green eucalypt capsules and has learnt to tap a variety of introduced and cultivated plants to extend its food supply. To orchards in the South-West which adjoin forest country it is a pest, like the associated King Parrot, feeding on pip-fruits, almonds, etc., and, with the Smoker, it is also troublesome in the wheat areas. Therefore it was declared "vermin" under the Vermin Act throughout the State when that measure was enacted in 1921 and this obtained for many years until the declaration was restricted to the districts of Dalwallinu, Armadale-Kelmscott and Bridgetown-Greenbushes some time prior to 1963. Despite the numbers killed annually the Twenty-eight shows no sign of diminution, and throughout the South-West one meets flocks of them, often disturbed from the roadsides whence they fly in undulatory glides ahead of a car.

There is considerable plumage variation in the species. In the wetter areas of the South-West corner occurs the all-green form with a prominent red forehead (the race *semitorquatus)*. Almost all birds south-west of a line from just north of Perth, through the Darling Range country to just west of Albany belong to this colour type. East of a line from about Moora, Goomalling, Wickepin and Nyabing all the birds have a broad yellow breast-band and only a trace of red, if any at all, on the forehead. This is the form often called the Port Lincoln Parrot (the nominate race *zonarius*), which extends over most of the inland area into South Australia. In between these lines, east of Perth and comprising most of the Wheatbelt, is a transitional area where the two colour types grade into each other. In the western portion of this intermediate zone there are many all-green birds as well as various degrees of hybridisation between the two forms. The situation has arisen, as mentioned on p. 53, through the geographical isolation of the original population in the South-West corner during an earlier arid period, and its differentiation into what is now the all-green *semitorquatus* type. However, full specific development was not attained (as in the comparable case of the White-breasted Robin) so that when subsequently the climate ameliorated and the yellow-breasted Port Lincoln type re-invaded the South-West secondary contact between the two forms allowed inter-breeding, producing the variable hybrid population now existing in the transition zone.

In the North-West has developed a fairly distinctive race, *occidentalis,* which is smaller and has a bluish tone in the upper plumage, lighter blue on the cheeks and the yellow on the under parts extending as far as the vent.

The call notes also vary regionally. The well-known double-noted "twenty-eight" call is confined to the races *semitorquatus* of the South-West and *occidentalis* in the North-West, where the notes are more drawn-out. Elsewhere the call can be represented by the syllables "tweet-tweet."

Mulga Parrot (Varied Parrot, Many-coloured Parrot)

Psephotus varius

Native names: Jooleen (Wooleen); Choogin (Yandanooka); Tu-lee (Paynes Find); Pullal-Pullal (Cundeelee).

Upper parts and breast a brilliant green with a bluish tone, lighter on the rump, which is followed by a band of dark green and then a narrower band of red. The forehead is orange-yellow and there is a red patch on the nape. There is a light orange-yellow band on the shoulder and the wing quills are blue, edged black. The tail is green with the outer feathers blue, largely tipped white. On the under surface, below the green breast, the abdomen is red and the under tail coverts yellow. Iris, dark brown; beak, slate-blue; legs, lead-grey. Length, 11-12¾ in. Wing span, 16 in.

The female is very much duller, greyish-green; with a brick-red band on the shoulder and lacking the red abdominal patch.

Distribution: Generally distributed, north to the Pilbara district and excluding the South-West corner (jarrah forest belt), west and south of a line joining Moora, Grass Valley, Quairading, Kweda, Lake Grace and Dundas.

Nesting: Nest usually in a hollow limb of a eucalypt and 4 to 5, occasionally 6-8, eggs. Average size of 10 eggs from Morawa and Yalgoo, 22.7 x 18.4 mm., varying from 21.7-23.8 x 17.0-19.3 mm. The eggs are placed on wood dust at the bottom of the hole. The male acts as sentinel in the vicinity of the tree in which the female is sitting and becomes greatly disturbed if the tree is approached. Incubation lasts 21 days.

Blue Bonnet (Oak Parrot, Naretha Parrot)

Psephotus hœmatogaster

Back, nape and breast pale brown; mantle olive-green; rump olive-yellow; tail green, outer feathers blue with white tips; shoulder bright blue with the wing coverts red and yellow, flight quills blue. The brownish breast grades into the yellowish abdomen; under tail coverts bright red. The female is not so brightly coloured as the male. Iris, brown; beak, bluish grey; legs, grey-brown. Length, 10¼-11½ in.

Distribution: This parrot has a very restricted distribution in Western Australia, being confined to the myall *(Acacia sowdenii)* country that fringes the north-western edge of the Nullarbor Plain.

Recently (January 1966) J. R. Ford has observed it in the myall belt south of the Nullarbor Plain, from Mundrabilla to Eucla and I. C. Carnaby has seen it in South Australia 10 miles east of Eucla. In the northern myall belt it has been proved to occur from Kitchener to 80 miles north of Loongana, a zone some 180 miles in length and probably not more than 20 miles in width. It may extend even further east, but it does not occur north in the mallee and mulga country of the Great Victoria Desert.

This little-known form is the Western Australian representative (the race *narethae)* of the better-known bird which inhabits a large part of south-eastern Australia, on the inland side of the Great Dividing Range and north to southern Queensland. The Western Australian race is a very distinctive one. It differs from the eastern birds in being more yellowish-green on the mantle and rump, paler grey on the chest, and with no red on the abdomen which is wholly yellow. The under tail coverts are red. Joseph M. Forshaw also draws attention to the two-tone facial colouring (blue, with the lower ear coverts edged pale yellow) not found in the nominate race. The local bird is also smaller, and the female is smaller than the male.

Nesting: The only nesting notes we have of this species in Western Australia are those of F. Lawson Whitlock who found altogether eight pairs nesting at Naretha. The nesting holes were all in desert sheoaks *(Casuarina cristata)* and the number of eggs laid varied from 4 to 6. They were rounded oval, shell close in grain, but slightly rough. Colour white, no gloss. Size, 22 x 18 mm.

Bourke Parrot (Night Parrot, Sundown Parrot)

Neophema bourkii

Back olive brown, head brown with a pink tone, cheeks brown streaked white, white around the eyes, forehead blue-grey. Edge of wing and quills blue, wing quills scalloped with yellow-buff. Flanks sky-blue. *Under parts pink,* paler and grey towards the throat and brighter towards the abdomen. Iris, brown; beak, brown; legs, grey-brown. Length, 8-9¼ in.

The blue is absent from the forehead in the female.

Distribution: An inland and central Australian species occurring north to the upper Ashburton River and Jigalong, and recorded as far south as Morawa, Wubin, Paynes Find, Leonora and Laverton.

Nesting: The Bourke Parrot lays its 4 or 5 white eggs in a hollow tree, at from 3 to 10 feet from the ground. They are deposited merely on the decayed woody material at the bottom of the hollow. A series of 21 eggs from Yalgoo averages 20.4 x 16.8 mm., varying from 19.4-21.8 x 15.6-17.8 mm. The eggs are laid on successive days and incubation starts with the first egg laid and lasts about 21 days. The young leave the nesting hollow in succession about 27 days after being hatched, and are fed by the parent birds for a week or ten days after leaving the nest. The male feeds the female by regurgitation during courtship, and whilst she is incubating, during which period she sits very closely, leaving the hollow but once a day and then only for short periods.

In common with several other species of its genus this parrot had become very rare over most parts of inland Australia where it formerly occurred in some numbers, the decline following, perhaps, on the subtle alteration of the environment due to stocking and the rabbit invasion. In recent years, however, it has become fairly plentiful in the mulga country between the Upper Ashburton and Murchison watersheds and eastwards beyond Laverton across the Gibson Desert to Central Australia. I. C. Carnaby has seen flocks of between 20 and 30 to 100 birds come to water at sheep troughs at Mulgul station, in the Ashburton divide in 1938. The explorer Charles Sturt was the first to note the distinctive habit of this parrot in coming to water after dark and before dawn, a habit leading to the general adoption of the name "Night Parrot," thus causing some confusion with the bird normally called by that name (see p. 291). However, birds do come to water points during the day also. The flight call is a mellow musical "chu-wee."

Scarlet-chested Parrot

Neophema splendida

Back and middle tail feathers green, outer tail feathers black and yellow; *head and throat royal blue* shading into lighter blue behind; *breast scarlet, abdomen and under tail feathers yellow.* Wing sky-blue on the shoulder, the quills dark brown edged deep blue and green. The female is duller and has the scarlet area of the chest replaced by green. Iris, brown; beak, dark grey; cere, black; legs, grey-brown. Length, 7½-8¾ in.

Distribution: An inland species, but until recently little was known of its usual range in this State. At the present time most of the records of its presence are from the eastern goldfields and eastwards into the Great Victoria Desert. There were several records of its occurrence in the early days. John Gould received a specimen from Western Australia in 1840, and in 1845 further specimens from the Moore River from Johnston Drummond. In 1854 Surveyor Robert Austin encountered the parrot on his expedition to the East Murchison country. Nothing was then heard of the species for many years. In 1939 there was an irruption in South Australia near Wynbring (on the trans-continental railway), when large numbers were trapped for the avicultural trade. In December 1941 H. M. Wilson saw one bird at Bullabulling, west of Coolgardie. In October 1948 L. R. M. Hunter observed a small party near Lake Cowan, 100 miles south of Kalgoorlie. In 1949 D. Marchant noted a small party near Kweda (between Pingelly and Corrigin). In September 1956 J. H. Calaby saw one bird at Coonana. In March 1957 and March 1958 K. Griffiths saw birds at Laverton. In January 1959 W. H. Butler saw a male and two females at Queen Victoria Spring, and in August 1960 a male bird at Maleura water hole in the Great Victoria Desert. In December 1959 G. M. Storr saw a pair at Coragina Rock, 40 miles south of Balladonia.

Nesting: There are no records of nesting within the State. It has been reported as following the usual practice of parrots and nests in a hollow in a tree. The eggs are white, 3 to 5 in number; size 23 x 19 mm. (A. J. North).

Elegant Grass-Parrot

Neophema elegans

Native Names: Gool-ye-der-ung, Gul-yidarang (South-West).

This grass-parrot's prevailing colours are green and yellow, with blue on the wing and tail. The basic body colour is bright olive-green, with yellow on the belly and sides of the under tail. The forehead is blue, with *yellow between the eye and the beak,* and extending towards the cheeks. The flight quills are black, with brilliant blue on the coverts and some of the secondaries. The tail is blue above and black below with canary yellow on the outer feathers. Iris, dark brown; beak, grey-horn; legs, grey. Length, 8¼-11¼ in.

The female is noticeably a duller green.

Distribution: Until recent years this parrot was known only in the South-West corner, north to Moora and east to Merredin and near Esperance. Formerly it was not found in the jarrah forest belt or the Swan coastal plain, but since 1937 the species has been undergoing a cycle of expansion, radially in the South-West and north to the pastoral country. It now extends to Moorine Rock, Lake King, Lake Grace, Albany, Boyup Brook, Bridgetown, Greenbushes, Collie, Brunswick Junction, Wandering and Pinjarra. It was first recorded from the Perth area (at Middle Swan) in 1945. In 1957 I. C. Carnaby observed it at Woodlands Station on the Gascoyne and in 1958 a specimen was received at the W.A. Museum from Hooley Station on the Fortescue. S. R. White recorded it at Mount Magnet in 1960 and T. E. Bush has seen it at Mullewa.

Nesting: The 4 or 5 eggs are dull white and almost round, and are laid in a hollow spout or in a hole in the trunk of a tree. A series of 5 eggs from Gnowangerup averages 20.7 x 17.6 mm., varying from 20.4-21.2 x 16.7-18.7 mm. They are distinctly smaller than those of the Rock Parrot.

Rock Parrot

Neophema petrophila

The Rock Parrot is a duller edition of the preceding species. The *under parts,* instead of being olive-green and yellow, almost lack any green hue and are a *smoky brownish-yellow.* There is blue on the forehead and lores, but no yellow, and the sides of the head are dull greenish-brown. Iris, dark brown; beak, bluish-horn; legs, mealy-brown. Length, 8½-9½ in.

Distribution: The Rock Parrot frequents coastal islands and mainland dunes and rocks (both granite and limestone) from the Great Australian Bight to Shark Bay.

Nesting: The 4 to 6 eggs are usually laid under a slab of rock on an islet, overlooking the ocean, above high-water mark, but often in such a position that spray is dashed over the rock in windy weather. A favourite site is a crevice or hollow in limestone rock. Often the site is well covered by overhanging vegetation *(Carpobrotus).* The eggs are white and rounded-oval. Average size of 14 eggs from Rottnest and Lancelin Islands, 25.2 x 19.8 mm., varying from 23.6-26.2 x 18.8-20.8 mm.

On the lower west coast the Rock Parrot is almost exclusively an inhabitant of islands and never goes on the mainlaind except when blown there by winter gales. However, from Cape Leeuwin eastwards it is commonly met with along the mainland shore, both among rocks and in the sand dunes. The same is true of some of the country north of Perth and in the vicinity of Lancelin Island, where the species occurs, flocks are to be seen on the mainland dunes. At Thirsty Point (15 miles south of Jurien Bay) G. M. Storr found these parrots especially plentiful in 1959, where they fed freely on the seeds of the introduced strand daisy *(Arctotheca nivea).* At the Albany railway yards birds not infrequently may be seen on the ground picking up spilled wheat grains. This is also the case at Esperance.

This parrot is entirely a seed-eater and has a partiality for the seeds of *Carpobrotus,* a plant which is plentiful on the islands which the birds normally frequent.

The species has been protected throughout the State since 1945.

Budgerygah (Shell Parrot)

Melopsittacus undulatus

Native names: Pa-thara (Coolawanyah); Padda-moora (North-West Cape); Coolyerrie (Barlee Range); Kol-yabiddy (Lyons River); Koolkillkaree (Jigalong); Kilykilykari (Warburton Range); Muddenburra (Wooleen); Din-gel-yer-ung (Avon River).

Head, nape and wing coverts narrowly barred with yellow and black; lower back brilliant green; wings grey-green; tail blue; the forehead and throat are yellow, with a number of blue spots on the lower cheek and black spots on the throat; rest of under parts bright green. Iris, white; beak, olive-grey; legs, leaden-blue. Length, 7-8½ in.

The sexes can be distinguished by the colour of the cere, which in the male is dark blue, and in the female pale blue in the non-breeding season and brown when in the breeding state. Immature birds of both sexes have the cere pale blue, like the non-breeding female. Varieties in which the green is replaced by blue and yellow are now commonly bred in captivity, and such colour phases occur occasionally in nature.

Distribution: All over the State, except in the South-West jarrah forest area. However, occasionally this highly nomadic species does penetrate into the South-West. Tom Carter recorded them as erratic visitors to Broomehill, and they reached the Williams River in August 1917. In the autumn and early winter of 1951, however, large numbers extensively invaded the South-West, reaching Gingin, Yanchep, Perth, various Darling Range localities (Bickley, Wooroloo, Baker's Hill), Clackline, Northam and Beverley.

Nesting: The 3 to 5, occasionally more, eggs are laid at the bottom of a hole in a tree, or in a spout of a tree, on the bare wood, no nesting material being used. Nine eggs (Barlee Range) average 18.2 x 14.5 mm., varying from 17.2-19.1 x 14.1-15.0 mm. In the northern areas the bird breeds at any time of the year after good rains. In the south, November is the usual month. Incubation takes 18 to 20 days.

Ground Parrot

Pezoporus wallicus

Native names: Boo-run-dur-dee (north of Perth); Djar-dong-garri, Djar-doon-gur-ree (Perth); Djul-bat-ta (south); Ky-lor-ing (Albany).

General colour grass-green with wavy barrings above and below of yellow and dark brown; *a prominent red forehead band;* no yellow on the cheeks; *a pale yellow wing stripe.* Iris, brown; beak, light horn colour; legs, long, flesh colour. Length, 12 in.

Young birds lack the red forehead band.

When flushed the Ground Parrot rises suddenly like a quail and flies off with a zig-zag flight, displaying the pale yellow wing stripe. It drops suddenly about 50 or 60 yards ahead, when it may be again flushed. The red forehead band is easily visible on birds which may be sighted on the ground.

Distribution: This species is now rare and of restricted distribution in Western Australia, but in the early days it occurred on the coastal plain from north of Perth to Albany. Up to recently the last individuals which appear to have been observed by naturalists in this State were noted by S. W. Jackson at Irwin's Inlet in 1912 and by F. Lawson Whitlock in the wet blackboy flats around Denmark in 1913. However, in December 1952 J. W. Baggs saw 4 birds at the Bow River, near Irwin's Inlet. In November 1963 members of the R.A.O.U. saw the birds at Cheyne Beach, where they had previously been observed by C. Allen in 1947.

Nesting: The nest is usually placed below some low bushy plant, where a circular depression is scratched out in the soil and lined with grasses. A nest found by Whitlock at Wilson's Inlet on November 20, 1913, had 3 fresh eggs; pure white, roundish in shape, fine and smooth with very little gloss. Size, 27 x 22 mm. Another nest found by the same ornithologist in the same locality on October 20, 1912, had two nestlings a few days old.

Night Parrot (Spinifex Parrot)

Pezoporus occidentalis

(formerly *Geopsittacus occidentalis*)

Similar in general colour pattern to the preceding species, but with *no red band on the forehead; bright yellow on cheeks and sides of neck;* abdomen and under wings yellow. Legs and toes long. The tail is much shorter than in the Ground Parrot. There is no information regarding the colour of the iris, beak and legs. Length, 11 in.

Distribution: A bird of the dry interior which in this State has been recorded west and south as far as the Cue district and north to the Peak Hill area.

Nesting: There is nothing on record of the nesting habits of this rare species in Western Australia. In 1883 F. W. Andrews, the collector for the South Australian Museum, recorded that the clutch consisted of 4 white eggs, laid in a hollow made by the bird in the inside of a tussock or bunch of porcupine grass *(Triodia).* In 1931 J. Neil McGilp, the South Australian ornithologist, obtained from

John McDonald, of Macumba Station, north-east of Oodnadatta, the information that the birds enter the nest through a tunnel starting low down on one side and almost under the bunch of porcupine grass. At the end of the tunnel there was an enlarged cavity in which the parent birds had placed some small sticks. Another nest, in a low dense samphire bush, was made by forming a runway, or tunnel, into the centre of the bush, where a fair number of small sticks made a platform between the central branches of the bush. In none of the 3 or 4 nests found was the nest actually on the ground itself. The eggs were compared in size to those of the Crested Pigeon.

This interesting and very rare parrot was first discovered by the Austin surveying expedition in 1854 (see p. 38), in granite breakaway country 8 miles south-west of Mt. Farmer in the Murchison district. It was subsequently observed in several localities in central and South Australia but has not been reported again with any certainty by any naturalist outside of Western Australia, since the middle 'nineties. H. M. Wilson has published observations made by Martin Burgoin, of the Public Works Department, on the occurrence of the parrot in Western Australia between 1912 and 1935 at Nichol Springs, Bolger's Soak, Pinyerinya Pool and Windich Spring. In all reported cases the bird frequented spinifex country where the surface water was fresh; it fed on the spinifex seeds and nested in its clumps. As to its decline Wilson wrote: "I have been impressed by the fact that the period of its comparative abundance extends only from 1875 to 1885, though much of its habitat was known before that. It is possible that some cause, such as climate, enabled the species to attain to abnormal numbers during that time, and that sbsequently unfavourable conditions greatly reduced the numbers and made the species extinct in the south-eastern portion of its habitat."

The type locality south-west of Mt. Farmer was visited by Dr. S. J. J. F. Davies in 1959. He found it to be an outlier of the arid country further east. It was granite and laterite breakaway country, an isolated remnant of erosion in sandplain, but differed from similar breakaways elsewhere in the Murchison drainage basin by the presence of spinifex. Davies knew of no other spinifex patch within a hundred miles. The locality is thus typical Night Parrot country from further inland and it is something of a coincidence that the Austin expedition was able to come across this restricted outlier with its interesting bird inhabitant.

Like the Bourke Parrot, which had suffered a similar decline, but is now apparently recovering, this species is nocturnal in its habits. It comes out to feed after dark and may visit water several times during the night, alighting right on the edge of the pool or spring. When watering, the birds give a long drawn-out mournful whistle, and when flushed utter a squeak. The flight is direct, about eight feet above the spinifex, and not undulating like the Twenty-eight Parrot. It does not drop suddenly when alighting, as does the Ground Parrot, but glides down, and makes off along the ground at right angles.

As a step towards preserving any survivors of this very scarce species it was placed on the totally protected list under the Game Act in 1937, which was continued under the Fauna Protection Act of 1950.

CUCKOO-LIKE BIRDS: Order Cuculiformes
CUCKOOS: Family Cuculidæ

Pallid Cuckoo (Rain-bird, Weather-bird, Scale-bird)
Cuculus pallidus

Native names: Booladoodoo (Jigalong); Dju-dur-run, Djudarran (South-West).

Upper parts a light uniform *brownish-grey,* with the inner webs of the flight quills spotted with white and the *outer tail feathers toothed with white* on both webs, above and below. The under parts are light grey, the under wing being grey-brown barred with white. As with many of the cuckoos, there is a small white patch on the leading wing-edge, near the shoulder. The female differs by being plentifully spotted and marked with chestnut and buff on the head; back and upper wing coverts and the tail feathers have chestnut as well as white toothings. The young are dark brown and white below, the throat being comparatively dark. Iris, brown with yellow eyelids; beak, brownish-black; legs, grey-black. Length, 11-13¼ in.

In the field the spotted female is readily distinguishable from the uniformly coloured male, and the calls are different.

Distribution: Found all over the State. The movements of this cuckoo are obscure. In some years, apparently when there are bountiful rains in the North-West few, if any, birds may visit the South-West; 1973 was such a year. Its song is first heard, in the Perth district, on its arrival about the end of May, or beginning of June and it is to be heard until towards the end of November, though odd birds may continue as late as mid-December. T. Carter found that the birds arrive in the Gascoyne district also about May, or with the first rains.

Nesting: The Pallid Cuckoo uses nests of birds constructing open, cup-shaped nests for the purpose of depositing its eggs. Honeyeaters are common fosterers, the species most parasitised in the South-West being Red and Little Wattle-birds, Yellow-throated Miners, Spiny-cheeked Honeyeaters, Singing Honeyeaters, Tawny-crowned Honeyeaters, Brown-headed Honeyeaters, Spinebills and White-naped Honeyeaters. C. L. E. Orton and P. T. Sandland have recorded that they have observed the Pallid Cuckoo in the Moora district using nests of the Willy Wagtail, Black-faced Wood-Swallow, White-winged Triller and Magpie-Lark. We have also information, from further south, that eggs of this cuckoo have been found in nests of the Restless Flycatcher, Brown Flycatcher, Scarlet Robin, Hooded Robin, and the Dusky Wood-Swallow, all of which build open nests. There is also a record of the Pallid Cuckoo using a nest of a Golden Whistler at Williams and a Pipit at Kojonup. East of Nullagine I. C. Carnaby has found its egg in a Grey-headed Honey-eater's nest. T. E. Bush has found eggs of this cuckoo in nests of the Wedgebill and Black-throated Butcher-bird. The egg is a very pale pink, or flesh-colour, without any markings, except that some specimens have a few spots of a darker hue scattered over the shell. Size, 24 x 18 mm.

A common sight in the bush in the spring and early summer is that of the young cuckoo being fed by its foster-parents. The gross youngster, often much larger than they, keeps up a sustained "tzwit, tzwit" begging call, as it waits for its seemingly endless succession of meals. The call resembles very much that of the nestling Red Wattle-bird.

The song of the male bird, a scale-like whistle, is perhaps the cuckoo call best known to the general public. The call of the female is a hoarse single-noted "cheer" or "kheer."

Oriental Cuckoo

Cuculus saturatus

Resembling the European Cuckoo *(Cuculus canorus)* in general appearance. Upper parts grey with a metallic glint; tail black spotted with white; throat grey, *breast white barred bronze-brown;* abdomen buff and similarly barred. Edge of wing shoulder pure white. Like the European Cuckoo this species also has a *brown (hepatic) phase* barred rufous and black above and below, including the throat. Iris, yellow, eyelid yellow; beak brown legs, olive brown. Length, 11-13 in.

Distribution: A non-breeding migrant from Asia to northern and north-eastern Australia. The only record so far known, south of the Kimberley Division, is of a bird seen by John Darnell at Mt. Goldsworthy, in the Pilbara, on March 19, 1973.

Fan-tailed Cuckoo

Cacomantis pyrrhophanus
(formerly *Cacomantis flabelliformis*)

Native names: Du-laar, Dyular (South-West).

Upper *dark slate-grey,* with a white patch near the forward angle of the wing, below the shoulder, and the *tail feathers,* except the central pair, *toothed and tipped with white. Under parts light rufous-brown,* except the throat and middle of the abdomen which are grey. The under tail is more strongly toothed with white than it is above. The immature bird is quite different in appearance, the upper parts being dark blackish-brown, barred lighter brown, with the tail black, toothed with brown on the outer feathers. The under parts are mottled black and grey, with brown in addition on the chest. The under tail is barred black and buff. Iris, dark brown, with yellow eyelids; beak, black, light at base of lower mandible; legs, yellowish-olive. Length, 10½ in.

Distribution: Confined to the South-West, north to a line joining Mt. Curious, Morawa, Westonia, Moorine Rock, Mt. Holland, Norseman and Israelite Bay. It is very plentiful in the lower South-West and in the Darling Range, and rather rare on the coastal plain at the latitude of Perth. These cuckoos do not make large-scale migrations in this State. What movement there is appears to be a seasonal dispersal from the lower South-West, where many birds remain throughout the year, radially towards the periphery of the geographical range. It begins in mid-March or early April. After breeding the birds retire south and south-westwards again. The karri country appears to be their headquarters.

Nesting: The Fan-tailed Cuckoo, although a much larger bird than the birds whose nests it selects in which to deposit its eggs, uses dome-shaped or hooded nests. A common selection is the nest of the Broad-tailed Thornbill, but at the northern limit of its range it has been found using the nest of the Redthroat, and in the south-east the Shy Ground-Wren. On rare occasions eggs are deposited in cup-shaped nests. Thus J. R. Ford found an egg in a Silvereye's nest near Fremantle in November 1952 and I. C. Carnaby one in a Reed-Warbler's nest at Belmont in November 1960. This cuckoo is ordinarily very rare in these districts and perhaps being out of its normal habitat breeding aberration resulted. The egg is white, marked with small freckles or small blotches of chestnut-brown, purplish-brown and light purple, tending to form a band around the upper half of the egg. Size, 22 x 15 mm.

The call is a strong trill which may be syllabised as "pee-e-er," repeated several times, and possessing that plaintive quality common to all cuckoo songs.

Black-eared Cuckoo

Chrysococcyx osculans
(formerly *Misocalius osculans*)

Upper parts *greyish-brown with a metallic glint,* rump light grey. Tail almost black tipped white and barred white on the outer feathers. A white patch on the leading edge of the wing, near the shoulder. *Under parts buff-brown, or light rufous.* A prominent white eyebrow and a black line from the base of the beak, widening into a *black patch* behind the eye. In the young the under parts are ashy white. Iris, brown; beak, black; legs, grey. Length, 5¾-8 in.

Distribution: Distributed over the State generally, except the heavy-forested country of the South-West corner. It rarely penetrates into the northern part of the Swan coastal plain. There is one record from Perth, and it was also seen at Coolup in the winter of 1955. It is a not uncommon inland cuckoo.

Nesting: There is little information available regarding the breeding habits of this bird in Western Australia. The egg, which is a uniform brownish-chocolate in colour and 19-22 x 14-15 mm. in size, has been found in the nest of the Redthroat, which lays a similarly dark-coloured egg. There is reason to believe that this cuckoo also makes use of the nest of the Shy Ground-Wren. It also occasionally parasitises the Yellow-rumped Thornbill. The nearest breeding record to Perth is from the mouth of the Moore River (1948).

The call is a melancholy whistle and may be rendered as "peer" (repeated from six to eighteen times), "pee-o-wit-pee-o-weer" (repeated about three times).

Narrow-billed Bronze Cuckoo (Horsfield Bronze Cuckoo)

Chrysococcyx basalis
(formerly *Chalcites basalis*)

Native names: Djuritch, Gutuban (South-West).

Upper parts bronze-brown, with metallic green reflections. *Under parts dull white with strong bronze-brown barrings,* usually incomplete in the middle and *ill-defined* on the throat, which is mottled grey and white. There is a dull white eyebrow. The basal halves of the *outer tail feathers are a prominent red-brown.* The tail ends in a black bar with white spots on the inner webs, and the outermost quill is toothed blackish-brown and white. Young birds lack the ventral bars but possess the red-brown patches at the base of the tail. Iris, red-brown; beak, black; legs, grey. Length, 5¾-6½ in.

Distribution: This is the most widespread of the two bronze cuckoos, being found over almost the whole State, and it breeds everywhere throughout that range. It tends to avoid, however, the jarrah belt of the South-West where the common bronze cuckoo is the next species. On the coastal plain it extends to some distance south of Pinjarra.

Nesting: The egg is white, minutely speckled and spotted with pinkish-red. Size, 18 x 13 mm. This cuckoo does not restrict its selection of nests for deposition of its eggs to either open or domed types. A variety of host species has been recorded, those building open nests including the Scarlet, Red-capped and Hooded Robins, White-fronted and Crimson Chats, Black-capped Sittella, Western Silvereye, Spinebill and Tawny-crowned Honeyeater and, among the domed nests, those of the Rufous-crowned Emu-Wren, Banded Whiteface, Western Warbler, Banded Blue Wren, White-winged, Black-and-white and Variegated Wrens, Broad-tailed, Slate-backed and Yellow-rumped Thornbills, Shy Ground-Wren and Field Wren. Not all of these hosts are equally successful, from the cuckoo's intention, as fosterers. Overseas ornithologists distinguish between "egg-hosts", such as most of the preceding listed species, from those of them, the "biological hosts", which actually rear the young cuckoo to the fledgling stage. In this case the cuckoo's egg frequently mimics the host's eggs in appearance. Stephen Marchant has applied this concept to cuckoos and their hosts in the Canberra area and elsewhere. He believes that the "biological hosts" of the Narrow-billed Bronze Cuckoo are the blue wrens *(Malurus)*, Thornbills (*Acanthiza*, except the Yellow-rumped Thornbill) and warblers *(Gerygone)*. These are the species most frequently parasitised, and when an occasional cuckoo lays in a "wrong" nest it pays the penalty by leaving few or no progeny. The method by which these cuckoos deposit their eggs in the domed nests of their hosts is still a matter of controversy. There is scope here for fruitful field observations, as well as extending or modifying Marchant's lists of "biological hosts."

Cuckoo embryos appear to be more viable than those of the host species; Dr. A. M. Morgan (South Australia) found an embryo of this cuckoo still alive in the egg 24 hours after taking it, whereas those of the host's eggs *(Malurus)* were dead.

Although this bird is closely similar in appearance to the Golden Bronze Cuckoo, the two species can be readily separated by the call-notes. That of this species resembles the final long-drawn-out, leisurely note of the Golden Bronze, "pee-er," "here", or "sphew," but lacks the preliminary "pee-pee-pee" sequences. In the south the bird is a migrant, the call being first heard, in the Perth area, about mid-June. What migratory movement occurs elsewhere in Australia is unknown, but some individuals, at any rate, are visitors to Java.

Golden Bronze Cuckoo

Chrysococcyx lucidus
(formerly *Chalcites plagosus*)

Native names: The same as for the previous species.

Upper parts pale bronze-greenish-brown with green and purple reflections, brighter than the previous species. The under surface is white with prominent *complete cross bars of bronze-brown, including the throat.* The outer tail feathers are brownish-black, spotted with white along their entire length, there being *no rufous on the tail.* Iris, brown; beak and legs, black. Length, 5½-6¾ in.

Distribution: This cuckoo is very rarely encountered in the northern part of the State, though specimens have been collected at Carnarvon (in May and August) and near Roebourne (July), being, perhaps, then in transit. Most of the breeding records are confined to the South-West corner, north to Geraldton and south-east to the Dundas area. However, in April 1934 I. C. Carnaby found the characteristic egg in a nest of the Variegated Wren at Landor Station, on the Gascoyne River. This was in an exceptionally good season. At the Royal Australasian Ornithologists' Union camp-out at the mouth of the Murchison River in September 1948, several individuals of this species were observed, and, though they were heard singing, no other direct evidence of nesting came under notice. D. P. Scot-Smith heard it calling at Mileura Station at the end of June 1965.

Nesting: The egg is pale olive, without any markings but this colour can be removed with water, after which the shell is found to be light bluish. Size, 19 x 12 mm. Almost invariably this cuckoo uses a dome-shaped nest in which to deposit its eggs, and the chief host is the Yellow-rumped Thornbill. Stephen Marchant considers this species to be the "biological host" of the Golden Bronze Cuckoo and explains the lack of resemblance between the eggs of host and cuckoo to the construction of the host's nest. It is so "complicatedly domed" that the inside of the nest chamber is so dark that the host is unable to appreciate the difference in egg colour. Those of most other builders of domed nests are much less light-proof and so the Narrow-billed Bronze Cuckoo, for example, "has found it advantageous to develop egg-mimcry." Nests of the Banded Blue Wren, Spotted Scrub-Wren and the Broad-tailed Thornbill are also commonly used by the Golden Bronze Cuckoo. On rare occasions eggs of the two bronze cuckoos are found in the one nest and we have found two eggs of this species of bronze cuckoo in a Yellow-rumped Thornbill's nest. A. Kalnins has found the egg of this cuckoo in the open nest of the Silvereye at Katanning.

The call-note, a sequence of repeated "pee-pee-pee," ending in a long drawn-out "pee-er" is uttered from a high perch, usually a dead branch in the topmost part of a tree. It is first heard, in the Perth area, in June, the earliest record of a call being May 4 (in 1928, in King's Park). The bird leaves about February.

The species migrates to the Lesser Sunda Islands (Lombok, Sumbawa, Flores and Wetar); eastern States individuals go largely to New Guinea.

Koel

Eudynamys scolopacea

The male is a large black bird with metallic iridescent plumage, long tail. The female is very different: head, black; back and wings, dark brown, barred white, tail similar but with broader off-white bars. Chin black, buff stripe below eye; rest of under white, narrowly barred with dark brown; buffy wash on upper chest. In young birds the forehead is extensively white and the dark head is blotched with chestnut and light buff. Otherwise like the female. Iris, red. Beak, olive grey. Legs, leaden blue. Length, 16 in.; wing span, 22 in.; weight, 7 oz.

Distribution: Normally in the eastern States and in the Kimberley Division, migrating after breeding to the islands north of Australia. Present in Australia between September and April.

The only record from the area treated in this book is of an immature female which was taken in an exhausted condition at Wamenusking, near Quairading, about 100 mlies easterly of Perth, on April 4, 1971. Its occurrence is interpreted as a case of "wrong-way migration", probably under the influence of a cyclone.

Pheasant-Coucal (North-West Pheasant)

Centropus phasianinus

Native name: Bul-bul (Ashburton district).

Of pheasant-like appearance and plumage colour this bird exists in two colour phases. (a) General colour *black,* with the wings a bright chestnut variegated with buff and black barrings. The long, wedge-shaped tail is black, barred buff and narrowly tipped white. (b) Entirely light coloured by comparison. The upper parts are dark brown with the tail barred buff, particularly towards the base. The wings are chestnut barred with buff and black, the buff predominating and the upper wing coverts and the scapulars pronouncedly streaked with buff. The under parts are buff. Iris, chestnut-brown; beak, black;; legs, slaty-blue: Length, male, 24 in.; female, 20-27 in. Wing span, 31 in. Weight, 1 lb 2 oz.

Distribution: A northern tropical species, extending through the coastal North-West as far south as the Minilya River; inland to Glen Florrie and Coongan.

Nesting: This is the only member of the cuckoo group in Australia which builds its own nest and incubates its own eggs like other birds. The nest is a globular structure about 14 inches in height and the same in diameter, situated usually in a tussock of grass and constructed of grass or leafy twigs. The entrance is through an oval cavity at the side and the actual cup for the eggs measures about 7 inches in diameter. Usually the birds construct a pathway, some 18 inches in length of green leaves, which they appear to renew as they become brown and shrivelled. The 3 to 5 eggs are pure white in colour and roundish in shape; size, 35-38 x 27-30 mm. Both sexes assist in incubation duties. Nesting may take place at odd times of the year but the usual period is from November to February.

The bird lives in rush beds and among coarse grass and is fairly plentiful along most of the North-West rivers and creeks. The call, which is mostly uttered in the early morning, is a repeated "coop-coop-coop," with a pause after the first two notes and the rest run together and uttered quickly. The alarm call is a harsh scolding note.

OWLS: Order Strigiformes

HAWK-OWLS: Family Strigidæ

Boobook Owl

Ninox novæseelandiæ

Native names: Koor-goo-mah-loo (Coolawanyah); Korlba (Lyons River); Goorgoor (Jigalong); Koo-go-burt (Toodyay); Goor-goor-da, Gurgurda, Gu-gu-mit (South-West).

Upper parts brown, marked with buff and spotted with white on the wing coverts. Tail obscurely barred buff on the outer feathers.

Fig. 48.—Boobook Owl.

A blackish patch behind the eye and white on the chin and forehead. Under parts buff-brown, longitudinally streaked lighter. Iris, yellow (adult) or brown (young birds); beak, slate blue or lead-grey; legs feathered to the toes, which are light blue-grey.

Length, 12-14½ in. Wing span, 27-29 in. Weight, 8-12 oz.

Distribution: All over the State.

Nesting: The Boobook Owl makes no nest but usually lays its eggs, 2 to 3 in number, in a hollow spout in a tree, the eggs being deposited on the wood-dust at the bottom of the hollow. It has also been observed using a disused raven's nest—there being apparently a local shortage of nesting holes. The eggs are, as is usual with all owls, rounded and pure white. Size, 42 x 35 mm.

The species name "boobook" is the aboriginal rendering, in the eastern States, of its call notes, which are usually syllabised as "more-pork", "mopoke" or "wook-wook". To Englishmen the note has a distinct resemblance to the call of the European Cuckoo and in the early days the similarity of notes gave rise to the saying that in Australia, that land of many paradoxes, the cuckoo sang at night. This most familiar of night sounds is by most people attributed to the Tawny Frogmouth (p. 304). It appears to be the owl's territorial call and is uttered mainly in the nesting season, being first heard, in the South-West, chiefly in August.

Barking Owl (Winking Owl)

Ninox connivens

Native names: Wool-boo-gle, Wulbugli (South-West); Goor-de-mul (York); Woo-rup (Toodyay).

Upper dark grey, unmarked except for prominent white spots on the wings. Under parts boldly streaked dark brown and whitish. Iris, bright yellow; beak, grey; legs feathered to the toes, which are yellow. Length, 15-17 in. Wing span, 42 in. Weight, 2 lb.

This owl differs from the Boobook Owl in its *much greater size,* looking about twice the bulk of the smaller bird, the disparity not being suggested sufficiently by the mere difference in linear measurements. In colour it *lacks the rufous tone* of the Boobook Owl, though the population in the Kimberley Division is brighter brown, particularly on the under parts. Generally it is a paler bird on the back with conspicuous white wing spots. It is more alert than the Boobook Owl.

Distribution: In Western Australia this species is found only in the South-West corner and in the Kimberley Division (and coastally into the North-West, south to the Ashburton River), but not in between, or in the inland.

Nesting: A hole in a trunk of a tree, or a hollow spout, constitutes a nesting place and it is often used year after year if the birds are not disturbed. No nesting material is placed in the hole selected, the 2 or 3 eggs which constitute the clutch being laid on the decayed wood at the bottom. They are white and round, measuring 46 x 41 mm.

Definitely rarer than the Boobook, this owl is only occasionally reported and its voice not often heard. It has a double-noted call which resembles the bark of a dog, and the vernacular name now coming into use was first used by the ornithologist John Gilbert and the botanist James Drummond when they heard one calling near Lake Coogee, Spearwood as they passed through on an expedition to Augusta on October 25, 1842. Another of its cries is a loud blood-curdling shriek, the so-called "screaming woman" call which for many years was attributed in the eastern States to the Powerful Owl *(Ninox strenua)*, a large owl which does not occur in our State. It is not a frequently uttered call. The native name "Nuelum" or "Newulum," of the Gnowangerup natives and associated with a legendary monster uttering a loud and frightening nocturnal call, probably refers to this owl.

BARN OWLS: Family Tytonidæ

Barn Owl

Tyto alba

(Plate 3)

Native names: Cogadoo (Well 42, Canning Stock Route); Winjahna (Coolawanyah); Jarli (Lyons River); Binar (South-West); Yon-ja (Perth); Minnar (Pallinup River).

Upper parts *grey beautifully mottled with black and white,* the wings marked also with buff; tail grey and barred with buff. Under parts white, very sparingly marked with black spots. A prominent round white facial disc is outlined with buff and dark brown. The legs are feathered almost to the toes. Iris, slate-blue to black; beak, ivory white; legs, whitish-grey. Length, 12½-14 in. Weight, 9-13 oz. The sexes are similar in size.

Distribution: Generally distributed.

Nesting: The 3-5 eggs are merely laid in a hollow in a tree and are, as is usual with owls' eggs, round and pure white. In July 1967 I. C. Carnaby found a nest at Lake Grace containing 7 young; there was a mice plague at the time. Size of eggs, 42-45 x 32-33 mm.

This is almost a cosmopolitan species. According to the specimens received at the Western Australian Museum since 1920, this

Plate 3.—Left: Masked Owl. Right: Barn Owl.

species outnumbers the Masked Owl by ten to one. In hard winters dead birds are not infrequently found. Dr. G. F. Mees, in a more detailed examination of Museum records, has shown that the Masked Owl appears to be more frequent in the heavier forested country of the South-West corner, and the Barn Owl an inhabitant of more open country.

Masked Owl (Cave Owl)

Tyto novæhollandiæ

(Plate 3)

Larger than the Barn Owl, *darker* and more strongly marked on the back and the facial disc surrounded with darker brown feathers, almost black in some cases. The facial disc itself is white in the male and with a pronounced brownish tone in the female. Under parts white spotted with black. The legs are fully feathered to the toes. Iris, yellow; beak, white. The sexes are dissimilar in size. Male, length, 14-17 in.; female, 17½-18 in. Weight, 1 lb. 13 oz. (male); 2 lb. 2 oz. (female).

Distribution: Occurs in this State in three widely separated areas: (1) in the Kimberley Division, (2) in the South-West corner, probably to the Mulga-Eucalypt line, and (3) in the Nullarbor Plain (a cave-dwelling population west to Cocklebiddy), an extension of the Eastern States distribution.

Nesting: The egg is white and rounded. There are few data available regarding the size and the number laid, but a typical egg would appear to measure 46 x 36 mm. and 2 eggs seem to be the number most usually laid. They are deposited in a hollow of a tree.

Eastern Grass-Owl

Tyto longimembris

Resembling the Masked Owl in general appearance. Upper parts brown, mottled with yellow and sparingly spotted with white. Tail yellow, barred and tipped brown. Breast dull white, abdomen white, with a few spots of dark brown. Facial disc dull white, outlined with buff; dark brown spot in front of eye. The *legs (about 3 in.) are long, with the lower portions of the tarsus almost bare,* yellowish. Iris, bluish-black; beak, yellowish. Length, 14½ in.

Distribution: At present this must be regarded as the rarest of our owls, for only one certainly identified record is known: namely a specimen picked up dead at Cranbrook in June 1944. It is a rare owl anywhere in Australia, and this, added to the difficulty of systematically observing nocturnal birds, no doubt accounts for this late recognition of its presence in Western Australia.

Nesting: In Queensland this owl has been recorded as making a nest under a clump of big, coarse grass; 4 to 6 eggs are laid; pure white, slightly glossy, slightly ovate, and finely pitted with limy nodules irregularly distributed. Size, 43 x 32 mm.

FROGMOUTHS and NIGHTJARS
Order Caprimulgiformes
FROGMOUTHS: Family Podargidæ

Tawny Frogmouth

Podargus strigoides

Native names: Tooli (Coolawanyah); Toywee (Lyons River); Peewee (Jigalong and Warburton Range); Joo-ee Wooleen); Du-leer, Gambigorn (South-West); Cumbine (Pallinup River); Combiac (Albany).

A beautiful mottled brown and grey, with a predominant dark brown marbling on the back and a lighter streaked pattern below. The wing quills have small white spots along the leading edges. The *great wide beak* is almost hidden under feathers, while there are strong bristles over the nostrils. Iris, golden-yellow; beak, brown; inside of the mouth, when the bird gapes, white; legs, pale brown. Length, 15-16 in. Weight, 9-10 oz. Wing span, 30-31 in.

Distribution: Generally distributed.

Nesting: The Tawny Frogmouth makes a frail, platform-like nest of sticks in a horizontal fork of a tree in which to lay its 2 white eggs, elliptical and slightly glossy. The nest, about 8 in. in diameter, is so flat that the young on occasions fall or are blown out of it. A series of 8 eggs averages 42.0 x 30.2 mm., varying from 39.2-45.0 x 29.1-32.1 mm. Incubation and nestling periods, both about 31 days. The nestlings have pure white down. Probably only the female broods but both sexes share equally in feeding the young (E. Lindgren).

This giant aberrant nightjar sleeps during the day with its body elongated and inert, simulating in a remarkable way a broken off branch or "spout" of a tree. When taken unawares in the daylight hours it will "freeze" into this attitude, its large yellow eyes elongating into narrow slits which intently watch the intruder. At dusk the bird awakens into life, and the "dead branch" reaction is not given if it is disturbed, the body being carried normally. Contrary to popular belief, the cavernous mouth is not used to capture flying insects on the wing. The Frogmouth feeds almost entirely on the ground. It watches for prospective prey from a vantage post not far above the ground and then noiselessly flies down to pick up a mouse, scorpion, centipede or maybe a snail. Moths and similar night-flying insects form practically no part of its prey.

Another popular misconception about the bird concerns its normal call, which is very widely held to be that well-known nocturnal sound, the double-noted "mopoke" or "more-pork." As has already been stated however, this is uttered by the Boobook Owl, the Frogmouth's call being "oom-oom-oom" repeated about a dozen times. However, the Frogmouth does occasionally give a call resembling that of the Boobook Owl. This may, possibly, be an example of mimicry.

It is a curious fact that the Frogmouth is perhaps the most frequent of all wild birds to be found dead in the bush. On roads it is not uncommonly knocked down by motor cars when it is feeding at night, and hawks and foxes appear to find it an easy prey.

OWLET-NIGHTJARS: Family Ægothelidæ

Owlet-Nightjar

Ægotheles cristatus

Native names: Orta-goota (Godfrey's Tank, Canning Stock Route); Kukubert (South-West).

A dumpy bird, with ash-grey upper parts, head mottled with black and with a nuchal collar of black and a narrower ring of buff. Tail and wings barred with brown. The under parts are finely barred grey, white and brown. The beak is very short and broad, well furnished with bristles. Inside of mouth pink. Iris, brown; beak, brownish-black; feet, pale flesh. Length, 8-9½ in.; wing span, 14-15 in. Weight, 1½-2 oz.

Distribution: Found all over the State, but apparently much commoner in the north.

Nesting: The Owlet-Nightjar either makes no nest or fashions a platform of dried leaves at the bottom of a hollow in a tree. The eggs are white, and 3 or 4 in number. They are more pointed at one end than at the other and measure 28-32 x 22-23 mm. in size.

This nocturnal bird is definitely less well known than the Frogmouth and often its presence is unsuspected until the domestic cat brings home a specimen or an injured bird is picked up. Its feeding habits appear to be intermediate between those of the Frogmouth and typical Nightjars, but are more akin to the former, as ground-obtained prey, such as weevils, millipedes, ants, etc., predominate in the diet. Aerial hawking for flying insects has been described, and most hunting is done from the ground, the bird rising to take flying insects.

The description given above applies essentially to the southern birds, but specimens from further north become progessively more rufous on the back and more white below. Specimens belonging to the South-West race, *cristatus,* with the dark grey upper plumage, have been identified north and inland to Pithara, Dukin and Narembeen. In the Murchison country occurs the race *murchisonianus,* predominantly grey but of a lighter tone and with rufous wash. In the Ashburton a very rufous variety occurs *(rufus),* the ground colour being a rufous brown, barred with grey. In the Kimberley Division is found the race *leucogaster,* with the under parts white. Sufficient specimens are not available in collections to indicate to us the boundaries between these forms, where they grade into each other.

Fletcher Moore records that the Aborigines believed this bird to have the power of afflicting human beings with sore eyes and that it bestowed the sense of smell on kangaroos, which originally lacked that attribute.

NIGHTJARS: Family Caprimulgidæ

Spotted Nightjar

Eurostopodus guttatus

Native names: Kal-ga (Perth); Quoin-kur-dong (Toodyay and Moore River).

A richly marbled plumage pattern of dark brown marked with chestnut and grey. Underneath chestnut finely barred brown, with a sharply defined *white, arrow-shaped throat patch,* resembling a clerical collar. In the wing towards the tip is a *large roundish white spot,* about one and a half inches in diameter. The beak is very small and delicately formed, but has a wide gape. Iris, grey; beak black, inside of the mouth pink; legs, slate. Length, 12-13 in. Wing span, 20 in. Weight, 3 oz.

Distribution: It is found all over the State, but is rarely observed owing to its nocturnal habits. In the daytime birds may sometimes be flushed from the ground, or perhaps a specimen seen on the road which has met misadventure through passing traffic. The species seems to be much more frequently met with in the north than in the South-West.

Nesting: The single egg is merely laid on the ground, usually amongst fallen leaves or other debris and its colours assimilate so well with this debris that it is extremely difficult to locate. It is variable in colour, being usually yellowish-green, spotted and blotched with purplish-brown, or brownish-black. Sometimes the ground-colour is pale olive-green. Size, 34 x 25 mm.; oval in shape and rather shiny.

SWIFTS: Order Apodiformes
SWIFTS: Family Apodidæ

Fork-tailed Swift
Apus pacificus

A large swallow-like bird with the upper parts a brownish-black, *a white rump* and the *tail long and strongly forked.* The throat is white and the rest of the under parts brown, each feather edged white. The wings are long and narrow, far over-reaching the tail. Iris, brown; beak and legs, black. Length, 7¼-7½ in. Wing span, 16-17 in.

Distribution: Swifts are likely to be seen as summer migrants anywhere in the State, particularly during stormy and humid weather caused by northern low-pressure systems. The earliest record of their appearance in the south is October 7 (1925, at Geraldton) and the latest June 5, 1964 (Crawley). Most of the visits occur during January and February. The birds hawk and circle usually at a high altitude, and may be accompanied by Tree-Martins, which look very diminutive in comparison. They are never seen to alight or perch in trees or buildings.

The Fork-tailed Swift breeds in eastern Siberia, Mongolia, Manchuria, Korea, north China and Japan.

Spine-tailed Swift
Chætura caudacuta

A slightly larger bird than the Fork-tailed Swift, differing in having a light brown mantle and *white under tail coverts,* a short *unforked tail,* and the feather shafts protruding as fine spines. The wings have a greenish gloss. Iris, dark brown; beak, black; legs, bluish-brown. Length, 7½-8¼ in. Wing span, 19½ in.

Distribution: This species, so common in the eastern States, has not so far been collected in Western Australia and in the past occasional sight reports of its presence were thought to be errors in identification. However, recent careful observations indicate that the Spine-tailed Swift does occasionally migrate down the western side of Australia. On March 23, 1944, W. H. Loaring saw a flock at Bickley, east of the Darling Range scarp, the short square tails being clearly seen. On February 7, 1956, John Warham saw one bird flying over Eclipse Is., south of Albany, the short squat tail and white below the tail being noted. Other published records of the species are too uncertain to be acceptable.

The Spine-tailed Swift breeds in eastern Siberia, Mongolia, Manchuria and Japan.

ROLLER-LIKE BIRDS: Order Coraciiformes
KINGFISHERS: Family Alcedinidæ

Laughing Kookaburra

Dacelo gigas

Upper parts and wings dark brown, with some pale blue feathers on the shoulder and a white patch at the base of the wing quills, very conspicuous in flight. The head is a very light brown with dark streaks on the crown and a dark brown line through the level of the eye. Behind the head is a broad white nuchal band. *The tail is brown, barred with black,* outer feathers largely white, barred black, but some males have blue rumps. Under parts white. Iris, light greyish-brown; beak, long and robust, black above and white below; legs, pale brown. Length 16½-18 in.; wing span, 29-31 in. Weight, 9-14 oz.

Distribution: The Laughing Kookaburra is not an indigenous bird but was introduced from Victoria by the then director of the Zoological Gardens, South Perth (Lt.-Col. E. A. Le Souef) from 1897 onward. It is now firmly established in the South-West forested country north to Jurien Bay, south to the Albany district and east to Bremer Bay. There is a colony on Bald Island. Eastward it has not spread, as a permanent inhabitant, beyond Moora, Bolgart and the Great Southern railway. However, odd birds constantly penetrate further inland but apparently never form established colonies. Such temporary invasions have been noted at Kellerberrin, Merredin, Nangeenan, Dangin (perhaps now resident), Kweda (perhaps now resident), Lake Grace, Lake King, Holt Rock, Gnowangerup, Borden and Ravensthorpe. The bird appears to have been independently introduced to the Mingenew and Irwin districts and is spreading; it it has been seen in recent years at Arramel (20 miles south of Dongara), Stockyard Gully and Cockleshell Gully.

Nesting: No nest is made but the bird lays its eggs in a hollow spout in a tall tree, or in a hole in the bole of the tree. The 2 to 4 eggs are white, glossy, and rather roundish. A series of 10 eggs from the South-West averages 46 x 36 mm., varying from 43-49 x 35-37 mm. Incubation period, 24-26 days; nestling period, 27-33 days (from studies on a Victorian population by Veronica Parry). Veronica Parry found an unusual social system in the birds she investigated. Breeding pairs were often assisted in incubation, care of the young, and defence of territory by former offspring, who remained as "helpers at the nest' or "auxiliaries." These were defined as "non-breeding adult birds whose capacity to breed was suppressed by their subordinate position within the family hierarchy." Later when it succeeded in obtaining a territory of its own (often when one of its parents or a breeding neighbour died) the auxiliary will breed. Kookaburras are capable of breeding when a year old.

Blue-winged Kookaburra

Dacelo leachii

Native names: Karoo-wolkun (Coolawanyah); Pooka, Pookar (North-West Cape); Jarrooroo (Lyons River).

Head streaked brown and light buff; nuchal collar white, mantle brown; *upper tail coverts sky-blue; wing quills purple-blue with a white stripe;* wing coverts enamel-blue and brown; throat white and the *rest of the under parts pale buff, lightly barred brown.* The tail in the female resembles that of the Laughing Kookaburra, and is a rich chestnut-brown, barred blue-black and tipped white. In the *male* the *tail is a rich purple-blue,* with the outer feathers barred and tipped with white. Iris, white; beak, upper mandible dark brown, lower dull white; legs, grey. Length, 16-17½ in. Weight, 9½-12½ oz.

This Kookaburra differs from the more familiar species of the south in the sky-blue rump, blue and white wing, and the blue tail in the male.

Distribution: A north Australian species which ranges throughout the North-West as far south as Shark Bay and the Wooramel River, and inland to Landor, Mt. Clere and Nullagine and the upper reaches of the large rivers of the North-West.

Nesting: The 3 or 4 white eggs are laid in a hole in a tree. Measurements, 41-46 x 34-36 mm. Breeding season, August and September.

The call is a loud cacophany, after the manner of the Laughing Kookaburra, but if the latter's is likened to a jovial human chuckle that of the Blue-winged Kookaburra can only be compared to a fiendish laugh, one apt to make the hair on the back of one's neck rise. The bird is very vociferous in the early morning and in the evening — during the day it rests quietly in the foliage of the trees. "One bird seems to start the chorus by a loud 'klock klock' and laugh-like chuckles and chatterings, and the other birds keep joining in until there is a medley of 'klock klock' and laugh-like chuckles and chatterings" (T. Carter). According to Carter, the name "Pookar" means "stinking," and is applied to the bird on account of the evil-smelling excrement and fish-remains found in the nesting-hole and at the surroundings.

Sacred Kingfisher

Halcyon sancta

Native names: Perina (Lyons River); Kun-yee-nuk, Kang-innak, Kanyinnak, Kandimak (South-West).

Dark greenish-blue above and *whitish-buff below.* Around the back of the neck is a black and buff collar. Young birds have a dusky clouding on the chest. Iris, brown; the long beak is black; feet, brown. Length, 8-9 in.; wing span, 13 in. Weight, 1-2 oz.

Distribution: Occurs all over the State, except in the far inland portion. In the south it is found inland to Tammin, Lake Grace, and Norseman, while in the North-West it penetrates along the rivers to their upper reaches, frequenting the eucalypt belts.

Nesting: Although in the eastern States the Sacred Kingfisher may excavate a tunnel in the bank of a creek as a nesting site, here in Western Australia it almost invariably uses a hollow spout or a suitable hole in the trunk of a tree. The selected site is often at a considerable height (40 to 50 feet) from the ground. No nest is made, the 4 or 5 eggs being laid only on decayed wood at the bottom of the hollow or spout. Similarly, when a hole in a bank is used, no nesting material is added. The eggs are shining white when fresh, but as incubation proceeds they lose their lustre and become discoloured. They are almost spherical. A set of 5 taken in the South-West averages 25 x 22 mm. in size.

In the southern portion of the State the Sacred Kingfisher is a migrant, arriving during the first half of September when it insistently reveals its presence by its "ki-ki-ki" note — one of the characteristic sounds of the spring and summer bushland. The majority of birds leave again in early March and most have gone by the first week in April, but a few individuals contrive to remain behind for the winter; these, however, are usually silent. Some of our birds migrate to Timor and adjoining islands of Indonesia. During migration (October) resting birds have been seen at Browse Island and Ashmore Reef, in the Timor Sea. A fledgling ringed at the Chain of Ponds, South Australia, on December 31, 1964, was found dead at Tuart Hill on March 16, 1968 — after its normal northern migration it had evidently made considerable westing on one or other of its return migrations. Though this species, one of the wood or forest kingfishers, is normally an insect-eater, the ancestral habit of aquatic feeding may be called forth when the appropriate stimulus is present. Thus it is that an occasional Sacred Kingfisher may prove to be a nuisance at ornamental fish ponds in suburban gardens.

Red-backed Kingfisher

Halcyon pyrrhopygia

Native names: Loorn (Godfrey's Tank, Canning Stock Route); Lawn (Jigalong).

Like the Sacred Kingfisher in general appearance, but slightly larger and differing in having the *head streaked dull green and white,* and the *lower back and upper tail coverts chestnut-brown.* The nuchal collar and entire under surface are pure white, lacking the buff characteristics of adult specimens of the Sacred Kingfisher. Iris, dark brown; beak, black; legs, grey. Length, $7\frac{3}{4}$-$9\frac{1}{2}$ in.; wing span, $13\frac{1}{2}$ in. Weight, $1\frac{1}{2}$ oz.

Distribution: Found all over the State, but normally excepting the South-West corner, to the Geraldton district, Kellerberrin and Norseman. In October 1967 Ian Edgar and Michael Morecombe found a nest at Grass Valley. One individual has been reported from South Belmont (seen in June 1956 by E. Lindgren and between October and December 1957 by P. Slater).

Nesting: It burrows a hole in banks of creeks or cliffs and the only situations in trees in which it nests are the arboreal termite mounds into which it drills. The 4 or 5 eggs are white, and as with other kingfisher eggs, rounded. Measurements, 24-26 x 21-22 mm. Weight, 6 gm.

This is the characteristic kingfisher of the inland and arid areas. Its voice is distinct from that of the Sacred Kingfisher and is a mournful whistle.

Mangrove Kingfisher

Halcyon chloris

Differs from the Sacred Kingfisher, which it much resembles, in *lacking any buff colouring on the ventral parts.* There is also no buff on the nuchal collar. *Two small white loral spots* at base of the beak. The plumage is duller, upper parts dusky green, with a brownish glint on the scapulars. Rump, light blue; iris, brown; beak, black, base of lower mandible, white; legs, blackish-brown. Length, 10 in.

Distribution: A northern species, which extends down the mangrove belt of the North-West coast as far south as Carnarvon. Occasionally it is seen at Denham, Shark Bay.

Nesting: The Mangrove Kingfisher nests in a hole in a tree (often high up) or in a hole in a termite's nest. The 3 white eggs measure 31-32 x 26-27 mm. (H. L. White).

Besides mangroves this species, in the far north, frequents the under-structure of jetties where the tidal rise and fall is great, and at the settlement at Cockatoo Island, Yampi Sound, it uses telephone poles and house tops as sally points. The call is a double-noted "kik-kik," repeated two or three times. It is rather like the call of the Sacred Kingfisher but sufficiently distinct to be easily recognisable.

BEE-EATERS: Family Meropidæ

Bee-eater (Rainbow-bird, Gold-digger, Gold-miner, Golden-Swallow, Berrin-berrin)

Merops ornatus

Native names: Berulberul (Lyons River); Dirroon dirroon (Jigalong); Bee-roo-bee-roo-long (Perth); Berrin-berrin (Avon River); Birunbirun (South-West); Birringo (Katanning); Perinbo (Pallinup River).

A brilliantly plumaged bird, mainly green in colour, with *sky-blue* and *orange-brown.* The mantle, wing coverts and forehead are green; nape and wing quills orange-brown, latter tipped black; lower back and upper tail coverts sky-blue; tail, black, the middle feathers, which are glossed with green, being *produced as thin filaments projecting for 1½-2 in. beyond the rest of the tail in the male, and for about ½ to 1 in. in the female.* The chin is orange-yellow; a patch on the throat black; breast green, becoming bluish towards the abdomen; under tail coverts sky-blue and the under wings chestnut. In the young the black throat patch is absent and the chin is paler. Iris, red; the long and curved beak is black and the legs, fleshy-grey. Length, 7½-10 in.; wing span, 13 in. Weight, 1 oz.

Distribution: All over the State, except the southern, more densely forested, parts of the Darling Range plateau and the extreme south coast to Albany. It extends down the western coastal plain to Cape Naturaliste and along the coastal strip to Cape Leeuwin.

Nesting: The nest is a tunnel drilled at a slight angle in flat ground. In the coastal plain the birds nest freely in the open country of the compact loamy soil near the foot of the Darling Range scarp. They occasionally nest in King's Park and a pair used to nest in the central park strip of Parry Street, Swanbourne. Nesting still takes place at the "quarries" on the Nedlands foreshore and at the Point Resolution Reserve. Dimensions of some nests measured by us are: (a) length of tunnel, 2 ft. 6 in.; terminal chamber, 7½ in. long, 6 in. wide, 5½ in. high, and the floor level 10 in. from surface. (b) tunnel, 2 ft. 6 in. long and of a width of 2¼ in.; chamber, 12 in. long, 7½ in. wide, 4¼ in. high, floor 10 in. from surface. (c) tunnel, 2 ft. 9 in. long and 1¾ in. wide; chamber 11 in. long, 7 in. wide and 4¼ in. high. The "mullock" is scraped out by the birds' strong feet and two marks, like railway tracks leading into a tunnel, indicate the method of excavation. Some loose soil is retained on the floor of the chamber on which the eggs are laid — these, usually 5 in number (but sometimes 4 or 6), are pearly-white and almost circular; 15 eggs from the South-West avearge 22.0 x 18.7 mm., varying from 21.0-25.0 x 18.0-19.5 mm.

When the eggs hatch and insects are brought in to feed the nestlings, the chamber becomes rapidly filled with insect remains. The contents of one such nest at Wembley, collected by L. J. McHugh, were sent for examination to K. McKeown, of the Australian Museum, Sydney. He found the material to comprise 2,753 head capsules of various insects. Of these 1,747 were Hymenoptera, mostly those of hive bees *(Apis mellifera)* but included quite a number of native bees. There were also 973 heads of Diptera, mostly those of blowflies *(Calliphora)*, while the remainder were of beetles. Dragonflies also are eaten. Bee-eaters' nests harbour among this mass of insect remains a dipterous parasite, a fly of the genus *Passeromyia*, whose maggots attack the young birds.

The Bee-eater is a strict migrant to the southern parts of the State, usually arriving in the Perth district during the first week of October, though very rarely some may be seen at the end of September. In 1954 the birds arrived a full month earlier than normal. Unlike the majority of migrants, which arrive silently or unobtrusively, the Bee-eaters appear in compact flocks which are very vocal, the call "pirr-pirr-pirr" (the i short as in pit) accounting for the onomatopoeic native names. Nesting activity follows immediately. The return movement to the north starts in the last week of January and the last stragglers have left by the third week in March. There is no known instance of any individual Bee-eaters wintering in the south, such as is the case with almost every other local bird migrant, even with those which cross the equator to the northern hemisphere.

Some Bee-eaters are found throughout the year from the Gascoyne and upper Murchison Rivers (Mileura Station) northwards to the Kimberleys, but until data from ringing become available details of the migration will not be perfectly understood. It is likely that "leap frog" migration occurs and the birds from the extreme south migrate furthest north. Large flocks, exceeding 100 individuals, have been seen passing northwards over the Kimberley Research Station (70 miles south-east of Wyndham) in early April (K. Immelmann). The "wintering area" in the extreme north of the species' range extends from Bali to New Guinea, and Celebes. In early October returning migrants have been seen at Browse I. and Ashmore Reef. in the Timor Sea, but these are probably laggards of the main movement.

Bee-eaters frequently perch on telephone lines and fences, which they use a sally posts from which to dart after a passing insect, and as they glide gracefully from perch to perch they display in turn the brilliant orange, bronze and sky-blue of their variegated plumage; the black tail, brown wings, and greenish-blue body make striking contrasts. They have a habit, occasionally, of concentrated roosting in clusters in tree tops. Dust-bathing is indulged in.

ROLLERS: Family Coraciidæ

Dollar-bird

Eurystomus orientalis

A gaudily coloured bird. Head dark brown, merging into a metallic green back. Wing coverts blue and quills black. Throat violet blue, rest of under parts metallic green. In the wing is a concealed *patch of pale blue* which becomes apparent as a *large "dollar" spot* in the extended under wing. Iris, dark brown, ring around eye red; *beak and legs, bright red.* Length, 10½-11½ in. The beak is duller in young birds.

Distribution: Northern and eastern Australia, migrating to New Guinea, the Moluccas and Celebes. In Western Australia it is rarely encountered south of the Kimberley Division, and so far there are only two records. On June 14, 1953, J. H. Calaby saw a bird at a eucalypt-lined creek at Winning Pool (perhaps a "wrong-way" migrant in view of the time of the year). In December 1972 John Darnell saw one bird at Shay Gap, 50 miles inland of Mt. Goldsworthy, in the Pilbara.

Nesting: Not recorded in our area. The bird lays 3-5, usually 4, eggs in the hollow of a eucalypt; white; varying from 28-29 x 32-36 mm. in size.

The birds arrive in Australia in mid-September, breed from October to January and leave in May. J. P. Rogers found that in the Kimberley their favourite perch is the topmost branch of a dead tree, always near a creek or waterhole. They were scattered along Parry's Creek every half-mile or so. G. F. Hill found them feeding on the winged females of the green tree ant.

PERCHING or SONG-BIRDS: Order Passeriformes

SCRUB-BIRDS: Family Atrichornithidæ

Noisy Scrub-bird

Atrichornis clamosus

Native name: Jee-mul-uk (King George's Sound).

Upper parts brown; in closer view lightly cross-barred darker brown. Throat white; chest blackish; breast dull white and abdomen bright buffy-brown. Tail long and fan-shaped; wings short. The female differs in lacking the black markings on the chest. Iris, dark

Fig. 49.—Noisy Scrub-bird.

brown; beak, "upper mandible reddish-brown, lower mandible greenish-white"; legs, "nearly greenish-white, becoming greenish-olive downwards" (Gilbert). Length, 9½ in. Weight, 2 oz.

Distribution: Known only from Drake's Brook, Waroona (just within the Darling Range); at Wallcliffe (Margaret River); Augusta and at King George's Sound. Recently ornithologists had come to resign themselves to the belief that this species was extinct as no sightings of it had been published since 1889. In December 1961 H. O. Webster created a sensation by identifying a singing male at Two People Bay, east of Albany. Detailed surveys have shown that the species occurs only in the Mt. Gardner area of Two People Bay, in the heavily vegetated gullies or in dense patches of eucalypts in the heath between the gullies.

Nesting: The first description of a supposed nest of the Noisy Scrub-bird was that of William Webb who found it at Torbay Inlet, some time prior to 1895. However, its authenticity has been doubted. The nest was on the ground, under a low spreading bush, was very loose and of slovenly construction and "composed mostly of thin strips of paper bark, wire grass and a few feathers." The one egg it contained was white, slightly blotched with chestnut-brown, and Webb gave the size as slightly smaller than the eggs of the Brown Quail [=28-32 x 21-23 mm.].

Another doubtful nesting record was made in October 1897, in a locality 25 miles east of Albany, on the way to Warriup. The scrub among which this nest was found consisted of a dwarf species of tea-tree *(Melaleuca)* which, "growing to a height of from 2 to 4 feet, forms comparatively open spaces in the midst of the dense 'marlock' or 'ma-lock' scrub that extends over great areas of this district" (T. Carter). The nest was found by A. Y. Hassell and his son, J. C. Hassell, and was described as situated on the ground, alongside the root of a eucalypt tree, which was about 7 inches out of the ground; dome-shaped and constructed of grass and rootlets, with a few leaves, and lined with a white downy substance. Size, 8½ in. high, 5¾ in. broad. It contained two slightly incubated eggs, which were swollen oval in shape, ground-colour of reddish-white, with purplish-brown markings, more numerous on the larger end, where they were confluent. Size, 22.8 x 17.5 mm. and 21.5 x 17 mm. The nest may have been that of a Western Grass-Wren (p. 369).

The first authentic nest, an uncompleted one, was found at Two People Bay by Lindsay Sedgwick on January 8, 1963, and described by Mr. Webster. It was a rather large globular structure, made of the flat leaves of the cut-rush, *Lepidosperma persecans,* the dimensions being 6 x 5½ in., and 6½ in. high, with a side entrance 1½ in. in diameter. Inside was a cardboard-like lining on the bottom, hard and smooth, but thinner and more brittle than the material reported in the nest of the Rufous Scrub-bird. The whole structure rested in a clump of *Juncus* on a rough platform of the same material as the nest, about 8 inches from the ground.

Subsequent detailed studies by F. N. Robinson and Graeme Smith, of the C.S.I.R.O. Division of Wildlife Research, have provided extensive information on the nesting of the species. Based on an examination of 40 nests the average has been found to be roughly spherical in shape, 6½ in. high, 5 in. wide and 6 in. deep, with a side entrance (1½ in. in diameter) usually opening into a short runway, 4-12 in. long. Some nests do not have a runway. The nest cavity is 2¾ in. in diameter, the bottom of which is lined with a hard cardboard-like material made of decayed rush. The nests are built on an average of 9 in. above ground, and ranging from 4½ in. to 4½ ft. They are usually constructed of rush leaves but a few are made with the leaves of *Dryandra formosa.* A small number of thin twigs and eucalyptus leaves are also used. All nests so far found have been in areas of eucalypts, the jarrah *(Eucalyptus marginata),* marri *(E. calophylla)* and bullich *(E. macrocarpa),* often close to a stream. The nest is usually built in a clump of rush *(Lepidosperma)* but may be occasionally placed in a clump of another rush *(Anarthia prolifera),* or at the base of a small shrub. It is unlikely that males have any influence on the selection of a nest site; they do not assist in the building of a nest. Females tend to build their nests, in successive seasons, in the same area, often within a few yards of each other.

Only one egg forms the clutch. It is pale cream buff in ground colour, with pale brown flecks, more concentrated at the larger end, oval in shape, slightly tapered towards one end. Average size of seven eggs, 28.4 x 19.9 mm., ranging from 26.0-30.8 x 19.4-20.6 mm. The laying season is from late May to late September. If an egg is lost before hatching the female may build another nest and lay a second egg. Incubation period is 36-38 days. Fledgling period is about 3-4 weeks. The male does not assist either in incubation or feeding the chick. Chicks are fed on invertebrates, including spiders, millipedes, isopods, crickets, cockroaches, earthworms and a variety of larvae.

During the first week after hatching the female broods the chick for a short period after each feed; the frequency of brooding thereafter decreases rapidly and during the last two weeks the female does not brood the chick at all during the day. However, it spends the night with the chick throughout the fledgling period. During this period the male may occasionally visit the vicinity of the nest at dusk. Nest sanitation is maintained by the chick excreting a faecal pellet which is taken directly from the chick by the female and deposited in a stream pool within a few yards of the nest. If there is no stream nearby the pellets are dropped on the ground. After fledging the female and the chick stay together for at least two weeks, but do not return to the nest at night.

Prior to December 1961 the Noisy Scrub-bird had been observed and collected by only four naturalists. John Gilbert, in 1842, obtained four or five specimens (one the type-specimen) at Drake's Brook and at King George's Sound, and he observed the bird at Augusta. George Masters, in 1866, 1868 and 1869, obtained seven specimens at King George's Sound, and William Webb, over a number of years, collected about eight specimens in the same area. A. J. Campbell obtained one specimen, the last to be collected, at Torbay in early October 1889 (now on exhibition at the National Museum, Melbourne), and recorded having heard the species calling at Boodjidup Lake, near Wallcliffe on November 4. Only 21 or 22 specimens survive in the world's museums.

From 1904 onwards intensive searches were conducted in all likely places in the South-West to re-discover the Noisy Scrub-bird but were unsuccessful until Mr. Webster's recent find.

Gilbert noted the Scrub-bird as an inhabitant of "the densest and rankest vegetation on the side of hills and the thick grass around swamps or small running streams." Masters stated that the species inhabited "dense masses of vegetation consisting of tall reedy grass and thick-growing low bushy shrubs" and Webb recorded that it was to be found "on the margins of fresh-water swamps." He thought that it probably occurred as far west as Cape Leeuwin and eastward to Cape Riche and added that he had never seen it inland.

However some surprising information on its inland distribution came to light in 1973. Ederic Slater, of the CSIRO, when searchng old documents at the British Museum (Natural History) in South Kensington, found reports by John Gilbert stating that in 1843 he had encountered the species, but in decreasing numbers, as far north as Mt. Barker.

As to its voice, Gilbert remarked that it had "extraordinary loud notes, many of which are sweet and melodious," which it uttered as it ran along the ground with the utmost rapidity. "Its notes are so exceedingly loud and shrill," Gilbert added, "as to produce a ringing sensation in the ears, precisely the effect produced when a shrill whistle is blown in a small room." Gilbert considered it to be "without exception the loudest of all the song-birds inhabiting Western Australia." James Drummond, who was Gilbert's companion when the species was first discovered at Drake's Brook, described it as "singing sweetly with loud clear notes." Masters stated that the bird was ventriloquial, the note "first seeming to come from one side, then from another, then sounding loud and clear a few feet in front." Webb gave a detailed description of the voice—loud, clear and piercing and something like "cheap-cheap-cheap-cheap," the first note being short, the second, third and fourth being each a little longer and lower. On the other hand, Campbell stated that the note was a loud sharp whistle *repeated eight or nine times* rapidly with *crescendo,* concluding in a sharp crack "that makes the woods resound." Mr. Webster has confirmed the general accuracy of these accounts and added the information that, like its congener in the eastern States (the Rufous Scrub-bird), it is capable of mimicry of the songs of other species in its vicinity. These include the Grey Butcher-bird, Willy Wagtail, Spotted Scrub-Wren and New Holland and Tawny-crowned Honeyeaters. Some observers, however, doubt whether there is any real mimicry involved. In its rich, vibrant quality the song recalls that of the Reed-Warbler, or, to European ears, the Nightingale. There is also some similarity to the ringing notes of the male Rufous Whistler. It is possible that it was the Noisy Scrub-bird which was heard by the Dutch navigator, Vlamingh, in January 1697, when he reported hearing the Nightingale on the banks of the Swan River.

The bird is a ground dweller and its powers of flight are feeble. Gilbert found it very elusive and it was only by mere chance that a glimpse of it could be obtained. He stated that "when running on the ground the tail is carried very erect with the head thrown up at the full stretch of the neck." According to Gilbert and Webb its food consisted of small beetles and other insects.

At Two People Bay the birds are reasonably plentiful in the densely vegetated ravines of gullies on the slopes of the Mt. Gardner promontory. The songs of the males can be readily heard, so freely that it might be imagined at times that the species is the most common small bird in the area. However, actually to *see* the birds themselves requires endless patience.

Population studies by Dr. Graeme Smith show that the breeding males are widely dispersed and defend their territories with the loud penetrating territorial song already mentioned. Due to the paucity of boundary interactions by the birds the size of the territories is hard to estimate. They are possibly of the order of 10 to 20 acres. However the observed movements are restricted to a small area of the territory, usually from ¼ to 5 acres. The actual nesting areas are on the periphery of the male's territories and well away from the area the males usually frequent. Males in occupation of territories sing throughout the year. However at the start of the breeding season there is a rapid increase in the frequency of calling which is maintained untl the end of the season. Mr. Robinson found that there was a general correlation between the maximum song output and the time of egg-laying. During the breeding season a male without a territory may attempt to establish one on the margin of an established territory, or in a small clump of eucalypts in the heath. The nature of the pair bond is not known, but probably the birds are monogamous. However in 1972 Dr. Smith noted one case where a male mated with two females — his own and a female in an adjacent territory which had lost her mate before the start of the 1971 breeding season. Similar reports (of occasional polygamy) have been made in regard to lyrebirds.

There can be little doubt that these two very distinctive genera, *Atrichornis* and *Menura,* are more nearly related to each other than either is to any other group. Their common characteristics, apart from anatomical considerations, include very strong voices and addiction to vocal mimicry, displayng with elevated tail and drooping wings, restriction of nest-building and brooding to the female, the dropping of juvenile faecal pellets into water, and of course, food seeking and nest-building on or near the earth. Moreover, as now appears, *A. clamosus* agrees with lyrebirds in breeding during winter and having only one egg to a clutch. This dual fact contrasts somewhat with *A. rufescens,* which breeds during spring-summer and normally has a clutch of two eggs. The explanation here, no doubt, is that the habitat of the Rufous Scrub-bird (mostly mountainous) is colder and much more fertile than that of its western relative. For the same reason, the eastern bird uses a greater amount of "cardboard" lining in its nest, simply because damp decayed wood is more readily available in rain-forests than it is in the drier areas of the west. A further point of contrast between the two scrub-birds, based on habitat, is that the eastern species is the more persistent and accomplished vocal mimic, a circumstance due, no doubt, to its having a greater "supply" to draw upon.

(The above paragraph has been kindly furnished by Mr. A. H. Chisholm of Sydney, who has written extensively on both scrub-birds and lyrebirds and has studied the two members of each genus in the field.).

The closely allied Rufous Scrub-bird *(Atrichornis rufescens)* still lingers in the sub-tropical brushes of coastal northern New South Wales and southern Queensland (from the Chichester River north to the Mistake Range). It is smaller (length about 7 inches) but generally similar in colour. Its habitat is the most impenetrable part of the scrub, among the jungle tangle, vines and fallen debris, under which it creeps and runs on the ground, never going up into a bush. It very rarely flies and then only with an awkward flutter, a few inches to a few feet above the ground. The long tail is often cocked, especially when the bird is agitated, and may even be held erect, as in the blue wrens; the small wings are drooped. With its strong feet it does a lot of scratching on the ground and feeds on snails and their eggs, worms, insects and land crustaceans *(Talitrus)*.

The song of the male is extraordinarily loud and penetrating, being a succession of notes sounding like "chirp-chirp-chirp-chirp" or "chip-chip-chip-chip." The male also mimics other birds to perfection and these imitations have a ventriloquial property; birds mimicked include the Spine-tailed Logrunner, Scrub-Wrens *(Sericornis)*, Yellow Robin *(Eopsaltria)*, Golden Whistler, Grey Fantail and honeyeaters. It is silent after the nesting season. The female has a soft squeaking call and a faint note resembling "tit-tit-tit-tit."

The dome-shaped nest is placed about 6 inches above the ground among ferns or grass and is loosely made of dead leaves and old dried grass; it measures about 7 inches in length and 4½ inches in width with a small side entrance 1¼ inches in diameter. The only lining is a peculiar tough composition resembling coarse cardboard, apparently made of soft decayed wood, worked up when wet and plastered over the inside of the nest. The clutch consists of 2 eggs, pinkish white in ground-colour, speckled with brownish-red, especially at the broad end. Size, 23 x 18 mm. Egg dates, October.

Anatomically the bird is interesting because of the very varied opinions on its relationships. It bears some resemblance to the lyrebirds *(Menura)*, particularly in the musculature of the syrinx, and of the shoulder girdle. It is the only known song-bird which lacks a furcula ("wishbone"), the clavicles being reduced to mere splints, evidently a consequence of its virtual flightlessness. The anatomy of the western bird has not been studied but no doubt resembles that of the eastern species. John Gilbert remarked on the extraordinarily muscular development of the legs and the great breadth and strength of the posterior parts generally.

LARKS: Family Alaudidæ

Singing Bushlark (Horsfield Bushlark)

Mirafra javanica

Like a pipit in general appearance but of richer colouring, tending to *more rufous in the plumage.* The beak is brown, feathers edged lighter brown and rich chestnut on the flight feathers. The under parts are buff, chin lighter and brown streakings on the chest. The sides of the *short tail* are white, as in the Pipit. The pale brown beak is distinctly un-pipit-like in form, being much more robust, resembling that of a finch. Iris, brown; legs, fleshy-white. Length, 5½-6 in.

It can be distinguished from the Pipit in the field by its *richer brown colour* and the *shorter tail* which gives it a less slender appearance. The *flight is less undulatory* than the Pipit's and it usually alights behind cover, from which it runs out to take a peep at the intruder. It runs nimbly, but not so quickly as the Pipit and it does not have the habit of the latter in wagging its tail up and down (Whitlock).

Distribution: A northern species extending through the North-West usually as far south as the Minilya River. In October 1972 M. G. Brooker encountered it on Callagiddy Station, 20 miles south-east of Carnarvon. That the species may be extending its range is suggested by a specimen found at Duketon, north-east of Leonora, in December 1974.

Nesting: The nest is a hollow scraped in the ground and lined thickly with grass. It is usually situated under the shelter of a tuft of grass. The 3 or 4 eggs are greyish-white, freckled all over with dark grey or greyish-brown. The freckles are often more numerous at the larger end and tend to form a dark cap. Size 18 x 14 mm. They are glossy. Sometimes the grass forming the nest, which measures about 4 inches across, is pulled forward and forms a sort of canopy over the nest.

It sings high in the air, like the European Skylark *(Alauda arvensis),* remaining poised with rapidly vibrating wings.

SWALLOWS and MARTINS: Family Hirundinidæ

White-backed Swallow

Cheramœca leucosternum

Native names: Chadichadi (Godfrey's Tank, Canning Stock Route); Boo-de-boo-de, Budibudi (Avon River).

One of the daintiest members of the swallow tribe. The *back is pure white,* the wings and tail black, and the head brown mottled with white. Ventrally, *the throat and chest are white* and the abdomen black. The tail is strongly forked. Iris, dark brown; beak, dark brown; legs, chocolate-brown. Length, 4¾-6¼ in.

Distribution: Generally distributed, except in the South-West forested area. The species is not usually found south of Mogumber or west of the Great Southern railway. The nearest observation to Perth until recently was made at Lake Yanchep (1902). It is found in the Northam district. However, in recent years a penetration southwards has been taking place and birds have been seen at Spencer's Brook, Muresk and between Bannister and Williams. Nesting has been reported south to Wanneroo (1969).

Nesting: For nesting purposes the White-backed Swallow drills a tunnel of about two inches diameter, up to arm's length, into a sandy bank, such as those of railway cuttings, mine workings and pits of various kinds, including those excavated under the joints of the pipeline of the Goldfields water supply. In the nesting chamber at the end of the tunnel is constructed a shallow saucer-shaped nest of fine grasses, fine rootlets and leaves. The 4 (occasionally 5) eggs are pure white and slightly glossy. Size, 16-18 x 12-13 mm. Nesting season, September to December.

The call is a subdued "check," uttered on the wing.

During intervals of cold, inclement weather in the winter the birds may cluster in the nesting burrows and go into temporary torpidity, lasting several days. Peter Congreve, describing four such birds, accidentally disturbed from a burrow at Yanchep during a cold snap in June 1971, stated they were inert and appeared dead, cold to the touch, eyes closed and wings closely folded. They recovered after finding, during the warm part of the day, and flew about as though nothing had happened. This type of behaviour has only recently been published, though Walter Gable noted it at Spencer's Brook during the winter of 1936, when some 16-20 birds were found in a burrow in a state of suspended animation. This temporary torpidity, now demonstrated to occur in an Australian species of swallow, is reminiscent of a belief held by the English naturalist Gilbert White, who wrote in 1772: "I am more and more induced to believe that many of the swallow kind do not depart from this island, but lay themselves up in holes and caverns; and do, insect-like and bat-like, come forth at mild times, and then retire again to their latebrae." This notion has since been dismissed as invalid in favour of the theory of total migration from England and Europe during the winter.

Welcome Swallow

Hirundo neoxena

Native names: Kun-na-meet, Kannamit (Perth); Cabbridge (Pallinup River).

Head and back a metallic black; wings and tail dark brown; *forehead, throat and chest rufous;* abdomen greyish-white. The inner feathers of the *deeply forked tail* have a sub-terminal row of white spots. Iris, dark brown; beak, black; legs, brown. Length, 6 in.

Distribution: Generally distributed, north along the coast to about Port Hedland, but inland no further north than Meekatharra and Wiluna. Found also on the islands along the coast, including the off-shore ones of Montebello, Barrow and Abrolhos Islands and the Archipelago of the Recherche.

Nesting: The nest is in the form of a half of a bowl, constructed of mud or clay, intermixed with grass and measures when first built about 4 in. in width and 3½ in. in depth. The half bowl is completed by a pair of birds often within three days and a further two or three days are spent in gathering the large quantity of feathers, grass and hair or fur, with which the nest is warmly lined. The same nest is usually occupied by the same pair of birds in successive years, each season more mud being added until the whole structure may become nearly twice the original size and eventually falls through its own weight. The clutch consists of 3, sometimes only 2, and on occasions as many as 4 eggs. They are white and speckled and blotched, in varying degree, with shades of brown and underlying markings of lavender. The markings may be spread generally over the surface of the shell but usually form a zone around the broad end. In shape they also vary, some being blunt ovals and others tapering towards the smaller end. Size, 18-19 x 13-14 mm. Two broods are raised annually and both parents share in the incubation, which takes 15 days. The young remain in the nest for a period of about a month.

These birds seem to have profited very greatly as the result of settlement, mainly through the provision of nesting sites. Nowadays one virtually never sees a Swallow's nest other than in artifical situations. Favourite sites are the verandahs of country railway stations, shops and hotels. The nests may also be built under bridges and jetties and also inside the kerosene tins mounted on posts so often used as outback mail-boxes. In the Peel Inlet, Mandurah, such tins used to be placed on spit posts to make them more conspicuous and they were frequently tenanted by Swallows. In some cases they were so far from shore that the young would inevitably drown on taking their first flight. To discourage such hopeless nesting ventures the fishermen would destroy the beginnings of nests but the birds were persistent. The most unusual instance of nesting is that reported by T. Carter from Point Cloates. One or two pairs nested on a boat used for station work whenever she lay any length of time at her moorings, and if the boat sailed when a nest contained eggs or young the old birds would accompany it, in one instance following her a distance of 35 miles and back.

Under natural conditions, before settlement, Swallows attached their nests to the insides of hollow trees and to the rock-faces in the cavernous coastal limestone. These situations are still used where human habitations are not available.

Swallows are non-migratory in this State, though in the winter they may gather into large flocks and roost together.

Barn Swallow

Hirundo rustica

Similar to the Welcome Swallow in general appearance, but with a *well-defined black collar* separating the chestnut throat from the pure white breast and abdomen. The tail streamers are also very much longer. In young birds, however, they are short.

Distribution: The Welcome Swallow does not occur in the northern parts of the State but, during the summer, parties of the Northern Hemisphere Barn Swallow visit the western parts of the Kimberley Division on migration. These belong to the Asiatic-breeding race, *H. r. gutturalis.* Up to recently there was only one positive record, an immature specimen collected by Dr. F. M. Rayner, in H.M.S. *Herald,* somewhere off the north coast of Australia, on October 18, 1860. It was almost 100 years later before the next observation of the species in Australia was made—Peter Slater saw a flock in Derby between January and March 1960. The birds were seen again during the summer of 1960-61. A flock of about 300 was seen just before the exodus migration in 1961. No swallows were seen at Wyndham or the Kimberley Research Station at Kununurra.

How widespread is the Australian range during migration is unknown. But it certainly extends into the area covered by this book. John Darnell has seen single birds at the sewage ponds at Mt. Goldsworthy, in the Pilbara in early November 1973 and at Cape Keraudren in October 1973 and 1974.

Tree-Martin

Petrochelidon nigricans

Native names: Kintillka (Jigalong); Gab-by-kal-lan-goo-rong, Gabbi Kallangorong (Perth); Kilbaltkilbalt (King George's Sound).

Upper parts steely black, except the *rump which is dull white* and there is a slight chestnut tinge on the forehead. The throat and chest are grey and the rest of the under parts are dull white. The tail is very slightly forked. When viewed from below the brown of the under surface of the tail contrasts with the white of the remainder of the under parts. Iris, brown; beak, black; legs, blackish-brown. Length, 5¼ in.

Distribution: Generally distributed all over the State.

Nesting: The eggs are usually laid in a hollow spout in a tree, often in hollows in dead trees, and, if necessary, the bird reduces the size of the entrance with mud. A foundation of dry leaves is built up at the bottom of the hollow, or feathers may be used. Although normally the Tree-Martin nests always in trees, for some years an interesting modification of the nesting habit has been developed in the city of Perth where the birds had taken to nesting behind the ventilators of the Public Library and of the Treasury Buildings in Barrack Street, becoming at times a nuisance to the civil servants working there. Occasionally mud nests are built under wooden shelters and verandahs, the structures resembling the nests of the Welcome Swallow. This habit has been noted at Northampton, Morawa and Perenjori. The eggs, usually 4 in number, are white, spotted and blotched with pale brown and pale mauve, mostly at the larger end where the markings tend to form a cap. Size, 17-20 x 13-14 mm. Nesting dates (South-West), August to November. The bird is a persistent nester. A set of 4 eggs was removed from a nest at Busselton on August 25. Another set was taken from the same nest on September 8, and a further set of 4 was noticed in the same nest on October 10. The first set differed greatly in marking from the next two.

This is the commonest and most widespread of the swallow tribe. Though it may be seen at all times of the year in the South-West and in the Perth district, there is some evidence of a long drawn-out passage movement. Parties are observed travelling north in the autumn and winter and south in the spring and summer.

On summer evenings large flocks may sometimes be seen congregating around various points, the individual birds hovering for a moment and then darting off again with vigorous wing-beats. Such flocks have a liking for roosting in the beds of rushes in swamps. It is a striking spectacle as the birds assemble over a swamp just before sundown. They fly in from all points of the compass and, after circling for a while, accumulate in a dense swarm, like bees. Then the whole mass of birds suddenly dives into the rushes with considerable speed, appearing as if pouring in through an invisible funnel. The din from the concerted twitterings of so many birds is almost overpowering, the restless birds shifting about actively and noisily changing roosting positions. The clamour subsides as dusk steals over the scene. Lake Yanchep Shenton Park Lake, Jackadder Lake (Woodlands) and Kogolup Lake (near Jandakot) are known to have been used as such roosting places.

Other night roosts are in groves of trees. A well-known one in the Perth area is at the south end of Crawley, towards Pelican

Point, where the aerial evolutions of the birds before dropping down to roost in the marri trees provide a spectacle for local residents. Sometimes the birds roost in neighbouring street trees and gardens. They leave by the end of April or the first few days in May.

Fletcher Moore derives the native name from *gabbi,* water, *kallan,* to collect, and *gorang,* to turn or twist.

Fairy Martin

Petrochelidon ariel

Native name: Tjoontpjangalankoo (Jigalong).

This species is very similar to the preceding, except that the *head is rusty-red instead of black,* but this is hard to detect unless the bird is very close. The call has a different quality from that of the Tree-Martin, being higher pitched. Iris, beak and legs, black. Length, 4¾ in.

Distribution: Occurs all over the State, except in the South-West forested area. Until recently it was not found, except sporadically, south of Gingin or west of the Great Southern railway. Recent nesting records in the South-West are detailed below.

Nesting: The Fairy Martin nests in small colonies and constructs a bottle- or retort-shaped nest of mud against the wall of a cave, a cliff-face, the arch of a culvert, or a building. Nests are to be found often in the vicinity of water—doubtless due to wet mud being available for construction. E. H. Sedgwick has calculated that a nest may be constructed of about 1,300 pellets of mud, each representing one beakful of mud brought in by the birds. The bowl is nearly round and up to about 6 inches in diameter; the spout-like entrance is about an inch in diameter and usually 2 to 3 inches long, but may be much longer—up to 9 inches. The bottom of the bowl is lined with feathers or grass. Two typical nests weighed 1 lb. 6 oz., and 1 lb. 8 oz. respectively, but one weighing 2¾ lb. has been recorded. The 3-4, rarely 5, eggs are white, with, usually, a few small specks of reddish-brown. Average size of 13 eggs from Carnarvon, 16.7 x 12.0 mm., varying from 16.2-17.0 x 11.6-12.3 mm.

The distinctive mud nests of this swallow often reveal the presence of the species in a locality when the birds themselves may not be noted. It was first known to nest in the Perth district in October 1937, at a pottery works at Belmont, but otherwise the nesting sites nearest to Perth included Gingin (One Tree Hill), and in an eroded granite cave formation at Gwambigyne (near York). In 1964 V. N. Serventy found old nests in old clay pits at Midvale (two miles east of Midland). Between September and November of the same year Bro. Matthew Amedy, of De La Salle College, found nests at disused clay pits at Middle Swan and under a culvert two miles south of Bullsbrook. There is one record of nesting from the mouth of the Frankland River, on the south coast, in December 1912 (S. W. Jackson). In September 1961 a colony of about 30 nests was discovered under a bridge at Benger (R. Taylor). H. Hassell has informed us that in 1961 Fairy Martins appeared for the first time at Warriup, east of Albany, and nested in the farm buildings (about 50 nests).

PIPITS: Family Motacillidæ

Australian Pipit (Groundlark, Grundy, Richard's Pipit)

Anthus novæseelandiæ

Native names: Pooree pooree (Jigalong); War-ra-joo-lon, Warrajudong (Perth); Cherruck (Katanning); Perit Perit (Pallinup River).

Upper parts brown, streaked darker; outer tail feathers white. Throat and abdomen, white; breast streaked brown and light buff. Iris, brown; beak, brown; legs, light brown. Length, 6-6½ in.

The Pipit has the same plumage pattern as the Bushlark (p. 321), but is usually less rufous and the beak is longer and finer.

Distribution: Found all over the State in open country, including many of the larger islands off the coast.

Nesting: The nest is a cup-shaped structure of soft grasses constructed in a hollow in the ground under the shelter of a plant, tuft of grass or a stone. A typical nest measures about 2½ inches in diameter and about 1½ inches in depth. The top of the rim is usually flush with the ground. A count of 22 nests in various parts of the State gave the number of eggs forming the clutch as 3 in 14 instances, 2 in 7 instances and 4 on one occasion. The eggs are greyish-white marked all over with small spots of pale umber and slaty-grey, particularly over the larger end. Average size of 15 eggs, 22.3 x 15.7 mm., varying from 21.0-23.7 x 15.1-16.6 mm. Eggs may be found in nests from August to January. Incubation period, 13 days; fledging 14 days (Ray Garstone).

The Pipit is one of the familiar birds of the settled countryside, as the clearing of the forest areas has notably increased its habitat and the bird nests freely in pasture paddocks and cultivation. Usually unobtrusive, it attracts attention in the nesting season by a vivacious sky song, like that of a lark. Pipits in Western Australia are not infrequently seen perching on fences, telephone wires and even trees, a habit which is not shared by the birds in eastern Australia. We have noticed a considerable diminution of the species over the last few years in the lower South-West — due possibly to the depredations at nesting time of the fox.

Yellow Wagtail

Motacilla flava

Above olive-brown with yellowish wash on upper tail coverts. Wing quills edged lighter. Eyebrow line white; lores and line behind eye dark brown. *Underparts yellow. Long tail* dark brown, outer quills white. Iris, dark brown; beak, greyish black; legs, black. Length, 6½ in.

Distribution: A very rare migrant to the Kimberley Division from the northern hemisphere. There are two records from the North-West. John Darnell saw a bird at the sewage ponds at Mt. Goldsworthy, in the Pilbara, in mid-December 1971 and early November 1972, the latter bird remaining for four days. Both of these birds were immature, characterised by duller upper parts than in the adult, under parts slightly off-white and a dark bib on the lower throat and upper breast, and slightly yellowish on the under tail coverts.

The crown colour varies according to the race, from almost concolour with the back, through blues, greys to almost black. Almost any of the Asiatic races might visit Australia.

Mr. Darnell recorded the call as a single-noted "tsip."

Pied Wagtail (White Wagtail)

Motacilla alba

Head and tail black, back grey; wing quills edged with white. Forehead, face and throat, white. Black band across the chest. *Remainder of underparts white.* The long tail *(wagged up and down* not sideways as in the Willy Wagtail) has the outer quills white. Iris, dark brown; beak and legs, black. Length, 7 in.

Distribution: A very rare migrant from the northern hemisphere. The only known record from Australia is of an individual seen in the Katanning district in the autumn of 1971 by officers of the Fisheries and Wildlife Department. Their description was identified as of this species by Dr. G. M. Storr, of the W.A. Museum. Normally the southern limit of migration is north-western India and mainland China, and rarely to Indo-China and the Philippines.

CUCKOO-SHRIKES: Family Campephagidæ

Ground Cuckoo-Shrike

Pteropodocys maxima

Native names: Wollpootatjarra (Jigalong); Pee-orr-ba (Cundeelee); Goo-ra-ling (York).

Head and upper back light dove-grey; *upper wings and tail black.* Lower back and upper tail coverts barred *white* and blackish.

Throat and chest grey; breast and flanks white barred blackish; abdomen white. *The black tail is forked* and the outermost feather is broadly tipped white. Iris, yellow-white; beak, black; legs, very dark slate. Length, 14½ in.

Distribution: Occurring generally over most of the State, usually well inland and north through the upper Ashburton River country to the southern part of the Kimberley Division, around Hall's Creek. It is sporadic in the South-West, and visits the Swan coastal plain in small numbers, usually on the open clay flats. It does not, however, appear to be really plentiful anywhere in its range and seems to be local in occurrence.

Nesting: The nest is a well-built structure, about 6-7 in. in diameter and 2-3 in. deep situated on the horizontal fork of a tree, and may be as low as 10 ft. from the ground. It is made of stems, flowers and grasses, bound together with cobweb. The clutch is usually of 2, occasionally of 3 eggs, which are normally plain olive-green, sometimes variegated, being marked all over with spots and blotches of olive-brown. Average size of 8 eggs from the North-West, 31.0 x 23.1 mm., varying from 28.7-31.8 x 22.6-23.7 mm.

The birds, in small flocks, spend much of their time on the ground and run about freely. These small flocks appear to consist of the parent birds and their progeny which remain together till the next nesting season, if not longer. The call resembles that made by the Kestrel, "chee-er-chee-er," and the alarm note is a piping "queel."

Black-faced Cuckoo-Shrike (Blue Pigeon, Blue Jay, Summer-bird)

Coracina novæhollandiæ

Native names: Te-by-by (Pilbara district); Eelooroo (Jigalong); Nu-lar-go (Perth); Carebumdulah (Pallinup River).

The general coloration is a *delicate dove-grey*. The *forehead, side of head, throat and upper breast are black,* as are the wing quills and end of tail which, however, is tipped white. The lower abdomen and under tail are white. Iris, blackish-brown; beak and legs, black. Length, 13 in. Weight, 4 oz.

Young birds have less black about the head, this being confined to a broad black bar passing from the beak past the eye to the ear region; the throat is grey and there are narrow white edgings to the black wing quills. Such birds, immatures of the pale northern race, were described as a separate species by Edwin Ashby, the so-called Gascoyne Cuckoo-shrike *(C. gascoynensis)*.

Distribution: This is one of the birds which is common all over the State and in a wide variety of country.

Nesting: The nest is usually constructed on the top of a horizontal fork of a tree — usually on a branch which is at least twenty feet from the ground. Sometimes the nests are quite low down, as low as 4 ft. The nest fits closely to the branch and is difficult to locate. A typical nest observed by us in November, east of Jerramongup, was on a horizontal fork of a yate *(Eucalyptus cornuta)* about 20 feet from the ground and was made of bark and pieces of stinkwood *(Jacksonia furcellata)* matted together with spiders' web. Outside measurements, 5½ inches by 4½ inches. The egg cavity was shallow and 3½ inches in diameter. There was no lining but the base of the egg cavity was free from spiders' webs and constructed only of stinkwood. The clutch is of 2 or 3 eggs, which are olive with spots of brown and some underlying ones of red. A series of 13 eggs from the Barlee Range and Carnarvon averages 32.7 x 23.0 mm., varying from 29.9-35.9 x 22.1-23.7 mm. Both sexes share in the building of the nest and incubation of the eggs. They are very shy and will suspend incubation at the slightest interference, the nesting material will then often be removed piecemeal and a new nest built on another site. Nestling period, 25 days (E. H. Sedgwick).

The flight is distinctive and quite cuckoo-like as the bird flies and glides alternately in an undulating manner, uttering a soft "churring" note. On perching the bird engages in a curious wing motion, one wing being lifted alternately with the other. It is often seen about orchards and has a penchant for fruits such as mulberries.

No migratory movements have been detected in this State, such as take place in south-eastern Australia, but there appears to be a strong tendency to flocking in mid and late summer. Graeme Smith has seen flocks of up to 20 birds in the winters of 1971 and 1972 at Two People Bay on the south coast.

White-winged Triller (White-shouldered Caterpillar-eater, Lincoln)

Lalage sueurii

Native name: Goo-mul-cul-long (Avon River).

The adult male in full breeding plumage has the *head, mantle, flight quills and tail a glistening black;* the lower back and rump are grey; there is *a broad white shoulder patch* and the edges of the inner flight quills (secondaries) are white. The *entire under surface is white.* The female is brown above and there is a brown wash on the breast. Iris, dark brown; beak, black; legs, grey-black. Length, 6-7¼ in. The young bird just out of the nest has brown streaks on the white under parts, the head and nape are freckled brown and white, and the brown wing feathers have prominent buff edges. In late summer and autumn the ventral streakings and head frecklings are lost through a partial moult and the plumage resembles that of the female.

The bird is interesting in being the only member of its family to have two moults in the year, details of which were first worked out by Ernst Mayr. The normal post-nuptial moult in the autumn produces an eclipse plumage which in the male differs strikingly from that just described. The wings and tail remain black and the rump grey but the head and mantle become brown as in the female. In transitional plumages the brown is intermixed with black. Between July and November—varying in different parts of the State—a partial, pre-nuptial moult occurs in which the brown feathers of the eclipse plumage are replaced by black and some of the wing coverts and the central tail feathers are also changed. The female does not undergo any colour changes.

Distribution: The White-winged Triller is found all over the State, except the heavier jarrah forest areas and the lower South-West. The bird occurs, however, within the forested areas after considerable clearing has taken place, as at Bridgetown.

Nesting: The nest is small for the size of the bird, is of shallow cup form, 2¼ inches in internal diameter, made of pliant creeper, with small strips of bark on the outside covered with spider web and placed at a horizontal fork of a tree. The nest is so shallow that it is often quite difficult to see from the ground. The clutch is of 1 or 2, sometimes 3 eggs. J. A. Estbergs found a nest near Carnarvon with 6 eggs, undoubtedly a combination clutch. It was in the same tree as another nest of the species which contained 3 eggs. The ground colour is green, heavily streaked and blotched, especially at the broad end, with brown. Size, 21 x 16 mm. Both sexes share in the work of constructing the nest and both take turns in the incubation of the eggs. Incubation takes about 14 days and the nestling remains in the nest a further 12 days. During the nesting period, the male bird sings vivaciously.

The Triller is a strict migrant in the southern parts of the State, arriving during the latter half of September or beginning of October, after it has attained its nuptial plumage, and departing by early March. So far as is known no birds in the male eclipse plumage have been seen in the south. The extent of the migration is not yet known but the birds remain throughout the winter, including males in eclipse and transitional dress, in the East Murchison area. Males in the eclipse plumage have been seen at Leonora. In the Kimberleys K. Immelman has seen breeding males and males in eclipse plumage throughout the summer; nesting occurs during the whole of this period. The Triller is mostly a ground feeder.

THRUSHES: Family Turdidæ

Southern Scrub-Robin

Drymodes brunneopygia

Upper parts brown, with the long tail rufous-brown, tipped white on the outer feathers. Under parts buff-brown. The two rows of wing coverts are tipped white, giving a white striped effect in this region. Iris, brown; beak, black; legs, dark grey. Length, 8¼ in.

The rufous lower back and tail, with white wing stripes, are conspicuous field marks.

Distribution: Has been recorded north to a line joining Peron Peninsula (Shark Bay), Perenjori, Wubin, Dowerin, Southern Cross, Boorabbin and Lake Dundas. The northern boundary will probably prove to approximate to the mulga-eucalypt line (p. 53). The Scrub-Robin does not occur in the heavy forested part of the South-West (west of the Great Southern railway and west of the Midland railway).

Nesting: The Scrub-Robin nests on the ground in thick scrub. The single egg is laid in a slight hollow about three inches broad, and is lined with grass, bark or rootlets. Invariably a number of short, thick sticks are criss-crossed around the nest. The egg is greenish-grey in colour with usually a zone at the larger end of brown spots and blotches. Occasionally the markings are more evenly distributed over the whole surface of the egg. An average sized egg measures 25 x 18 mm.

The Scrub-Robin is one of the characteristic inhabitants of the mallee country and on sandplain with some tall scrub and dwarfed eucalypt. Though unobtrusive and shy it may, with patience, be approached quite closely. Usually it keeps just beyond one's reach, and a fleeting glimpse is caught of an alert brown bird walking along the ground between thickets, with the tail partially cocked. Most often, however, its presence is revealed only by its notes, which include a loud "did you *wick*," and a clear piping "pee-pee-pee,' or "pee-e-e," usually uttered from cover. The territorial challenge song is a persistent "chip-per-per-p'ree," uttered from a low perch.

QUAIL-THRUSHES and BABBLERS: Family Timaliidæ

Chestnut Quail-Thrush (Chestnut Ground-bird, Copperback)

Cinclosoma castanotum

Native name: Boone-yung (Avon River).

The male is a strikingly patterned bird in *chestnut, black and white*. The head and upper back are brown and the lower back and rump are chestnut-brown, the black tail being tipped white. The wings are brown and black and on the shoulder black and spotted with white. The *throat and chest are a velvety black,* with a white line extending out from the side of the beak and above this is a black line beneath the eye. The belly is white with black towards the thighs. The female lacks the black band on the breast, has less black on the shoulders and has a chestnut patch on the side of the neck. Iris, brown; beak, black; legs, lead-grey. Length, 9¾ in.

Distribution: The South-West, north to the mulga-eucalypt line, is the range of the principal form of the species, but excluding the heavy forested area of the extreme south-west corner (see Fig. 7). It is not usually recorded west of the Great Southern railway line, but John Gilbert in 1839 noted it on the old York Road, about 43 miles from Perth at a place then called "Banks' huts." This spot is still occupied as a farm and though Quail-Thrushes have not been reported in the locality in recent years an outlying colony of the White-browed Babbler is settled near there. It has also been observed at Dryandra and Cranbrook. Julian Ford has positively identified it north to Callagiddy, Butcher's Track (Muggan area), Pindar, Crows Nest Range, Davyhurst, 12 miles south of Menzies and thence into the Great Victoria Desert (north to Wanjarri, the Warburton area, and the Rawlinson Range), and south to the Nullarbor Plain.

H. T. Condon, of the South Australian Museum, has shown that a well-marked, brightly coloured race *(clarum)* occurs widely in the arid areas of Australia within the same area of occurrence as the next species; in Western Australia he recorded examples of it from Separation Well, Callion and Kalgoorlie. The Harold Hall expedition of the British Museum also collected a specimen at the Rawlinson Range. J. D. MacDonald of the British Museum has pointed out that the specific characteristics in these ground-frequenting birds must be sought in the ventral anterior aspects, from chin to breast, which are the parts individual birds see of each other. The upper parts reflect only the results of selection for camouflage and so the desert form of the present species (the race *clarum*) is hard to distinguish, when viewed from above, from the races of the next species. Another distinctive race of the Chestnut Quail-Thrush has evolved in the Nullarbor Plain (the race *alisteri);* its back including the head, is a rich russet. It is sometimes separated out as a full species, the Nullarbor Quail-Thrush.

Nesting: Nest a depression in the ground, usually under a bush or tuft of grass, or close to a fallen branch. Lined with grass, dead leaves or strips of bark. Eggs, 2, white, spotted with various shades of brown, most numerously at the larger end. They average 28-32 x 21 mm. in size.

The Chestnut Quail-Thrush, as its name might imply, is more given to running than flying when disturbed. It never flies very far when flushed and, on rising, its wings make a distinct whirring noise. "These birds have a peculiar low piping whistle, which is usually uttered as they are running away from one on the ground." (T. Carter).

Cinnamon Quail-Thrush (Chestnut-breasted Ground-bird)

Cinclosoma cinnamomeum

Native name: Milka (Warburton Range).

In the male the *back and upper wings are a rich chestnut;* the middle tail feathers rich chestnut and the outer ones black broadly tipped white. The wing coverts are black tipped white and the flight quills brown. The crown of the head is brown and there is a white eyebrow line. The lores and throat are a glistening black with chestnut behind the eye and a white streak below this. The *chest and flanks are chestnut;* middle of the breast black and the abdomen white with black blotches. The female is similar but has no black markings on the head and under parts. Iris, reddish-brown; beak, black; legs, grey. Length, 8-9 in. Weight, 2½ oz.

It differs from the nominate race of the Chestnut Quail-Thrush in the rich rufous colour of the back and from all its races in that the black of the ventral parts is interrupted by the rufous chest.

Distribution: This Quail-Thrush is found in the mid-north; south to the mulga-eucalypt line (Fig. 4) and north to the Ashburton River and Jigalong (the range of the race, *castaneothorax).* Julian Ford has made positive identifications north to Old Ullawarra, Talawana, Karara Well (Well 24 on the Canning Stock Route), Windy Corner (Gibson Desert) and Mt. Leisler (Kintore Range in the Northern Territory, just across the State border), and south to Muggan Station, Tallering Peak, Thundelarra, 12 miles south of Menzies, breakaways 16 miles east of Neale Junction, Hann Breakaways, Mt. Cassidy (Warburton area) and the Rawlinson Range.

Nesting: The nest is a moderately deep depression in the ground, probably scratched out by the female, and lined with strips of bark, a few acacia leaves, or leaves of the gidgie tree. It is located under the shelter of a bush, or under a fallen branch. Eggs, 2 or 3, pale buff in ground colour, marked with numerous small irregular-shaped dots of various shades of brown, with a few underlying spots of neutral tint. The markings are more numerous at the large end of the egg. Size, 27-31 x 20-21 mm.

F. Lawson Whitlock, who studied the species in the Wiluna district, states that soon after sunrise the male perches himself on some dead branch, or other point of vantage, and utters his rather plaintive and monotonous call. This may be represented by the letters *i* and *e*. The *i* is uttered short, and is rather rapidly repeated, the final *e* being long-drawn and a half-note higher in tone. It may, perhaps, be expressed as follows "i-i-i-i-i-e." This call is repeated at short intervals and two or three males will often reply to one another. Derek Goodwin renders the call as a di-syllabic "feeeu, feeeu." The bird runs. The tail is flicked up and down when the bird is alarmed. Besides insects these birds feed extensively on seeds, including mulga seeds. They are unhusked.

Grey-crowned Babbler

Pomatostomus temporalis

Native names: Ca-cac-co (Pilbara district); Doyndoyn (Lyons River); Turdoo wardoo (Jigalong).

Back and wings dark brown; flight quills with chestnut band; tail blackish-brown, broadly tipped with white on all but the central feathers. From the beak extends a *broad greyish-white streak over each eye,* leaving between only a narrow grey-brown strip on the crown of the head, which is continuous with the back. There is a broad dark brown streak from the lores through the eye to the sides of the neck. The throat is white, *chest and breast a rich chestnut, abdomen and flanks dark brown.* Iris, yellow to brown; beak long and down-curved, brown; legs, greyish-black. Length, 9¾-10 in.; wing span, 12¼ in. Weight, 2½ oz.

Distribution: A northern species known to extend south at least as far as the Wooramel River, the upper Murchison (Mileura Station, Karalundi Mission) and Wiluna. However, it may range even further as in October 1965 A. G. Mathews came on a solitary individual in mulga between Wurarga and Gullewa. T. E. Bush recorded it breeding 20 miles north of Pindar.

Nesting: The birds construct large dome-shaped nests of strong sticks and rootlets loosely put together. The favourite sites are lone patches of mulga, where the nest is placed on the topmost branches, securely woven into an upright triple fork, though nests are also built in river gums along creeks, up to 40 feet from the ground. A typical nest is about 2 feet high and 15 inches wide, with a spout-like entrance at the top, made of longer sticks, and about 2 to 2½ inches in diameter. The nesting chamber, 9 to 10 inches in diameter, is about 9 inches from the mouth of the spout and is lined with fine shredded bark, grass, fur or pieces of wool. The bottom lining is usually of pieces of the leaves of flannel bush *(Solanum lasiophyllum),* among which the eggs are found hidden if the nest is unattended but the eggs are exposed if the bird is suddenly flushed from it. Although new nests are usually built each season old ones may be renovated and additional material built into the structure; some of these old nests may be up to 3 feet in height and 2 feet in width. During heavy rain the nests become sodden and may require relining, or even rebuilding. Although there might be two or more nests ready for eggs in the one colony, sometimes as many as 6 nests in the one tree, only one will be in actual use for nesting. Nest-building appears to be a prolonged process, the birds spending two or three months on the job, all of the company, who share in its construction, sheltering in the structure at nights, it being used as a community roosting place. However, once egg-laying begins the birds cease being conspicuous about the site and are never seen near the particular nest unless flushed (observations by A. H. Robinson).

The eggs, 2 or 3 in a clutch, sometimes only one, are distinctive, being brown and streaked all over with hair-like markings of darker brown. Average measurements, 27 x 18 mm. Only one female lays in the nest, but when the young are hatched the entire group assists in feeding them. The birds queue up in front of the nest to feed the chicks in turn. The egg shells are removed by the parent. The young are entirely naked at hatching, there is no nestling down and a completely adult plumage is aquired when the young leave the nest. Incubation period, 17 days (Klaus Immelmann).

The birds associate in flocks of about 6 individuals, though there might be up to 12 at times. Social preening is very common. The courtship display includes food begging as in young birds. The parties usually are very local and haunt the same patch of trees year after year.

The Western Australian form (the race *rubeculus)* is more extensively reddish on the underparts than the eastern States bird, and was formerly ranked as a separate species, with the vernacular name of Red-breasted Babbler.

White-browed Babbler (Apostle-bird, Cat-bird)

Pomatostomus superciliosus

Native name: Gnow-un (Avon River).

Upper parts brown, including the tail which is broadly tipped white on all but the central feathers. *Under parts white,* becoming brownish towards the abdomen. There is a well-defined and conspicuous white eyebrow line, reaching from the base of the beak to far behind the eye. Below this line the brown of the upper parts extends to well below eye level. Iris, bright brown; *beak, long and down-curved;* dark-brown; legs, grey. Length, 7-8¾ in.

Distribution: Resident and breeding in the southern part of the State, north to the Tropic of Capricorn, extending eastwards to central and south Australia. It does not occur, however, in the South-West corner much west of Toodyay, Dryandra, Williams, Kulikup and the southern end of the Great Southern railway, except for a few isolated colonies on the south coast (Busselton, Pemberton, the lower reaches of the Warren River, Irwin's Inlet, Denmark, see Fig. 8). Out of Toodyay towards Perth, nesting colonies are known as near as 43 miles from Perth, and also at Clackline.

Nesting: The nesting arrangements of this bird are similar to those of the preceding species and it has the same habit of using "roosting-nests." The eggs, usually 3 in number, are also similar to those of the Grey-crowned Babbler, but are smaller in size. A series of 29 eggs from different parts of the State averages 23.8 x 16.8 mm., varying from 21.7-25.6 x 15.7-18.4 mm. The measurements are highly variable but there do not appear to be any constant regional differences as once thought by T. Carter.

These babblers are mostly found in small companies, whose noisy caterwauling is responsible for one of their local names. The calls vary, the commonest being a reedy "churr-r-r"; another, which seems to be the alarm call, is a strident "wit-wit." On the ground they move by a series of hops, but so rapidly that they almost appear to be running. Besides insects, they feed on seeds (unhusked).

OLD WORLD WARBLERS: Family Sylviidæ

Spinifex-bird (Carter's Desert-bird)

Eremiornis carteri

The upper parts are a *uniform bright brown* without any pattern of streakings. Crown of head rufous. The *long and wedge-shaped tail* becomes darker towards the tip. The under parts are dull white with a buff wash towards the flanks. The brown under tail feathers are edged with light buff. Iris, brown; beak, brown; legs, light brown. Length, 5¾-6 in.

Distribution: This is an inhabitant of the arid country of northern central Australia, from the Cloncurry district of Queensland west to Hall's Creek and the Fitzroy River and south through the North-West to the Minilya River. It also occurs on some of the islands off the North-West coast, the Dampier and Montebello Groups (actually reported on Barrow, Hermite and Thevenard Islands).

Nesting: A nest found by F. Lawson Whitlock on Barrow Island on August 20, 1918, was open and cup-shaped, substantially built of finely shredded grass and herbage, and lined with fine roots. It was placed in a bunch of spinifex *(Triodia)* and measured 5½ inches across and 2½ inches deep. The egg cavity was 2¼ inches by 1¼ inches. The 2 or 3 eggs were oval in shape, surface of shell very fine and slightly glossy Ground colour, pinkish-white, closely covered all over, and particularly at the large ends, with minute markings of pale lilac and purplish-brown, and from light to dark reddish-brown. Average measurements, 17-18 x 12 mm. The female alone appears to build the nest, but the male assists in feeding the young.

In the nesting season the male sings conspicuously from the top of a bush and Whitlock has recorded that the liquid notes cannot be rendered better than by the French words "je suis a vous" (pronounced "je swee, ah voo"). Sometimes only the first syllables are uttered, when the call closely resembles that of the Pipit. Simultaneously the female adds two or three grating notes, sounding like "jut-jut-jut." In flight the Spinifex-bird seems encumbered by its long broad tail which is carried in a semi- drooping fashion. When perched, on *Triodia* clumps or other bushes, it elevates its tail. On Thevenard Island, off Onslow, it shelters in the clumps formed by the inter-twining stems of the curious vine *(Sarcostemma australe)*, common on most of these islands.

The Spinifex-bird was first made known to science in 1900 by A. J. North from specimens obtained by T. Carter near North-West Cape.

Golden-headed Fantail-Warbler

Cisticola exilis

This is one of the comparatively few species of Passerine birds which have a double moult and rather strikingly different seasonal plumages. The *male in the summer* has the *crown of the head a golden-brown* colour, the rump somewhat similar and the *back streaked dark and light brown*. The *tail is short*. In the *winter* plumage the male is strongly streaked black and light brown, with a brownish collar around the neck and the *tail becomes long and wedged-shaped*. On the under surface the grey tail quills have a row of conspicuous black spots near the tip. The under parts are uniform buffy-white. The female throughout the year resembles the male in winter dress. Iris, light brown; beak, dark brown; legs, buff-brown. Length, male in winter, 4¼ in., in summer, 3½ in.

Distribution: An eastern and northern Australian species which only penetrates into the northern portion of our area from the Kimberley Division. It has been recorded as far south as Millstream Pool, on the Fortescue River.

Nesting: The Fantail-Warbler makes a small dome-shaped nest with an entrance near the top. Tall grass or a clump of rushes is the usual site; very fine grasses are used for the body of the nest and the outside is covered with plant-down or leaves stitched on to the nest with cobwebs. Soft plant-down is used as a lining to the egg-cavity. The 3 or 4 eggs are small (18 x 12 mm), and are bright, light turquoise-blue ovals, spotted and blotched with rust-red to rust-brown and purple, sometimes dark to blackness, with a few secondary markings of grey, and the marks frequently form a zone at the larger end.

The Fantail-Warbler has become the most common bird in the irrigated rice plots in the Ord River scheme in the East Kimberley and it breeds earlier (October-November) in these plots than in the surrounding bush (January-March). The grass in the native bushland starts to grow only after rain, which is far later than in the artificially irrigated rice plots. The view of tall grass appears to be the breeding releaser as is the case with several species of Reed-Warblers *(Acrocephalus)* in Europe (Klaus Immelmann).

Little Grass-bird

Megalurus gramineus

Upper parts *brown streaked black;* the tail brown with black along the shafts. Under parts dull white, with *black spots and streaks on the throat and breast.* A whitish line above the eye. Iris, pale brown; beak and legs, brown. Length, 6½ in.

The Little Grass-bird is much smaller than the Reed-Warbler, whose habitat it often shares, and differs from it in having a streaked pattern both above and below and lacking any chestnut tone in the plumage.

Distribution: Found in the South-West, from Carnarvon in the north to Shark Lake, near Esperance Bay, in the south-east. Occasionally it may wander outside these limits. Thus in August 1962 J. R. Ford saw one bird and heard it call, in the rushy margins of Windich Spring on the Canning Stock Route, 80 miles N.N.E. of Wiluna. However, the species was not recorded on later visits there by ornithologists (February and August 1964).

Nesting: The nest is deep cup-shaped made of coarse grasses, and placed in the centre of a clump of rushes growing in water. It is warmly lined with feathers and usually a couple of feathers of the Swamphen are so worked into the rim as to form an arched canopy. The eggs are 3 in number, occasionally 4, and rarely 5, and in the lower South-West are to be found in nests from September onwards. They are white in ground colour, rather oval in shape, and are covered with small reddish spots and specks, which sometimes predominate at the larger end. Average measurements of 6 eggs from the lower South-West, 18 x 14 mm.

The Little Grass-bird frequents not only fresh water swamps but also occurs in the rushes and samphires which margin the coastal estuaries. Curiously, however, it has not infrequently been observed both there and in the eastern States, in the apparently unfavourable environment of coastal islands; A. J. Campbell recorded it at Breaksea Island. It is a very secretive species and its presence would be mostly undetected but for its rather mournful-sounding bi-syllabic note of "pee-pee-e." It may still be commonly heard on the Swan River in the Causeway marshes, despite the inroads of the reclamation works, and another metropolitan stronghold is the salt marsh in the Lucky Bay area. It still nests in the rush beds at Pelican Point. The species also survives in the dense growth of yanget in the middle of Herdsman's Lake.

Reed Warbler ("Swamp-Tit")

Acrocephalus stentoreus

Native names: Goor-jee-goor-jee, Gurji-gurji (Perth).

Upper parts bright brown, wing and tail quills dark brown. A line from the base of the beak to above the eye is buff. The under parts are buff or light chestnut brown, the throat and middle of the abdomen being white. Iris, brown; beak, brown; legs, grey. Length, 7 in.

Distribution: Generally distributed throughout the State, whereever there are fresh water swamps.

Nesting: The nest is beautifully adapted for the special stresses of its environment. The bird peels off strips from the rushes and winds them in and out around three or four standing rushes till a deep cup-shaped structure is formed. No further lining is added. The lip of the cup is narrowed in, presumably to prevent the eggs or young from rolling out should the rushes be bent over by the wind. As a further safety measure the rushes carrying the nest are permitted to slide up or down through the fabric as the whole sways in the wind. In the Perth area nests are sometimes built in the swamp paperbark *(Melaleuca rhaphiophylla),* rigidly held in the multiple forks of the slender branches. Usually 4 eggs are laid—pale greyish-white or very pale yellowish-brown in ground colour, spotted and blotched with light brown and lavender. Two sets of 4 eggs each, taken in the South-West, average 21 x 15 mm. in size. Egg dates, from September to February in the South-West.

The bird is confined to rush-beds in fresh water swamps and is never seen out of them. Most rushy patches of any extent have their quota of Reed-Warblers, even in such restricted places as the Jolimont and Shenton Park Lakes in the metropolitan area. Within this haven the bird leads an active existence; wary and alert. It is hard to pick up when at rest owing to its concealing colouring and the casual observer mostly sees it as a richly plumaged brown form flying swiftly through the rushes. Its presence, however, is made obvious by the vigorous and melodious calls of "Twitchee-twitchee-twitchee-quarty-quarty-quarty" which may be heard for a considerable distance. This song is similar to but more powerful than that of the Brown Honeyeater. The ship's company of the Dutch navigator Vlamingh thought they heard "Nightingales" and these may have been either Reed-Warblers or Noisy Scrub-birds singing in the rush-grown lakes which were Perth and its suburbs in the seventeenth century.

The seasonal movements of Reed-Warblers are incompletely known. During the winter months the species certainly vacates the lower South-West districts in April and does not return until August. In the neighbourhood of Perth, however, the birds appear to be resident throughout the year.

Brown Songlark ("Corn Crake")

Cinclorhamphus cruralis

Native name: Ye-jul-lup (Avon River).

Upper parts dark brown, streaked paler. Entire under parts dark brown, with the *throat almost black in the male and sparingly spotted with buff.* Females and young birds lack the dark under parts, having a white throat and the remainder of the lower surface dull white, spotted and strongly marked with brown. Iris, brown; beak, black; legs, brown. Length, male, 9¾-10½ in.; weight, 2½-3 oz.; female, 7½ in.; weight, 1 oz. The great size disparity between the sexes, in both this and the next species, is unique among Australian song-birds.

Distribution: Generally distributed, but not found in the South-West jarrah forest area. However, it extends as far south as Busselton on the coastal plain, west of the Darling Range scarp. On the south coast corridor (see p. 59) it occurs at Albany and Walpole.

Nesting: The Brown Songlark makes a cup-shaped nest in a hollow in the ground under protection of a tuft of grass. The birds probably scratch out the hollow themselves. The nest is like an enlarged edition of that of the Pipit and, like the nest of that species, is constructed of grasses and lined with softer material. The 3 or 4 eggs are salmon-pink, speckled all over, particularly at the larger end, with pinkish-red. Average size of 8 eggs (Barlee Range), 23.1 x 17.1 mm., varying from 21.0-24.8 x 16.2-17.7 mm. Nesting has been reported at Busselton.

As they soar—often gliding without perceptible wing movements—and from vantage points on a dead tree, these birds utter a curious call, "witchy-weedle" or "whittaker-whittaker," which sometimes has a harsh tone and may be likened to a wheel turning on a rusty axle. When perched they may elevate the tail almost to an upright position.

The species appears to be a spring and summer visitor to the southern parts of the State, and open clearings, such as pasture paddocks and aerodromes, are an attraction to them. They may even be seen at times on the open grassy areas at the eastern end of the Perth Esplanade. C. F. H. Jenkins records that they arrive at Northam about the end of July and depart in January. At Kellerberrin they arrive in June and July, nest in August and September, and are usually gone by December (Bruce W. Leake). A. L. Milhinch has recorded odd birds remaining throughout the winter at Northam (as happened in 1955 and 1964). D. G. Bathgate has observed both species of Songlark only during the winter, April to July, at Shark Bay. These birds feed on both insects and seeds.

Rufous Songlark

Cinclorhamphus mathewsi

Native name: E-role-del (Avon River).

Upper parts greyish-brown, streaked dark brown; *rump, chestnut-brown;* tail, darker brown. White eyebrow. Under parts dull white with a brown clouding and freckling across the upper breast. Iris, brown; beak, black; legs, grey. Length, male 7½-8¼ in.; weight, 1-1½ oz.; female, 6¾-7¼ in.; weight, slightly under 1 oz.

In the field it looks like a large pipit but lacks the white outer tail feathers, and differs from the Brown Songlark in its smaller size, chestnut rump and light under surface.

Distribution: Found all over the State except the South-West corner (Darling Range forested area). Occasionally it visits the coastal plain as far south as the Bunbury district. It now occurs in the Boyup Brook and Bridgetown districts in cleared country, apparently penetrating westwards from the Kojonup and Arthur River country (G. A. Lodge).

Nesting: The nest of the Rufous Songlark conforms to the type constructed by the Brown Songlark, and is found in similar situations. It measures 4½ inches in diameter, 3 inches deep, with the egg cavity 2½ inches by 2 inches deep. The 3 or 4 eggs are white, speckled all over, particularly at the larger end, with reddish spots. Average size of 7 eggs (Ashburton River) 21.5 x 15.3 mm., varying from 20.8-22.5 x 14.7-15.6 mm.

The species favours grassland areas with trees, including ringbarked country, and, like the Brown Songlark, is given to perching in trees and on stumps. The call is "witchy-weedle," repeated several times and often preceded by a trill and sometimes followed by an unelaborated trill which may be varied in pitch. It has not the harsh, grating quality of the Brown Songlark. The bird is a spring and summer migrant to the southern portions of the State. C. F. H. Jenkins has recorded it arriving at Northam in August and leaving in December, the numbers varying annually. In 1965 A. L. Milhinch

Plate 4.—Left: Banded Blue Wren. Right: Turquoise Wren.

recorded the first arrival at Northam on June 29. At Boyup Brook it arrives in September, becomes silent at the end of December, after breeding, and leaves by early March (G. A. Lodge).

AUSTRALIAN WARBLERS: Family Maluridæ

Banded Blue Wren (Splendid Wren)

Malurus splendens

(Plate 4)

Native names: Djarjilya, Djur-jeel-ya, Deldillia (Perth), Jeer-jal (Avon River).

This and the following species are the only local blue wrens in which *the fully plumaged male is entirely blue and black,* with *no white or red in the colour scheme.* The head, back, wing coverts and under parts are a brilliant violet-blue, with the forehead and ear coverts lighter blue and the wing quills a greenish-blue and black. A band across the breast, and a band across the nape behind the head, from eye to eye, are black. Iris, brown; beak, black; legs, brown to black. Length, 5¼ in. The female is brownish-grey on the upper parts and whitish below, with a dull blue tail, the beak and a small area around the eye being chestnut. The young male is similar and in the transition plumage will also have blue wings. Adult males in eclipse plumage retain the black beak and blue wings (W. H. Loaring).

Distribution: This species is most common in the South-West corner. from Geraldton to Esperance, but it extends far to the north and inland, however, as apparently sporadic colonies. Thus it has been reported north to Shark Bay (Hamelin Pool, Edel Land), the upper Ashburton River, Wiluna and Leonora.

Nesting: The nest is an oval structure of dry grasses, with a circular entrance at the side lined with feathers, the whole having a rather untidy than neat appearance. It is placed in a bush or a creeper at no great height from the ground — 4 feet being a not unusual height. A typical nest measured 5½ inches x 3¼ inches, nesting cavity, 2½ inches deep, entrance 1½ inches in diameter. The eggs, 3 in number, sometimes 4, are white, spotted with reddish-brown often forming a zone at the larger end. On one occasion we found six eggs in a nest — evidently a combined clutch of two different females. A set of 3 from the Bridgetown district measure 17 x 14 mm. In the lower South-West the birds breed late into the year, eggs having been noted from October to the third week in January. Two broods are often raised in a season. W. H. Loaring has found the incubation period to vary from 14 to 16 days. P. Slater, from observations at many nests, has determined the usual period to be 13-14 days and the nestling period 12 to 14 days. The male takes no part in building the nest or in brooding, but assists in feeding the young.

The birds are usually seen, even in the nesting season, in small groups, there being an apparent excess of females, though some of these may be immature males which cannot be distinguished as such. This character has earned for the bird the title of "mormon-wren," as for the kindred species in the eastern States. The explanation has also been given that the apparent extra females may be young of a previous hatching, still in attendance on the parents. A. H. Robinson puts forward the suggestion, however, that blue wrens may have social habits similar to those he has found to hold for the Western Magpie (see p. 452), where *groups* of birds defend territory, instead of individual pairs as is the case with the majority of birds. In the North-West he has found several nests of *Malurus lamberti* side by side. C. F. H. Jenkins has seen two fully-plumaged females feeding one particular nestling.

Turquoise Wren

Malurus callainus

(Plate 4)

Very similar in plumage pattern to the preceding species but differing from it in having *a broad black bar across the lower back,* separating the blue of the upper back and the tail. The head and upper back, and the upper tail coverts, are a brilliant light blue, *much lighter* than in the Banded Blue Wren and altogether *lacking the violet tone* of that species. The throat is a similar violet-blue but the ear-coverts are a light silvery blue, not pale violet-blue. The abdomen differs from that of the Banded Blue-Wren in being a lighter blue with no violet tone. Iris, brown; beak, black; legs, brown to black. Length, 4¾-5¼ in. Female similar to that of the preceding species.

Distribution: Until 1960 this exquisitely coloured wren was considered to be restricted to South and Central Australia (occurring from near Port Augusta to Ooldea and north to the Macdonnell Ranges). However, in August 1960 W. H. Butler encountered the species in mulga country in the Great Victoria Desert, at Maloora waterhole, about 70 miles west of the South Australian border. He came on a party of two females and a male. Subsequently it has been established that the species occurs over a wide area, from the Rawlinson and Warburton Ranges westwards to the Gibson Desert, at least to Mungilli Claypan (100 miles E.N.E. of Carnegie Homestead). The birds here appear to be transitional with the Banded Blue Wren (lacking the black bar across the lower back).

Nesting: So far not observed in Western Australia. The nest and eggs resemble those of the Banded Blue Wren.

Blue-and-White Wren (White-winged Wren)

Malurus leuconotus

Native names: Wee-rirr-eer-ri (Coolawanyah, all species); Yerreeyarro (Lyons River, all species); Tjilltjilka (Jigalong, all species).

The male is a *brilliant blue* above and below, with *prominent silky white shoulder* patches. The female is brown above, white below and with a bluish tinge on the grey tail: it can be distinguished from the female of *Malurus lamberti* by the absence of the chestnut lores and orbital ring. Iris, brown; beak, black in the male, brown in the female; legs, pale brown. Length, 4½-5 in.

Distribution: This is a widely distributed species, recorded from central Australia to the coast, north as far as the Pilbara district. It probably extends to the southern boundary of the Kimberley Division as Fauna Warden V. T. Nelson sighted a family party at Louisa Downs station, W.S.W. of Hall's Creek. The general southern boundary is a line joining Mogumber, Corrigin and Norseman. On the coastal plain it has a broken distributon south to Perth (cf. Fig. 9). It is a characteristic species of the driest areas and occurs on scrubby sandplain, *Casuarina*-clad ironstone slopes and the samphire flats around salt lakes and estuaries.

Until comparatively recently it was not thought that this species occurred further south than Mogumber, on the Moore River, north of Perth. However, in May 1931, W. H. Loaring observed it in the thickets behind Burns Beach, west of Wanneroo, and in August 1949 it was seen behind the beach dunes at City Beach. In June 1948 it was most unexpectedly met with actually within the city limits in the samphire marshes at the Heirisson Islands, alongside the west end of the Perth Causeway. This discovery was made during a survey by naturalists of the fauna and flora of that area before its transformation by reclamation works. Occasionally birds are seen in the shoreline vegetation at Pelican Point, in the Swan River estuary and before their reclamation in 1963-69, in the Causeway marshes east of Heirrisson Island. Further inland the nearest record to Perth is at Lake Pinjar (seen by Miss D. A. Rook, October 1962).

Nesting: Like other members of the genus *Malurus,* the Blue-and-white Wren builds a domed nest, with an entrance near the top, of dried grasses. The nesting cavity is lined with feathers, wool or plant down. Average dimensions are 4½ inches in height and 3 inches in breadth. It is usually placed in a low bush a few feet from the ground. The 3 or 4 eggs (size 15 x 12 mm.) are white, with purplish or reddish-brown markings, especially at the larger end.

F. Lawson Whitlock has described the voice of this species as follows:— "If one imagines a toy sewing-machine being run at a high rate of speed the resultant sound would resemble the sustained reeling notes of these wrens." The native name is onomatopoeic.

Black-and-White Wren

Malurus leucopterus

Very similar to the preceding, but the blue in the plumage pattern is replaced by *black,* shot with blue. When viewed in certain lights is could be described as deep blue-black. Tail, dark blue. Iris, dark brown; beak, black in the male, brown in female; legs, brown. Length, 4¾-5 in. Females brown above and white below as in preceding species.

Distribution: This wren is found only on Dirk Hartog and Barrow Islands.

These two isolated colonies, over 350 miles apart, have aroused considerable conjecture as to their origin and their relationships with the Blue-and-white Wren. Are these island populations to be considered as survivors or relicts of a form which has now disappeared from the mainland, or, alternatively, have they evolved independently from Blue-and-white ancestors cut off on these islands? The latter explanation demands a coincidence in the direction of mutation of the controlling gene or genes, but this need not be considered improbable. A similar phenomenon has been noted in wrens and honeyeaters in Flinders and King Islands, in Bass Strait, and it would appear that in all these cases the same factor, or series of factors, is operating independently in the island environments to produce, by selection, a similar result in each. In other words, ecotypic variation is taking place. The explanation involving relict populations is supported to some extent by the occurrence of dark individuals on the mainland and F. Lawson Whitlock considers that coastal populations of the Blue-and-white Wren are darker than those further inland. On one occasion he observed a bird at Onslow which he thought to be a black-and-white bird and only after collection was it found to be an unusually dark specimen of *M. leuconotus*. Edwin Ashby collected a specimen of the same species, with almost black feathers intermingled with the blue, on the coast hills 160 miles north of Perth, and on Peron Peninsula, Shark Bay, he took a specimen which was so dark in colour that it appeared black in some lights. L. E. Sedgwick saw a "decidedly black and white" bird at Monkey Mia, Peron Peninsula in September 1966. Both Whitlock and T. Carter have collected specimens of the Black-and-white Wren on Dirk Hartog with some blue feathers in the plumage.

Nesting: Whitlock is our authority on the nesting habits of this bird, both on Dirk Hartog Island and on Barrow Island. The birds appear to favour the big, fairly open flats, where salt-bushes of various species are abundant and the scrub generally is of a more dwarfed character. A prickly *Hakea* is a favourite site for the nest, which is placed close to the ground, a foot being the average. A typical nest is strongly constructed of very fine, dry grasses, matted with spiders' webs, cocoons, and small pieces of wool. It is of the usual domed style as built by all the blue wrens, there being an entrance at the side. A nest is 4-5 inches high and about 2¾-3 inches broad, with the entrance hole 1¾ inches in diameter. The portion containing the eggs is lined with fine grasses, cocoons, and wool, and the eggs number 3 or 4. Like the eggs of other species of blue wren, the colour is white, marked with small dots and splashes of various shades of brown and red dispersed over the whole surface, at other times concentrated mainly towards the larger end, even to the extent of making a distinct cap. Measurements, 15 x 12 mm. Weight, 1 gm. Nesting dates: Barrow Island, mid-July to mid-September; Dirk Hartog Island, end of August. This is one of the species parasitised by the Narrow-billed Bronze Cuckoo. Males may breed before they attain the typical adult plumage.

The reeling song is weaker than that of the Blue-and-white Wren.

Variegated Wren (Purple-backed Wren)

Malurus lamberti

(Plate 5)

The resplendent male is blue and black, with *conspicuous chestnut shoulder patches* and a *white abdomen.* The head is bright enamel purple-blue, behind which is a black nuchal band and behind this again a violet-blue mantle. The lower back is black and the tail dull white. Apart from the vivid rufous shoulder patch the wings are brown. The ear coverts are a sky-blue. The *throat and chest are a velvety black.* Iris, brown; beak, black; legs, brown. The female is grey-brown above with a dull blue tail, the under parts being light buffy-white. The beak is light chestnut in the female and the lores and a narrow ring of feathers around the eye are also chestnut, contrasting with the olive-grey of the head plumage. Length, 5¼-5½ in.

Distribution: This is the most northerly ranging of three similar-looking species, the others being *M. pulcherrimus* and *M. elegans.* It is found north to the Kimberley Division and south to a line joining Geraldton, Carnamah and about Kalgoorlie. Eastwards it ranges into Central Australia, thus linking up with the eastern Australian populations. Recently J. R. Ford has discovered an unsuspected penetration of this species along the coast south of Geraldton and almost to Fremantle, the birds occurring in the thickets of the coastal dunes. In 1975 a party was seen at Pelican Point in the Swan River estuary. This narrow "tongue" south of Geraldton overlaps considerably the distribution in this sector of the Blue-breasted Wren.

Nesting: It builds a dome-shaped nest, with an entrance at the side near the top, and constructed of pale-looking dry grasses and stalks of flowering plants lined inside with plant down. A typical nest measures 5 inches high and 4 inches broad. The 3 (sometimes 4) eggs are like those of the Blue-and-white Wren—white, with purplish-red spots and specks which are often mainly at the larger end. Size, 17 x 12 mm. Incubation period, 14 days; nestling period, 12 days (observations by N. Ives on captive birds).

The song consists of a few brief rattling notes and a high-pitched alarm note like that of the Western Grass Wren *(Amytornis textilis)*. It is a very silent and unobtrusive bird (Whitlock).

Blue-breasted Wren

Malurus pulcherrimus

(Plate 5)

This species is very similar to the preceding and it is very difficult to tell them apart in the field, unless one is very close and is able to view the adult males with a good glass in bright light. The principal difference is the colour of the *throat and chest which in this bird is a deep dark blue,* appearing almost black. The females are indistinguishable.

Distribution: This species inhabits a comparatively narrow belt generally south of the range of the Variegated Wren and inland of that of the Red-winged Wren. On the west coast it occurs from Tamala (south of Shark Bay) south to the mouth of the Namban River (south of Jurien Bay), where, as J. R. Ford has shown, it overlaps the range of the Variegated Wren over a distance of 280 miles. The south-western boundary is not certainly known, but it probably trends south-eastwards from the Namban River towards the Wheatbelt, the southern portion of the boundary approximating to the Great Southern railway. Its westernmost limit on the south coast is at Warriup. The north-eastern boundary is similarly undefined but known localities are 21 miles north of Galena, Mingenew, Bunjil, Wongan Hills and Norseman, narrowly overlapping the southern boundary of the Variegated Wren. It has been collected south-eastwards as far as the coast south of Mundrabilla.

Nesting: The nest is usually placed low in a small bush and is similar to that made by other red-shouldered wrens. The eggs, 3 in number, are white, blotched and spotted with reddish-brown. They measure 16 x 13 mm. T. E. Bush has found this species to be an earlier nester than the other blue wrens in the South-West, breeding starting in August.

Plate 5.—The chestnut-shouldered Wrens.

Blue-breasted Wren | Variegated Wren

Red-winged Wren

Red-winged Wren (Marsh Wren)

Malurus elegans

(Plate 5)

Native name: Djur-jeel-ya (Perth, not differentiated from the Banded Blue Wren).

The plumage pattern of the male is similar to that in the two preceding species. It has the same *deep bluish-black breast* (edged with black towards the abdomen) as the Blue-breasted Wren, but is readily recognisable from both it and the Variegated Wren by the *lighter blue of the head and mantle.* This blue is of a light silvery hue, with no purple in it, as in the other two species of red-shouldered wrens, and the bases of the mantle feathers are white. The females are similar to those of the Variegated and Blue-breasted Wrens, but the beak is dull black (shining black in the males). There is also a greater contrast between the grey head and the mantle (which is rather browner) than in the related species. Length, $5\frac{1}{2}$-$6\frac{1}{4}$ in.

Distribution: This wren is confined to the South-West corner from about Moora along the western part of the Darling Range to Pemberton, along the south coast to Warriup, with a possible outlier in the Stirling Range. The wren occurs in the Porongorups Range and C. J. O. Harrison and P. R. Colston in 1966 encountered it in small swampy areas at Kiriwina (east of Mt. Barker) and between the Stirling and Porongorups Ranges. The birds may prove to be widely distributed in such scattered patches of habitat.

Nesting: The usual situation for a nest of the Red-winged Wren is close to the ground in the base of a clump of sword-grass *(Lepidosperma longitudinale)* on the edge of a brook. Such situations, are however, not always selected by the birds. For example, H. O. Webster, during observations over four nesting seasons at Karridale and Smith's Brook, noted ten nests built close to water in rushes and eight built well away from rushes and water. In one instance a nest was observed some three to four hundred yards from a brook in the limbs of a dead karri *(Eucalyptus diversicolor)* sapling, while another was still further distant from water and was located practically on the ground among grass, ferns and Scotch thistles growing in a clover paddock. A typical nest is constructed of grasses, strips of *Leptospermum* bark, and leaves of the flooded gum *(Eucalyptus rudis).* The nest is well-lined with the vegetable down from the cones of *Banksia,* feathers, shredded bark and moss. It is globular in shape and is a little larger and of more sturdy construction than that of the Banded Blue Wren. It has been suggested that the leaves of the flooded gum are included by the birds in the components of the nest because of the odour acting as a deterrent to the myriads of midges, mosquitoes and other stinging insects which frequent the localities used by the birds for nesting purposes. The eggs are white in colour and are spotted and blotched sparingly with rufous with a tendency for the markings to concentrate at the larger end and so to form a cap or zone. Size, 15 x 12 mm. Of 18 nests observed by Mr. Webster, 3 eggs occurred on six occasions, 2 eggs ten times and one nest contained one young. From his observations it appears that the female alone constructs the nest and incubates the eggs — incubation lasting 15 to 16 days. When the eggs are hatched the female is assisted in the feeding of the young by her mate and also by what appear to be the progeny of the previous season's rearing. The young remain in the nest for a period of 12 to 13 days. October and November are the nesting months.

This and the other two species of chestnut-shouldered blue wrens are so much alike that some modern ornithologists combine them together as one species. Apart from the physical similarity between them this view is supported by the general facts of distribution. Their ranges on the whole are mutually exclusive, suggesting that the three species are really geographically replacing forms of each other. However, recent collecting has shown that there is some overlap in their distributions at the margins of their respective ranges, indicating that the three forms have differentiated sufficiently to be considered valid species. Additionally, J. R. Ford has shown that there are statistically significant size differences between the two most closely similar forms — the Blue-breasted and Variegated Wrens, the former being the larger.

There are decided differences in habitat between the South-West form, the Red-winged Wren, as compared with the other two. Within its comparatively restricted range the Red-winged Wren is almost entirely confined to the neighbourhood of swampy areas, either in the margins of swamps or in the sword-grass marshy sections of brooks and streams. In fact F. Lawson Whitlock has suggested that Marsh Wren would be an appropriate vernacular name for it. However, the species does forage, and even nest, some little distance from the swampy thickets themselves. It is a bird which, like the ecologically equivalent race of *M. lamberti* in the Sydney district, cannot tolerate too much interference with its habitat by settlement and so withdraws steadily away from the populated areas. Thus it has long since disappeared from the vicinity of Perth, though once it nested at Herdsman's Lake, within the Perth suburban area, the last known specimen having been collected there for the Western Australian Museum in January 1901. In 1907 Whitlock saw it in thickets at Lake Jandakot. It is doubtful whether the species now occurs anywhere in the swamps on the coastal plain between Perth and Bunbury, although it still exists in the swampy margins of the Gingin Brook west of Gingin. It survives in the gullies of the Darling Range, such as at Bickley, the vicinity of the Mundaring and Victoria Reservoirs and Roleystone. It has been observed as far east in the Darling Range as the 53-mile peg on the Perth-Williams road (K. G. Buller). It is fairly plentiful in the lower South-West in swampy localities, such as those near Bridgetown, at Pemberton, along the Warren River and south-east to Warriup, beyond Albany. These southern districts are now its strongholds, and it appears to be the common blue wren at Albany.

The other two forms, *pulcherrimus* and *lamberti,* live in quite different situations, among mallee, mulga, salt-bush and scrubby sandplain country, though in most books and papers *M. pulcherrimus* erroneously has been given the reputation of a rare species of severely circumscribed range. Despite the fact that the Red-winged Wren may be found in the same geographical areas as the Blue-breasted Wren in certain parts of their range (e.g., the Stirling Range and Warriup), the two remain strictly in their sharply different habitats. On the other hand the Blue-breasted and Variegated Wrens may occupy similar habitats in their areas of overlap and hence difficult field identification problems arise. This has happened in recent years when the problem of the identification of the red-shouldered wrens in the coastal strip between Geraldton and Fremantle was under discussion and varying opinions were published in the natural history journals. However, the two species never mingle, or occur close together.

Red-backed Wren

Malurus melanocephalus

Breeding males are *velvety black,* including the tail; the mantle and humeral coverts are a *brilliant red;* wing quills brown. Males in the non-breeding plumage lack the red and the black parts become dark brown. Females similar to the non-breeding males. Iris, dark brown; beak, black in male, brown in female; legs, light brown. Length, 4¼-5 in. Weight, 6-6.5 gm.

Distribution: Northern and north-eastern Australia. In this State it had formerly been regarded as confined to the Kimberley Division, but recently it has been found to extend to Cape Keraudren in the North-West. John Darnell in 1972 found a small population there which appeared to be resident in the coastal dunes. In June 1974 Ron Johnstone collected two specimens there for the W.A. Museum; they were in sub-adult plumage and were from a party of five feeding in *Spinifex longifolius* at the edge of mangroves.

Southern Emu-Wren

Stipiturus malachurus

(Plate 8)

Native names: Djur-jeel-ya, Jar-jil-ya, Jirjil-ya (South West).

This small-bodied bird is in some respects one of the most remarkable of our song-birds in appearance. No other species has such a curious tail; this is considerably longer than the body, being about 4 to 5 inches long and consists of six delicate grey plumes, of emu-feather pattern. The upper parts of the body are streaked grey and black, becoming brighter on the head. The *throat and chest and a line from the beak to above the eye are a light blue,* the remainder of the under parts being bright buff. The ear coverts are streaked grey and white. The female is plainer garbed and lacks any blue colour, being entirely chestnut below. Iris, brown; beak, dark brown; legs, yellow-brown. Length, including tail, 7-7¾ in.

As they fly rapidly from bush to bush, Emu-Wrens bear a strong resemblance to giant insects, with the tail streaming behind the body. The call has a general likeness to that of the blue wrens, but is sharper and more high-pitched.

Distribution: In general the distribution of this species is in the South-West within the area from Shark Bay (Dirk Hartog Island), the Wongan Hills, Lake Grace and on the south-east coast to Israelite Bay. Whether it occurs continuously through this region cannot be stated on present knowledge as the bird is easily overlooked. It is one of those forms which avoids the heavy jarrah country, but it extends all round the south coast and along the west coast plain, where it is found in local colonies, on damp heathy country and on the scrubby vegetation of the coastal dunes. It is quite common on the Bunbury dunes, for instance, and is present at Lake Jilbup (south of Fremantle), south of Mandurah and from the mouth of the Moore River northwards. On the coastal plain it has been recorded at Cookernup, on the wet *Melaleuca* flats west of Coolup and near Maida Vale in the Perth district. In the Darling Range gullies it has been seen at the scarp at Gosnells. Inland it is to be found on the scrubby ironstone hills and on dry sandplain country.

Nesting: The nest is a loosely built oval structure with an entrance at the side; usually made of grasses and lined with the same material. Built close to the ground it is difficult to locate in the thick herbage which is usually chosen by the birds as a building-site. The usual number of eggs laid in Western Australia is 3, and the months in which they may be found in nests are September and October. This applies to the Bunbury district; birds nesting further north, or inland, commence nesting operations earlier. The eggs are white, freckled and spotted with reddish markings, particularly around the larger end. Average measurements, 17 x 12 mm.

For a long period the bird was overlooked in the Perth area though in 1839 John Gilbert found it nesting there. That it had not really become locally extinct was proved when W. H. Loaring, in 1931, found parties of the birds in the bottle-brush flats on the Welshpool Road, two miles from the foothills.

It will be seen that the Western Australian bird has much wider ecological tastes than the same species in the eastern States, where it is strictly confined to sub-swampy areas along the coast and is never found far inland. The inland areas, in the mallee districts of Victoria and South Australia, are occupied by a distinct species, not represented at all in Western Australia, the Mallee Emu-Wren *(S. mallee)* which is more nearly allied to the Rufous-crowned Emu-Wren described below. As might be expected, the Western Australian birds exhibit more geographical variation than is shown in the eastern States. The population inhabiting the wetter south coast region consists of larger birds with coarser black streakings on the head and back (the race *westernensis*). The race occupying the drier, more inland areas *(media)* is smaller and has narrower black streakings; specimens have been identified as far south as Tambellup. On Dirk Hartog Island there exists a still smaller race *(hartogi)* which is more grey in tone and has the black streakings further reduced.

Rufous-crowned Emu-Wren

Stipiturus ruficeps

This species is similar to the preceding, but differs in that the tail is relatively shorter and the plumes coarser. The head is bright rufous-brown, without any streakings, these being confined to the back. The ear coverts, instead of being grey and white, are *blue* and white, the blue being continuous with that of the eyebrow above and the throat below, thus entirely encircling the eye. The female lacks any blue and is entirely buff-white below. Iris, brown; beak, brown; legs, pale brown. Length, 4¾-5 in.

Distribution: This Emu-Wren is found in the mid-north and the interior, north to the Pilbara district and Godfrey's Tank at the northern end of the Canning Stock Route. The present known southern boundary is approximately a line joining North-West Cape, Wiluna and Naretha. Thence the species extends on a broad front into central Australia. It is the most widely distributed of the emu-wrens.

Nesting: A nest found at North-West Cape by T. Carter, in October 1916, was an oval structure with an entrance at the side;

composed of fine bark, grasses and spiders' webs, and lined with feathers and the flowers of shrubs. It was placed a foot from the ground in the middle of a small bush. The 2 eggs were white, well marked all over, particularly at the larger ends, with spots and specks of reddish-brown, with a few others of light umber. The usual clutch size has been found to be 3. Size, 17 x 12 mm.

The species frequents spinifex *(Triodia)* country which is dotted with larger bushes or small trees. It rarely flies unless hard pressed, but runs from one bush or clump of spinifex to another, meanwhile uttering its faint yet rather shrill squeak.

THORNBILLS and THEIR ALLIES: Family Acanthizidæ

Western Warbler (Sleepy Dick)

Gerygone fusca

Native names: War-ryle-bur-dang, Warryl-bardang (Perth).

Upper parts greyish-brown; under surface entirely dull white. There is a small dull white patch at the base of the beak. The tail is dark brown, the outer feathers with a *broad white band towards the base and conspicuously spotted with white at the tips.* Iris, red-brown; beak and legs, black. Length, 3½ in.

The members of this genus are distinguished from the closely allied *Acanthiza* by the absence of frecklings and other markings on the forehead and ear-coverts.

Distribution: This species occurs from the south coast north to the southern boundary of the Kimberley Division, but is limited in its extension inland. It is most plentiful in the forested parts of the South-West and there are no records east of a line from Perenjori, Merredin, Newdegate, the Stirling Range and Two People Bay. In the North-West it has been observed inland to Cue, Wiluna and the upper reaches of the Ashburton and Oakover Rivers. Another population widely separated geographically occurs in the mountain ranges of central Australia, and this extends westwards into Western Australia in the Rawlinson Ranges and the Gibson Desert as far as the Van der Linden Lakes. Birds have been seen by M. Brooker 50 miles N.E. of Rawlinna in July 1970 and May 1972.

Nesting: The Western Warbler makes an oval nest, with an entrance at the side, and it is usually to be found suspended in the outer foliage of a sapling 10 to 25 feet from the ground. It is composed of fine bark and dry grasses closely woven with spiders' webs and the cup for the eggs is warmly lined with feathers. The 2 or 3, very rarely 4, eggs are pinkish-white, speckled, particularly at the larger end, with reddish-brown. Size, 16-18 x 11-12 mm. The bird in the lower South-West is a late breeder, November, December and January being the usual months during which eggs may be found.

The Western Warbler is essentially an inhabitant of the foliage canopies of the tree-tops among which it is in active movement. It is far better known by its song than by its appearance; the notes have a pensive melancholy quality and the bird never seems to finish its song. It is one of the distinctive songs of King's Park and the city gardens in spring and summer. An English ornithologist, Alan Bell, when hearing it for the first time, thus described it: "I have never heard sounds so plaintively microscopic, so clear and yet scarcely perceptible. The ghost of a kitten's mew—the echo of dwarf violins played in the moon—these were the bird's notes." In the Wheatbelt (according to observations by E. H. Sedgwick at Caron, Bilbarin, Nangeenan and Dangin) the song is heard only from May to September and may indicate that the birds leave the area in the summer months.

Perth has the privilege of being the only capital city to "own" this songster—in the east it is an inland species.

Dusky Warbler

Gerygone tenebrosa

Upper parts grey-brown, with a *pale buff eyebrow;* under parts dull white. Tail brown, with a black sub-terminal band, and obscure white spots near the tips of the outer feathers, but relatively inconspicuous compared with the preceding species. Iris, white; beak and legs, black. Length, 4½ in.

This warbler is readily distinguished from the Western Warbler by the presence of the buff eyebrow line and the *virtual absence of white on the tail.* The songs of the two species are somewhat similar, but that of the Dusky Warbler is more lively.

Distribution: A northern bird, recorded down the North-West coast in the mangrove belt, as far south as Bush Bay, 24 miles south of Carnarvon.

Nesting: A nest collected at Port Hedland by F. Lawson Whitlock on October 28, 1914, was dome-shaped and composed of shreds of bark matted together with spiders' webs. It was situated in a mangrove tree, 8 feet above the mud, on the tide-line. It took the birds 16 days to construct. The 2 eggs are white with reddish-brown spots and splashes which are mainly concentrated at the larger ends. Size, 20 x 12 mm.

Broad-tailed Thornbill (Red-tailed Thornbill, formerly Brown Thornbill)

Acanthiza apicalis

(Plate 6)

Native names: Djool-be-djool-bung, Djulbidjulbang (Perth).

Upper parts brownish-olive, *rump brown to reddish-brown,* a black sub-terminal tail band with slight white tipping to the outer tail feathers. The forehead and ear covert feathers are freckled or "scalloped" with light brown crescents. The *throat and breast are strongly streaked black and white.* Iris, reddish brown; beak, black; legs, dark brown. Length, 3¾-4½ in.

The striated breast and tawny rump are diagnostic field-marks, this being the only Western Australian Thornbill with a streaked breast, it having a grey appearance when seen at a distance. The other local thornbills with reddish on the rump *(A. robustirostris* and *A. uropygialis)* have unmarked, white under parts. The distinctive habit of carrying its tail cocked at an angle, like a blue wren, also marks off the Broad-tailed Thornbill from any other.

In our previous editions we combined this species with the eastern States Brown Thornbill *(Acanthiza pusilla)*. It is now considered that though they are very similar they are specifically distinct, They were kept separate by the earlier ornithologists.

Distribution: Generally distributed; north to Shark Bay and Carnarvon, the upper reaches of the Gascoyne and Ashburton Rivers, and to about the Ophthalmia Range, but not found in the country drained by the lower courses of the Ashburton. It occurs on Bald Island, east of Albany.

Nesting: The nest is a compact oval structure, with an entrance in the side near the top, usually made of grasses and a few leaves matted with spiders' webs and cocoons and snugly lined with feathers and plant down. The top of the nest is continued outwards to form a hood over the entrance. It is often within a couple of feet of the ground and a prickly mimosa is often selected. In one instance, in the Bridgetown district, the birds had lined their nest with feathers from a dead Owlet-Nightjar which was lying close by. The nest measures 4¼ inches in height and is about 3¼ inches in diameter. The entrance hole is about an inch in diameter. The eggs, usually 3 in number, are white, spotted and blotched, mostly at the larger end, and often to the extent of forming a cap, with reddish brown. Size, 17 x 13 mm. Incubation period, 18-19 days; nestling period, 17 days (W. H. Loaring).

This is a tree-haunting thornbill and not a ground-feeder, though it may forage over quite low bushes and saplings. Unlike most other thornbills it does not tend to associate in flocks in the non-breeding season, but remains in pairs. It is a pleasing songster and is an accomplished mimic. It has a more penetrating call than the Brown Thornbill of south-eastern Australia and persons who are hard of hearing can hear the Broad-tailed Thornbill when they cannot the Brown Thornbill.

Plate 6.—Thornbills (1).
Slate-backed Thornbill
Broad-tailed Thornbill
Chestnut-rumped Thornbill

Slate-backed Thornbill (Robust-billed Thornbill)

Acanthiza robustirostris

(Plate 6)

Back grey with a *slaty or bluish tone,* crown streaked grey and black, *Rump chestnut;* under parts plain greyish-white, lighter on the abdomen. *Iris, reddish-brown;* beak and legs, black. Length, 3¼-4¼ in.

It can be readily distinguished in the field from the Chestnut-tailed Thornbill by its blue-grey back, streaked crown and dark eye.

Distribution: A central Australian species found south and west as far as Broad Arrow, Yalgoo and Wurarga, and north to the Peak Hill district and the Ophthalmia Range.

Nesting: A favourite situation for the nest is a medium-sized, or even small, narrow-leaved mulga, growing in company with two or three similar bushes, but often quite isolated (Whitlock). The nest itself is globular, with an entrance at the side, the upper portion more fragile than the lower, and the whole made of dry grasses held together with spiders' webs and lichen. The lower portion, which contains the 3 eggs—the usual number laid—is lined with hair-like material. It is usually situated between 4 and 6 feet from the ground, but nests have been observed as high as 15 feet. The bird rears two broods in favourable seasons—from July to October. The eggs are white, with a very faint tinge of pink, and are speckled with minute markings of reddish-brown, which tend to form a cap over the larger end. Size, 17 x 12 mm.

The song is a high-pitched twitter, the alarm call being a harsh "thrip-thrip."

F. Lawson Whitlock, the discoverer of the species, found it to be associated with a species of poverty-bush *(Eremophila pantoni),* which is always preferred as a nesting site. It is a common bird in the desert country from Cosmo Newbery to the Warburton Ranges.

Chestnut-rumped Thornbill

Acanthiza uropygialis

(Plate 6)

Upper parts greyish-brown, *rump a rich chestnut-brown,* much brighter than in the Broad-tailed Thornbill; tail, black slightly tipped with white on the outer feathers. *Under parts, white.* Forehead and ear coverts freckled with black and whitish, with a trace of buff on the forehead. *Iris, white;* beak and legs, black. Length, 3½-4 in.

Differs from the other chestnut or tawny-rumped Thornbills by its unmarked under parts (from the Broad-tailed Thornbill) and grey-brown instead of grey-blue back (from the Slate-backed Thornbill). From the latter it can also be distinguished by its white eye and the presence of buff on the freckled, not streaked, forehead.

Distribution: Generally distributed north to the Fortescue River, but, until recently, not found in the South-West corner, south-west of a line joining Moora, Bolgart, Meenar, Dangin, Corrigin, Kulin, Lake Grace, Lake Magenta and Esperance. The species may be extending its range into the South-West. A. L. Milhinch has recorded occasional birds at Seabrook (four miles S.E. of Northam) since 1955. Dr. G. F. Mees saw birds 10 miles east of Brookton in 1960, between Beverley and Brookton in 1961, and at Marley Pool, east of York, in 1963.

Nesting: This Thornbill is an exception in its nesting habits to other local members of the genus (excepting the Western Thornbill) in that it builds its nest inside a dead hollow branch often with the entrance through a knot hole. On occasions it will build in a hollow in a fence post. A dome-shaped nest of grasses is, however, constructed in the hollow and it is profusely lined with fur, feathers or other soft material. The clutch comprises 2 to 4 eggs, which are white spotted with reddish-brown, mostly at the larger end. Average size of 11 eggs, 16.0 x 12.0 mm., varying from 15.3-16.8 x 11.3-12.5 mm. Two broods are usually raised in a season.

Western Thornbill

Acanthiza inornata

(Plate 7)

Native name: Djo-bul-djo-bul (Perth).

This is the plainest coloured bird of the genus and has no special distinguishing markings. The back is greyish or brownish-olive and the tail has a black sub-terminal band. The under parts are bright buff. On the forehead and behind the eye are brown frecklings. Iris, light coloured, whitish or cream; beak and legs black. Length, 4 in.

Distribution: Confined to the South-West heavy forest belt, north to the mouths of Cockleshell Gully and Hill River, and inland to Moora and Mogumber, the Great Southern Railway and the Stirling and Porongorups Ranges. It is plentiful on the Swan coastal plain and the Perth area, being a permanent resident of Kings Park.

Nesting: The nest is an oval structure with an entrance hole at the side. More loosely put together than is the nest of the Broad-tailed Thornbill, it is made of grasses and lined with feathers. The nest is usually placed in the centre of a bushy tree against the trunk or in a hollow limb with the entrance through a knothole, or stuffed shapelessly behind a piece of loose bark. It is often placed in the dry

Plate 7.—Thornbills (2).
Yellow-rumped Thornbill
Samphire Thornbill
Western Thornbill

"beard" or in the crown of the blackboy *(Xanthorrhoea preissii),* particularly in branched specimens. The clutch is of 3 eggs, which are pinkish-white, spotted with reddish and purplish-brown, the spotting usually tending to form a zone round the larger end. Size, 6-17 x 12-13 mm. According to observations by W. H. Loaring the male does not assist with the nest-building or in the incubation of the eggs, but takes a full share in feeding the young. The eggs are laid at two-day intervals and the incubation period is 20-21 days.

A frequenter of various types of woodland habitats, this is one of the thornbills which usually keeps to the foliage, though it may descend to quite low levels such as saplings and the tops of fallen trees and bushes. It also forages on the ground. It is quiet and unobtrusive and may be easily overlooked, but can be readily "called up" to within a foot or so of the observer. Mimicry of the song of other birds is practised to some extent and among the calls imitated are those of the Western Rosella, the Grey Fantail and the Squeaker. In the non-breeding season the birds gather into small flocks and move about in the forest.

Birds from the extreme south coast have a richer plumage than those from further north, with brownish-olive upper parts and buff under parts. These have been described as a separate race, *A. i. mastersi.* The more northern birds *(A. i. inornata)* are greyish-olive above and more whitish below. The differences are most marked in the newly-acquired plumage after the annual moult, which occurs in February-March. As wear and fraying progress the worn plumage of *mastersi* comes to resemble that of the northern population.

This species is not found in the eastern States, its nearest relative there being the Buff-rumped Thornbill *(A. reguloides).*

Samphire Thornbill (Slender-billed Thornbill)

Acanthiza iredalei

(Plate 7)

Upper parts brownish-olive, *rump buff-yellow,* tail black with grey tip. The under parts are buff with grey clouding on the breast. The forehead and ear coverts are freckled with light brown. Iris, creamy-white; beak, brown; legs, black. Length, 4¼ in.

In the field, from its tail pattern, this thornbill looks like a bleached form of the Yellow-rumped Thornbill, but not only is the rump much more drab, this species differs in the absence of the spotted forehead markings.

Distribution: A characteristic inhabitant of the samphire margins of the inland salt lakes, but it also occurs in mangroves along the coast (as at Carnarvon and the Wooramel River) and in the scrubby vegetation of the Nullarbor Plain. It is not found south-west of a line joining Shark Bay, Yalgoo and Coolgardie, nor has it been reported north of Carnarvon and Wiluna. It occurs at Lake Throssell, north-east of Laverton.

Nesting: A nest collected at Lake Austin was dome-shaped, with an entrance at the side. It was loosely constructed of bark of a stringy nature and portion of a soft plant, which were woven together with spiders' webs—the whole measuring 5 x 3 inches. The egg-cavity was lined with the whitish vegetable-down of *Dicrastyles fulva,* but the birds also use rabbit fur. The nest is mostly found on the top of a clump of samphire. The eggs, 3 in number, are of the usual thornbill type—white, and speckled with small spots of reddish-brown, particularly at the larger end. Size, 14 x 10 mm.

The Samphire Thornbill was first found to occur in Western Australia in 1903, when F. Lawson Whitlock collected specimens at Lake Austin. The call notes resemble the syllables "tsip-tsip-tsip."

Yellow-rumped Thornbill (Tom-tit, Chigaree)

Acanthiza chrysorrhoa

(Plate 7)

Native names: Jee-da, Jida (Perth); Jit-burt (Avon River).

Upper parts brownish-olive with greyish-black head, and *a bright yellow rump,* the tail with a broad black terminal band and the feathers very slightly tipped with white. The under parts are dull white, with buff along the flanks. The blackish feathers of the forehead are *boldly spotted with white;* there is a white line over the eye and the ear coverts are freckled in the usual thornbill manner. Iris, light greyish; beak and legs, black. Length, 4¾ in.

Distribution: Generally distributed north to the Murchison River, but more inland it extends further north, reaching the upper Ashburton River and the Tropic of Capricorn. Eastwards it ranges into central Australia.

Nesting: This thornbill makes a nest unlike that of any other bird. It is usually built in the drooping foliage of a sapling, from six to twelve feet from the ground, and at times may be built into the bulky structure of the nests of larger birds, especially the raptores. It is an untidy oval with an entrance low down on one side and so constructed that it is covered with a small hood, making it difficult, at first glance, to see how the bird enters the nest. The nesting chamber inside is well lined with feathers or plant-down. The peculiarity of the nest is, however, the construction on the top of one or more additional open, cup-shaped nests which do not appear to be used by the bird and whose function has aroused controversy.

One theory is that the extra nest is used by the male bird at night while the female is incubating the eggs, but there is no evidence in support. A further theory is that it is intended to delude the Bronze Cuckoos, which parasitise the nests of this species, into placing their eggs in the dummy nest. Again, however, there is no evidence to support this belief. The most plausible explanation, put forward by K. A. Hindwood of Sydney, is that it is due to the persistence of the male bird's nest-building impulses long after the essential nest is completed. The male, though sharing in the building of the nest and later in caring for the young, takes no part in the incubation of the eggs. This view is supported by the intensive observations of K. Immelmann who never found thornbills occupying these upper chambers at night, though birds of another species (the Zebra Finch) occasionally did so. The clutch size varies from 2 to 5, with 3 most frequent, and occasionally 4. They are white in colour, but sometimes have a few faint spots of reddish-brown, especially over the larger end. They vary considerably in shape, and a series of 13 from the South-West averages 17.8 x 12.5 mm., varying from 16.0-19.6 x 12.0-13.1 mm. The bird is one of the earliest to nest in the lower South-West, eggs being found in June. Incubation lasts 18-20 days and the nestling period 17-19 days. Two broods are usually reared in the season, but up to four (each successfully fledged) are known. The young birds of the first brood assist in feeding the progeny of later clutches.

This is one of the commonest small birds of the South-Western bush, preferring open or lightly wooded country to dense forest. When nesting is over, neighbouring families unite into foraging flocks which sometimes invade Perth suburban streets. Birds of other species often form mixed flocks with them. By the end of July most pairs have separated out and selected nesting territories.

Spotted Scrub-Wren

Sericornis maculatus

Native names: Goor-gal (Avon River); Girgal (South-West); Pin-ki-mum (King George's Sound).

A lively but secretive bird whose characteristics are rather difficult to pick out as it skulks in the thickets. The upper parts are brownish-grey with a black sub-terminal band across the tail. The under parts are white, *strongly streaked with black* on the throat and breast and the flanks brown. The wing coverts have *a prominent black and white mark*. There is a *white line above the eye* to the base of the beak and the lores below this are black. The beak is rather long for the size of the birds in this family. The female is less strongly marked than the male, having brown lores instead of black, less pronounced streaks on the under parts and a fainter tail band. Iris, white with yellow tinge; beak, dark brown; legs, brown. Length, 4½-5¼ in.

The birds from the humid South-West corner are pale yellow on the middle of the abdomen but this colour is lacking in birds from the Abrolhos, the Recherche and the northern, inland and south-eastern parts of the range (E. Mayr and R. Wolk).

Fig. 50.—Spotted Scrub-Wren.

Distribution: A resident of the South-West inside a line joining Shark Bay, Mullewa, Perenjpori, Bunketch (40 miles north-east of Wongan Hills), near Kweda (between Pingelly and Corrigin), Newdegate, Lake Magenta, and along the south coast strip to Eucla. It is found on some of the adjoining islands including Bernier Island, Dirk Hartog Island, the Abrolhos (East and West Wallaby Islands), Rottnest Island, Bald Island, and the Archipelago of the Recherche (Mondrain, Middle and Salisbury Islands). The northernmost limit on the mainland is Callagiddy Station, Carnarvon.

Nesting: The Spotted Scrub-Wren builds a dome-shaped nest of about 6 inches deep and 4 inches wide of grass, leaves, bark, etc., and lines the bottom with feathers. There is a circular entrance in the side. The nest is situated close to the ground and no particular situation can be indicated. It may be in a low bush, or in a collection of fallen branches, or in the leaves of an overhanging branch of a tree. The 3 (sometimes 2) eggs are white and spotted, usually at the larger end only, with purplish-brown and lavender. Average measurements 20 x 15 mm.

Near Perth this species is most plentiful in the thickets on the coastal dunes. Further inland the bird may be found in the dense undergrowth lining paper-bark swamps and rivers, but the distribution does not seem to be continuous. Along extensive stretches on the south coast it may be the only small bird—other than the ubiquitous Silvereye—occurring in the dune thickets.

Although it is a secretive bird it can readily be "called up" from its leafy fastnesses. It emits a rapid "tsi-tsi-tsi" call, a loud and rapid succession of notes which may be repeated up to 20 times. A variation of this note is "tseer-tseer-tseer," also frequently repeated. The warning call appears to be "tsi" or "tsa", uttered as a single note.

Redthroat

Pyrrholœmus brunneus

Native name: Ber-rit-ber-rit (Avon River).

General coloration grey; *a conspicuous chestnut throat patch* in the male; lighter grey breast. Forehead freckled with whitish crescents, lores and eyebrow line dull white. The blackish *outer tail feathers tipped white,* which are conspicuous when the bird fans out its tail as it breaks cover when flying from shrub to shrub. The female differs in the absence of the rufous chin and closely resembles the Dusky Warbler in appearance but may be distinguished by the much more extensive white lores. Iris, chestnut-brown; beak, brown; legs, brown. Length, 4-5 in.

Distribution: An inland and mid-west species, north to the Fortescue River and south and west to Moora, Cunderdin, Kweda, Lake Grace, Lake King, Salmon Gums and Coragina Rocks (40 miles south of Balladonia) and along the coastal vegetation to Eucla. Generally it inhabits the country north and east of the range of the Spotted Scrub-Wren.

Nesting: The Redthroat builds a globular nest of strips of bark and grass, with an entrance at the side. The egg cavity is lined usually with feathers. It is placed in a bush, such as tea-tree or saltbush, a few feet from the ground, but F. Lawson Whitlock found one nest in the burrow of a lizard. The 3 (sometimes 2, rarely 4) eggs are wholly brown in colour, darker at the larger ends. A series of 10 eggs averages 19.2 x 14.4 mm., varying from 18.4-20.4 x 13.4-15.6 mm. The Black-eared Cuckoo makes use of the nest of this species.

The Redthroat is an accomplished mimic and has a varied repertoire. Some of its songs have been rendered by E. H. Sedgwick as follows: "weet-weet, winit, winit wer-re"; "cirra-re-chee-chee"; "whrr-whrr-whrr-whrr, cheer-cheer"; "wir-peter-peter, wrr, eddle-d'roo"; "wee-er, wee-er, wee-er, wee-er" (second two notes faster than the first two); "chit-chit-chit, wee-er, wee-er, wee-er."

It makes a peculiar fluttering noise with its wings when flying.

Weebill

Smicrornis brevirostris

Native names: Geah-ter-but, Gi-aterbat (Avon River).

Back bright olive, yellowish-olive on the rump and the head grey-brown. The tail is brown with a broad black sub-terminal band, the outer feathers with a white spot near the tip. Breast grey and the *rest of the under parts pale yellow.* Iris, pale-yellowish-white; the short beak, brown; legs, brown. Length, 3-3¾ in.

Distribution: Fairly generally distributed throughout the State. In the south-west, though it occurs plentifully in the Perth district and the coastal plain south to Busselton, it is rare in the lower Darling Range jarrah forest area and is absent south of a line joining Cape Naturaliste, Granite Peak and Albany.

Nesting: The Weebill's nest is a small rounded purse-like structure, with a hooded entrance at one side, near the top. The birds first construct a cup-shaped nest and then add the superstructure which makes the whole into a sphere. It is constructed of fine grass and spiders' web and decorated on the outside with cocoons. Usually suspended in a drooping branch at heights from 4 feet from the ground upwards, and is usually also well hidden in foliage. Mostly 3 eggs are laid, sometimes 2, which are cream in colour, with a zone of faint brownish spots around the larger end. Size, 15-17 x 10-12 mm. Weight, 1 gm. August to October are the main nesting months in the south.

The distinctive song has been syllabised by the words "Winnie Wieldt" and "pee-pee p'wee-weep."

Shy Ground-Wren

Hylacola cauta

Upper plumage a uniform dark brown, with a rufous wash on the lower back, deep chestnut on the rump and the tail tipped with white on the outer feathers. Under parts white with the throat and breast heavily streaked with black. White line above the eye. Iris, yellowish-brown; beak, blackish-brown; legs, fleshy-brown. Length, 4¼-4¾ in.

The Ground-Wren can be recognised from the Field-Wren, which occupies somewhat similar situations, by the *absence of a streaked pattern* on its *darker back,* its *rich rufous-red rump* and *no yellow wash below.*

Distribution: A southern species found north to the mouth of the Murchison River, Perenjori, Bunketch (40 miles north-east of Wongan Hills) and Kalgoorlie, but excluding the forested South-

West corner (west of a line joining Dongara, Wongan Hills, Dumbleyung and the Stirling Range). It is a denizen of the mallee and sandplain country.

Nesting: The Shy Ground-Wren builds a domed nest of bark, lined with feathers, and placed on the ground at the base of a bush or under fallen timber. A typical nest measures about 6 inches high and 4 inches broad. The 3 eggs are dark grey, spotted brown, particularly at the larger end, where some examples may be solid jet black. They measure 20-21 x 14-15 mm.

These birds hop very nimbly from tussock to tussock and always contrive to have stems and branches between the observer and themselves. As they run the tail is held erect. The typical melodious call of the Ground-Wren is rendered by the words "chee-chee-chick-a-dee."

Field Wren (Sandplain Wren)

Calamanthus fuliginosus

Native name: Nee-an-tee (North-West Cape)

Upper parts varying from olive-brown (in more southern birds) to rufous-brown (in the north-west) *and coarsely marked with dark brown, longitudinal streaks* (narrower in the north); crown rufous-brown with dark streaks; a white eyebrow; tail chestnut-brown to rufous, with a black sub-terminal band and outer feathers tipped white. Under parts whitish, *strongly streaked dark brown* (in southern birds the breast and abdomen are yellowish-buff). Iris, straw-yellow; beak and legs, brown. Length, 4-5½ in.

In our earlier editions we treated this bird as two separate species — the Striated Field-Wren *(Calamanthus fuliginosus (= montanellus))* of the southern part of the State, and the Rufous Field-Wren *(C. campestris)*, of the north. However, the various populations are now regarded as geographically varying components of one species. H. T. Condon, of the South Australian Museum, has stated (1951) that "examination of a good series of skins from Tasmania, Victoria, South Australia and Western Australia has shown that there is a perfect gradation in size, plumage pattern and plumage coloration from east to west and north to south, with intermediates between the various forms occurring at several places".

Distribution: The northern boundary of the species' range extends from North-West Cape to Wiluna and eastwards to Central Australia. The species avoids the heavy forested south-west corner and it is not found south-west of a line running from Lancelin to Bolgart, Wongan Hills and south to Kellerberrin, Kweda, Kojonup, Cranbrook, Mt. Manypeaks and Two People Bay. In May, 1973, M. K.

Tarburton saw a bird in the coastal scrub at Jimmy Newells Harbour near Albany. It occurs on the Shark Bay islands (Dirk Hartog and Dorre).

Nesting: The Field-Wren builds its nest on the open sandplain often in a slight hollow. It is globular, with an entry at the side, and measures 4 to 5 inches in diameter. Made of grasses, leaves and flower-stalks, and lined with feathers. The eggs are usually three in number, light brown in colour, clouded with markings of a darker shade. Measurements, 19-22 x 15-18 mm.

In the southern parts of the State the motorist, driving in sandplain country, will frequently flush Field-Wrens from the sides of the road and some, usually immature birds, may fail to avoid a collision. On a 36-mile stretch of sandplain, on a motor run between Ravensthorpe and Jerramungup in November 1945, one of us counted no fewer than 29 Field-Wrens flushed from the roadside and the strip of scrub along the centre of the road. From observations on a farm being developed from virgin bush at Gibb Rock, E. H. Sedgwick found that this species will persist until the last vestige of cover remains. Shelter belts provide an adequate habitat.

These birds appear to maintain territories throughout the year and the song is uttered from the top of a bush. It may be described as a loud, clear "whirr-whirr-chick-chick-whirr-ree-ree", the final sequence being a mellow liquid sound.

Southern Whiteface

Aphelocephala leucopsis

Native name: Beetunkara (Warburton Range).

Back brown, with slightly rufous rump; tail black, tipped white on outer feathers. *Forehead, white;* under parts, white; *chestnut flanks;* no black breast band. Iris, white; beak and legs, black. Length, 4-5 in.

The white forehead and chestnut flanks are conspicuous field marks of the western population (the race *castaneiventris*). The central Australian race *(whitei)* extends to the Rawlinson and Warburton Ranges and south-westwards into the Victoria Desert and the Nullarbor Plain, at least as far as Naretha. The flanks in this race are brownish-buff instead of rufous-chestnut.

Distribution: An inhabitant of the mid-north and inland areas, north to the Tropic of Capricorn, and excluding the south-west (from Mullewa, Kellerberrin, Widgiemooltha and Cocklebiddy). This avoidance of the south-west is in contrast to the distribution of the species in the eastern States, where it ranges far into the higher rainfall area, being regularly present to the Great Dividing Range and sometimes appearing close to the coast, as near Sydney.

Nesting: The domed nest, of small sticks or rootlets, well-lined with feathers is usually built in a hollow branch or tree trunk, entrance being gained through a knot-hole or crack. The bird, however, accommodates itself to any nesting-site. It is partial to using the nests of the White-browed Babbler, and F. Lawson Whitlock has found the nest in a variety of other situations—in a half-opened fruit tin, recumbent logs, holes in dead timber and, in one instance, in the roof of a tool-shed. The eggs—usually 3 or 4, but up to 6 in the clutch—are pinkish-brown and purplish-grey. The larger clutches are laid in good seasons. Sometimes the markings form a zone around one end. Size, 16 x 12 mm.

This small bird may be met with in parties in lightly wooded country. In those areas in which it overlaps the range of the Banded Whiteface, it favours mulga thickets rather than the open plains in which the other species is found. The song is composed of two or three plaintive but musical notes, a repeated "wit".

In addition to insects this species feeds extensively on small seeds, unhusked.

Banded Whiteface

Aphelocephala nigricincta

Upper parts chestnut; head brown with *white forehead;* tail black, tipped white; flight quills brown; under parts white with a *narrow black band across the breast;* flanks, chestnut. Iris, white; beak and legs, black. Length, 4½-5 in.

Distribution: A central Australian species extending across the Gibson Desert west to about the junction of the Gascoyne and Lyons Rivers (sometimes as far west as Callagidey, near Carnarvon), north to Jigalong and Godfrey's Tank (on the Canning Stock Route) and south to the Murchison district (recorded at Cue). In the south-east a specimen has been collected in the Great Victoria Desert, 120 miles north of Neale Junction (M. Brooker).

Nesting: The nesting situation is entirely different from that of the preceding species. The nest is a somewhat similar domed structure but is never built within a hollow, but always in an open bush. It is constructed of black sticks, has a long spout-like entrance and is amply lined with feathers (emu, cockatoo, etc.). The 3 or 4 eggs are not distinguishable from those of the preceding species, but are a little larger. A series of 13 eggs from Mulgul and Milgun Stations averages 17.6 x 12.8 mm., varying from 16.0-18.9 x 12.2-13.7 mm.

This Whiteface frequents bare clay flats or stony plains, covered with odd mulga, curara *(Acacia tetragonophylla)* or wax-bush *(Eremophila fraseri)*. However, sometimes the two species of whiteface may be found together. On the ground they hop, both sedately and rapidly. Unhusked seeds form a large part of the food, as well as insects, and a strong gizzard is developed in the alimentary canal.

This species was first described from specimens collected near Hermannsburg mission in central Australia by the Horn Expedition in 1894, and was not thought to extend into Western Australia until Otto Lipfert collected specimens along the Canning Stock Route in 1930-31. Subsequently I. C. Carnaby observed it much further to the westward and we are indebted to him for the nesting notes.

Striated Grass-Wren (Rufous Grass-Wren)

Amytornis striatus

Native name: Nyandi (Lyons River).

Upper parts bright chestnut, each feather streaked longitudinally with white. *Chin white,* rest of under parts buff-white. The flanks chestnut in female. Tail brown. There is a *black facial streak,* from the base of the beak to behind the eye, and the ear coverts are streaked black and white. Iris, beak and legs, brown. Length, 6-7 in.

Distribution: A mid-northern species ranging north to the Pilbara district and Godfrey's Tank near the north end of the Canning Stock Route, and south to the east Murchison country (recorded between Sandstone and Wiluna, and Kathleen Valley). Eastwards it continues across the Gibson Desert into central and eastern Australia. One specimen was collected in the Great Victoria Desert, 80 miles north of Neale Junction (M. Brooker).

Nesting: The Rufous Grass-Wren makes a globular nest of fine grasses and stems of spinifex *(Triodia)* and builds it in the top of a bunch of the spinifex. The entrance is at the side, low down and the egg cavity is lined with vegetable down. The clutch comprises 3 eggs, which are white, finely flecked all over with small spots of pale red, umber and grey. Size, 20 x 15 mm.

The male has a rather pleasing song, and on calm, sunny days he will perch at a height of 3 or 4 feet and pour it forth. He may keep up the performance for ten minutes at a time if undisturbed, and in this respect he resembles a Field-Wren. The common call note is a clear, liquid and musical "tu-tu-tu" and this is frequently responded to by other males (and possibly his own female) within earshot. These notes are incorporated in the song and are followed by other more rippling notes, which again are followed by further sounds difficult to describe on paper. The whole effect is distinctly pleasing (Whitlock).

Fig. 51.—Striated Grass-Wren.

Fig. 52.—Western Grass-Wren.

Western Grass-Wren (Thick-billed Grass-Wren, Rock Wren)

Amytornis textilis

Native names: Nyern-de and Jee-ra (Avon River); Tcher-ree (Broomehill).

Upper parts brown, each feather streaked longitudinally with white, with some black streakings on the head also. The *sides of the head, throat and breast are similarly streaked,* but there is no black facial streak, as in the preceding species. The abdomen is buff-white. The female differs from the male in having rufous flanks and axillaries. The tail is long and often carried erect, as in the blue wrens. Iris, beak and legs, brown. Length, 6½-7½ in.

This bird is distinguished from the preceding species by its larger size, the presence of streakings on the throat and breast, absence of rufous on the back, and no black face stripe.

Distribution: This Grass-Wren has been observed over a wide area, but generally rather south of that occupied by the previous species. It has been recorded north to Point Cloates, Wiluna and the Rawlinson Range and south to the south coast but absent from the south-west corner (west of the Great Southern Railway).

In recent years it has not been reported south of Shark Bay and Kalgoorlie. The ornithologist Tom Carter collected specimens in marlock country at Broomehill up to 1910, and a grass-wren probably of this species was noted near Esperance in 1921. In the Murchison country F. Lawson Whitlock found it confined to the salt-bush. At

Shark Bay it occurs in *Acacia* thickets. It is still very plentiful in the Rawlinson Range east of the Gibson Desert, where it leads an agile existence among the crags, a veritable rock-wren. Tourists visiting Standley Chasm, near Alice Springs, can get close views of it.

Some modern ornithologists separate out the birds from the Rawlinson Range and other central Australian ranges as belonging to a distinct species, the Dusky or Thin-billed Grass-Wren *(Amytornis purnelli)*, characterised by the beak being more pointed, longer and slenderer.

Nesting: It builds a globular nest of grasses, but the "roof" is not as dense as the remainder of the nest, and is more of a "hood". The grasses are rather loosely put together in the construction, and no lining, other than finer grasses, is used. It is usually built in a low, thick bush within a couple of feet of the ground. A prickly bush is often selected. The 2 (sometimes 3) eggs are rounded oval and are white, or faintly reddish-white, marked with spots and blotches of red or purplish-red and a few of lilac-grey which appear as if below the surface of the shell. The markings tend to form a zone or cap, at the larger end. Size, 23 x 17 mm.

The bird hops very rapidly over rocks and stones, with the tail held up nearly vertical most of the time. When agitated it moves with drooping wings, head down and the tail held out far behind. The earlier observers (e.g. Whitlock) questioned its vocal powers and doubted whether it had a song. However, at the Rawlinson Range, Derek Goodwin, of the British Museum, heard males utter short song phrases incorporating one or more calls with a reeling or sibilant trilling character, suggestive of a blue wren. The female often replied with a sibilant "sree" like the reeling trill of a blue wren but more sibilant and "shivery'. The alarm note is a still small voice, mouse-like in quality. Sometimes it is a loud "tsip" or a long drawn-out and high-pitched "tseet".

The bird has a hard grinding gizzard and besides insects it feeds considerably on seeds, which are unhusked. Quite large seeds including mulga, may be eaten.

Western Bristle-bird

Dasyornis longirostris

Native names: Djyr-dal-ya, Djardal-ya (Perth); Booteritch (King George's Sound).

Upper parts bright brown, head duller, brown spotted with grey. Under parts grey, flanks and abdomen brown. The wings are short and rounded and the *tail long, with graduated feathers*. Iris, orange-red; beak, dark brown; legs olive-grey. Length, 6¾ in.

Distribution: One of our rarest birds. It formerly occurred around the South West coastal belt from Perth to the east of Albany, but it is now extinct over much of this range. It survives in some numbers east of Albany (Mt. Gardner, Two People Bay, Waychinicup River and the Fitzgerald River National Park).

It was first discovered in Western Australia in 1839 when John Gilbert collected specimens in the vicinity of Perth, but has never been observed there since. Some thirty years after Gilbert's original discovery George Masters in 1868 collected ten specimens at King George's Sound and also a nest and two eggs (September). Further specimens were obtained by William Webb in the same area in the 1880's. Thereafter there were no reports until 1907 when F. Lawson Whitlock located several individuals near Wilson's Inlet. The birds continued to be observed there until 1914 when a bushfire apparently wiped out the little colony. Nothing more was heard of the species until February 1945 when K. G. Buller collected a specimen at Two People Bay, east of Albany. The species is still plentiful there, and is one of the several ornithological attractions of the fauna reserve created recently for the Noisy Scrub-bird.

The Western Bristle-bird is closely related to the larger and paler Eastern Bristle-bird, *Dasyornis brachypterus,* with which it was combined in our earlier editions.

Nesting: The nest is formed of dry wiry grass, without any lining, more globular than those made by blue wrens but like them in having a side opening. It is built concealed in clumps of coarse grass within a foot or so of the ground, in such situations as dwarf banksias. The dimensions of one nest found near Albany were 4 inches in height, 6 inches deep and 5 inches broad. The 2 eggs are dull brownish-white, blotched and freckled with purplish-brown, some of the blotches, particularly at the larger end where they are most numerous, being sub-surface markings. Size, 26 x 19 mm. As with the Rufous Bristle-bird, it is possible that often one of the two eggs laid by this bird is infertile. At any rate, in the case of a nest found by Whitlock in 1907, an infertile egg was found in what was, presumably, an old nest. Territories range in size from 2½-7 acres (Graeme Smith).

Gilbert found the bird to be remarkably shy and he had great difficulty in obtaining even a glimpse of it. Its actions were extremely quick and the bird ran over the ground with its tail erect, sometimes ascending to the top of a bush to sing. Buller also observed that the bird he collected carried its tail erect and at intervals fanned it out. All its movements were very jerky. However Graeme Smith, who studied the bird recently in the Two People Bay reserve, found that the tail is normally held horizontal when the bird is feeding and is elevated only when it is disturbed. Usually it is very difficult to see in the thick vegetation. Most of the singing is done from the ground and only rarely will a bird sing from the top of a bush and then only the head is visible. It is very active when feeding, moving quickly over the ground. At sunrise and towards sunset it occasionally feeds on open tracks at which times it may be approached quite closely and followed for some distance. The flight is weak, usually just skimming the tops of the vegetation, and rarely for more than ten yards.

When consigning to England the nest and eggs he had collected, Gilbert labelled them as those of the "Reed-wren", thus suggesting the habitat in which the species appears to have occurred in the Perth area. On the other hand the country in which Buller found the species frequenting was a large belt of low, heathy scrub, where the thick undergrowth was covered with a tangle of wire-like creepers. The country in which Whitlock found the species consisted of blackboy *(Xanthorrhoea)* flats, which in winter and spring are sloppy, extending to shallow pools after every fall of rain. These flats were interspersed with gravel ridges which are clothed with a dense growth of flowering vegetation reaching to the knees. Graeme Smith's habitat at Two People Bay was dense heath, three feet high, with abundant sword rush *(Anarthria scabra* and *A. prolifera)* among the shrubs. After bushfires the heath takes about eight years to become suitable habitat again for the bristle-birds.

Gilbert described the notes uttered by the bird as "loud, clear, and extremely varied." Whitlock syllabised the song of the male as a quick "chip-pee-tee-peetle-pet," usually uttered in the early morning or towards sunset. He thought the female emitted only a metallic "tink." Graeme Smith's detailed studies indicate that Whitlock's "chip-pee" sequence is apparently the territorial song of the male; it is a variable song in which the number of notes and, less frequently, the order of the notes, may be changed. Song syllables are short, varying from two to thirteen and averaging six. This song is responded to by another bird, presumably the female, and may be syllabised as "quick more beer", a sharp, three-noted whistle, starting after and ending before the close of the male's song. These songs are given throughout the year, usually in the early morning and late afternoon. Both male and female have two alarm calls, a loud "zee-eek", usually given when the bird is flushed and a soft "zit" which is often given when the bird becomes aware of an observer but not disturbed enough to move away. It may also function as a short distance contact call.

Rufous Bristle-bird

Dasyornis broadbenti

Back greyish-brown, but lacking any grey on the wings and tail, the rump chestnut-brown, and the head a bright rufous brown. The throat and breast have a speckled appearance due to each feather having a black centre and a greyish-white edging. Belly greyish-brown. The wings are short and rounded, and the tail long and graduated. Iris, brown; beak, brown, lower mandible paler; legs, brown. Length, 10½ in.

Distribution: This rare species has an extraordinarily restricted distribution in Western Australia being confined to a narrow strip of coastal country from Cape Naturaliste to Cape Leeuwin. In the eastern States it is much commoner and occurs in the coastal dune thickets from the mouth of the Murray River in South Australia to Marlo in eastern Victoria.

Nesting: There are no records available of nests found in this State. Captain S. A. White has stated that in South Australia the species is a late breeder, often not nesting till late in December. The nest is a large oval structure of bent and twisted blades of sword-grass, or rootlets, with a side entrance and neatly lined with fine grasses. It measures about 7 inches high by 4 inches in breadth. A. G. Campbell has recorded that it is remarkable that one of the 2 eggs, which is the normal number laid, is as a rule infertile. Size, 27 x 22 mm. The female alone appears to build the nest and is the only one of the pair that incubates, but the male brings her food and also assists in the feeding of the young. The eggs are dull pinkish-white, well marked all over with spots and dots of reddish-brown, purplish-grey, and umber, especially at the larger end.

The species was first discovered in Western Australia by A. W. Milligan in October 1901, at Ellensbrook. C. P. Conigrave found it at Cape Naturaliste in July 1904 and again in 1908, and the light-house keeper there, P. Baird, collected other specimens in December 1906. Nothing more was heard of it until L. Glauert, of the Western Australian Museum, saw a bristle-bird in the coast thickets at The Fishing Place, west of the Mammoth Cave, in March, 1940, and which, from the location and habitat, was probably an individual of this species.

OLD WORLD FLYCATCHERS: Family Muscicapidæ

Brown Flycatcher (Jacky Winter)

Micrœca leucophæa

Upper parts *grey-brown,* wing quills and tail dark brown, *outer tail feathers white.* Under parts dull white and some grey clouding on the breast. Iris, brown; beak, dark brown; legs, black. Length, 4¾-5½ in.

Distribution: Found mainly south of the mulga-eucalpyt line but extending to Leonora and eastwards to White Cliffs and the Great Victoria Desert. It is absent, however, from the South-West corner, south of the Moore River and generally west of the Great Southern railway. However, there is a record 9 miles east of Williams. There is a slight penetration along the south coast corridor (Fig. 8) as the species has been recorded at Nornalup (Royal Australasian Ornithologists' Union field camp, 1927).

Nesting: The nest is a tiny cup-shaped structure, about two inches in diameter, of grass-stems held together with spiders' web, built close into a horizontal fork of a tree up to 20 feet from the ground. Eggs, 2, pale greenish-blue with spots and blotches of purplish-brown and underlying markings of greyish-lilac. In some cases the markings tend to form a zone at the larger end. Average size, 20 x 14 mm.

Scarlet Robin

Petroica multicolor

Native names: Goo-ba (South-West); Men-de-ete (King George's Sound); Demclick (Pallinup River).

The male is recognisable instantly by the *vivid scarlet breast,* the upper parts, head and chest being black. There is a *small white patch on the forehead,* above the beak, a prominent white wing splash and the outer tail feathers are white also. The female is grey in those parts of the plumage where the male is black and the breast is a paler red which is less extensive in area. The abdomen is white. Iris, hazel; beak and legs, black. Length, 4¾-5 in.

Distribution: Found in the forested country of the South-West corner, from Cockleshell Gully (near Jurien Bay) in the north to the east of Esperance and inland to New Norcia (rare), Bolgart, Northam (rare), near Kweda (between Pingelly and Corrigin), Wickepin, Lake Grace and thence to the south coast at the Gairdner River, near Bremer Bay. Eastwards it has been seen at Jerdacuttup Creek and Stokes Inlet, and the distribution here may be discontinuous.

Nesting: The nest of the Scarlet Robin is a beautiful structure, cup-shaped, with thick walls, and cosily lined with hair, feathers or sheep's wool. It is usually placed in a vertical fork of a tree, not infrequently in the trunk fork of a blackboy, sometimes on a horizontal branch and occasionally in a shallow hollow in the side of a tree trunk. It is not usually more than 15 feet from the ground, but may be up to 50 or 60 feet high. Made of strips of bark thickly matted with spiders' web, a nest measured 3 inches in diameter by 2¼ inches deep. The egg cavity was 2 inches deep and the same in breadth. The 3 (sometimes 2) eggs are greyish or greenish white and are thickly spotted all over with pale brown specks and blotches. Size 18-19 x 13-14 mm. The species is an early breeder and we have found a full complement of eggs in a nest as early as July 28 in the Bridgetown district. Courtship feeding is indulged in by the male, which, however, takes no part in nest-building, but may occasionally assist in incubation. It takes a full share in feeding the young. Incubation period, 15 days; nestling period, 16-18 days (W. H. Loaring).

Albino males, with red breasts, are reported occasionally, and they breed successfully.

The cheery, trilling warble of the male bird is one of the pleasant sounds of the bush. In the lower South-West the birds are often seen on the roadsides or perched on the telephone wires and they are trustful visitors to the very doors of country homes and the quieter streets of townships. Around Perth they have become scarcer with the years and though once regularly present in King's Park they are now extremely scarce there.

The vernacular name is derived from the English Robin Redbreast. However, not only does the bird belong to a different family, but its appearance and habits are really unlike.

Red-capped Robin

Petroica goodenovii

Native names: Tjulily-tjulilypa (Warburton Range); Me-ne-gedang, Minidang, Minijidang (Avon River).

The male of this red-breasted robin can be distinguished from the previous species by the *brilliant red patch on the forehead,* from the beak to the crown. The female, unlike that of the Scarlet Robin, lacks any red on the breast, which is grey, while the forehead patch is an inconspicuous reddish-brown. Iris, dark brown; beak and legs, black. Length, 4½ in.

Distribution: Found throughout our area, being very common inland and visiting the coastal areas in winter. It is, however, only a rare visitor in the South-West jarrah area—the headquarters of the Scarlet Robin. In parts of the western Darling Range small numbers appear in most winters and some individuals penetrate occasionally on to the coastal plain near Perth. An isolated colony is resident on Rottnest Island, probably a relic of a former dry climatic cycle when it may have been the common species on the adjacent mainland.

There are in Australia five species of robins with the breast of some shade of red or pink, but only the Scarlet and Red-capped Robins occur in this State. Of the whole five the Red-capped species has the widest distribution. However, as indicated above, this range does not coincide with the more populous portion of the State. Hence the erroneous impression is gained that the Scarlet Robin, though actually of very limited range by comparison, is the more common of the two.

The Red-capped Robin is closely related to the Scarlet Robin and, in general, the two species have mutually exclusive breeding ranges. There is a narrow region of marginal overlap where both robins are found nesting together. At Narrogin, for example, where such is the case, B. V. Teague found the Red-capped Robin "tied" to jam *(Acacia acuminata)* country and it preferred jam and *Hakea* trees for nesting. The Scarlet Robin there was characteristic of the tongues of marri *(Eucalyptus calophylla)* and *Casuarina* country, but might nest in jams. At Woodanilling Ray Garstone found the two species nesting together, the Scarlet Robin nesting exclusively in the wandoo country. The Red-capped Robin may occasionally nest in the South-West, at the close of its winter incursions there. Such instances have been reported at Wooroloo (E. H. Sedgwick, October, 1952), Bibra Lake (J. R. Ford, September, 1949), and Mundijong (C. F. H. Jenkins, December, 1954). No cases of interbreeding between the two species have been seen in Western Australia but one has been reported in Victoria.

Nesting: The Red-capped Robin builds a beautiful cup-shaped nest of green moss and lichen and lines it with rabbit-fur, etc. Usually to be found in a horizontal fork of a tree near (about four feet) the ground. It is a very neat and small nest—about 2½ inches in diameter and the same in depth. The 2 (occasionally 3) eggs are roundish and pale green, spotted, particularly at the larger end, with umber and grey. Size, 14-16 x 12-13 mm. We have seen one egg which was almost globular, 14 x 13 mm. July to October are the nesting months. The normal incubation period is about 12 days. Both parents assist in the care of the young. The male may breed before it has assumed the full adult plumage.

The call resembles the syllables "tick-tick-tick-tick," and the territorial song "tick-it-ter-rri-ri-ri."

Hooded Robin

Petroica cucullata
(formerly *Melanodryas cucullata*)

Native names: Waid tjulka-tjulka (Warburton Range); Goo-ba-mogin (Perth).

The general colour is *black,* including the head, throat and upper chest. The remainder of the *under parts are white,* as are also the shoulders, edge of the wing, and a wing stripe. Also the lateral tail feathers are white at the base. The female is dark grey or brown above and the under parts are dull white with grey clouding on the breast. Immature birds are streaked black and whitish, both above and below, with the abdomen whitish. Iris, dark brown; beak, black; legs, black. Length 5¾ in.

Distribution: The species is generally distributed but occurs sparsely in the heavier forested areas of the South-West, except in open country and clearings.

Nesting: The nest is a cup, made sometimes of grasses, at other times of strips of bark, or both of these matted with spiders' webs. Lined with fine grasses, Banksia down, hair, etc., it is placed in a fork in a tree, or in a ledge or hole in the side of a tree, usually within ten feet of the ground. A nest observed in the Bridgetown district, on October 23, measured 3¼ inches across and 2¼ inches in depth, the inside measurement of the nesting cavity being 2¼ inches in diameter and one inch deep. As with the White-breasted Robin, the 2 eggs (the number usually laid) vary between themselves in ground-colour. One may be distinctly bluish, and the other greenish, while in some cases one may be even brownish. They have a few faint markings of liver-colour which usually take the form of an indefinite cap over the larger end of the egg. The other end is more pointed than is generally the case with the eggs of these Australian robins. Average size of 10 eggs, 21.0 x 15.6 mm., varying from 19.5-22.4 x 15.0-16.0 mm.

Western Yellow Robin

Eopsaltria griseogularis

Native names: Bam-boore, Bambun (Perth).

The most prominent feature in the coloration of this species is the *brilliant yellow lower breast and abdomen.* The chest and upper breast are grey and the throat white. The head, back and tail are grey and rump yellow. There is an inconspicuous white wing stripe, which, as in the following species, shows up as a small shoulder patch when the bird is at rest. Iris, black-brown; beak and legs, black. Length, 5½-5¾ in.

This is the Western Australian derivative of the Yellow Robin *(E. australis)* of south-eastern Australia, and differs in having a grey breast (yellow in the eastern birds).

Distribution: Restricted to the South-West corner, south and west of a line passing from Peron Peninsula (Shark Bay), south-eastwards to Mullewa, Bonnie Rock, Mt. Jackson, 50 miles south of Karonie, Balladonia and Madura.

Nesting: The Western Yellow Robin builds a deep cup-shaped nest (about 3 inches in height and 2½ inches broad) in a perpendicular fork of a jarrah sapling, usually about five to twelve feet from the ground but sometimes higher. It is beautifully made of jarrah bark kept together with spiders' web and the egg cavity invariably has three or four dry jarrah leaves placed loosely in it on which the eggs lie. The bird hangs strips of jarrah bark on the outside of the nest in so effective a manner that the whole structure blends well with the trunk of the sapling in which it is built and is easily overlooked when the nest is searched for. When, on occasion, the bird makes use of the fork of a blackboy *(Xanthorrhoea)* or of a marri (red gum) sapling as a site for the nest, it camouflages it in the same manner. The 2 oval eggs are buff (occasionally olive) in colour with brownish blotches which often form a collar around one end. Size, 22 x 16 mm. September and October are the breeding months in the jarrah area, but we have seen eggs in nests as late as December on the coastal plain at Bunbury.

This is another of the species which exhibits a certain amount of local migration of the kind described for the Grey Fantail (p. 382). The Yellow Robin is, or rather used to be, a common inhabitant of the Darling Range country, being far less shy than the White-breasted Robin and occupying most types of woodland habitats. It visits the coastal plain near Perth in the winter months but not so commonly as the Grey Fantail. These visits have become less frequent in recent years and though once the bird used to appear regularly each year in Kings Park it has not been observed there since the winter of 1935. A solitary individual was seen on Carnac Island, off Fremantle, by J. A. L. Watson in January 1956, possibly a bird which had flown to the island the previous winter.

The decrease in the visits to the coastal plain seems to be correlated with a declining population strength in the Darling Range country. W. H. Loaring confirms observations elsewhere that the Yellow Robin has become a scarce bird at Bickley, the decline having set in as long as 40 years ago. The forest habitat has not been noticeably despoiled and it is difficult to attribute a likely cause.

The bird has a habit of clinging sideways to the trunk of a tree.

The notes uttered by the Yellow Robin are varied, but all rather mournful in tone. The alarm note is a rather harsh "chit chit," which may be varied to a "chit, chit," short pause, "chit, chair."

White-breasted Robin

Eopsaltria georgiana

Native name: Boi-jeel (Augusta).

This is a quietly garbed but attractive bird. The upper parts are *a dark bluish-grey,* wings black with an inconspicuous white stripe and the tail quills black, slightly tipped white. The *under parts are white* with a greyish clouding on the breast. When the bird is at rest a white patch on the wing, just below the shoulder, is noticeable. Iris, brown; beak, strongly built, black; legs, black. Length, 6-6½ in.

Distribution: Confined to the south-west. Formerly it was thought to be restricted, as a common species, to the lower south-west from Cape Naturaliste to Albany, with isolated occurrences in gullies in the Darling Range north to about the level of Pinjarra. However, in October 1957 J. R. Ford and B. V. Teague made the surprising discovery that it occurred in the coastal thickets near Dongara, and subsequent surveys by Ford showed that along the coast it ranged from Geraldton in the north to Lancelin in the south in the thickets of coast wattle, *Acacia rostellifera,* and *Melaleuca cardiophylla.* That it formerly may have extended farther south of this habitat is suggested by the observation of a bird by W. B. Alexander in July 1919 in a patch of coast wattle near what is now Floreat Park. J. R. Ford has also found the bird in Darling Range gullies north to the Canning Dam area.

The southern population is most plentiful in the karri country and is largely confined to the luxuriant vegetation especially *Trymalium spathulatum,* growing along streams. G. M. Storr has recorded isolated populations in the karri outlier in the Porongorups and also at Bald Island. Ian Abbott has seen it on Michaelmas Island and Ray Garstone has observed nesting in the Stirling Range (Toolbrunup Peak).

Nesting: The White-breasted Robin makes a cup-shaped nest of dry grasses and rootlets and camouflages it on the outside by placing lichen, etc., around it to make it agree with the appearance of the tree or bush in which it is built. No attempt at lining the nest is made by the bird. Usually selecting a site within a few feet off the ground, the bird builds in a *Leptospermum,* or in a paperbark, but on occasions a jarrah is used. We have even seen a nest built on the top of a bracken fern. The eggs, 2 in number, vary between each other in colour. They may be light olive, light brown or pale blue. A casual inspection of the eggs of the White-breasted Robin gives the impression that they are without markings, but a close examination, especially with a magnifying glass, discloses that they have faint markings, spots and streaks, of silver colour. Average size of 9 eggs, 21.3 x 15.6 mm., varying from 20.8-21.6 x 15.0-16.0 mm. As incubation proceeds, they tend to become brownish in colour.

The breeding season is protracted, being from June to December in the south-west forested area. Two or more broods may be reared.

The existence of a nest or the presence of very young birds is invariably disclosed by the warning note uttered by the birds, which may be heard between September and November. When one bird is sitting on the nest, warning of the approach of an intruder is invariably given by the bird "off duty", which utters a rather harsh "chit-chit". This is the only note, song or otherwise, that we have heard the birds utter. However, in the northern population J. R. Ford has heard a whistling call rendered as "wee-oh". The male feeds the female on the nest. Both birds will react to the presence of an intruder, when young are in a nest, by falling to the ground and trailing a wing in the approved decoy fashion.

This is one of the few birds found only in the south-west and in no other part of Australia. It is now considered (J. R. Ford and J. A. Keast) to be a modified descendant, through the loss of yellow pigment, of an earlier colonisation of the Yellow Robin group in southern Western Australia. The Western Yellow Robin appears to be evolving in a similar direction as the upper breast, in western birds, has lost its yellow colour and become grey. This instance of "double invasion" of a species group into the south-west is discussed in more detail on p. 52.

The White-breasted Robin has the reputation of being a very shy and secretive species. However, if an observer sits down quietly in a gully or thicket it frequents, it will probably come and settle on a twig within a few feet of his face and remain there for a considerable period. The slightest movement and it is off in a flash, the sombre-coloured plumage making it difficult to follow in the dim light of the undergrowth. In quiet gardens built in the bush the bird may be a frequent visitor; we have watched a White-breasted Robin from the verandah of a guest-house at Nornalup where at the time some thirty guests were at lunch within. The bird perched on a fruit tree and then flitted to a post-and-rail fence. It is much easier to observe during the nesting season, at which time it is really confiding and may be "called up" close to an observer if a squeaking noise is made with the lips. In general habits the species retains its similarity to the Western Yellow Robin and the same mannerisms of clinging to the side of a tree and remaining motionless, and of jerking its tail and wings.

Mangrove Robin

Eopsaltria pulverulenta

(formerly *Quoyornis leucurus* and *Peneœnanthe pulverulenta*)

Similar in general coloration to the preceding species, except in the tail pattern. The *outer tail quills have a broad white band towards their bases,* there being no white on the tips. The wings are brown with an inconspicuous white stripe, and the head has a brown tone. The tail is black. Iris, brown; beak and legs black. Length, 6¼ in.

Distribution: A northern species, occurring in the mangrove belt south to Exmouth Gulf. F. Lawson Whitlock collected it at Cossack and T. Carter obtained specimens near North-West Cape.

Nesting: There appear to be no descriptions extant of nests or eggs observed in Western Australia. In the eastern portion of its range the Mangrove Robin builds an open cup-shaped nest which it constructs principally of bark, fibre and spiders' webs, and is usually placed in a fork of a mangrove from five to fifteen feet from the mud in which the tree grows. It lines the nest with rootlets and fibre and decorates it externally with long fragments of bark. The eggs, 2 in number, are green, conspicuously spotted with brown or reddish-brown, intermixed with faint lilac markings. Size, 18-19 x 15-16 mm. Nesting occurs during the rainy season.

FANTAILS: Family Rhipiduridæ

Grey Fantail

Rhipidura fuliginosa

(formerly *Rhipidura flabellifera*)

Native name: Gadjinnak (South-West).

Upper grey, under buff, with the throat grey and the chin white. There are two white bars on the wing and the outer parts of the *long fan tail* are white. It has a short white eyebrow. Iris, dark brown; beak, black; legs, dark brown. Length, 6 in.

Distribution: Likely to be found anywhere in the State, except in the far interior. It is, however, most abundant in the heavier forests of the South-West and is rare and local north and inland of a line joining the mouth of the Murchison River, Perenjori, Merredin, Lake Grace and Hopetoun. In the North-West, south to Shark Bay, the species shows a distinct partiality to mangroves.

Nesting: The Grey Fantail makes a cup-shaped nest which is a small edition of that of the Willy Wagtail, but to the bottom of the nest is added a "tail" which hangs down and gives to the whole a "wine-glass" effect. Usually attached to a thin branch of a tree at no great height from the ground, it is composed of dry grasses and shredded bark matted with spiders' webs. It is about 2½ inches in diameter and 1½ inches deep; the tail-piece varies in length, sometimes being as much as 4 inches. The eggs, 2 or 3 in number, are white marked with specks and blotches of brown and pale lavender, which are chiefly at the larger end and tend to form a distinct zone around the broadest part. A set of three from Bunbury averages 16 x 12 mm. in size.

Both parents assist in building the nest and in the care of the young. They are very sensitive to interference before the eggs are hatched and are likely to desert the nest if it is closely inspected. Incubation and nestling periods, each 15 days (W. H. Loaring). After the young are hatched the parents are roused to great excitement if the nest is approached.

In many parts of its area of occurrence the Grey Fantail is a local migrant. In the Darling Range, for instance, and the wetter portions of the South-West it is plentiful throughout the year. From these favoured areas some individuals disperse into neighbouring country in the winter months, between April and October. In the Perth district and surrounding parts of the coastal plain, Grey Fantails first appear at the onset of the winter rains and continue to be met with until spring. Similar movements have been noted at Moora, Broomehill, Nangeenan, Dangin, Bilbarin, Caron, and elsewhere in the Wheatbelt where continuous observations have been made by ornithologists. At Moora the birds appear towards the end of April, remain during May, and disappear again towards the end of June.

In its general behaviour the Grey Fantail resembles the Willy Wagtail and is one of the most confiding birds in the bush. More than any other bird it will enter camps and can be "called up" to within a few feet. The South-West Aborigines, however, disliked this intimacy, regarding it as a bird of ill omen, and would abandon an undertaking if one of these fantails appeared in the camp immediately beforehand. Curiously, the Grey Fantail does not continue its association with man in more sophisticated surroundings and when it visits the Perth suburbs in winter it shows far less familiarity than the Willy Wagtail.

Willy Wagtail (Black-and-white Fantail)

Rhipidura leucophrys

Native names: Jind-ril (Godfrey's Tank, Canning Stock Route); Chitti-chitti (Coolawanyah); Chee-ri-gal (North West Cape); Jindy-jindy (Lyons River); Tjittee tjittee (Jigalong); Chindi-chindil (Warburton Range); Wil-ar-ing, Willaring (Perth); Jit-te-jit-te (Avon River); Chitti-chitti (Pallinup River).

Head, throat and upper parts glistening black; wing quills blackish-brown; *breast and abdomen white.* Eyebrow white. Iris, dark brown; beak and legs, black. Length, 8-8¾ in.

Young birds have the eyebrow and narrow bars on the wing-coverts brown.

Distribution: Found all over the State, but absent in the Darling Range (jarrah forest area) except where the country has been opened up by settlement.

Nesting: This bird makes a beautiful cup nest, about 2¾ in. broad and 1¾ in. deep, of grass, which it binds so firmly together with spiders' webs that it becomes so closely matted as to be difficult to tear apart. The nest is placed usually on a horizontal branch and a full set of eggs is 4 in number, but in recent years only 3 are usually laid. They are greyish, spotted with brown, which markings tend to concentrate around the thickest part of the egg and form a distinct ring. As incubation proceeds the eggs assume a brownish tinge. They average 18-20 x 14-15 mm. in size. Eggs are laid on successive days and incubation begins on the second day. Several broods are raised in the season. The first clutch of eggs is laid in the South-West about the third week in August, the second about the first half of October, the third about the beginning of November, and the fourth, if any, late in December. Incubation takes about 13 days and the young remain in the nest another 13 days. The juveniles are driven off just prior to, or at the hatching of, the next brood (E. H. Sedgwick).

This familiar species has endeared itself to man and has profited greatly as the result of settlement. It has comfortably adapted itself to living with man in towns and suburbs and has been able to spread into forest country, which it normally avoids, after clearing. It is strongly territorial and drives off birds much bigger than itself, including Brown Hawks, Magpies and other birds, and its aggressiveness leads it at times into molesting quite harmless species such as Thornbills and Silvereyes. It is a friendly associate of cattle and horses and farm-stock generally. A common sight in the paddocks is that of a Wagtail using the back of a cow or a horse as a vantage point and thence sallying to and fro in pursuit of insects.

MONARCH FLYCATCHERS: Family Monarchidæ

Restless Flycatcher (Scissors-grinder)

Myiagra inquieta

(formerly *Seisura inquieta*)

Native names: Jit-tee-gnut, Jitti-ngat (South-West).

Upper steely blue and the *entire under parts white.* In colour pattern it closely resembles the Willy Wagtail, but the white of the under parts extends right up to the beak. The feathers of the crown are capable of being erected into a *semi-crest*. The beak is long and broad at the base. Iris, dark brown; beak and legs, black. Length, 8 in.

Distribution: A South-West species, occurring south and west of a line passing through Moora, Southern Cross and Fraser's Range. It is, however, only a very occasional visitor to the Perth district and the coastal plain, but has been known to nest at Claremont and Midland Junction.

Nesting: The nest is usually built towards the outer extremity of a horizontal branch of a tree at a considerable height from the ground — up to 50 feet. It is a neat cup-shaped structure, composed of fine grasses and strips of bark compactly bound together with spiders' webs. It is lined with finer material, usually hair. In order to harmonise the nest with the branch, the bird camouflages the outside with spiders' egg sacs and lichen. C. L. E. Orton and P. T. Sandland state they have noticed that at Moora the bird always nests in trees growing in swamps. The 3, rarely 4, eggs are white and have well-defined bands of spots of purplish-brown and purple around the larger end. A few spots of the same colours occur elsewhere on the surface of the eggs. Measurements, 18-21 x 14-16 mm.

A distinctive habit of this bird is its peculiar hovering behavour, with the body held arched and the wings quivering rapidly, the bird uttering the while a harsh metallic call which has earned for it the name of "Scissors-grinder." Another note is a whistling "kheer, kheer." It favours the neighbourhood of swamps and creeks.

Satin Flycatcher

Myiagra cyanoleuca

Sexes different in appearance. In the male the upper parts, and the throat and upper breast, are a glossy greenish-black, with velvety-black on the lores and forehead; rest of under white. In the female the upper parts are brownish blue and the chin, throat and foreneck are orange rufous, rest of under white. Iris, dark brown; the wide beak is leaden black; legs, black. Length, 6½ in.

Distribution: An eastern States species which migrates north after breeding in the south, including Tasmania. During the northern, post-breeding, migration some individuals may deviate widely from the normal route. There is one record from New Zealand (June 12, 1963) and on March 29, 1964, a female bird was found at Langhorne Creek, south-east of Adelaide, the first record of the species in South Australia. On April 12, 1973 M. G. Brooker collected an immature female near Twilight Cove, south of Cocklebiddy, the only record of the occurrence of the species in this State.

WHISTLERS and SHRIKE-THRUSHES
Family Pachycephalidæ

Golden Whistler
Pachycephala pectoralis

Native names: Pe-dil-me-dung, Pidilmidang (Perth).

The brilliantly coloured male first arrests attention by the *bright yellow lower breast and abdomen.* The *throat is white,* separated from the yellow of the breast by a well-defined *narrow black band* which arches up and is continuous with the black of the head. The yellow of the breast also continues upwards as a narrow nuchal collar. The remainder of the back is a bright olive-green and the wings and tail quills are grey and black. The female, like the immature male, has no yellow in the plumage (except for a concealed tuft of yellow feathers over the vent, but these cannot be seen unless the bird is handled), being grey above and buff on the lower breast. The throat and upper breast are grey. Iris, red-brown; beak, black; legs, grey. Length, 6¼-7 in.

Distribution: This species is restricted to the South-West corner, probably reaching the Mulga-Eucalypt line, recorded limits being Gnow's Nest Range (south-east of Yalgoo), Bonnie Rock, Widgiemooltha, 50 miles south of Karonie, Eyre and Madura. It occurs on some islands of the Archipelago of the Recherche (recorded on Sandy Hook, Woody, Remark, Long, Mondrain, North Twin Peaks and Middle Islands).

Nesting: The nest, shallow, cup-shaped, is built in a vertical fork of a sapling and is made of grasses, shreds of bark and leaves, and lined with fine grass. The 2 eggs have a yellowish ground-colour and a broad ring of umber spots around the larger end, interspersed with underlying spots of grey. Size 23 x 17 mm. The male assists in the building of the nest and in brooding. Incubation period, 15 days (W. H. Loaring).

One of its most frequent calls may be rendered by the words "we-wi-wi-*wit.*"

The bird is an inhabitant of densely wooded country, including thickets of no great height.

Mangrove Golden Whistler (Black-tailed Whistler)
Pachycephala melanura

Very similar to the previous species. Base of tail greenish yellow. Western Australian birds are much smaller than the Golden Whistler, with relatively longer and heavier beaks. Edges of wing quills grey (olive to grey in the Golden Whistler), base and tip of tail olive (not grey). The female is yellowish on the belly. Length, 6½ in.

Distribution: A northern tropical form restricted to mangroves; in this State as far south as Exmouth Gulf. It was formerly regarded as a subspecies of the Golden Whistler, and it was so treated in our earlier editions.

Nesting: Similar to the previous species.

Rufous Whistler

Pachycephala rufiventris

The plumage pattern is somewhat similar to that of the Golden Whistler, but the adult is *chestnut or rufous* where the other is yellow. The chin is white, and the black band between the white and the chestnut is continued up as a broad line passing through the eye to the beak. The rest of the head and the back are grey. The *female* and the immature male are *longitudinally streaked black and white on the under parts,* there being a buff tone on the abdomen. The young male before it fully assumes the adult dress has a lighter chestnut breast with a grey chest band and the white throat is lightly streaked with black. Iris, red-brown; beak, black; legs dark slate-grey. Length, 6¼ in.

Distribution: Found all over the State, though rare in the heavier forested country of the lower South-West (Cape Naturaliste to Albany). It is found on Rottnest Island.

Nesting: The nest is a fragile structure——cup-shaped, formed of grasses and rootlets, and lined with finer material. It is placed in an upright fork, up to 25 feet from the ground and its frail construction often permits of the eggs being seen from the ground. The 2 or 3 eggs are olive, spotted and blotched with dark brown. Average measurements, 22 x 17 mm. Both sexes share in the incubation of the eggs and the brooding of the young, the female taking the night shift. Nesting period, 15 days.

This bird is not migratory as it is in the eastern States but in the winter months may be somewhat nomadic. From life-history studies made by A. H. Robinson, of Coolup, it appears that the male Rufous Whistler establishes his nesting territory about July in the South-West, when the spring song starts. It also appears that once a territory has been acquired some males, at least, defend it by song indefinitely, even during the non-breeding season. When he is joined by a female the male bird's song becomes richer, the familiar "e-chong" and whip-crack songs being part of the territorial and sexual displays. The following is a rendering of a full territory defence song: very rapid: "Pee-pee-pee-pee-pee-pee-pee"; rapid: "joey-joey-joey-joey-joey"; deliberate, and with a whip-crack effect: "e-chong e-chong e-chong e-chong e-chong e-chong." There may be variations; the final whip-crack whistle may be given only once or twice, and we have counted no fewer than 39 repetitions of the "joey-joey" call in the song of one particular male. The mating or sexual song is a low song composed of single notes followed by a trill. In both types of songs the bird does a bowing display. Mrs. E. Erickson, of Bolgart, has recorded that duet singing occurs. The female intersperses the notes "pee-pee" with the male's notes in the preliminary sequence, and gently answers "chong" to his ringing "e-chong" but never gives the "joey-joey" notes. Mrs. Erickson has found that though the male establishes the territory, the female will assist the male in its defence. The territorial songs of this bird and of the Brown Honeyeater provide some of the finest and most spirited bird-music in Perth parks and gardens in the spring, but probably the sweetest song of the Rufous Whistler is its autumn song.

It favours more open country than the Golden Whistler and as the South-West forested area is cleared it is becoming more plentiful and tending to replace the other species. It was first recorded in Bridgetown in 1939 and is now found at Augusta, all the way from Cape Naturaliste in the west coast corridor.

Gilbert Whistler

Pachycephala inornata

The plumage pattern is different from any other of our local whistlers. The male is an entirely grey bird, with a rufous or *chestnut throat and upper breast* and buff on the abdomen. There is a black line from the beak to the eye. The female lacks the chestnut markings and is dark grey ventrally, streaked dark brown longitudinally, from the throat to the abdomen. Iris, reddish-brown; beak, black; legs, slaty-black. Length, 7-8 inches.

Distribution: Found in the mallee and lightly timbered country of the southern part of the State; not occurring in the heavily forested area west of the Great Southern railway, or in the vicinity of Perth. M. Brooker mist-netted one 50 miles N.E. of Rawlinna in July 1970, and saw it at Kanandah (near Naretha) in January 1971.

Nesting: The nest is deep, cup-shaped and constructed of twigs, with the egg-cavity thickly lined with shredded bark or merely with finer twigs and grass. A typical nest measures about 5 inches in diameter and 4 inches in depth, and is usually placed within a few feet of the ground, in a vertical fork of a densely-leaved bush. The 2, sometimes 3, eggs are white to light buff, spotted with irregular markings of different shades of brown, with sometimes others of black, and some underlying ones of lavender. These markings tend to form a zone over the larger end. The eggs are subject to variations in shape, but a typical clutch measures about 23 x 17 mm. They are rather like those of the Dusky Wood-Swallow.

The Gilbert Whistler has a chuckling "chop chop" note and also a sibilant whistling call, not unlike that of the Southern Scrub-Robin.

White-breasted Whistler

Pachycephala lanioides

Black slaty-grey; tail brownish-black; wing quills black, edged whitish-grey. Head shining black with brown nuchal collar. *Throat white, followed by a broader black and a narrower chestnut band* continued upwards to join the head and nuchal collar respectively; abdomen white. In the female and immature male the upper parts are grey-brown and the under parts white, streaked with dark brown. The wing quills and tail are darker grey than the back. Iris, light brown; beak, black; legs, grey. Length, 8 in. Weight, 1½ oz.

Distribution: A northern species extending south along the coastal mangrove belt as far as Carnarvon.

Nesting: It nests in mangroves, making a flat, flimsy nest, open, and about six inches in diameter. It is constructed of fine rootlets, and placed as high as 15 feet in a fork of a mangrove. The 2 eggs conform to the type laid by other members of the genus *Pachycephala,* being olive in colour marked at the broader end with a zone of olive-brown tinted spots and splashes. They are glossy. Average measurements, 27 x 18 mm.

F. Lawson Whitlock, who found the species nesting in mangroves at Condon in August, states that, as a rule, the male sings at no great distance from the nest but takes no part in building, or in the care of the young.

The call note is a soft, clear whistle and the alarm note, with which the male warns the female of any approaching danger, is sharp and harsh. In addition the male has a very loud song which is clear and musical and, as with other whistlers, is impetuous and joyous.

Western Shrike-Thrush (Thrush)

Colluricincla rufiventris

Native names: Goo-de-lung, Gudilang (South-West).

Upper *dark grey* with a slight brownish tone; under grey, *lower belly and under tail coverts buff.* There is a slight whitish area between the beak and eye. In the female there is a more rufescent tone in the plumage, the buff, which in the male is confined to the lower abdomen, is more extensive; the breast is strongly streaked; there is a buff eyebrow and the back is paler, being greyish-brown. Iris, brown; beak, black in the adult, brown in immature birds; legs, grey. Length, 9-9¾ in. Weight, 2 oz.

Distribution: Generally distributed, coastal and inland, north to the Fortescue River, east to the Great Victoria Desert, and along the coastal vegetation to Eucla.

Nesting: Nest bowl-shaped, made of bark and fibre and placed either in the centre of thick foliage or in the end of a broken spout of a tree, often at a considerable height from the ground. Mrs. Rosemary Balmford found a nest in breakaway country at Kathleen Valley which was in a hole in the rock, some 10 ft. from the ground. The hole was just big enough to accommodate the nest. Eggs, 2 to 4, usually 3, are white with brown and bluish-grey spots. Average size of 16 eggs, 28.3 x 20.7 mm., varying from 26.7-32.0 x 20.0-21.7 mm.

This species ranks as one of the most melodious songsters in the bush, but is heard only in the nesting season, renderings being: "purr-purr-purr quee yule," "yoh-ho-ee," 'pip-pip-pip-pip-ho-ee," "purr-purr-purr-he-whew-it," "e-all, ee-all, queel," and "Cooee, who are you." The winter call is a sharp "yor-ick," or "war-ick," loud and unmusical. Though very similar to the Grey Shrike-Thrush *(C. harmonica)* of the eastern States, it lacks that bird's tolerance of the inroads of civilisation. The eastern bird will frequent suburban areas and even nest in the verandahs of homes. The western bird has none of this familiarity and, though not shy, will live only in the relatively little-altered bush. Thus it has long since forsaken the immediate neighbourhood of Perth (W. B. Alexander in 1919 found them plentiful at the Swanbourne rifle range, Herdsman's Lake and Perry Lakes), though still found a few miles north and south on the plain and common enough in the nearer parts of the Darling Range. It occasionally visits King's Park.

SHRIKE-TITS: Family Falcunculidæ

Crested Shrike-Tit

Falcunculus frontatus

Native names: Jil-le-e-lee (Avon River); Djoon-dool-goo-roon (South-West); Goore-beet-goore-beet, Gurbitgurbit (South-West, coastal).

Back, wings and tail bright olive-green; flight quills blackish-grey. The head and throat are black with a prominent broad *white stripe over the eye and another from the beak across the cheek.* The *breast is bright yellow,* the abdomen white and the under tail coverts yellow. The *crown feathers are capable of erection into a conspicuous upright crest.* Female birds have a greenish throat in place of the black and the crest feathers are very much shorter. Immature birds are brown above and have the entire under parts yellow. Iris, brown; beak, strong and massive; legs, leaden-blue. Length, 7 in.

Distribution: Found in the South-West to Morawa, Moora, Kellerberrin, Kalgoorlie and Norseman, favouring the lighter timbered wandoo, salmon gum and jam country, but not venturing far into the heavy jarrah forest areas. It occurs at Dryandra and has been seen in karri country 7 miles south of Shannon. We have a record from Winnijup, September 1932 and G. A. Lodge has recorded them haunting flooded gums along the Blackwood River at Boyup Brook. Occasionally birds appear in the western coastal strip (one record from Wanneroo in 1907 and another from Karridale in 1961). There is also some penetration along the south coast corridor (Fig. 8) and S. W. Jackson recorded the species in the vicinity of Irwin's Inlet on numerous occasions in the summer of 1912-13, and C. J. O. Harrison and P. R. Colston met it at the Scott River in February 1966.

Nesting: The nest is deep and cup-shaped and is composed of the bark of trees matted together with spiders' webs. It is lined with fine grass. Usually placed high up in a vertical fork at the extremity of a branch of a tree. The clutch is usually of 2, rarely 3, eggs, which are white, speckled with dark brown, especially over the larger end. Size, 24 x 15 mm.

The call is a soft, clear, single-noted "ko" or "koo," or "too-ee, too-ee," repeated at minute intervals.

Crested Bellbird

Oreoica gutturalis

Native names: Jindangie (Well 45, Canning Stock route); Pan-pan-panella, Bokkun, Bokkun-bokkun (North-West Cape); Bagobago (Lyons River); Wunnbunnpullalla (Jigalong); Panpanpalala (Warburton Range); Baacka-baacka (Yandanooka); Bokanbokan (South-West).

Upper parts brown, the head being black with *an erect crest.* The wing and tail are darker and there is a chestnut shade on the rump. The *lores and chin are white,* chest black and abdomen white.

The female has practically no black in the plumage. The chest is brown and there is no white around the head. Iris, male, bright orange; female, yellow orange. Beak, male, black; female, grey with a greenish tinge. Legs, grey. Length, 8-9 in.

Distribution: Generally distributed, north to the southern portion of the Kimberley Division. Until recently it was not usually found in the South-West corner south and west of Mogumber and the Great Southern railway (cf. Fig. 7). However, the species appears now to be penetrating into the northern and eastern parts of the South-West block of forest country. In 1954 Crested Bellbirds were recorded at Beermullah. In August 1957 H. A. Atkinson and J. R. Ford found them six miles north of Yanchep. Miss D. A. Rook found the species at Lake Pinjar in 1962, and R. H. Taylor had noted it still further south, near the Gnangara pine plantation in 1960. More recently Mrs. Bruce Leake and Miss Olive Leake heard the characteristic song at Upper Swan. In the eastern sector B. V. Teague found them 10 miles north of Williams in the winter of 1945. R. Job observed birds at Dryandra in December 1967. Ray Garstone has recorded the species in the Woodanilling district, west to the Beaufort River crossing.

Nesting: The nest is deep cup-shaped, composed of strips of bark and lined with finer material, grass, leaves and bark, and the bird has the unusual habit of placing live caterpillars on its edge. It is to be found in a variety of situations; in a vertical fork of a tree, in a hollow in a stump, or in the hanging bark of a tree. The caterpillars are usually of the hairy variety and the bird appears so to treat them, probably by pinching, that they are in a state of semi-torpor. The eggs, 2 or 3, sometimes 4 in the clutch, are whitish, blotched sparingly with dark brown and grey. They are rather like those of the Shrike-Thrush, as is also the nest, other than for the presence of caterpillars. Average measurements of 12 eggs, 27.4 x 20.3 mm., varying from 25.9-28.1 x 19.9-20.6 mm.

The celebrated ventriloquial call, which consists of two slow notes followed by three quicker ones, is well syllabised by the bird's aboriginal, onomatopoeic, names such as "pan-pan-panella." These notes sound deceptively like a distant cattle bell. Whilst moving over the ground, under dense bushes, it utters a call resembling "chuck-a-chuck-chuck."

Wedgebill (Wheelbarrow-bird, Waggon-bird)

Sphenostoma cristatum

Native names: Won-bon-alah (Godfrey's Tank, Canning Stock Route), Jeedadoo (Barlee Range); Chidadoo (Wooleen); Geetardo (Yandanooka).

Upper light brown; pale below; a *white line along* outer edge of the wing. The head bears *an erect crest,* and the *outer tail feathers are tipped white.* As the tail is frequently outspread this white flash is conspicuous and readily distinguishes the Wedgebill from the only other crested bird of the dry areas—the Crested Bellbird. Iris, brown; beak, black; legs, black. Length, 7½-8½ in.

It also differs from the Crested Bellbird in *running* instead of hopping along the ground and in its flocking habit.

Distribution: Found in the mid-north and centre; north to the Ashburton River and Godfrey's Tank on the Canning Stock Route, and south to the Mullewa district, Yandanooka and Thundelarra Station (48 miles south of Yalgoo) being the southernmost recorded localities. Probably the mulga-eucalypt line (see Fig. 4) will prove to be the main southern boundary.

Nesting: The nest, of flattened cup-shape, is fairly loosely constructed of small sticks and is built in scrubby trees, or in clumps of saltbush from 4 to 10 ft. from the ground. Curara *(Acacia tetragonophylla)* is often chosen. The nest is usually lined with grass roots. The clutch consists of 2 or 3 eggs, which resemble those of the Western Whipbird in colour. They are blue, spotted with black and greyish-purple. In some cases the greyish-purple spots are lacking. In another, comparatively rare, colour variant the markings take the form of a distinct ring of spots and hieroglyphic lines around the broad end. Average size of 31 eggs from the Barlee Range and Exmouth Gulf, 24.3 x 17.4 mm., varying from 23.0-25.8 x 16.4-18.2 mm. In the Barlee Range eggs have been found in nests from February to May, and incubation lasts 17 days (A. H. Robinson).

This ground-frequenting bird is one of the characteristic inhatants of the mulga country of the north and one least likely to be overlooked. The song is remarkable and carries some considerable distance, being clear, ringing and bell-like. It lends itself to imitation. The native names, like those of the Crested Bellbird, are onomatopoeic. The most frequently uttered notes have been variously syllabised by Europeans, from the sentimental rendering "sweet Kitty Lintoff" to the earthy "did you get drunk?" whilst the French words "je suis a vous" (je swee ah voo) used for the rendering of the song of the Spinifex-bird (see p. 338) are also a good imitation of that of the Wedgebill. The song is repeated over and over again with maddening monotony and though not really ventriloquial, as is the song of the Crested Bellbird, does resemble that species in its habit of starting off in a low tone and gradually becoming louder, sounding as if the bird were coming closer. The final note, which is emphasised, has a distinctive metallic ring, as if it were sounding from a rainwater tank.

The bird sings from the top of a shrub, but is rather wary and does not allow too close an approach. It often sings on moonlight nights. To the bird-lover from the south, the Wedgebill's song, like that of many other birds of the Eremea, has a sweetly-sad quality, somehow appropriately in tune with the solitude and rigour of the environment.

Western Whipbird

Psophodes nigrogularis

Native name: Tar-ding (King George's Sound).

Upper surface a quiet *olive-green,* with the long tail tipped black and white on the outer feathers only, both above and below. Underneath grey, *the throat black with a white stripe on each side* behind the beak. The forehead feathers can be erected into a *crest.* Iris, red; beak, black; legs, slaty black. Length, 9½-10 in.

Young birds lack the black throat.

Distribution: A resident and breeding inhabitant of the South-West mallee areas. In former times it was known from the dune thickets near Perth, Busselton, Margaret River, Irwin's Inlet and King George's Sound. It was also recorded at Wongan Hills but recent records are only from the south-east—Lake Grace (22 miles north), Gnowangerup, Borden, the Stirling Range, Mt. Taylor, Two People Bay, Pallinup estuary, Ravensthorpe (11 miles west), the Fitzgerald River National Park and Hopetoun.

Nesting: In the Gnowangerup district the Western Whipbird nests in the prickly *Banksia caleyi* and makes a shallow cup-shaped nest of very fine twigs, strips of bark and a few dry grass-stems, lining it with dry grass-stems of a finer quality than those in the body of the nest. The egg-cavity measures about 3½ inches and is about 2 inches deep. The nest is usually placed a few feet from the ground in the centre of the *Banksia.* Nests have also been found in low shrubs, about a foot from the ground in *Daviesia, Agonis* and *Hakea.* It is an early breeder, eggs being found in nests from mid-July to the end of October. The clutch consists of 2 eggs, which are blue, sparingly spotted with black. They closely resemble those of the Wedgebill. Six eggs in the South Australian Museum collection average 26.6 x 18.9 mm. and vary from 25.7-27.7 x 18.0-19.6 mm (H. T Condon). Incubation period, 21 days; nestling period 9-12 days. Observations suggest that only one sex probably does most of the incubation; however both parents feed the fledged chicks. These probably stay with the parents until the next breeding season. Territories range from 7-14 acres (Graeme Smith).

Until is was proved in 1939 that this bird still survived in some strength in the mallee country near Gnowangerup it was feared that the species might have become extinct in Western Australia. Prior to the Gnowangerup records the most recent observations had been made in 1901 at the mouth of the Margaret River by A. W. Milligan and in 1912 at Irwin's Inlet by S. W. Jackson. John Gilbert found it at the Wongan Hills in 1842 and the recent discovery of his notebook has revealed the highly interesting fact, first made known in 1950, that he had actually met with the bird near Perth in October 1842. Gilbert wrote: "This bird, although assimilating in many particulars to the Jeemuluk (Noisy Scrub-bird) and Djyrdalya (Bristle-bird), differs from both in being much more widely dispersed, and the dissimilar places it inhabits; in the neighbourhood of the Vasse and Augusta, for instance, it is only found in the rankest vegetation growing around swamps or small running streams; near Perth I found it on the dry sandhills immediately adjacent to the beach, and at the [King George's] Sound in similar situations in groves of the *Metrosideros* [= tea-tree], while in the interior from the settled districts [Wongan Hills], it was confined to the dense dry thickets, where for a very great portion of the year not a drop of surface water is to be met with for several miles around."

The bird is excessively shy. It spends most of its time on the ground and runs rather than flies when disturbed.

John Gilbert described the song as harsh and grating, one which "it is quite impossible to describe, and which is so different from that of every other bird I ever heard or am acquainted with, that I shall have no difficulty in recognising it again wherever I may hear it." Gilbert compared its chattering notes with the calls of a crake and found that these were only uttered when a pair were together, being sung as a duet. From the Victorian mallee F. E. Howe and J. A. Ross have recorded that there the song is strange, rattling and staccato, and inclined to be harsh. It consists of twelve distinct utterances of four groups of three notes each, produced fairly quickly with a slightly perceptible pause after the third, sixth and ninth. On the other hand, in the Gnowangerup district no call has been heard from this bird other than a "chitter-chitter," like that of the Southern Scrub-Robin when disturbed at its nest. E. Lindgren, who located a pair of birds 75 miles east of Borden in August 1957, heard a duet song made up of "6 to 8 whistling notes, the first call being 3 or 4 notes, immediately answered by a call of similar duration and quality but consisting of different notes." H. O. Webster, at the Two People Bay faunal reserve, has recorded this duet song, the male singing a phrase which may be rendered by the words "It's for *tea*cher," and the female follows immediately with "Pick-it-up." They vary the loudness of their song, "beginning so softly that the observer is led to believe the singer is some distance away. The song becomes much louder and it is amazing to discover how close the bird actually is."

NUTHATCHES: Family Neosittidæ

Black-capped Sittella (Nuthatch)

Neositta pileata

Native names: Goo-mal-de-dite, Gumalbidyt (South-West)

A brightly-patterned bird with *head black,* except for the white forehead, the back light grey-brown, the rump white and a short white-tipped tail. The wings are black with a *conspicuous orange "window';* under parts white. The female differs in lacking the white forehead and the black on the head being more extensive. Iris, reddish-brown, eyelid, yellow; beak, long with a distinct upward tilt, yellow at the base and shading into red-brown towards the tip; legs, yellow. Length, 4-4$\frac{3}{4}$ in.

Distribution: Generally distributed north to the Pilbara district and extending into central Australia.

Nesting: The Black-capped Sittella builds its nest in a small vertical, and usually dead, fork of a limb of a tree, usually at a considerable height from the ground—up to 50 feet. It is a beautifully made nest, deep cup-shaped, of spiders' web and cocoons woven together and the outside decorated, or camouflaged with small pieces of bark or lichen. It measures about 3 inches in height and 2½ inches in breadth. The 3, rarely 4, eggs are pale bluish-white, spotted and blotched, mainly at the broader portion, with olive and dark grey markings. Size, 17 x 14 mm.

Sittellas are woodland birds and in the non-breeding season they move in small flocks restlessly and noisily from tree to tree, foraging for insects. They are very numerous in the South-West, often visiting King's Park and occasionally moving along the street trees of the better wooded suburbs.

In the Kimberley Division, and extending far south into the dry country beyond Mount Alexander, is a rather similar species, the White-winged Sittella, *N. leucoptera.* It is distinguished by possessing *black streakings* on the back and a *white* instead of an orange "window" in the wing. It is also smaller.

AUSTRALIAN TREE-CREEPERS: Family Climacteridæ

Rufous Tree-creeper ("Woodpecker")

Climacteris rufa

Native names: Jin-nee, Jinni (Avon River).

A *rich dark brown above, reddish-brown below,* streaked with black and buff markings on the upper breast. There is a reddish-brown eyebrow and *a chestnut wing stripe,* which is very conspicuous in flight. The tail has a black sub-terminal band. Iris, reddish-brown; beak, brown; legs, black. Length, 5¾-7 in.

Distribution: Occurs generally south of the mulga-eucalypt line (Fig. 4), with its headquarters in the jarrah and karri forest belt, and wandoo and salmon gum country. J. R. Ford has recorded it in the *Eucalyptus gongylocarpa* and *Triodia* savannah woodland in the Great Victoria Desert. It is the Western Australian representative of the Brown Tree-creeper *(C. picumnus)* of the eastern States.

Nesting: As with the other forms of Australian tree-creepers, the western bird uses as a nesting site a hollow spout in a tree, at heights up to 25 feet, though mostly under 10 feet. On occasions a hollow in a log lying on the ground is used. The cavity selected is a deep one and much nesting material—grasses, plant down, soft bark, and even sheep's wool, rabbits' fur and dung is used to make a base on which the eggs are laid. The eggs are white, speckled all over with reddish spots and splashes. Clutch size, 2 or 3 (Darling Range); 2 (Pingelly-Corrigin district); 1 (Lake Grace). Measurements, 21-24 x 18-19 mm.

The call is a rather shrill "tsoo-tsoo-tsoo," or "tsit-tsit-tsit."

The birds may be seen spiralling up the trunks of trees in their search for insects and a favourite feeding situation is on the ground, among fallen timber. The species rapidly withdraws in the face of closer settlement and it has long since disappeared from the coastal plain. Near Perth it can still be seen, however, in the jarrah forest plateau of the Darling Range, though even here its numbers are diminishing rapidly.

White-browed Tree-creeper

Climacteris affinis

Native name: Tiru-tiru (Warburton Range).

Upper parts brown, with a black sub-terminal tail band and a brownish orange band in the wing. There is a white eyebrow line. The *throat and breast are grey, sharply marked off from the streaked abdomen* and under tail coverts, a strongly-patterned black and white. The females have slight rufous on the throat and a narrow rufous line above the white eyebrow. Iris, light brown; beak and legs black. Length, 5-5½ in.

This tree-creeper resembles the preceding species in its upper coloration, including the chestnut wing "window", but differs from it by the replacement of the rich rufous ventral coloration by the black and white streakings.

Distribution: A mid-northern species, extending north to the upper Gascoyne and Peak Hill districts, eastwards into central Australia and south to a line joining Mingenew, Southern Cross and Kalgoorlie. It is an inhabitant of the mulga country and usually it is exclusive of the areas in which the Rufous Tree-creeper occurs, but may overlap the range of the Black-tailed Tree-creeper.

Nesting: The nest consists of a collection of grass, plant-down, soft bark, or fur and hair, placed at the bottom of a hollow in a limb or in the trunk of a tree. The 2 or 3 eggs are pinkish-white, covered all over with spots of pinkish-red and purplish-red. Size, 22 x 15.16 mm.

These birds feed extensively on ants and they may be often seen feeding on the ground at the base of trees, hopping about sideways, and apparently gulping down ants.

Black-tailed Tree-creeper

Climacteris melanura

Native name: Chinin-chinin (Pilbara district).

Upper parts dark-brown, blackish-brown on the tail; *throat black streaked white;* rest of under parts bright brown. A chestnut *patch or "window"* in the wing. The *female* differs in having a *white throat,* below which there are *white striations on a rufous-ground.* The under tail coverts in both sexes are dark brown barred lighter. Iris, brown; beak and legs, dark brown. Length, 6½-7½ in.

Distribution: A northern species extending south to the Gascoyne River. In older reference books the Western Australian form, south of the Kimberley Division, is given as a distinct species, the Allied Tree-creeper, *Climacteris wellsi.*

Nesting: The nest is similar to that of the White-browed Tree-creeper and may be found also in a hollow in a limb of a tree, or in a hollow in a trunk. The clutch consists of 2 eggs, which are similar to those of the White-browed Tree-creeper, but are larger, averaging 24 x 17 mm.

This tree-creeper is found among the river gums (*Eucalyptus camaldulensis*) of the North-West rivers and creeks and is rarely seen away from the gum country. The call is a loud sharp whistle.

FLOWER-PECKERS: Family Dicæidæ

Mistletoe-bird

Dicæum hirundinaceum

Native name: Moo-ne-je-tang (Perth).

The male has the *upper parts a gleaming black with blue metallic reflections.* Below, the *throat, chest* and *rump are a brilliant red,* and the remainder white with a black median longitudinal stripe which does not, however, extend as far as the abdomen. Iris, chocolate-brown; beak, grey-black; legs, black. Length, 3¾ in. The female has entirely dark brown upper parts except for the shining black tail; it is greyish-brown and the under tail coverts are a dull red. The juvenile has a red beak.

Distribution: Distributed all over the State. It is uncommon in the jarrah forest country of the South-West, but is occasionally met with in the Perth district, and in some local areas, such as parts of Nedlands, a pair or two may occur regularly.

Nesting: The Mistletoe-bird constructs a type of nest which is unique among the birds of Western Australia. It is so constructed that the materials which are used by the birds—spiders' webs and cocoons, together with fluffy seeds—are so firmly bound together that it is almost felt-like in its consisency. It is pear-shaped, and is suspended from a twig in a bushy tree from about 4 feet up to about 50 feet. The entrance is slit-like and about 1¼ x ½ in. The whole nest measures about 3 inches deep and 2 inches broad. A nest brought back by F. Lawson Whitlock from central Australia was found to weigh, including a portion of the twig to which it was attached, only 7 grains. The 3 eggs are pure white and are long ovals in shape. Size 17 x 11 mm.

The bird, as its name implies, is intimately associated with mistletoe *(Loranthus)* and is the plant's most active disseminator. As an adaptation to this specialised diet the muscular portion of the stomach (the gizzard) has virtually disappeared and the alimentary canal (including the oesophagus, glandular portion of the stomach and the intestine) appears superficially to be an even duct with no enlargements. Another characteristic is that the bird defaecates sitting along, and not across, a branch so that seeds become attached to the branch.

Its high-pitched calls have been syllabised as "swiz-it, swiz-it"; "weet-weet-swizet," "weet-weet-swizet-it," and "per-swizit-per-swizit-it."

DIAMOND BIRDS (PARDALOTES): Family Pardalotidæ

Spotted Diamond-bird (Pardalote)

Pardalotus punctatus

Native name: We-dup-we-dup (Perth).

Crown, wings and tail, *black, spotted with white* (yellow in female). The feathers of the back buff broadly edged with black, giving a spotted effect. *Rump chestnut-brown,* becoming red at the tail coverts. There is a white eyebrow and the side of the head is grey finely barred with white. Underneath the throat and under-tail coverts are bright yellow (buffy in female) and the rest of the under parts buffy-brown. Iris, grey; beak, short and thick, black; legs, fleshy. Length, 3¾ in.

Distinguished from the Red-tipped Diamond-bird by most of the upper parts, including the back, being spotted, no yellow between the eye and beak, and no white wing line.

Distribution: Restricted to the South-West corner, King's Head (20 miles north of Cockleshell Gully, near Jurien Bay) and Moora to the Stirling Range.

Nesting: The Spotted Diamond-bird always nests in a bank in the earth clinging to the roots of some large tree that may have been blown down, or even in the ground itself—it never nests in a hole in a tree as does the Red-tipped Diamond-bird. A tunnel is excavated and at the end a space is enlarged in which the birds construct a globular nest of fine strips of bark or grasses, with an entrance at the side opposite the tunnel leading to it. The tunnel, which is often as much as 18 inches in length, can be recognised by there always being two longitudinal depressions at the bottom, formed by the scratching of the birds when excavating. The 4 eggs, roundish in shape, are pure, glossy white. Size, 17 x 13 mm.

The call, "sleepy-baby," is uttered so frequently that it becomes monotonous and has led settlers to apply the name "headache bird" to the species. The voice may easily be mistaken for that of the Western Warbler.

The Spotted Diamond-bird is plentiful in the forests of the South-West, but is less common in the Perth area. There appears to be an annual autumn movement from the south to the Perth district which in some years may be quite considerable. The birds appear during May and stay to nest in the spring.

Yellow-rumped Diamond-bird

Pardalotus xanthopygus

Native name: Bilyabiet (Pallinup River).

This bird is similar to the preceding species, except that the *rump is bright yellow* instead of chestnut brown.

Distribution: Confined to the mallee country of the south-east; west to Tambellup and Lake Grace and Warriup, north to Norseman and east to the Eucla region. In this area it replaces the Spotted Diamond-bird.

Nesting: The nesting habits of this bird are generally the same as with the preceding species, except that the nesting tunnel is drilled into flat or gently sloping ground and not into a bank. A set of 4 eggs taken at Lake Grace measures 17 x 13 mm.

Red-tipped Diamond-bird (Mannakin, Pardalote)

Pardalotus substriatus

(Plate 8)

Native names: Bawalilly (Lyons River, all species); We-dup-wee-dup, Widap-widap, Wida-wida, Widgi-winji (Perth); We-dee-we-due (Avon River).

Head black, hinder part marked with short white stripes; back grey; rump, tawny-brown; short tail black, slightly tipped white; wing black with white stripes along several of the outer quills forming a conspicuous *white band* at the head of which is a *red spot* (which may rarely be yellow). From the base of the beak to above the eye is a bright yellow line which is continued backwards as a white eyebrow. Chin and throat bright yellow. Remainder of under surface buff. Iris, cream; beak, short and heavy, black; legs, brown. Length, 4-4½ in.

Distribution: The most widely distributed diamond-bird, ranging into central Australia and north to the Fitzroy River. It is very common in the South-West.

Nesting: This diamond-bird is different from other members of the genus in that it does not build its nest in an excavation in a bank, but constructs a dome-shaped nest of grasses or thin strips of bark, sometimes lined with feathers, in a hollow or hole in a tree. The site is often at a considerable height from the ground and we have seen a nest well over 50 feet high in a marri *(Eucalyptus calophylla)*. The bird is also known to nest in buildings and for several years a pair nested in a hole in the old gaol building of the Western Australian Museum, Perth. The 3

Plate 8.—Upper: Southern Emu-Wren, male and female.
Lower: Red-tipped Diamond-bird.

eggs (on occasion 4 or 5) are roundish in shape, pure white and very glossy. Size, 17-19 x 14 mm.

Until recently it was thought that this common species was virtually sedentary, or at the most subject to only minor nomadic movements. However E. H. Sedgwick has demonstrated that the species is strongly migratory. Many individuals may vacate the heavily-timbered forest block in the South-West between March and May and radiate out into the drier parts of the State, even as far as the Pilbara, where they are winter visitors. On February 27, 1972 an immature bird flew exhausted on to a fishing boat 16 miles at sea, off Lancelin. Between August and September the southward movement is in progress, but by October the species has still not fully penetrated into the deep South-West. Some individuals, however, do remain behind in the south. The birds may not breed in the north but full data on the distribution of breeding birds are lacking. This may fluctuate according to seasonal conditions.

The call, "pick-wick," "be-quick" or "wit-a-witta," is far more often heard than the singer is seen, as it is a denizen of the tree-tops. It is not, however, confined to these situations and may at times join the mixed flocks of thornbills and other small birds in foraging in the lower branches of trees and even near the ground.

The native names are onomatopoeic, and one, "widji-winji," means "Where is the emu?" (Fletcher Moore).

A somewhat similar species, the Black-headed Diamond-bird *(Pardalotus melanocephalus)*, which entirely lacks the white stripes on the head, occurs in the Kimberley area and F. Lawson Whitlock recorded that he observed it in the Pilbara district in 1909. However, as no specimen was collected, nor has any specimen been collected south of the Kimberley Division, we do not include the species in this book.

Red-browed Diamond-bird

Pardalotus rubricatus

Native names: Be-lul (Well 33, Canning Stock Route); Wunnbunnbullalla (Jigalong); Dia-dia (Warburton Range).

Rather like the Red-tipped Diamond-bird in general appearance, with a black head which is *spotted* with white. Back brown, with yellow on the rump. Tail black, slightly spotted with white. The eyebrow line commences as a reddish-orange spot near the beak and continues as a buff line. The wing has no red spot, but the outer wing quills have the outer vanes yellow. The under parts are greyish-white with a yellow patch on the breast. Iris, pale yellow; beak, dark brown; legs, grey. Length, 4½ in.

Distribution: A northern and central Australian species, extending south as far as the Gascoyne River and then trending south-easterly to the Great Victoria Desert.

Nesting: The Red-browed Diamond-bird drills into a bank and at the end of the tunnel so made, which is on occasions as long as 2 feet, excavates a round chamber of about 4 inches in depth and height. Within this chamber it constructs a cup-shaped nest of strips of bark (cajaput) which it lines with fine grasses or paper-bark. The 2 or 3 eggs (sometimes only 1 or as many as 4) are nearly round, pure dull white. Average size of 11 eggs, 17.7 x 13.9 mm., varying from 16.6-18.9 x 13.4-14.2 mm.

The call is a soft, monotonous note and is uttered by both sexes, rendered as "chirgajip" by E. Lindgren. Derek Goodwin renders it as a liquid "pwer, pwee-ti-ti-tit."

SILVEREYES: Family Zosteropidæ

Western Silvereye (Greenie)

Zosterops gouldi

(formerly *Zosterops australasiæ*)

Native names: Jule-we-de-lung, Julwidilang Perth); Poang (Pallinup River).

Entire upper parts a bright olive-green; wing and tail quills grey edged green. Throat and under tail coverts yellow-green and rest of under parts grey. *Around the eye a circlet of white feathers.* Lores and below eye dark grey. Iris, light brown; beak, dark brown; legs, slate-grey. Length, 4-5 in. The male is brighter yellow on the throat than the female.

Distribution: A South-West form extending to Shark Bay, Carnarvon and Boolathanna Station (rarely as far as Point Cloates and the De Grey River in the winter) and inland to Mingenew, Wongan Hills, Kellerberrin (irregular), Lake Grace, Peak Charles, Norseman and Newman Rocks; it occurs in the coastal vegetation as far east as Eucla. It has now appeared in Kalgoorlie. It is also found on most of the neighbouring islands, including those in Shark Bay, the Abrolhos and the Recherche Archipelago.

Nesting: The Silvereye chooses usually a shrub in which it builds a suspended cup-shaped nest of grasses, binding the grasses with cobwebs and lining the whole with finer grasses, sheep's wool or horsehair. The cup is deep and about 2 inches broad. The clutch consists of 2 or 3, occasionally 4, pale blue eggs. Average size of 13 eggs from Perth, 16.8 x 12.4 mm., varying from 15.6-17.8 x 12.0-12.8

mm. Incubation takes 10 to 12 days, nestling period 12 days (W. H. Loaring).

This is perhaps the commonest small bird in the Perth area and over much of the South-West. After the nesting season, by January, the birds gather into foraging flocks, which are noisily on the move until the pairs separate out again next spring. In the city and suburbs they play the role of the Sparrow *(Passer domesticus)* in the eastern States, or the tits *(Parus)* in Europe, visiting gardens, shrubberies and even the backyard fowl-run. In the fruit season they cause some trouble to figs, grapes and other soft fruits. Orchardists adjoining forest country and running water-courses may find them a serious pest in varying degree. How great their numbers are, and illustrating the futility of keeping them down by mere destruction of individuals, are the examples of attempted control quoted by B. H. Woodward in the *Western Australian Year-book for 1900-01:* "In one orchard on the Swan River, near Perth, at least 20,000 birds have been shot. In an orchard near Bunbury 1,200 of these birds were shot in one day."

This Western Australian species differs from its eastern Australian ally *(lateralis)* in that the whole of the upper parts are green (the back is grey in the eastern birds). It also lacks the pre-nuptial moult which characterises the eastern populations. Nevertheless, some ornithologists combine them as one species.

Yellow Silvereye

Zosterops lutea

Differs from the preceding species in having the *whole of the under surface bright yellow,* and the whole of the upper surface has a more yellow-green tone. The head looks almost entirely yellow. Iris, reddish-brown; beak, black; legs, dark grey. Length, 3¾-4½ in.

Distribution: A northern species extending as far south as Shark Bay; it does not go far inland.

Nesting: The nest is a neat deeply woven cup, with the edge woven over the supporting branches, usually of mangroves. Made of grass stems with cobweb on the outside. Overall diameter, 2½ inches, the nesting cup being 1¾ inches in diameter and 1½ inches deep. Clutch, 3 or 4 eggs, and of a paler blue than the preceding species. Size, 17 x 12 mm.

The Yellow Silvereye is practically confined to mangroves, which, on the mainland, reach their southern limit in Shark Bay (except for an isolated colony in the Leschenault Inlet, Bunbury). It does not occur on the mangroves at the Abrolhos Islands. At a small mangrove lagoon on Peron Peninsula, Shark Bay, in July

1944, we had a striking illustration of how the two species of Silvereye remain attached to their preferred habitats. Whilst one of us was observing the birdlife in the mangroves, finding the Yellow species quite abundant, the other, a few chains away found the Western Silvereye plentiful in the low scrub to the very edge of the lagoon. The two species existed here within earshot of each other, yet not intermingling. T. Carter stated however, that in the North-West Cape area flocks of considerable size leave the mangrove area in winter and feed in the scrubby coastal sandhills and more open country. At Carnarvon we have seen the species at such times intermingling with the Western Silvereye. At Pelican Island, Shark Bay, where there are no mangroves, the Yellow Silvereye lives in the thickets of nitrebush *(Nitraria schoberi)* and it has no competition from the Western Silvereye, which is absent from the islands. Similarly, on Thevenard, Airlie, Large and North Sandy Isands, off Onslow, we have seen the Yellow Silvereye in the acacia thickets. The species occurs generally, in mangroves and elsewhere, on Lowendall Island and the Montebellos.

HONEYEATERS: Family Meliphagidæ

Brown Honeyeater

Lichmera indistincta
(formerly *Gliciphila indistincta*)

Native name: Jin-jo-gour (Avon River).

A small modestly plumaged honeyeater, greyish-brown above and grey below, with olive-green on the wings. Behind the eye is a tiny patch of minute yellow and white feathers. Iris, brown; beak, long, rather robust and curved, dark brown; legs, grey. Length, 5-5½ in.

Distribution: Found all over the State, coastal and inland, but it is not permanently resident in the jarrah forest belt of the South-West except where settlement has opened up the country, as for example, Bickley in the Darling Ranges and the Bridgetown area. However, it invades the karri belt (as at Pemberton) when the trees are heavy in blossom. In the northern part of the jarrah belt, such as along the Toodyay and York Roads, it is generally distributed, favouring the vicinity of swampy areas. In the semi-arid eucalypt woodland zone it is restricted to the dense shrubbery around granite outcrops and to towns where there are street trees and gardens. In the far north it is common alike in the coastal mangroves and wooded country and scrub inland.

Nesting: The nest of the Brown Honeyeater is small and cup-shaped, it measures about 2½ inches in diameter and nearly the same in depth. Constructed of shredded bark and lined with the reddish down of the zamia palm, it is usually found suspended in the twigs of a small tree or bush. The 2 eggs are blunt ovals, white in colour with a few reddish-brown spots. Sometimes the spots are altogether absent. Size, 17 x 13 mm. Though it is one of the smallest honeyeaters, it is sometimes parasitised by our largest cuckoo, the Pallid *(Cuculus pallidus)*.

This bird calls attention to itself by its loud spirited song; the ordinary warning and signal notes are harsh, but the song in the nesting season is remarkably similar to that of the Reed-Warbler. The Reed-Warbler's song, however, is more melodious, louder and uttered with greater vigour.

In the South-West the Brown Honeyeater is particularly noticeable in gardens when the tree-lucerne is in blossom. In Perth it is familiar in street trees and gardens in the better-wooded suburbs, and may be seen probing in the blossoms of the coral tree *(Erythrina indica)*.

Honeyeaters are a characteristic Australian family and, as their name implies, feed on the nectar of flowers, for which purpose their beaks and tongues are adapted. The beak is lengthened and curved for probing into flowers and the horny tip of the tongue frayed out into a brush, while the sides are curled upwards to form a more or less closed tube. In their visits to flowers the birds also capture insects, which form a substantial part of their diet. In consequence of their feeding habits these birds may play an important role in pollination and many species of Australian plants depend on honeyeaters and other birds as effective pollinators. O. H. Sargent, the Western Australian botanist, considered that honeyeaters are the regular pollinators of the genus *Eucalyptus,* and that adaptation to bird demands has been important in the evolution of these and other plants, including *Banksia, Templetonia,* the Sturt Pea *(Clianthus speciosus)* and kangaroo paws *(Anigozanthos)*. However, more recently, Professor L. D. Pryor, of Canberra, has expressed the view that pollination in *Eucalyptus* is largely by insects, and perhaps to some extent by birds.

Pied Honeyeater

Certhionyx variegatus

A *black-and-white honeyeater,* head, throat, and upper parts black, under white. There is a broad white stripe on the wing and the tail is black in the centre and at the tips, the rest of the outer tail being white. The female and immature male are quite different. The upper parts are brownish-grey with buff-white markings on the wing coverts. Throat and chest brownish-grey, marked brown; abdomen white. The male bird at first appearance looks like a Hooded Robin or male Triller and the female like a female Triller. Iris, bright brown; a prominent sky blue lobed wattle dependent from the lower eyelid, much reduced in the female; upper eyelid black; beak, slaty-blue tipped black, downcurved and rather long; legs, leaden grey. Length, 5¼-7¼ in.

Distribution: This is a mid-northern form, ranging north to about the Fortescue River and south to about Geraldton, extending eastwards to the interior and Central Australia (Fig 6). Recorded south-western limits of its distribution are Nangetty Station (Irwin River), Wongan Hills, Wubin, Coonana and Naretha (on the edge of the Nullarbor Plain).

Nesting: The Pied Honeyeater's nest is cup-shaped and is usually placed in a fork in a low tree or large-sized bush, often up to 15 feet from the ground. The red-flowered *Hakea multilineata* is often used as nest site. The nest is strongly made of grasses kept well together with spiders' webs; size, about 4 inches broad. The eggs are white (occasionally buff white), marked all over with dark brown and grey spots and blotches. They are quite unlike the usual type of eggs laid by the honeyeaters and resemble in appearance eggs of the Dusky Wood-Swallow, but are smaller. Average size of 20 eggs, 22.1 x 16.0 mm., varying from 21.1-23.5 x 15.3-17.3 mm. The species is also unusual in laying a greater clutch than any other local honeyeater, excepting the Miner. The clutch consists of 2 or 3 eggs, and rarely 4 may be found. In the Murchison district the principal breeding month is September.

Both sexes share in the building of the nest and the incubation of the eggs. A. Fewster and M. T. Millard established the incubation period as "not more than 12 days 21 hours." The female has been observed to react with the "broken wing" response when disturbed at the nest containing young.

The male has a distinctive territory-advertising display. After rising vertically into the air, it descends with aerial evolutions, the tail fanned out, and often to the accompaniment of a piercing call which has been syllabised as "te-titee-tee-tee"; when not displaying the call has a mournful cuckoo-like quality and bears a distinct resemblance to the monotonous call of the Little Grass-bird. The sharply contrasting black and white plumage pattern arrests attention during these displays. Such behaviour has been noted by us at Shark Bay in early August and in inland New South Wales later in the same month (Dr. A. Chenery). Apparently it is not indulged in after mating has started when the birds become silent and undemonstrative.

At many parts of its range the species is regarded as a winter migrant, arriving after the first winter rain at Point Cloates and in July at Erlistoun. On migration flocks of birds have been reported.

The Pied Honeyeater was first recorded in Western Australia when John Gilbert sent specimens to Gould in January 1843 during his second collecting expedition to the State. These specimens came from Wongan Hills.

Fig. 53.—Pied Honeyeater, male.

Fig. 54.—Black Honeyeater, male.

Black Honeyeater

Myzomela nigra

Native name: Dwer-da-ngok-ngun-nin (Avon River).

The upper parts, head, chest and a line along the middle of the abdomen blackish-brown; the remainder of the under parts are white. The female differs strikingly from the male, being brown above with a white eyebrow, throat brown and greyish-white, and the abdomen white. Iris, brown; beak, long, down-curved, black; legs, black. Length, $3\frac{3}{4}$-$4\frac{3}{4}$ in.

Distribution: Generally distributed north to the Pilbara district, but avoiding the South-West corner (in this case the known limit is a line joining Moora, Aldersyde and Dumbleyung. Occasionally birds may wander into the outer fringes of the South-West forest belt (E. H. Sedgwick recorded an invasion into Wooroloo in February 1952).

Nesting: Nests found by F. Lawson Whitlock at Wiluna in August were loosely built, small, open, and rather shallow structures of small dead twigs and pieces of dry grass, etc., matted with cobwebs, and placed in the fork of a small horizontal limb, within a few feet of the ground. Size, 2 inches broad and 1½ inches deep. The 2, occasionally, 3, eggs are yellowish-buff, with a zone at the larger end of minute spots and cloudy markings of pale slate and umber. Size, 15-16 x 11-12 mm. Whitlock noted that the female does all the work of nest-building, the male perching near at hand on some conspicu- fluttered down to the nest. C. G. Gibson observed that the male ous twig and uttering a monotonous call, with an occasional erratic flight around. When the female arrived with building material he shared in the incubation of the eggs; he also assists in feeding the young.

Grey Honeyeater

Lacustroica whitei

Upper parts brownish-grey; wing quills dark brown with outer vanes slightly tinged olive-yellow; tail dark brown, slightly tipped white; a ring of ash-grey feathers around the eye; under parts white, breast clouded with grey. Immature birds have a yellow wash on the cheeks and throat and a ring of buff-tinted or golden feathers around the eye. Iris, brown; beak, down-curved but very short for a honeyeater, black; legs, lead-grey. Length, 4-4½ in.

Distribution: Restricted to the mid-north; north to the Minilya River, the upper Ashburton and Mundiwindi, south to about Geraldton and Mingenew, then eastwards to central Australia and northern South Australia.

Nesting: A nest found by F. Lawson Whitlock at Lake Way was attached to the extremity of a horizontal branch of a narrow-leaved mulga. Both birds assisted in the building of the nest which was a frail structure and lacked neatness and finish in its architecture. It was about 7 feet from the ground and was constructed of long horsehairs which were held in position with spiders' webs and cocoons and the whole was attached to the extremity of a slender branch by spiders' webs. The walls were very thin and the whole nest measured 2¾ inches x 2 inches x 1 inch in depth. The 2 eggs were plainly visible from below and were swollen oval in shape, fine in texture, and slightly glossy. Colour white, moderately marked with rich reddish-brown spots, while some of dull purplish-grey were also at the larger ends of the eggs and appeared as if beneath the surface of the shell. Size, 17 x 12 mm., and appear large for the size of the bird.

In October and November A. G. and B. A. Wells found nests in the Yalgoo district in narrow-leaved mulga and curara, 3-8 feet from the ground (eggs and nestlings). Both parents shared in the incubation and feeding of the chicks.

This little-known and distinctly un-honeyeater-like species was first discovered by Whitlock at Lake Austin in 1903, but was not scientifically described until the same collector obtained specimens at Wiluna in 1909. In the field the bird looks a uniform grey and it has a superficial likeness to the Silvereye, which it also resembles in habits and notes, according to Whitlock. It does not appear to favour the eucalypts and seems to be a creature of mulga and acacia woodland.

Singing Honeyeater (Grape-eater, Grey Peter)

Meliphaga virescens

(Plate 1)

Native names: Doo-rum-doo-rum (Perth); Gool-boort (Avon River); Darandura (Lake Guli, Canning Stock Route).

Greyish-brown above, *streaked buff and grey below.* A conspicuous *black line passes from the beak past the eye,* below which is *a prominent yellow ear patch, tipped with white.* Iris, brown; beak, rather short and down-curved, black; legs, light grey-blue. Length, 7½-8¾ in.

Distribution: Found all over the State, though it avoids the block of heavy-forested, mainly jarrah country in the South-West corner, but since settlement and consequent clearing it has appeared in many new localites such as Bridgetown and along the coastal strip of country as far south as Hamelin Bay. It is a familiar dry-country honeyeater and is well established on most of the islands off the coast—even the smallest. It does not, however, occur on the Abrolhos Islands, which group is entirely lacking in honeyeaters.

Nesting: The Singing Honeyeater builds a suspended cup-shaped nest of green grasses and lines it with banksia, or other flower down, rabbits' fur, sheep's wool, or even kangaroos' hair. It is usually situated in the centre of a bush or tree and well-hidden in the foliage. The 2 or 3 eggs vary from pale buffy-white to pale-salmon, darker at the larger end, and sometimes have a few specks of reddish-brown towards the larger end. Average size of 11 eggs, 22.2 x 16.0 mm., varying from 21.0-25.0 x 15.3-17.2 mm.

This honeyeater is a common Perth species and has made itself at home in the parks and street trees. The common call is an unmusical "queek" or "sheek," and another is an animated "terric,

terric, terric," resembling the notes of the Bee-eater. A very rapid, rippling call, "rer-er-er-er-er-er," or "te-e-e-e-e-e" may be an alarm or warning call. Where there is a dense population of the birds, as at Shark Bay and the Perth metropolitan area, their continuous singing is a melodious feature of the early dawn chorus in spring and early summer. This is an aggressive bird towards smaller species. K. Immelmann has observed it ransacking nests of Zebra Finches at York, to feed on the eggs. In the Kimberleys he saw these honeyeaters plundering the nests of Zebra and Double-bar Finches.

Yellow-plumed Honeyeater

Meliphaga ornata

(Plate 1)

Native name: Miamit (South-West).

Back brown, with olive on the wings and tail and brighter olive-green on the head; under parts whitish-grey, *strongly streaked* with brown. There is an indistinct brown line through the eye. A *bright yellow plume* behind the ear coverts, with *no black markings* in this region. Iris, hazel; beak, black; legs, flesh-colour. Length 5¼-6½ in.

Distribution: A South-West species probably distributed north to the mulga-eucalypt line (recorded northern limits are the mouth of the Murchison River, Yandanooka, Callion, Coolgardie, Queen Victoria Spring and Kanandah), and eastward along the coastal mallee fringe to Eucla. It avoids the heavy jarrah forest tableland; but is common in the coastal tuart belt south to Busselton (Figs. 6 and 9). Besides the tuart it has a partiality to wandoo forest.

The clearing of the forest is enabling the species to penetrate westward. G. A. Lodge observed it on flowering flooded gums between Boyup Brook and Bridgetown in November 1962. In January 1963 it was recorded breeding in wandoo country at Mayanup.

Nesting: The nest is cup-shaped but not deep, and is composed of dry grasses without any other lining. A typical nest measures about 2½ inches x 1½ inches. It is usually suspended in the drooping leaves of a sapling but may be placed in a bush as low as three feet from the ground. The 2 or 3 eggs are white with a pinkish-buff tinge, and are spotted and blotched with pinkish-brown with some of dull grey which look as if beneath the surface of the shell. The markings tend to form a zone at the larger end. Size, 19-21 x 14 mm. The species is a late breeder, the nesting season being from October to December.

The Yellow-plumed Honeyeater is a bird which is heard calling at a very early hour—well before sunrise. The call is loud and clear, but harsh and sharp and is associated with steep aerial flights as in other honeyeaters. Its strong warning call when predators are about is taken up by other individuals of the species and other birds also react to it. It is a bird of pugnacious character and, when feeding at the flowers of a tree, the birds quarrel much among themselves for positions. Should a bird of another species arrive at the feeding site, the honeyeaters usually make an organised attack on it.

Yellow-fronted Honeyeater

Meliphaga plumula

(Plate 1)

Resembles generally the Yellow-plumed Honeyeater but can be readily distinguished from it by the *much less conspicuous streakings* on the under parts and the presence of a *brownish-black mark above the yellow ear tuft.* The back is brown with olive on the wings and tail and brighter olive-green on the head. The under parts are whitish-grey with faint brown markings and a yellow wash. Iris, dark brown; beak and legs, brown. Length, 5-6¼ in.

Distribution: A northern and central Australian species which occurs south to about Carnamah and thence south-easterly to Kalgoorlie and the Great Victoria Desert. During his visit to Western Australia (1839-40) John Gilbert obtained specimens in the "York district," but the species has not since been recorded from there; in actual fact they were probably collected on his visit to the Wongan Hills.

Nesting: This bird builds a suspended cup-shaped nest in a bush within a few feet of the ground. It is constructed of dry grasses. The 2 eggs are light salmon-colour and are marked with minute spots of pale reddish-brown which tend to form an irregular zone over the larger end. Size, 20 x 14 mm. In South Australia E. F. Boehm has recorded a nestling period of 15 days.

Mr. Boehm renders the call as "it-wirt, wirt, wirt wirt," and has heard a male sing beautifully with canary-like notes when perched on a tall shrub. At times the birds make erratic flights 20 feet or more above the scrub and emit low guttural calls on alighting.

Purple-gaped Honeyeater

Meliphaga cratitia

(Plate 1)

Upper olive-green, brighter green on the wings; head grey. Under parts light grey with yellow wash. A *black patch in front of and behind the eyes,* extending over the ear coverts; behind this there is a *small yellow plume.* There is a bright yellow stripe from the base of the lower mandible to below the eye. A narrow fleshy ridge or line of vivid lilac is exposed between the feathers from the angle of the jaw to the hinder part of the cheek, but this is not evident in the field unless the bird is at close range; its colour is yellow in young birds. Iris, brown; beak, black; feet, grey: Length, $7\frac{1}{4}$-$7\frac{1}{2}$ in.

Distribution: So far reported only from a restricted area in the Wheatbelt region, north to the Wongan Hills, east to Mt. Hampton (40 miles S.S.W. of Southern Cross) and Mt. Holland, and south-west to East Broomehill and the Stirling and Green Range. It does not occur in the heavily-forested South-West corner. It favours thickets of mallet, yate, and mort, and is fairly numerous in the southern portions of its circumscribed range. In the Hopetoun area, for instance, it is one of the most plentiful of the honeyeaters, occurring even in the street trees of the township.

Nesting: The species builds a suspended nest, cup-shaped, and made of thin strips of bark and lined with grass. It is usually suspended from a horizontal branch of a bush or small tree a few feet from the ground. The 2 eggs are white with a pinkish tinge and have reddish-brown spots and specks which tend to form a zone at the larger ends. Size, 21-22 x 15-16 mm.

Grey-headed Honeyeater

Meliphaga keartlandi

(Plate 1)

Native name: Widge-e-doo (Coolawanyah).

Upper parts greyish-brown with olive-green on wings and tail feathers; *crown of head grey;* under parts pale yellow, faintly streaked with light brown. There is a black-brown line from the base of the beak, through and around the eye, *widening behind into a dark slaty patch* which has a sheen in certain lights. Iris, brown; beak, dark brown, lower mandible yellow at base; legs, light grey. Length, 6-$6\frac{1}{4}$ in.

Distribution: A northern and central Australian species, extending through the North-West to Point Cloates and to the Hamersley Range. Along the Canning Stock Route it occurs as far south as Lake Disappointment. Its favourite haunts are rocky gullies in rugged range country and around springs, and it seldom encroaches on the river gums habitat of the White-plumed Honeyeater—the other common species of its general area of distribution.

Dr. G. F. Mees has shown that the species had been collected over 50 years before it was scientifically named. There is a specimen in the natural history museum at Leiden, Holland, probably collected at Depuch Island when H.M.S. *Beagle* visited it in 1840.

Nesting: This honeyeater usually builds low down and as a rule in some shrub or large plant growing in the bed of a creek or on a bank close at hand. No concealment is aimed at; a favourite site is a branch of a prickly grey-brown-leaved canjie bush *(Acacia pyrifolia)*. The nest is usually, but not always, suspended, and is constructed of fine strips of bark, spiders' webs, and vegetable down; it harmonises wonderfully with the foliage in which it is built (Whitlock). The 2 eggs are pale salmon or pale flesh-pink, or white, sparingly marked, particularly at the broader end, with light brown spots. Average size of 13 eggs from Exmouth Gulf, 20.0 x 14.3 mm., varying from 18.9-21.7 x 13.2-15.0 mm. The species appears to breed at any time of the year after good rains.

The notes uttered by this bird are very variable, some of them musical, but others, again, rather harsh. It can hardly be called a melodious singer, its efforts in that direction being far inferior to those of the White-plumed Honeyeater; but on the other hand in variety of note or call it far exceeds the latter (Whitlock).

White-plumed Honeyeater (Native Canary)

Meliphaga penicillata

(Plate 1)

Native names: Tui-dee (Pilbara district); Widge-e-doo (Coolawanyah); Indatjarra (Jigalong).

Upper parts greyish-brown with olive-green on the wings and tail. The *forehead and sides of the head are yellow,* which gradually merges into the brown of the back. There is a small white plume behind the ear coverts. The under parts are greyish with a yellow wash, particularly apparent on the throat and breast, and there are indistinct brown streakings as in the Yellow-fronted Honeyeater. Iris, dark brown; beak, black; palate, orange; legs, pale brown. Length, $6\frac{1}{4}$-$6\frac{3}{4}$ in.

Distribution: This most widespread Australian honeyeater occurs, in this State, north to the Pilbara district and south to Moora and Kalgoorlie and thence eastwards into central Australia (Fig. 6). In July 1969 a pair of birds were seen in a suburban garden at Joondanna, north of Perth. It is essentially a bird of the river gums *(Eucalyptus camaldulensis)* and is never seen in the dry scrub country. Its comparatively restricted distribution in Western

Australia contrasts rather strikingly with the wide range it enjoys in central Australia and the eastern States. It is a familiar bird in gardens at Adelaide and Melbourne, and has a habit of frequenting the street trees and gardens of our northern towns, such as Carnarvon.

Nesting: The nest, cup-shaped and so fragile that the eggs are often visible from below, is suspended in a bush or a tree, and often in the interior of a plant climbing over a tree. It is constructed of grasses or fine strips of bark (cajaput), plant-down and spiders' web, and is lined with wool or plant-down. Size, 3½ inches diameter and 2 inches in depth. The eggs are white and sparingly spotted, and then mainly over the larger end, with reddish-brown, pinkish-brown and purplish-grey. Average size of 10 eggs from the North-West, 20.2 x 14.8 mm., varying from 18.9-21.1 x 14.3-15.1 mm. Clutch, usually 2, occasionally 3.

Like the Yellow-plumed Honeyeater, this species has the habit of rising frequently into the air from the top of a tree and uttering its distinct note, which may be syllabised as "chick-oo-wee." It takes the place of the Yellow-plumed Honeyeater as the "watchman of the bush" in the North-West (E. Lindgren).

White-eared Honeyeater

Meliphaga leucotis

Native name: Duranduran (South-West).

General colour light olive-green, with the *head, throat and chest black,* and behind the eye a *white ear patch.* The wings and tail quills are greyish-brown, edged olive-green. The *under parts are light olive-green, tinged yellowish.* Iris, light brown; beak and legs, black. Length, 8¼ in.

Distribution: Found in the South-West, north to the mulga-eucalypt line, but excluding the South-West corner itself (namely west of Wongan Hills, Wickepin and the Stirling Range, see Fig. 6). On the south coast it occurs westwards to Two People Bay. Recorded inland and eastern limits are Queen Victoria Spring and Eucla. It is an inhabitant of the salmon gum and mallee country, and the tea-tree areas around salt lakes.

Nest: The nest is usually built in a bush or small tree at no great height from the ground—up to 10 feet. It is cup-shaped and composed of shredded bark or grass which is held together with spiders' web to make a compact cup. As lining for the egg cavity, the White-eared Honeyeater uses sheep's wool, kangaroos' fur or horsehair. The 2 or 3 eggs (size, 21 x 15 mm.) are white in colour and are sparsely spotted and blotched with red or reddish-brown, usually towards the larger ends. In some parts of the eastern States this bird has the reputation of being very tame during the nesting season and there are a number of records of birds alighting on the heads of human beings and endeavouring to tug out hair for nest-building purposes. In other cases they have attempted to pull wool from socks. In this State an instance of an individual pulling at the laces of an observer's shoe has been recorded.

This honeyeater has a loud ringing "chock" and other notes, and like most honeyeaters can be readily "called up" by imitative calls and studied close at hand.

White-naped Honeyeater

Melithreptus lunatus

Native names: Jin-gee (coastal); Ber-ril-ber-ril (Swan River); Bun-geen (Avon River); Banggin (South-West).

Back, wings and tail a bright olive-green, *around the rear of the black head passes a narrow white crescent.* The entire under parts are white with a small, inconspicuous patch of black feathers between the angle of the lower jaw. The bare skin around the eye is grey below and chalky white above. Iris, brown; beak, black; legs, flesh-brown. Length, 5½ in.

Distribution: Confined to the South-West corner, north to Moora, in the main jarrah forest belt and east to Broomehill, the Stirling Range and along the coast to Stokes Inlet. Its distribution between here and the mouth of the Gairdner River may not be continuous, but it has been observed at Hopetoun and the Jerdacuttup Creek.

Nesting: The White-naped Honeyeater builds a cup-shaped nest, usually of bark fibres, rootlets and dry grasses, which is slung from twigs of a sapling at a height of up to 30 feet from the ground. At times little else than dry grass is used in the construction of the nest, but in some cases, besides the items mentioned above, sheep's wool and vegetable down are employed. The clutch consists of 2 (sometimes 3) eggs, which are pale buff, marked mostly at the larger end with spots and blotches of reddish-brown and grey. Size, 18 x 14 mm. The bird is late in breeding—eggs may be found in nests from November to January. Incubation takes about 14 days and the young remain in the nest a further fortnight.

This honeyeater makes its presence known in the bush by its harsh grating notes, "sherp, sherp, sherp." Its usual call while foraging in the tree canopies is a single-noted "tsip" or "tsit," uttered continuously and sounding like insect calls.

Brown-headed Honeyeater

Melithreptus brevirostris

Upper parts olive-green; *head brown, around the back of which is a narrow buff crescent.* Under parts buff grey. The naked space around the eye is a light orange. Iris, brown; beak, dark brown; legs, light brown. Length, 4½-5½ in.

This species has a similar plumage pattern to the White-naped Honeyeater, but with duller colours; it is brown where the other is black, buff replaces the white and the green tone is reduced.

Distribution: Found in the South-West, the inland limits recorded so far being Tamala (south of Shark Bay), 50 miles north of Galena (on the Murchison River), Mullewa, Mulline, Menzies, 45 miles east of Norseman, and Madura. It appears to avoid the jarrah forest belt, but penetrates westward at times to about Wooroloo and south along the Swan coastal plain. In Woodanilling it is common and favours the wandoo country. Forest clearing has enabled it to penetrate to Boyup Brook and Bridgetown (first seen in 1962). It is rarely seen around Perth, its headquarters appearing to be the mallee and salmon gum country.

Nesting: The bird builds a small cup-shaped nest of strips of bark, grasses, and horse or cowhair, matted together with cobwebs and spiders' cocoons, and placed in the topmost twigs of a sapling. The clutch comprises 2, sometimes 3, eggs. Size, 16-17 x 13 mm. They are pale reddish-buff, spotted and speckled, especially at the larger end, with reddish-brown and purplish-grey. They are very like the eggs of the previous species, but slightly smaller.

Golden-backed Honeyeater

Melithreptus gularis

The *golden-yellow of the back and rump* is the most conspicuous feature of the plumage. The *head is black,* with a *white crescent mark* towards its hinder margin; wings and tail brown. The under parts are white but there is a slight but prominent *dark patch on the middle of the chin.* The bare skin around the eye is yellowish-green. Iris, dark brown; beak, black; legs, brown. Length, 6-6½ in.

Distribution: A northern and central Australian species extending through the North-West country to Exmouth Gulf. Our inland records are from Coolawanyah and Marble Bar. These birds belong to the race, *M. g. laetior.*

Nesting: No nesting data are available for the portion of the State covered by this book. Further north, at the junction

Plate 9.—Upper: Spinebill, male.
Lower: Tawny crowned Honeyeater.

of the Fitzroy and Margaret Rivers in the Kimberley Division, nests found by G. A. Keartland (1897) and E. J. Harris (1899), in January and February, were cup-shaped and suspended in the branches of *Acacia* and *Bauhinia trees.* A single egg appears to form the clutch and it is pale fleshy-buff which gradually passes into a warm reddish-buff at the larger end, where there are spots and blotches of a slightly darker hue, intermingled with underlying markings of faint purplish-buff. Size, 22 x 16 mm.

The call-note is single-noted and sharp, but it has an attractive song which Klaus Immelmann considers the most beautiful of all the honeyeaters, resembling that of the European Song-Thrush *(Turdus ericetorum).* It is repeated two to four times and both sexes sing.

Western Spinebill

Acanthorhynchus superciliosus

(Plate 9)

Native names: Bool-jeet, Buljit (Perth)

The most distinctive feature of this small, active honeyeater is the fine, *long down-curved beak,* and the *rich chestnut throat and chest* of the male bird; this chestnut is continued as a band across the nape of the neck. Below, the chestnut area is followed by *narrow white and black crescent bands* across the breast. The remainder of the under parts are buff, with a small white patch beneath the chin, extending out below the eyes and another behind the eye. Apart from the chestnut nuchal collar the upper parts are a quiet olive-grey. The *outer tail feathers are white,* very conspicuous in flight. The female is duller; lacking the bright chest and breast pattern. Iris, red; beak, black; legs, grey. Length, 6 in.

Distribution: Confined to the South-West corner, north to Cockleshell Gully (near Jurien Bay) and Moora (rare) and east generally to the Great Southern railway (but recorded near Kweda between Pingelly and Corrigin), the Stirling Range and along the south-east coast to Israelite Bay.

Nesting: The Spinebill makes a compact cup-shaped nest of strips of bark or fine stems, and mats them together with spiders' webs, lining the cup snugly with the flock from zamia palms. It may be built either in a low bush or in a small tree, up to as much as 12 feet from the ground. Early nesting birds usually have only one egg, but those nesting later in the year have 2 in the clutch. They are elongated oval in shape; ground-colour white with a buff tinge (sometimes a pink tinge), and sparingly spotted and blotched with chestnut-brown and purplish-brown. These markings sometimes are so concentrated at the larger end as to form a ring or crown. Size, 18 x 13 mm.

The female does most of the nest building and alone incubates the egg or eggs, but the male assists with the upbringing of the young. He defends the territory around the nest. In its territorial display flight it rises vertically to about 35 ft. uttering its call at the peak of the climb and for the first foot of the descent (K. Immelmann). Eggs may be found in nests from September to January.

Spinebills are among the most plentiful of honeyeaters in the South-West and are particularly partial to banksia country, where they may be seen probing the robust flowering cones of these trees. They add to the attraction of King's Park by feeding on the kangaroo paw flowers, thrusting their delicate beaks into the tubular flower whilst perched gracefully on the stem.

Tawny-crowned Honeyeater

Gliciphila melanops

(Plate 9)

Native names: Wy-ro-dju-dong (Perth); Poo-let (King George's Sound).

Upper parts brown, streaked darker; under parts dull white. The *crown is chestnut or rufous-brown,* and over the eye is a white line. A *dark band* commences at the beak and passes through and behind the eye and then downwards like a crescent, but just fails to meet below its fellow of the other side. Under wing, salmon. Iris, brown; beak, black; legs, light grey. Length, 5½-6¼ in.

Young birds lack the tawny crown and have a prominent yellow patch on the throat.

Distribution: An open country honeyeater, found in sandplains and heathy areas of the South-West, from the mouth of the Murchison River, inland to Mullewa, Wongan Hills, Southern Cross and Norseman, and eastwards along the coast beyond Israelite Bay to Point Culver. It avoids the heavy timber belt west of the Great Southern railway, but occurs in a narrow coastal corridor round most of the South-West corner.

Nesting: A deep cup nest, of coarse dry stems and grass, and lined with soft wool-like material. It is built in a low bush, usually within a foot of the ground. Eggs, 2 (sometimes 3) of white ground colour, sparingly blotched, especially at the larger end, with brown. Size, 19-23 x 13-15 mm. Egg dates, Dandarragan, August; Lake Grace, July to February; south coastal districts, November-January.

Though a creature of the open heaths, this honeyeater may occasionally feed on flowering gum trees and mallee thickets in the neighbourhood of open country. At times, too, it will penetrate many miles into heavy forest country, such as the dense karri, jarrah, and marri, when it is strongly in bloom.

The bird has a sweet bell or flute-like note, rather ventriloquial in quality, which it utters from the top of a bush, whence it has a habit of making sudden, restless flights. Its almost continuous gentle singing falls pleasantly on the ear of the observer in the sandplain and mallee when the plants are in blossom.

In the Wheatbelt, according to observations made at Nangeenan, Dangin, Bilbarin and Caron by E. H. Sedgwick, the Tawny-crowned Honeyeater appears to be a winter visitor (April to October) presumably from the wetter portions of the South-West.

White-fronted Honeyeater

Phylidonyris albifrons
(formerly *Gliciphila albifrons*)

Native name: Gool-be-gool-burn (Avon River).

In general this bird is somewhat like the New Holland Honeyeater in appearance and might be confused with it, having a similar *streaked black and white pattern.* The yellow wing patch is however, much less apparent. There is no white eyebrow, but a prominent broad white patch on each side of the face between the eye and the beak, and there is a white line below the eye. There is a tiny wattle of red flesh at the outer angle of the eye, but which cannot be noticed unless the bird is viewed very close at hand. Behind this is a silvery-slate ear patch. A circlet of white feathers surrounds the brown iris. The throat

Fig. 55.—White-fronted Honeyeater.

is wholly dark brown and the rest of the under parts are dull white. Beak, black; legs, grey-brown. Length, 7¼ in.

The sexes are similar in colour pattern, but the male is perceptibly darker than the female. The dark parts are blackish-brown in the male and brown in the female.

Distribution: Generally distributed north to the Fortescue River and inland to central Australia but absent from the South-West (west of Moora and the Great Southern railway—see Figs. 6 and 7). It penetrates at times further west and has been observed north-east of Gingin and at Wooroloo.

Nesting: The nest is usually built in the centre, and near the top, of a small bush and within a foot or two of the ground. It is cup-shaped and constructed of grass and shredded bark, and lined with plant down. Size, 3 x 3½ inches. The 2 or 3 eggs are pale pinkish-white, spotted with chestnut, mostly at the larger end. Size, 18-21 x 14-15 mm. The female alone incubates the eggs.

This bird has a variety of calls: A metallic "cheep-cheep," "chu-chu-chu," a trill, and other musical notes, and vocal sounds suggesting a shaking of quills. It appears to be a winter visitor to the Wheatbelt area, presumably from further north.

New Holland Honeyeater (Yellow-winged Honeyeater)

Phylidonyris novæhollandiæ

(formerly *Meliornis novæhollandiæ*)

Native names: Ban-dene, Bandin (Perth); Woo-re-ning (King George's Sound).

Mostly *black, streaked with white on the back and especially below.* Slightly white on the forehead. Narrow white feathers on the throat; of beard-like effect. The outer vanes of the wing feathers, towards their bases, are yellow, forming a conspicuous golden patch in the middle of the wing when it is folded. The outer vanes of the basal parts of the tail feathers are also yellow. The *tail feathers are tipped with white. Iris, white,* clearly visible at close quarters (dark in young birds); beak and legs, black. Length, 6¾-7½ in.

Distribution: Confined to the South-West corner, north to Dandarragan and east to Mogumber, Beverley, and on the south coast to Point Culver. It is plentiful on those islands of the Archipelago of the Recherche which have eucalypt and other thickets, such as Woody, Sandy Hook, Thomas, Mondrain and Middle Islands. It also lives on Bald Island.

Nesting: The nest is an open cup-shaped structure, usually placed in a bush at a height of about five feet from the ground. It is constructed of twigs, rootlets, strips of bark, or grasses, and zamia palms. The nest is not the neat, closely welded structure typical of most honeyeaters, but much more roughly made. The eggs, usually 2, sometimes 3, in number, are pinkish pale buff, sparsely spotted and blotched, often particularly at the larger end, with dark red and underlying markings of slate-grey. Average size of 20 eggs, 20.6 x 15.3 mm., ranging from 21.7-19.0 x 14.8-16.1 mm. Incubation period, 15 days; nestling period, 15-16 days (W. H. Loaring). It is known to be an out-of-season breeder and W. H. Loaring has a record of a nest with young being found in mid-April at Bickley. However, spring nesting is normal. At Two People Bay egg-laying takes place between June or July and late September.

In the northern parts of its range, as around Perth and Mandurah, this honeyeater is almost always found in the thickets around swamps, along water-courses and in garden areas such as at the University grounds, Nedlands. In the lower South-West, as at Bridgetown, where it is among the commonest of birds at certain seasons of the year, it frequents wooded country, such as banksia and eucalypts, not always close to water. Tree-lucerne when in blossom in gardens is also much visited by this honeyeater. The note is loud and shrill; when the vicinity of its nest is approached it utters loud, chattering notes. A restless bird, it is susceptible to being "called up" by chirping, but unlike such birds as thornbills, which will remain around after being called up, this honeyeater merely dashes up and then departs just as quickly.

Fig. 56.—New Holland Honeyeater.

White-cheeked Honeyeater

Phylidonyris nigra
(formerly *Meliornis niger*)

Native name: Ban-dene (Perth).

Closely resembles the preceding species in general colour pattern, but differs in the following features: There is a *broad white cheek patch* starting just below the eye; there is a white line above the eye, which widens slightly in front; there are *no white spots on the tail,* and the *iris is brown.* Beak and legs, black. Length, $6\frac{1}{2}$-$7\frac{1}{4}$ in. In young birds the white superciliary line widens slightly with a yellow patch on the lores.

Fig. 57.—White-cheeked Honeyeater.

Distribution: A South-West species of a similar distribution to the New Holland Honeyeater, but avoiding the jarrah forest area. In the north it occurs as far as the mouth of the Murchison River and south-east to Wickepin, Lake Grace and Esperance.

Nesting: The nest, cup-shaped, is usually built in a low bush and is very like that of the New Holland Honeyeater; the eggs (two are usually laid) also resemble those of that bird; they measure 21 x 15 mm.

A case of inter-breeding between the New Holland Honeyeater and this species was observed by B. V. Teague at Narrogin in August 1953; the pair successfully raised two young. In September 1973 Mr. A. Y. Pepper successfully mated a male New Holland with a female White-cheeked in his aviaries at Scarborough; two young were raised.

This bird is much less abundant than the New Holland Honeyeater and is apparently very local in its occurrence. Near

Perth it may be encountered in the woolly-bush *(Adenanthos)* thickets towards Forrestdale and east of Cannington. It is also plentiful in the coastal wattle thickets and *Banksia prionotes* between Sorrento and Burns Beach. Since 1970 it has become a fairly frequent visitor in late summer and autumn to gardens in some Perth metropolitan suburbs (such as Nedlands).

It has the habit, during the nesting season, of flying above its nesting site to a considerable height and then, with closed wings, descending rapidly until near the ground, when it again opens its wings and glides to a position on the top of a bush or tree. The call is a single-noted squeaky whistle. It also has a melodious warbling song which J. R. Ford has syllabised as a repeated "twee-ee-twee-ee," often uttered during its aerial evolutions. Another call is a rapid "hee-hee-hee-hee-hee-hee."

Yellow-throated Miner (Dusky Miner, Mickey Miner)

Manorina flavigula

(formerly *Myzantha flavigula*)

Native names: Beingy (Lyons River); Peen peen (Jigalong); Boon boon (Mileura); Bil-ya-goo-rong (coastal, near Perth); Bill-your-ga (Avon River); Puer (Pallinup River).

The Miner varies in colour according to locality. In the South-West the birds are larger and darker. The upper parts are grey, the rump lighter and the wing and tail quills dark grey, with olive-green on the outer feathers. The forehead also is washed with olive-green. The *beak and a patch of bare skin around and behind the eye are yellow.* The *lores and ear coverts are black,* with white at the extreme tip. The throat and breast are grey, barred with white, and the abdomen wholly white. Iris, brown; legs, brownish-yellow. Length, $10\frac{1}{2}$ in. Weight, $2\frac{1}{4}$-$2\frac{1}{2}$ oz.

Inland the birds become progressively lighter, the grey rump becoming almost pure white and the throat and a cheek line below the black ear patch pure white.

Distribution:: Miners inhabit the whole State, except the block of heavy-forested jarrah country in the South-West. However, they extend down the coastal plain somewhat south of Pinjarra (Figs. 6 and 9). Previously it was held that two distinct species of Miners occurred in Western Australia, the Dusky Miner *(M. obscura)* in the South-West and the Yellow-throated Miner in the remainder of the State. It is now considered that these are only geographical races of one species, and, as already stated, the various forms merge into each other.

Nesting: The nest is cup-shaped, made of grass and twigs, and is rather large for the size of the bird. It is lined with fine grass, hair, or rabbit fur, and is usually placed in a bush or small tree such as a drooping branch of the sheoak. The 3 or 4 eggs are salmon-buff, spotted, chiefly at the larger ends, with reddish- and purplish-brown. Average size of 20 eggs, 25.9 x 18.8 mm., varying from 23.9-28.5 x 18.0-19.9 mm. August to November are the chief breeding months in the south.

Red Wattle-bird (Gilly-Wattler, Barking Bird)

Anthochæra carunculata

Native names: Djung-gung, Djang-gang (South-West); Dongrruck (Katanning); Tungaluk (Pallinup River).

Brownish-grey, strongly streaked white, above and below. Tail and wing feathers tipped white. A *yellow patch* in the middle of the abdomen, and a *small red fleshy wattle pendent* behind the eye. Iris, red; beak, black; legs, brownish-grey. Length, 13¼-14½ in. Weight, 3½-4½ oz.

Distribution: Found in the South-West, north to the mulga-eucalypt line (Fig. 4), and eastwards to Eucla. In the Archipelago of the Recherche it has been reported on Middle Island.

Nesting: The Red Wattle-bird builds a saucer-shaped nest of twigs, in trees from a fairly low situation up to as high as 50 feet from the ground. A typical nest measures 7 or 8 inches in diameter. The clutch consists usually of 2 or sometimes 3 eggs, which are pinkish-buff, spotted with reddish-brown and purplish-grey, particularly at the larger end. Average size of 33 eggs, 30.7 x 21.8, varying from 28.0-33.8 x 19.5-23.3 mm.

In the Perth and lower South-West districts Red Wattle-birds are resident throughout the year. In the outer parts of its range the bird seems to be only a winter visitor (April to October or November), as at Coorow, Caron, Moora, Kellerberrin, Nangeenan, Bilbarin and Dangin (where the movements have been studied by ornithologists), withdrawing to the South-West corner in the summer after breeding. In the Perth district this migratory passage movement can be well observed. From about April to June small parties of up to 15, or sometimes 30, birds may be seen unobtrusively working their way north. The return movement is in evidence from about August to November or later. At Busselton on March 8, 1934, S. R. White observed hundreds passing east in parties of up to 17.

This is one of the commonest honeyeaters of the Perth area, the raucous notes being a feature of King's Park and city and suburban streets when the coral tree *(Erythrina indica)* is in flower, the birds probing the blossoms for insect and nectar. They have been reported as damaging snapdragons and fuchsias in gardens in their search for nectar.

The Red Wattle-bird has never been regarded as a game-bird in Western Australia as it used to be in the eastern States.

Little Wattle-bird

Anthochæra chrysoptera

Native names: Djung-gung, Djang-gang (Perth); Tur-dal-l (Avon River); Dun-gur-uk (King George's Sound).

Like the preceding wattle-bird in general appearance, but lighter on the head; *no wattle;* with small white spots on the throat and chest instead of streaks; *no yellow on the abdomen,* and no white tips to the wing feathers. The best distinguishing field character, however, is the presence of a broad patch of *reddish-chestnut in the wing,* well seen when the bird takes flight. Iris, bright reddish-brown; beak, dark brown; legs, slaty-black. Length, 10½-12¼ in. Weight, 2¼-2½ oz.

Distribution: Found in the South-West, from Geraldton around the coast to Israelite Bay, but not further inland than the Midland and Great Southern railways. It is rather local in distribution.

Nesting: The rather shallow cup-shaped nest is usually built in the fork of a perpendicularly growing branch of a tree—often a *Banksia.* It is small for the size of the bird, being 4 to 6 inches in diameter and 2 to 3 inches in depth. It is usually constructed of grass or small twigs, put loosely together and lined with soft grass, plant-down or wool. Although in the eastern States the clutch consists of 2 eggs, in Western Australia it is invariably limited to one. The colour is pinkish-buff, with spots, chiefly at the larger end, of reddish-brown and purplish-grey. Average size of 8 eggs 28.9 x 20.7 mm., varying from 27.5-29.8 x 19.3-21.6 mm.

As already mentioned, the bird is local in distribution, favouring *Banksia* country. In the vicinity of Perth it has declined in numbers in recent years. There is a "family resemblance" between the calls of this and the preceding species, but the notes of the Little Wattle-bird are quite distinctive.

Spiny-cheeked Honeyeater

Acanthagenys rufogularis

Upper parts streaked grey and dark brown, with a white rump, strongly streaked dark brown, and white tipping to the tail. There is a dull olive-green patch on the wing. The under parts are pale with the throat and chest salmon-brown and the belly streaked buff and brown. A black line passes through the eye, and from the base of the *parti-coloured beak*—which is black at the tip and deep flesh-pink at the base—extends a patch of purple-pink naked skin below and beyond the ear. Below this is a prominent white stripe which expands into a wide ear patch behind, marked with black. The lower border of this variegated and colourful head pattern is a dark streak marking it off from the salmon-coloured throat. Iris, grey-blue or brown; legs, dark grey. Length, 9-10½ in.; weight, 1½-2 oz.; females average slightly smaller than males. Weight, 2 oz. In the young the ear patch is plain yellow with no black markings.

Distribution: Generally distributed, north to Jigalong, the Fortescue River and Barrow Island, but absent from the South-West corner (west and south of a line passing from Dongara to Moora, Cunderdin, Kojonup, Broomehill and Bremer Bay, Figs. 6 and 7).

Fig. 58.—Spiny-cheeked Honeyeater.

Nesting: This bird makes a cup-shaped nest of green grasses held together with spider's webs, 4 to 5 inches in diameter and 3 to 4 inches in depth. It is usually suspended from the foliage of a bush, about three feet from the ground. The eggs, 2 or 3 in the clutch, are olive-brown and are spotted and blotched with various shades of reddish-brown, mostly at the larger end where the markings tend to form a cap. They are small for the size of the bird, as is also the nest. Average size of 8 eggs, 25.2 x 18.3 mm., varying from 23.8-25.9 x 17.9-18.8 mm.

The bird has a plaintive call—more melodious than the notes of the other wattle-birds.

AUSTRALIAN CHATS: Family Ephthianuridæ

White-fronted Chat (Tin Tack)

Ephthianura albifrons

Native name: Yaba-wilban (South-West)

Back grey; tail black with outer tail feathers tipped white; wings brown; *crown of head black; forehead, sides of head and throat white;* black band across chest; abdomen white. The female has no black on the head and the breast band is a dark brown and narrowed. Iris, light brown; beak and legs, black. Length, 4½ in.

Distribution: The bird is an inhabitant of the southern part of the State, north to Shark Bay, Milly Milly Station, Cue and the Eastern Goldfields. Occasionally it wanders further north still, and has been reported at Point Cloates.

Nesting: The White-fronted Chat builds a cup-shaped nest in a bush, usually at a height of about 3 feet from the ground. The nest is constructed of grass or twigs and is lined with hair. The white eggs, usually 3, sometimes 4, are spotted at the larger end with reddish-brown spots. Size, 19 x 15 mm. It is, in the lower south-west, an early breeder—eggs are often found in nests in the Busselton district in July. When eggs or young are in the nest the bird will readily attempt the broken wing reaction.

This gregarious bird favours open country, the samphire margins of estuaries and salt lakes, and even, occasionally, may be seen on ocean beaches. After the breeding season it gathers into small flocks —of up to as many as 40 individuals. These flocks then lead a nomadic life in their search for food. The call is a metallic note, similar to that of the Zebra Finch.

Crimson Chat

Ephthianura tricolor

Native names: Belidita (Godfrey's Tank, Canning Stock Route); Billy deedy (Jigalong).

Distribution: North to the Pilbara district and the southern drier portion of the Kimberley Division, and south to about the Moore River, Kellerberrin and Kalgoorlie. Occasionally individuals stray south along the Swan coastal plain, and have been seen in company with the White-fronted Chat on the Cottesloe golf links.

Nesting: It builds its nest in the top of a low bush, such as salt-bush, or in a bunch of spinifex *(Triodia)*. It is cup-shaped and constructed of fine twigs and grass, lined with fine grass, hairs and bark fibres. The 3, sometimes 4, eggs are white, spotted with reddish-purple. Size, 17 x 13 mm. The male assists the female in nest construction and both parents share in incubating the eggs, feeding the young and assisting with nest hygiene. At times the nests are grouped in communities, each, however, being the centre of a strongly defended territory, which in the Morawa district was observed to be approximately half an acre in area. Both sexes defend the territory by song and aggressive bill clicking displays, but after the young are fledged family parties flock together on common ground (S. R. White).

Mr. White has differentiated five types of call notes, as follows: (1) A sweet but shrill whistle usually uttered three times, "whee! whee! whee!" (2) A mournful whistle in two syllables "tee-whee . . . !" uttered by the male in territory establishment. (3) A low truculent chatter usually uttered when a territory holder attacks a straying neighbour. (4) A plaintive hissing made by birds attempting the use of the injury-feigning decoy. (5) A metallic "ting!" somewhat resembling the call of the White-fronted Chat and which appears to be used as a flock call.

Orange Chat

Ephthianura aurifrons

Back and wings yellowish-green streaked brown; wing quills grey; tail, black; tipped white; *crown of head and rump, orange-yellow; chin, black; remainder of under parts, orange-yellow.* The female is brown above, lower back and upper tail coverts yellow; under whitish with a yellow tinge, under tail coverts yellow. Iris, indian-red; beak and legs black. Length, $3\frac{1}{2}$-$4\frac{1}{2}$ in.

Distribution: A bird of the mid-north and interior; not recorded in the south-west, the known limits so far being Morawa, Haig and Eucla. To the north it has been recorded up to Roebourne and Jigalong.

Nesting: Nest cup-shaped, made of fine twigs and lined with rootlets. Generally placed in a small bush about three feet from the ground. Eggs are usually 3 in number, and both parents take part in incubation. Colour, white, spotted, particularly at the larger end with purplish-red. Size, 18 x 14 mm.

A favourite habitat of this bird is the samphire belt around salt lakes and marine marshes. The flight call is a mellow "cheek-cheek".

FINCHES and BUNTINGS: Family Fringillidæ

Goldfinch

Carduelis carduelis

Back, brown; *wings, black with bright yellow band* conspicuous in flight, white spots on feathers; distinctive *red, white and black markings on head,* formed by red forehead and chin, side of head and throat white, and back of head and nape black. Under parts, white with band on breast and flanks buffy-brown. The young are greyish-brown, streaked darker brown, but with the wing pattern like the adult. Sexes, similar. Iris, brown; beak, whitish, tipped darker; legs, pale flesh. Length, 4¾-5 in.

Distribution: Introduced, originally from Europe; and became established locally as escapees from aviaries. The species had been released from time to time on former occasions, but the present colonies originate from birds which were first noticed at the Government Gardens, Perth, in October 1933. Thence they slowly spread into the suburbs, and became thoroughly established but still not generally plentiful. After 1963 the species entered into a fairly sudden decline in the metropolitan area and it is now virtually absent. During the period of its presence foraging flocks were to be encountered at intervals at Floreat Park, King's Park, Subiaco, Hollywood, Nedlands, Claremont, Cottesloe, Fremantle, North Perth, East Perth, Mt. Lawley, Inglewood, Maylands, Bayswater, Beechboro, Belmont, Caversham, Herne Hill, South Perth, Canning Bridge, Victoria Park, South Belmont, Kenwick and Gosnells. In January 1957, the species was noticed at Albany, where it is now established.

Nesting: The nest is a neat cup, often built in pine trees. It is made of dried grass, woolly bush *(Adenanthos)* and other stems and lined with plant down and feathers. It is about 3½ in. in diameter, the nest cavity being 2-2½ in. wide and 1¼ in. deep. The eggs are bluish-white, marked with red-brown. Size 17 x 13 mm. Usually 3 eggs form the clutch; in N.S.W. the usual number is 3-4, and in England 5 is the average, but varying from 3 to 7.

Goldfinches are most numerous in waste land. They visit gardens to feed on sunflower, cosmos, thistles and other seeds and are not often seen in the virgin bush. They have a preference for roosting in the introduced pines and pencil pines in parks and private gardens, and, among native trees, in sheoaks *(Casuarina).*

GRASS-FINCHES: Family Estrildidæ

Red-eared Firetail

Emblema oculata
(formerly *Zonæginthus oculata*)

Native names: Jeer-ree, Jiri (Perth); Dwar-den-ngool-gnan-neer (Darling Range); Poir-een, Poor-een (King George's Sound).

Upper parts and tail olive-brown, finely barred black. *Rump crimson,* the crimson extending along the outer vanes of the outer tail feathers, giving the tail a brown streaked-red appearance when fanned out and a red-bordered appearance when relaxed. Lores and forehead black; *crimson patch behind eye;* a ring of naked skin around the eye pale blue. Breast buff-brown, finely barred black; lower breast, abdomen and under tail coverts spotted black and white. Iris, red; *beak, scarlet;* legs, dark brown. Length, 4½-4¾ in. Juvenile birds lack the red ear.

Distribution: Confined to the south-west corner, in the coastal belt and the Darling Range, south-easterly to the Hamersley estuary. After a gap it reappears at Lucky Bay (east of Esperance) and also occurs at Mississippi Bay and Duke of Orleans Bay (see p. 62). In the early colonial days it was found in the neighbourhood of Perth, and until about 1912 it could still be seen in the thickets along the banks of the Murray River at Pinjarra. As a result of extensive clearing of river and swamp frontages which has gone on in the coastal plain it has apparently disappeared entirely from the coastal plain. It still survives, however, in some of the Darling Range gullies north to the Mundaring reservoir, and has been recorded as far inland as the 53-mile peg on the Albany Highway (K. G. Buller).

Nesting: The nest is a bulky structure completely built of long grasses, with a spout-like entrance and thickly lined with feathers and plant-down. It is best described as retort-shaped and it is more neatly and closely constructed than is usual with the Australian finches. This stability is obtained through the inclusion of twining stems (such as *Thysanotus patersonii)* and tips of green bushes amongst the grass used in construction of the nest. These tips are apparently nipped off by the birds and carried to the nest under construction. In the forest areas it is usually placed in a small tree or sapling, but on the coastal areas the bird has a habit of building the nest in a prickly *Hakea.* A nest at Two People Bay was in a *Banksia coccinea.* In spite of its bulk the nest is not conspicuous, being usually well-hidden in the foliage. The number of eggs laid varies from 4 to 6; they are pure white, but the contents give them a beautiful salmon-pink tone. Size, 25 eggs in the South Australian Museum (McGilp collection) range from 16-19 x 11.5-12.3 mm. (H. T. Condon).

K. Immelmann, in a study of the species at the Wongong Gorge, south of Armadale, found that mated pairs inhabit permanent territories of 100-200 yards in diameter, which are weakly defended except in the immediate vicinity of the nest. The male selects the nest site and the female may accept or reject it. The male carries nesting material to the site but the female alone does the building. Both parents share the incubation of the eggs, in shifts of 1-1½ hours, and feed the young.

The Red-eared Firetail is a small bird which may easily be overlooked in the heavy-forested areas as it is inconspicuous in thick vegetation and, unlike the northern Australian grass-finches, never associates in flocks, hence no flock call has been developed. Pairs and aggregations of up to 6 or 7 birds are normal. It is also inconspicuous from its habit of feeding on the ground on grass and other seeds. At Wongong Dr. Immelmann found the birds feeding on the seeds of the sedge, *Lepidosperma angustatum.* Attention to the bird is usually drawn through its mournful note—a soft, low "wee-ee", or "oowee". This call enables a pair of mated birds to hold contact with each other. Dr. Immelmann recognised other calls associated with nesting and incubation relief. In open country, where it is found in the vicinity of paperbark swamps, the bird is easily located, when flushed in sunlight, by the conspicuous red colouring of the rump.

Painted Finch (Mountain Finch)

Emblema pictus

(formerly *Zonæginthus pictus*)

Native names: Nee-murri (Pilbara Diistrict); Koo-je-lan (Coolawanyah).

Back brown to auburn, rump crimson and tail dark brown. Scarlet on forehead, around the eye and on the chin. *Under parts black; spotted with white, with irregular scarlet patches.* Iris, white; beak, upper mandible, black tipped red, lower mandible, red; legs, flesh. Length, 4¼-4½ in.

Distribution: Until recent years this northern species was known to range only as far south as Boolathanna Station (north of Carnarvon) and the Barlee Range. It has now embarked on a spectacular spread southwards and south-eastwards, possibly because of the extensive inland rains in recent years. In September 1963 I. C. Carnaby saw a pair in breakaways 30 miles south of Yalgoo. In September 1968 an apparently isolated nesting colony was discovered by S. J. J. F. Davies and M. W. R. Beck at Mount Hale in the Murchison district. In August-September 1973 D. R. Reid encountered the bird in various localities from the Gahnda rockhole on the Laverton to Warburton Ranges road, eastwards to the Northern Territory. In 1974 he found it present in the Leonora and Menzies districts (37 and 50 miles west of Leonora, at Gwalia and 29 miles ENE of Menzies).

Nesting: The nest is built in spinifex *(Triodia)* and is very small, about 4 in. long, 4 in. high and 5 in. broad. There is no funnel entrance and underneath the nest there is a platform built of pieces of wood and bark, with a few small stones added. This platform is unique in the grass-finches and K. Immelmann considers it to be an adaptation for nesting in spiny *Triodia*. The outer part of the nest consists of stiff pieces of grass and the inner of soft grass, the lining being of plant down, hairy seeds and fronds of the fern *Cheilanthes*. Unlike most other Australian grass-finches, Painted Finches never roost in nests and do not build roosting nests. During incubation only the female sleeps in the nest, the male on a twig near the ground. The 3 or 4 eggs are of the usual grass-finch type, but are rounder and chunkier than eggs of other species. A series of 8 eggs from the Hamersley and Barlee Ranges average 13.9 x 11.1 mm., varying from 13.6-14.3 x 10.8-11.3 mm.

This finch, which F. Lawson Whitlock suggests might appropriately be called the Mountain Finch, occurs near water in the vicinity of rocky ranges, and generally frequents rough country among spinifex. It is abundant in such situations. It is specially adapted to ground life, mostly sitting on stones and rocks.

Dr. Immelmann, after a study of all Australian grass-finches, ranks the call of the Painted Finch as one of the loudest and harshest finch-calls. The call has one to five syllables, sounding like "truht" and "trai(r)ht", and is given in flight and on the ground. The song, too, is very loud and contains many flight-calls interwoven with softer sounds and trills. The female has a special nest-warning call if there is any danger around the nest containing eggs or young. It is a very sharp and harsh "terrai(r)ht" sound.

Zebra Finch (Chestnut-eared Finch, Pimp, "Java Sparrow")

Tæniopygia castanotis

Native names: Chiga (Godfrey's Tank, Canning Stock Route); Newmerri (Pilbara district); Yim-eye (Coolawanyah); Neamoora (North-West Cape); Nyinnyinka (Jigalong); Nyi-nyi (Warburton Range); Nye-nye (Cundeelee, the "ny" sound is nasalised, as "gn" in Gnow.)

Back grey, crown dark grey, rump white merging black and white at sides; central tail feathers black boldly barred white, outer tail feathers charcoal grey. Lores and cheeks white, bordered black. *A large chestnut patch behind the eye. Throat and breast finely barred black and white;* abdomen white; flanks chestnut, spotted white. In the female there are no chestnut ear patches and the breast and flanks are pale grey. Iris, red in the adult, dark brown to chestnut in the young; beak, red in the male, orange in the female, black in the young until between 8 and 11 weeks; legs, orange. Length, 4½ in.

Distribution: All over the State, except in the south-west corner (the general limits being the Midland and Great Southern railways). Zebra finches are resident in the Northam-York area, associated with thickets of needle-bush *(Hakea preissii)*. Occasionally it penetrates into the south-west forested country (T. Carter observed the species at Lake Muir in January 1916). Flocks occur at Lake Grace in the summer, extending their range following land clearing (I. C. Carnaby).

Nesting: The nest is usually a very small and untidy dome of stiff grass externally and softer grasses internally, lined with feathers, plant down and wool. If grass is unavailable it is built of small twigs and roots. There is a small entrance funnel which is damaged during incubation and the feeding of the young. The nest is usually built in a bush, but it may be found in a hollow spout of a tree. The bird sometimes chooses unusual nesting situations, not infrequently building in the bulky nests of larger birds, especially those of the larger birds of prey. The eggs are white and the number laid is from 3 to 7, usually 4 or 5, the latter being more frequent. There are many records of much larger clutches. F. Lawson Whitlock recorded 23 eggs in one nest at Marble Bar. These are combination clutches laid by more than one hen bird. Whitlock noted 13 occupied nests in one large *Hakea* bush in the Wiluna district. Average size of 10 eggs from the north-west, 15.2 x 10.6 mm., varying from 14.7-15.7 x 10.2-10.9 mm. The incubation period is 12½-14 days and the young remain in the nest for 20-21 days (K. Immelmann). The parent birds do not remove droppings and the lower part of the nest becomes almost solid with their accumulation.

Both parents share in the incubation and feeding of the young. During nest-building the male carries the pieces of grass to the nest while the female stays inside and arranges them (K. Immelmann). From observations of wild-trapped birds in captivity Miss C. A. Nicholls has noted that if a second clutch of eggs is laid soon after the first-clutch nestlings have left the nest, these are incubated principally by the female. The male stays with the nestlings for the greater part of his time, feeding them until they become independent, after which he usually assists the female with the incubation and rearing of the second clutch. The female will become increasingly hostile toward her first chicks, driving them as far from the nest as possible. The male rarely becomes so antagonistic. Young captive birds may begin to breed at 8 weeks of age. Under optimum aviary conditions up to 19 successive broods of young may be raised.

In the non-breeding season special roosting nests are built which are even more untidy than the breeding nests and may lack a roof. Old breeding nests may be relined and reroofed and used as roosting nests and other birds' nests, such as those of the Yellow-rumped Thornbill, may be used for this purpose. Zebra Finches never roost on a branch if they can find a nest.

The calls have been studied in detail by Immelmann. The loud identity call sounds like a small toy trumpet. It is very similar to the call of the White-fronted Chat and finches will answer to the chats' calls. The softer intimate call serves to keep a pair or a small group in contact. The male has a very high-pitched nesting call, sounding like the whining of a small child in the distance; it is uttered during the search for a suitable nesting site and in the nest itself, to attract the female to the site. Unlike other grass-finches the Zebra Finch has a special attacking call, given when one bird chases another, sounding like the sudden ripping of a piece of cotton cloth. During courtship the male, again unlike most other grass-finches, never carries a feather or a piece of grass in its beak but hops towards the female while singing. Social preening is very common in the species and one often sees a pair huddled together preening each other.

The Zebra Finch is the most widely distributed Australian finch and the commonest finch in the State, excluding the Kimberley Division, which is the true home of the family in Western Australia. There eleven species occur.

Star Finch (Red-faced Finch)

Neochmia ruficauda

(formerly *Bathilda ruficauda*)

Back, olive-green, tail brownish-red with light spots towards the rump. Scarlet on forehead, around the eye and on the chin. *Breast, olive-green, spotted with white;* abdomen, dull yellow. Iris, light orange; beak, coral-red; legs, yellow. Length, 4½-5 in.

Distribution: A northern species occurring as far south as the Ashburton district, the Barlee Range and Mount Augustus. M. Brooker encountered the species at Carnarvon, including Rocky Pool, in May-June 1970.

Nesting: A nest on Ullawarra Station, Barlee Range, was in a small bush amongst long grass and about 2 feet from the ground. Shaped like a bottle lying on its side, it was composed of green grass and lined with wool and feathers (A. H. Robinson). K. Immelmann, from observations in the Kimberley Division, considered this species to build one of the warmest and softest of all finch nests. No roosting nests are built. The clutch consists of 3 to 6 eggs; they are oval in shape and pure white. Size, 14-15 x 10-11 mm. Egg dates, Hardey River, March and May; young in nest, August.

The species is restricted to rush-grown areas around swamps and is rather local and uncommon.

The loud identity call is a high-pitched "sssit," the soft intimate call sounding like "tlit." The song of the male is very low and soft and can be heard only from close at hand. In the courtship dance the male carries a piece of green grass and hops up and down on a branch beside the female. The female engages in a special courtship invitation in which she flies around the male with a long piece of grass in her beak (K. Immelmann).

In the North-West the species is restricted to rush-grown areas around swamps and is rather local and uncommon. In the Kimberleys in the dry season (which is the non-breeding period) the birds assemble into flocks of several hundred individuals. K. Immelmann found in the Kimberleys that during the breeding season, in the "wet," the Star Finch became nearly insectivorous, feeding mostly on flying termites, which it caught on the wing.

Red-browed Finch

Ægintha temporalis

Upper parts dull olive-green, head grey, rump crimson, tail dark grey; a crimson line extends from the beak backwards over and beyond the eye; under parts grey, buff-grey towards the abdomen. Iris, reddish-brown; beak, red; legs, flesh. Length, 4½ in.

Distribution: Introduced (aviary escapee). The species occurs native in eastern and south-eastern Australia. Since the early 1960s colonies have become established in the Darling Range, between Kalamunda, the Mundaring Weir and Darlington.

Nesting: The nests are domed structures, about 4-6 in. high, with a 2-3 in. entrance spout, built of grass stems and leaves and bark, lined inside with feathers—especially white fowl feathers if available. Locally they are built in *Agonis linearifolia,* the prickly *Mirbelia dilatata* and *Melaleuca parviflora,* at heights of 4-20 ft. above the ground. The eggs are white, oval in shape and measure 15-17 x 11-12 mm. The clutch size in the Kalamunda colonies is 4, compared with an average of 5 (occasionally up to 8) in N.S.W. Incubation period, 14 days.

The presence of this Eastern States species in Western Australia (evidently as the result of aviary escapes) was first detected by Mr. John Dell, of Kalamunda, who has made a detailed study of the local groups at Hackett's Gully and the Mundaring Weir Road. In 1965 he estimated that the population, after at least 4 years, consisted of about 230 individuals. Another aggregation, of about 20 birds, was reported by Mrs. Marie Parton at Jungle Gully, Bickley, close to Mr. Dell's sites, the birds being first seen about 1960. Their preferred habitat was orchard land and heavily vegetated creeks and paper-bark swamps, areas where clearing and destruction of habitat had caused a decline of native species. They fed on the small seeds of introduced grasses chiefly *Poa annua* and *Digitaria sanguinalis*. There was no apparent inimical effect on native birds, the native Red-eared Firetail behaving aggressively to the newcomer.

OLD WORLD SPARROWS: Family Passeridæ

House Sparrow

Passer domesticus

Brown and grey. The head is dark grey on the crown, with a chestnut bar behind the eye and curving around on each side to form a chestnut nuchal collar. Upper back streaked black and brown; lower back grey; tail brown; wings brown, streaked black and with a white bar across the primary coverts. Throat and upper breast black; cheeks and rest of under parts grey. Iris, brown; beak, black; legs, light brown. Length, 6 in. Wing span, 8-9 in.

The female lacks the black and chestnut markings on the head and back. The head, neck and lower back are brown, the upper back and wings are streaked brown and black, and the wing stripe is less conspicuous. Under parts greyish-white. Iris, brown; beak, brown; legs, brown.

Distribution: This introduced bird has colonised every Australian State except Western Australia and the Northern Territory. However, there has been a penetration into this State near Eucla but the birds failed to establish themselves. About 1917 Sparrows reached the Western Australian border at Eucla and by the following year had reached Mundrabilla Station, 60 miles west. They died out during the summer of 1918-19 and there has been no further invasion along this route. The nearest colony in South Australia at the present time is at Coorabie, near Fowler's Bay, and near Tarcoola on the transcontinental railway. "One of the contributory checks to the Sparrow's western progress must undoubtedly be the replacement of the horse by the motor vehicle as the main means of transport in this country of widely-scattered settlement" (C. F. H. Jenkins).

Not infrequently Sparrows land at Fremantle from visiting ships from the Eastern States. In 1894 some were seen at Government Gardens, Perth, and destroyed. A pair escaped from a ship at Fremantle in 1927 and one was killed. Another bird was destroyed in 1930, one in 1946 and two in 1957 (C. F. H. Jenkins). In March 1964, nine birds were shot at the Fremantle railway station, seven were juveniles which could only recently have left the nest.

In September 1966 two birds shot on the Geraldton wharf proved to be Tree Sparrows *(Passer montana)*. They were reported to have accompanied a ship whose last port of call was in south-east Asia. In December 1970 another Tree Sparrow was shot by an Agriculture Protection Board officer at Palmyra, near Fremantle.

Nesting: There is no record that the Sparrow nested in Western Australia during its brief penetration in 1917-18. It builds a bulky domed nest of sticks and dried grass, cosily lined with feathers, often under eaves of houses or in cavities in buildings. In England the clutch is usually 4-5 eggs, but may be up to 6. In Sydney the usual clutch is 3-4 eggs, sometimes 5 and very rarely 6 (N. Fearnley). In South Australia 4-7 eggs have been recorded. The ground colour is pale bluish or greyish, sometimes white, marked and freckled with pale brown to blackish, more densely at a zone at the broader end (H. T. Condon). Size, 20-23 x 15 mm.

STARLINGS: Family Sturnidæ

Common Starling

Sturnus vulgaris

Black with *green and purple metallic iridescence,* spangled with pale spots. Body feathers long and pointed, especially on throat and breast. Young uniformly greyish brown. Iris, brown; *beak, yellow* to dull brown; legs, reddish brown. Length, 8½ in.

Distribution: Introduced from Europe (native home, Great Britain and continental Europe and coastal north Africa). In Australia the species now occurs widely in the south-east of the continent, where some individuals perform extensive migrations (eg. Tasmania to southern Queensland, proved by ringing). It appears sporadically in southern Western Australia, as detailed below, but has not yet become permanently established.

Nesting: The nest is an untidy mass of sticks and other plant material, roughly cup-shaped, and often heavily insect-infested. It may be built in rock crevices, tree hollows or under roofs of buildings. The pioneering flocks in the Nullarbor Plain in 1975 nested in a variety of situations: the tubular supports of micro-wave towers, walls of caves and in a 44-gallon petrol drum. Eggs, 5-7, pale light blue, averaging 21 x 30 mm., varying from 20-22 x 27-35 mm. (English data).

The first record of the species' presence in this State was in a report by W. B. Alexander, of the W.A. Museum, to a meeting of ornithologists in Adelaide on March 30, 1917, wherein he stated "That recently a starling had been sent to Perth for identification from the Albany district. . . . Upon an official being sent down to investigate he was told by a resident that a pair of birds had appeared, and had been under observation for some time, until they nested, when they were both shot, and one of them sent to Perth to be identified. Upon the tree being felled with the nest, it was found to contain without doubt the eggs of the starlings" (*South Australian Ornithologist,* 3, 1917, p. 56).

The next definite record is of a bird shot at Gingin in September 1936 by Ian Edgar. He saw a lone bird on his property, "Strathalbyn", and, though recognizing it as a Starling, sent it to the Museum for confirmation.

In October 1971 officers of the Agriculture Protection Board found a colony of Starlings established and breeding in a swamp just north of Cordingup, some 40 miles east of Esperance. A smaller colony was later located about 8 miles further east. Altogether about 40 birds were destroyed, but some were known to have survived.

In 1972 eight birds were shot in the Esperance area. In September 1973, following reports of Starlings about 20 miles west of Esperance, an intensive aerial and ground search was made by Agriculture Protection Board officers and Dr. S. J. J. F. Davies, of CSIRO. No Starlings were actually seen during these surveys.

In 1975 occurred the biggest invasion of Starlings into Western Australia from the Nullarbor Plain. They probably first appeared in August or September and the first report was from Mr. A. Nash, of Mundrabilla Station, who shot a Starling there on October 17. Agriculture Protection Board officers then found birds to be present also at Eucla and at Abrakurrie and Weebubbie Caves nearby, and at Warbla Cave across the State border north-east of Eucla. Starlings were found to have penetrated into Western Australia as far as the micro-wave tower about 20 miles east of Balladonia. On the South Australian side of the border there was a large concentration of starlings at Koonalda Station and birds were found to be at the Koonalda Cave and at Giant's Head Cave. A destruction programme was vigorously carried out and by the end of 1975 most of the Starlings known to be in Western Australia had been accounted for. By the end of March 1976 about 250 birds had been destroyed.

MAGPIE-LARKS: Family Grallinidæ

Magpie-Lark (Mudlark, Pee-wee, Pee-wit, Murray Magpie, Tilyabit)

Grallina cyanoleuca

Native names: Jillinberri (Lyons River); Chillin-birrin (North-West Cape); Yeellin-berra (Coolawanyah); Dee-will deewill (Jigalong; Chillenbidy (Wooleen); By-yoo-gool-ye-de, By-yu Gulyidi, Dil-a-but, Dilabit, Dilebert (South-West); Qwoolibruk (Katanning); Kulyaberuk (Pallinup River).

A study in *black and white.* In the male the head, back, wings, and the end of the tail, throat and breast are a gleaming black, and the eyebrow, a large patch below and behind the eye, the base of the tail, the shoulder, and the lower breast and abdomen are white. There is also slight white tipping to the secondary wing quills. The female differs in the head pattern; it has the forehead and throat white; but no white eyebrow. Young birds up to three months of age have the white throat marking of the female and the black forehead and white eyebrow of the male. Iris, cream-yellow; beak, whitish; legs, black. Length, 11¼-12 in. Weight, 4¼ oz.

Distribution: This species is likely to be found over most of the State, where it has access to surface water. It is absent, however, in the extreme South-West, south of a line joining Cape Naturaliste and Albany, but has recently penetrated south-west along the coastal strip as far as Augusta. In the country south of the mulga-eucalypt line it is hardly ever seen east of Southern Cross, Holt Rock, Lake Grace and Jerramongup. It has been seen near Karonie and Fraser's Range.

Nesting: As one of its names indicates, this bird makes a nest of mud which, to hold it together, it mats with grass—even so, many nests are destroyed by heavy rain. The nest is a bowl-shaped structure and is generally placed on a horizontal fork of a tall tree, and very often near water. A nest from the Bruce Rock district measured 6 inches in diameter and was 3½ inches deep, with walls ¾ inch thick. It weighed 2 lb. 1¼ oz. The eggs, 3 to 5, are white with violet and violet-brown blotches, but are very variable in coloration and shape. The size is also variable. A series of 28 eggs averages 28.4 x 20.6 mm., varying from 24.2-31.1 x 19.4-21.7 mm. Incubation period, 18-19 days. Nestling period, 17-18 days. Two or more broods are raised each nesting season. Both parents assist in the building of the nest, incubating of the eggs and the care of the young.

This species has been studied in considerable detail by A. H. Robinson, of Coolup, who has found that the birds pair for life and hold the same territory in successive seasons, its area in the Coolup district being from 15 to 20 acres. Both male and female share in the territory defence, but only against members of their own sex, and therefore both sing. This is one of the few species which indulge in a dual song, the familiar "te-he pee-o-wit." One bird calls "te-he" and the other follows with "pee-o-wit,' perfectly synchronised. Territory maintenance partially breaks down in the non-breeding season, during winter, when neighbouring families assemble into a central flock for roosting purposes at night, though this may not be a general practice. Some of these winter flocks may attain large proportions, as at Northam where aggregations of about 500 birds may be seen. In the spring the males begin to pay more and more attention to their own territory and are joined eventually by their own females, the roosting flock remaining only of immature unmated birds.

The tendency to form large winter flocks appears to vary in different parts of the country. It is not a feature of the Perth area at present, but when John Gilbert was studying the local birdlife in 1839-42 he found large congregations of Magpie Larks "on the banks and muddy flats of the lakes around Perth." Subsequently the species declined greatly in numbers around the metropolitan area and elsewhere, and at least from 1890 to the early 'twenties of the present century only two nesting areas were known, one at Guildford and the other at Cannington. In the following decade the species increased considerably, becoming a relatively common bird and has since maintained its numbers. There is some evidence that the increase has been due to the rise in the level of the water-table which occurred during the same period and which rejuvenated many swampy areas. The increase has also been observed in other parts of the South-West. The Magpie Lark first made its appearance at Lake Muir in 1919 and in the Bridgetown district in the mid-twenties. S. R. White first recorded its breeding in the Busselton district in 1934. Since then there has been a large increase in numbers in both of the latter districts.

White-winged Chough

Corcorax melanorhamphus

Sooty black. In flight *a prominent white wing patch* is revealed, formed by the white inner vanes in the middle of the primaries; these are even more conspicuous in the under wing. *Iris, red;* beak long and down-curved, black; legs, black. Length, 17-17½ in. Wing span, 24½-26 in. Weight, 11½-13 oz.

Distribution: Until recently this species was considered to be restricted to south-eastern Australia, extending westward from Eyre Peninsula as far as Fowler's Bay. In October 1958 Professor A. J. Marshall saw one bird between Cocklebiddy and Madura.

Nesting: White-winged Choughs associate in communal nesting groups of about a half-a-dozen birds, all assisting in building the nest. This is a large bowl-shaped structure of mud, lined with bark, feathers and fur, about 9 in. in diameter and 6 in. in depth, built on a horizontal branch. Usually 4 or 5 eggs are found in the nest, but there may be as few as 2 or as many as 9, apparently depending on the number of laying hens in the group. The eggs are pale whitish buff, boldly marked with brown blotches and submerged ones of pale lavender. Size 39 x 29 mm., and varying from 36-43 x 28-30 mm.

WOOD-SWALLOWS: Family Artamidæ

White-breasted Wood-Swallow

Artamus leucorhynchus

Upper parts entirely dark brown, except for a *broad white bar on the rump. Under parts white,* except for the head region where the throat and upper breast are like the back. Iris, brown. The short, thick beak, somewhat finch-like in appearance, is blue, with black at the tip. Legs, light grey. Length, 7 in.

Distribution: A northern species which extends down the mangrove belt of the North-West coast, reaching as far south as Carnarvon and occasionally even further south in Shark Bay. In the southern part of its range it is only a winter visitor. It is sometimes found further inland, but it is characteristic of mangroves. It also occurs on many of the islands along the North-West coast.

Nesting: The nest is a fairly substantial structure, compared with that made by other species of Wood-Swallows, built of grass, with no lining. It may be placed in the topmost fork of the highest mangrove in the immediate neighbourhood. One nest, at the Onslow jetty, was built for several successive seasons on a coil of rope inside a fire bucket. The nests are easily located as the male is generally perched near at hand and, when approached, utters a harsh alarm note to which the female usually responds and slips off the nest. The 3 or 4 eggs are creamy-white, with a zone at the larger end of spots and blotches of reddish-brown and slate. In some cases the markings are disposed over the general surface of the eggs instead of being confined to a zone at the larger end. Size, 22-23 x 17 mm., but measurements are inclined to vary.

Masked Wood-Swallow

Artamus personatus

Native names: Tjitta tjalbee (Jigalong); Jit-bung (Avon River).

Upper parts light dove-grey, head dark and the *face and throat black,* bordered behind by a narrow white line. The remainder of the *under parts are greyish-white.* As in all the local wood-swallows, except the preceding species, the tail feathers are tipped white. Iris, dark brown; beak blue, tipped black; legs, lead grey. Length 7-7¼ in.

Distribution: Found all over the State, except the extreme South-West corner (namely the jarrah forest tableland and the coastal plain). However, it occasionally penetrates into the outer parts of this area and it has been seen at Bannister and very rarely near Perth.

Nesting: The nest is generally built within a short distance of the ground—two to five feet, and is the usual loosely constructed saucer-shaped structure built by most Australian wood-swallows. It is constructed of grasses or twigs, and often lined only with finer pieces of the same material. The 2 or 3 eggs vary in colour, sometimes being pale brown, at other times greyish-white or even greenish-grey. They are marked with blotches of various shades of brown with underlying spots and blotches of grey. Average measurements of 17 eggs, 20.2 x 16.7 mm., varying from 19.4-22.8 x 15.6-18.1 mm. If the birds start nesting sufficiently early in the season, two broods are sometimes raised. Incubation takes 12 days and the young remain in the nest for a further 12 days.

Fig. 59.—White-breasted Wood-Swallow.

Fig. 60.—Masked Wood-Swallow

Although the Masked Wood-Swallow is distributed generally over the continent of Australia in some seasons invasions on a large scale occur in the southern portion of the State. Thus F. Lawson Whitlock in July and August 1909 observed large numbers passing through the east Murchison district and in the latter month a large flock became weather-bound on the spinifex plain at Bore Well. Some remained to breed, but by November not one bird was to be seen. From Broomehill, T. Carter recorded that a few were seen almost every summer, but occasionally in greater numbers. For example, in November 1912 the birds arrived suddenly in great flocks and immediately started to breed, leaving towards the end of the year when the young were strong on the wing. Similarly, in November 1943 we noticed flocks of several hundreds at Jerramungup, some 50 miles south-east of Broomehill, and I. C. Carnaby recorded the Masked Wood-Swallow as having arrived one year in November after heavy rain in the Lake Grace district and bred profusely a week or so later. In October 1972 large flocks, flying very high, were seen entering the Helena Valley near Gooseberry Hill from the north-west.

White-browed Wood-Swallow

Artamus superciliosus

The head and throat, and the entire upper parts greyish-black with a *conspicuous white eyebrow* which becomes broader behind the eye. The *under parts,* except for the throat, are *rufous-brown.* The outer tail feathers are tipped white. In the female the white eyebrow is less conspicuous. Iris, dark brown; beak, blue, tipped black; legs, grey. Length, 8½ in.

Distribution: This is a common Eastern States species but in Western Australia it appears to occur only in parts of the interior and the Kimberley Division, as an extension of South Australian and Northern Territory wandering populations. A few individuals accompany larger flocks of Masked Wood-Swallows.

For many years the only occurrence on record, within the area treated by this book, was Wiluna where in early August 1909 F. Lawson Whitlock collected four specimens among a large aggregation of Masked Wood-Swallows on the move. The next report came in 1964. On September 3 V. N. Serventy saw one bird 36 miles west of Mount Magnet. A few days later I. C. Carnaby came on several individuals among large numbers of Masked Wood-Swallows at Thundelarra Station, between Yalgoo and Paynes Find, and both species were breeding together in three separate localities. Six nests were found (September 14, 15).

On August 12, 1965, G. S. Cowles (with a British Museum expedition) saw two individuals east of the Warburton Range. On November 27, 1970 M. G. Brooker saw several in the Great Victoria Desert, 80 miles N. of Neale Junction. At Callagiddy Station, Carnarvon, on August 29, 1972, he saw two birds; Masked Wood-Swallows were very abundant in the locality at the time.

In the following year Mr. Carnaby found a nesting pair of a Masked Wood-Swallow and a hybrid Masked and White-browed at Mulgul Station. The hybrid had a white brow and a pinkish breast.

Nesting: The nests are open frail structures, rather saucer-shaped and shallow; size, about 4 inches in breadth and 2 inches deep, made of dried grasses, rootlets, strips of bark and thin twigs, etc., lined with fine grass or rootlets. At Thundelarra Mr. Carnaby found the nests built low in boggada *(Acacia linophylla),* in a species of prickly *Acacia* and in *Callistemon.* The clutches were of 2 or 3 eggs. Average size of 10 eggs, 22.3 x 16.6 mm., varying from 20.6-23.7 x 16.3-16.9 mm. The eggs are glossy, creamy-white, speckled and spotted all over with umber and pale slate, forming a well marked zone at the large end.

Black-faced Wood-Swallow

Artamus cinereus

Native names: Yalbonjarry (Lyons River, all species); Wokka tjalbee (Jigalong); Be-wo-wen, Biwoen, Bee-woo-een (Perth, Avon River).

General colour dark smoky-grey, darker on the wings and lighter below. The tail and *a patch around the eye to the beak, and the chin are black.* The outer tail feathers are tipped with white, The strongly-built beak is slaty-blue, black at the tip. Iris, brown; legs, light slate-grey. Length, 7-8 in.

Distribution: Found all over the State, except the jarrah forest tableland of the South-West corner. In former times its occurrence in the South-West was much more restricted than it is now, and though it extended along the coastal plain to the south of Perth, inland it probably did not range far south of the Great Eastern railway. Now with the habitat alteration in the Wheatbelt, among other causes, it has spread to Williams, Wagin, Dumbleyung and Ravensthorpe. On the coastal plain it is found as far as Busselton.

Nesting: The Black-faced Wood-Swallow builds a saucer-shaped nest of dried grass, rootlets and pliable stems lined with dried grass, of more compact form than the nest of the Dusky Wood-Swallow.

It is placed in a variety of situations, but usually in a branch of a small tree within a few feet of the ground; in the Perth area a *Hakea* bush is often chosen. The clutch consists of 3 or 4 eggs; they are glossy, white, spotted and blotched with chestnut-brown and light purple. In shape they are blunt-oval. Average size of 16 eggs, 22.4 x 16.6 mm., varying from 20.4-25.3 x 15.7-18.7 mm. In the South-West the birds nest between November and January but in the North-West they will breed at any time of the year following on rainfall, showing the most rapid response of any bird to rain.

These gracefully flying birds frequent open or lightly timbered country in small companies and in the Perth district are commonly to be seen in the open clay flats and bare coastal hills.

Fig. 61.—Black-faced Wood-Swallow.

Fig. 62.—Dusky Wood-Swallow.

Dusky Wood-Swallow

Artamus cyanopterus

Native names: Be-wo-wen, Biwoen, Bee-woo-een (South West, not differentiated from the previous species); Worle (Albany).

Greyish-brown all over, except for the back, wings and tail. There *is a white line edging the outer wing* and the tail is tipped white. Iris, dark brown; beak, blue, tipped black; legs, grey-black. Length, 7-7¼ in.

This species is similar in appearance to the Black-faced Wood-Swallow but is darker below and lacks the black facial marking. The best field recognition mark is, however, the white wing edging, but which despite its general conspicuousness, may be at times difficult to pick out.

Distribution: This was formerly the common wood-swallow of the South-West and it probably ranged regularly north to the mulga-eucalypt line; there are records of its occurrence north to Geraldton, Perenjori and Kalgoorlie. Its usual range is now very much reduced and, on the coastal plain and in the Wheatbelt it is giving way to the Black-faced Wood-Swallow. It is still fairly plentiful in the Darling Range and may be seen north to New Norcia, but it is only occasionally seen on the Swan coastal plain. In the Wheatbelt it is now rarely met with north of Wagin. Thence its present range extends eastwards to Newman Rocks and the Israelite Bay country. M. G. Brooker observed a party at Caiguna, on the Eyre Highway, in October 1973. Its seasonal movements are not fully understood and at times it may still wander towards the northern limits of its former range. Thus in June 1965 A. G. Mathews saw a flock of about 25 birds 45 miles S.W. of Paynes Find, and collected specimens.

Nesting: The Dusky Wood-Swallow is not particular in the selection of a site for its nest—a bush, a low tree, a piece of loose bark on the side of a tree, or even the top of a thistle may be chosen for the open, shallow, and loosely-put-together saucer-shaped nest. It is almost invariably constructed of twigs or rootlets, and generally lined with finer rootlets. The 3 or 4 eggs are stout oval in shape, glossy, and white or yellowish-white in colour, spotted and blotched with umber and various shades of grey, the markings sometimes forming a zone round the larger end of the eggs, sometimes round the centre. They measure, on an average, 22 x 18 mm.

This species inhabits much heavier timbered country than does the Black-faced, and it may be seen hawking for insects at a high altitude above the tall forest trees. It is partial to open tracts within the jarrah and karri forest and hence is commonly met with on farm clearings. At dusk, like most other wood-swallows, the birds cluster very closely together to roost.

Little Wood-Swallow

Artamus minor

Mainly dark brown, with black wings, rump and tail, the latter being tipped white on the outer feathers. Iris, brown; beak, blue, tipped black; legs, dark grey. Length, 5¾ in.

The Little Wood-Swallow is almost black in appearance, and looks only about half the size of the other wood-swallows. The distinctive call, somewhat resembling that of the Swallow, is a high-pitched "peep."

Distribution: A northern species ranging as far south as Kalbarri, the Murchison River gorge (downstream from Galena), Ellendale Gorge, Tallering Peak, Gnow's Nest Range (south of Yalgoo), Field's Find and Mt. Kenneth (west of Youanmi). In the south-east it has been seen at Hann's Tabletop, a mesa in the Great Victoria Desert.

Nesting: The Little Wood-Swallow haunts cliffs and breakaway country, where it nests in crevices among the rocks or on ledges in caverns. The flimsy nest is a small shallow saucer of dry mulga leaves. Occasionally the nest may be found in a hollow limb of a tree or on a tree-stump, when it is more strongly put together with rootlets. It is, however, never as compactly made as are the nests of the other species of wood-swallows. The 3, rarely 4, eggs are white, yellowish-white or creamy-white and have a zone at the larger end of spots of various shades of brown and slate. Size, 20 x 14 mm.

AUSTRALIAN MAGPIES: Family Cracticidæ

Squeaker (Grey Bell-Magpie, Jay, Currawong)

Strepera versicolor

Native names: Bella (Milligan); Dje-laak (Gilbert); Djillak, Tee-luk, Jilak (South-West); Bell (Broomehill); Jarburn, Jarbell (Pallinup River); Yimbalu (Fraser Range).

Almost a uniform *dark grey* in general plumage, *with a white "window" in the wing,* the under tail coverts white, and the secondary wing quills and tail narrowly tipped white. Iris, yellow; beak, heavy and massive, black; legs, black. Length, 20-20½ in.

The white wing "windows" are conspicuous as the bird flies with slow flapping wing-beats.

Distribution: A South-West species ranging north to the mulga-eucalypt line (Fig. 4) eastwards to Eucla; occasionally further north in places, following the river gums (e.g. at Laverton and near Wiluna). D. R. Reid has recorded it at Breaden's Bluff.

Nesting: The Squeaker builds a shallow cup-shaped nest of thin twigs in a vertical fork of a tree, often at a considerable height from the ground. It is lined with fine roots and coarse grasses. The eggs, 2 or 3 in the clutch, occasionally 4, may be found in nests from July to November. They are, in ground colour, light brown of varying shades, usually pale, and marked with spots and blotches of deeper brown and neutral tint. Average size of 24 eggs from the South-West, 43 x 31 mm., varying from 40-47 x 29-32 mm. The female sits closely but when the nest-site is approached she slips quietly away into the nearest cover. The young accompany the parents until the latter commence nesting again in the following year.

Up to 1949-50 the Squeaker had been diminishing in numbers in the South-West and was hardly ever seen in the near vicinity of Perth. Recently there has been an indication of a revival in numbers in the South-West and Wheatbelt. It is abundant in the mallee country of the south-east and in some of the country towns in this region, such as Lake Grace and Ravensthorpe, the birds may be seen in the street trees.

The bird is not popular in the fruit-growing areas on account of the damage it does, not only to stone-fruit but to apples.

Its usual note is a loud gong-like "tew-tew," which sounds clearly through the bush; it also has a peculiar call which has been likened to the mewing of a cat, or to the sound made on a tin trumpet.

Grey Butcher-bird (Derwent Jackass)

Cracticus torquatus

Native names: Wandoa (Lyons River); Wadda-wadda Wooleen); Wad-do-wad-ong, Waddo-wadong (Perth); Yoy-jeddie (Pallinup River).

Head, cheeks and tail black, back grey, rump white. Wings, brownish-grey. The *whole of the underparts are greyish-white.* Iris, reddish-brown; beak, strong, hooked, bluish-grey, becoming black towards the tip; legs slaty-black. Length, 9½-11½ in. Weight, 3½ oz.

Distribution: In the southern part of the State, north to the Minilya and Upper Ashburton Rivers and Jigalong. In the extreme south, however, it does not occur south of a line passing through Augusta, Nannup, Kirup, Cranbrook and Albany.

Nesting: The nest is usually placed in a vertical fork of a tree at no great height from the ground—usually from 6 to 20 feet, but sometimes higher. It is a shallow, cup-shaped structure, composed of dead twigs, roots, etc., and lined with grass or rootlets. The cup itself is neat, but viewed from below, owing to twigs being used in its construction, it appears rather untidy. A typical nest is about 9 inches in diameter. The lined cup for the eggs measures about 4 inches in diameter and 2 inches in depth. The clutch consists of 3 or 4 eggs. A set of three from Bunbury are pale brown with small reddish-brown blotches concentrated towards the larger end, in one egg forming a cap. A few black spots are distributed over the eggs mostly towards the larger ends. Size 30 x 32 x 22 x 24 mm. July to November are the chief breeding months. Incubation period, 23 days; nestling period, 25 or 26 days (W. H. Loaring).

These birds defend territories throughout the year and, as the female shares in the defence, she sings as vigorously as the male. The social organisation does not resemble that of the Western Magpie, as orthodox pairing occurs. The song, which is very spirited and contains some beautiful notes, may be heard the year round.

The butcher-bird's habit of impaling prey on twigs and thorn is modified in an interesting fashion in this species and in such a way as to bring it within the category of the very few birds which are tool users. E. H. Sedgwick has witnessed at Bilbarin a Grey Butcher-bird eating a nestling Brown Honey-eater by using the jagged end of a dead branch as a vice to assist it in tearing its prey apart. He has also recorded a similar observation of a Butcher-bird eating a phasmid. C. F. H. Jenkins has observed the birds wedging lizards into small forks of trees, which were used as larders for later feeding.

Fig. 63.—Black-throated Butcher-bird.

Fig. 64.—Grey Butcher-bird

Black-throated Butcher-bird (Pied Butcher-bird, Break o' Day Boy)

Cracticus nigrogularis

Native names: Cockaraboota (Jigalong); Coolburra (Wooleen); Cudgeego (Yandanooka).

This Butcher-bird is more Magpie-like in appearance than the preceding species—with its *sharply contrasted black and white plumage.* It has a *jet-black head, neck and breast,* as well as the upper back, tail, shoulders and outer wing quills. There is a white nuchal collar, a white wing bar, broad white tips to the outer tail feathers and white lower breast and abdomen. Iris, brown; beak, bluish-grey, blackish towards the tip, strongly hooked; legs, black. Length, 13-15 in. Weight, 5½ oz.

Distribution: Found all over the State, except the South-West corner, the southern boundary passing through Gingin, near Toodyay and Lake Grace. It is extremely plentiful on the road between Geraldton and Dongara. There are occasional records in the South-West, such as at Wanneroo, Woodmans Point, Carmel (in the Darling Range), Busselton and Tambellup, indicative of a steady evtension of range south-westwards, tending to the type of distribution illustrated in Fig. 9. It colonised the Moora district in 1912 (C. L. E. Orton and P. T. Sandland). However, it is still scarce south of a line joining Hill River, Walebing, Wongan Hills, Wyalkatchem, Merredin and Kalgoorlie. It is rather more of a wanderer than the Grey Butcher-bird.

Nesting: The nest is very similar to that of the preceding species but is usually higher in trees. The usual clutch is of 4 eggs, but varies from 3 to 5. Under drought conditions in the North-West a single egg might comprise the full clutch (I. C. Carnaby). The eggs are larger than those of the Grey Butcher-bird, the average size of 10 eggs being 32.9 x 24.8 mm., varying from 31.1-35.5 x 23.8-25.5 mm. Weight, 9 gm. They are usually a dark olive-brown covered all over with small markings of a deeper shade, especially towards the broad end, where there are also, as with the grey species, a few black spots. The ground-colour is subject to considerable variation in its shade of brown. August to December are the nesting months in the southern part of its range.

The bird is one of the finest of our songsters and has a wide variety of clear and mellow calls. Some may be rendered as "por-rk-it-tee," "quo-orrr" and "chu-eep." This Butcher-bird is also a noted mimic.

Western Magpie

Gymnorhina dorsalis

Native names: Goore-bat, Gurbat, Gurbal, Gooraba (Perth); Curabarty (Beverley); Gurbal, Toorakee, Crow-bardy (Broomehill); Coolbardie (Katanning); Kulbardie (Pallinup River); Yelyel (Fraser Range).

The plumage pattern is a sharply contrasted black and white. In the *adult male* the head, breast, abdomen, flight feathers and the tip of the tail are a shining black. *The upper parts are* white from the hindneck to the black tail-tip, as also are the shoulders of the wing, under tail coverts and the base of the under tail. The *female* differs strikingly in having the *back black,* with narrow white edgings to the mantle feathers and a white collar behind the head. *Young birds* resemble the female in general colour pattern, with the white edgings to the mantle feathers less developed or absent and with a brown tone to the feathers which progressively disappears as the birds grow older. Iris, bright brown; beak, bluish-white with a black tip in the adults of both sexes, darker in immature individuals. Length, 16-17 in. Weight, 10-12 oz.

The young male is four years old before it assumes the fully white back.

Distribution: Generally distributed north to the mulga-eucalypt line (p. 53). In the extreme South-West it is usually absent in the heavy forest country, south of a line joining Augusta, Nannup, Mount Barker and Albany, but it appears after extensive clearing. Prior to settlement it was unknown west of the Carbanup River, near Busselton (this was the case when T. Carter visited the area in 1887). In the north it is extremely sparse north of the mulga-eucalypt line and appears to be absent in the coastal country north of the Murchison River. It has been recorded at Seemore Downs and Eucla in the south-east. I. C. Carnaby has found it nesting between Madura and Eucla, and saw it 13 miles east of the border, in South Australia. Exact limits of distribution are uncertain owing to possible confusion with the Black-backed Magpie.

Nesting: The nest is a large bowl-shaped structure, constructed of sticks and twigs, with the inside lined with grass, bark, hair, wool or leaves. It is usually placed in a fork of a tree—often up to 40 or 50 feet from the ground. Nests built of wire are not uncommon. One such from Lake Grace examined by C. S. Hamilton, of the Education Department, weighed 5 lb. and was made up of 238 separate pieces of wire ranging from 4½ in. to 4 ft. 2 in. in length. These pieces totalled a length of 243 ft. of wire. A larger nest from Caron, weighing 13¾ lb., was examined by us and found to consist of 286 pieces of wire, including barbed wire of a similar size range to the previous nest, and totalling 338 feet in length.

The eggs are not so variable in colour and markings as are those of the other two species of Australian magpies, being bluish or greenish in colour, blotched, smeared and streaked all over with liver and brown. A set of 4 collected by A. H. Robinson at the Edmund River, in the Upper Gascoyne River district, had 2 eggs of the 4 with a blue ground with the submerged spottings of more normal eggs of the next species *(G. tibicen),* but smaller than the average of the northern eggs. The other 2 eggs were distinctly smaller and resembled the eggs of this species in ground colour and were heavily smeared and streaked with brown and purple, but also with submerged dark purple spots. Was this the result of hybridisation with the next species? The usual clutch is of 3 eggs, but may vary from 2 to 6. Average measurements of 25 eggs, 40.1 x 27.3 mm., varying from 38.0-43.0 x 25.5-29.0 mm.

Though this bird is so common and familiar to all, the intimate details of its peculiar life-story have only recently been revealed through the studies of A. H. Robinson and H. M. Wilson. Robinson has shown that instead of *pairs* of birds defending territory, as is normally the case, *groups* of Magpies form the territorial units. The group consists of six to twenty individuals, of varying sexual ratio, and the territory occupied by them is approximately 100 acres in the Coolup district. The birds are permanently resident on their territory and every bird, both male and female, adult and immature, participates in its defence, the adult males being the most aggressive. The beautiful carol is the communal territory song and it reaches its finest volume in the mating season before nesting begins. Magpies become relatively quiet during the incubation period.

Robinson found that no attachment is discernible between particular individuals in the nesting season, the various males being more or less promiscuous within their own group and mated to no particular female. The females alone build the nests and incubate the eggs, the males playing no part until the feeding of the young begins on their leaving the nest. The males then assume a share in their upbringing and protection, but the females continue to do most of the work. The nests of a group may be placed in the one tree, adjacent trees, or as far apart as a quarter of a mile.

The birds of a group as a rule will not stray away from their territory. However, frontier fights are not uncommon and there may occasionally be mass invasions of a territory by neighbours. Strange magpies, including pets introduced into a new area, will be severely molested by the resident magpies, though cases are known where a transported pet, after a time, is accepted in the local group. Magpies become very aggressive to other species, including human beings, in the nesting season and may swoop viciously at anyone intruding too close to the nests.

Population studies by Wilson show that the Swan coastal plain is more densely populated by magpies than any other part of the State, averaging 25 birds per square mile. The next densest area is the wandoo-jam country between Northampton and Katanning, with nine birds to the square mile. In the jarrah forest the average falls to five birds per square mile. He estimates that the total number of the species in Western Australia is between 450,000 and 900,000 birds.

Black-backed Magpie

Gymnorhina tibicen

Native names: Wundoolah (Coolawanyah); Koorburroo (Mulga Downs).

Similar to the preceding species, except that the sexes are not so dissimilar and resemble in plumage pattern the female of the Western Magpie. Both male and female are black-backed but the female is grey instead of white on the hindneck. Iris, brown; beak, slate, blue-black tip; legs, black. Length, 14½ in.

Distribution: Though widely distributed in the eastern States the Black-backed Magpie in Western Australia occurs only in the north, to as far south as Onslow and the middle Gascoyne (Landor and Milgun Stations). In the south-east it has been recorded at the Warburton Range and in the Great Victoria Desert (25 miles south of Neale Junction and Carlisle Lake). The exact range is, however, not certainly known owing to the confusion in the field with the Western Magpie and specimens in collections are scarce. The situation is further complicated by the fact that pastoralists are known to have introduced the southern bird on their stations in the Minilya and lower Gascoyne country, and possibly elsewhere.

Nesting: The eggs of this species are very variable in coloration throughout the whole of its range and no standard description can be set down. When reviewing the eggs of the species for G. M. Mathews' *Birds of Australia,* H. L. White found it necessary to give detailed descriptions of no fewer than twelve different clutches. Some clutches we have seen from the Barlee Range have the ground colour varying from grey to brown, marked with dark brown spots, larger at the broader end, also submerged purple spots. of dark blue or purple. The essence of the difference between the eggs of this species and those of the Western Magpie appears to be that the markings are *spottings* in this species instead of the *streakings* or smears in the Western Magpie. There is no difference between the nests and in the number of eggs laid. A series of 15 eggs from Ullawarra and Landor Stations average 39.8 x 27.8 mm., varying from 34.0-46.0 x 24.5-32.0 mm. In the Barlee Range A. H. Robinson found the favourite nesting tree to be the Mountain Wattle *(Acacia xiphophylla)*, the highest tree away from the creeks.

Magpies of either species in the north are much less plentiful than in the south. They have a scattered, patchy distribution, are much more shy and wary and are difficult to approach. They are also more silent and the song is rarely heard.

BOWER-BIRDS: Family Ptilonorhynchidæ

Spotted Bower-bird

Chlamydera maculata

Native names: Tarradee (Millstream); Thurra-ji (Coolawanyah); Ir-karka, Iligura (Warburton Range).

Back and wings, blackish-brown, *boldly spotted with orange-brown,* tail tipped orange-brown. The head is spotted grey and brown, having a metallic sheen, and on the nape is a beautiful *nuchal collar of vivid pink* with a silvery lilac hue in certain lights. The throat is blackish spotted orange-brown and the rest of the under flanks. The female has a much smaller nuchal collar. Iris, deep brown; parts are yellowish-brown, becoming chestnut-brown towards the beak, black; legs, green-grey. Length, 10¼-11¼ in. Weight, 4½ oz.

Distribution: A mid-northern and central Australian form extending north to the Pilbara district and south to the North-West Cape Range, the middle reaches of the Gascoyne River, Poona and Tuckabiana (near Cue), and Malcolm; east to Lake Carnegie. The Gibson Desert may separate the western from the central Australian population, which, however, extends westwards from the Alice Springs and Macdonnell Ranges areas and crosses the State boundary to the Rawlinson Range at Giles and to the Warburton Range. A third population inhabits eastern Australia, inland from the Great Dividing Range.

Nesting: While the nest has been described from other parts of the range of the bird in Australia, there is only a single description of one observed in Western Australia. This was a nest recorded by F. Lawson Whitlock about 20 miles north-east of Wiluna in October 1909. It was about 15 feet from the ground near the top of a casuarina and was a rough platform of dried and black twigs, with a fairly well-defined, but rather shallow cup in the centre which was lined with fine twigs and a few casuarina "needles." The 2 eggs are a delicate french- or greenish-grey and are covered with continuous lines of light brown which wind in and out in a more or less circular manner around the wider part of the shell. There are also irregular blotches of darker brown. In shape the eggs are slightly elongated ovals, with no very great difference between either end. Size, 32 x 26 mm.

From descriptions given by Whitlock and A. H. Robinson of the bower or playground, it appears that the bird first makes a foundation by collecting a mass of twigs of $\frac{1}{8}$ to $\frac{1}{4}$ inch diameter and of about 9 inches long to raise the floor of the bower some 6 to 8 inches from the ground. The walls of the arch of the bower are constructed by the nipping off by the bird of twigs from a low dense bush, the wood of which is very elastic, like a willow, and, in the green state, fairly soft. The general height of the sides of the bower is about 16 inches and the twigs forming them are erected until the sides are about 6 inches thick. The length of the completed bower is about 2 feet. Inch square flakes of white limestone, bones, green seed-pods, mulga beans, sandalwood nuts and green leaves are among the objects collected by the bird to ornament the floor of the bower.

The male in display at the bower puffs out his feathers, showing the beautiful pink tract of plumage on his neck to great advantage. With various harsh cries he advances into the middle of the playground and makes a vigorous attack on some object lying on the floor. He advances, backs, hops from side to side, jumps into the air and, with much apparent ferocity, rushes at his mate. Now and again she utters a short, harsh cry but otherwise seems to regard the demonstrations made by the male as very matter-of-fact and hardly worthy of notice. These demonstrations may last for over an hour. While the display is going on other birds act as audience. They squat rather than perch on the branches of the tree overhead and occasionally one or other of them utters a harsh cry. To close the display the male hops to the top of the tree and the party flies off.

The Bower-bird is much given to mimicry and will imitate the calls of other birds, such as the Whistling Eagle, Brown Hawk, Blue-winged Kookaburra and Butcher-bird. It also utters a cat-like cry which is probably its natural call.

The distribution of the species in Western Australia appears to be largely determined by that of the wild fig *(Ficus platypoda)*, on the fruits of which it feeds, and the southern "fig line" is also the main southern limit of occurrence of the Spotted Bower-bird. The wild fig grows in rough country, among rocks and cliffs, but is not uniformly distributed and the bower-bird likewise has a patchy distribution. When writing of his observations in the North-west Cape Range, T. Carter considered the bird very rare, but the troops stationed at Exmouth Gulf during World War II found it very abundant in the rugged ridges of the range.

CROWS and RAVENS: Family Corvidæ

Australian Raven (Crow)

Corvus coronoides

Native names: Kwaggum, Queg-gum, Par-dang (South-West); Wur-dang, Woordang, Warrdong (Broomehill and Pallinup River).

A glistening black, with the feathers having a blue-violet glint when viewed in certain lights. The *bases of the body feathers are dusky grey* (sometimes whitish-grey) and the hackles or *throat feathers are long and lanceolate in adult birds.* Young birds also differ from the adults in having brown eyes instead of white and in the inside of the mouth being flesh-pink instead of black. The heavy predatory beak is black. Legs, black. Length, 20-22 in. Weight, 1 lb. 4 oz.-1 lb. 7 oz.

Distribution: Confined to the South-West, from the mouth of the Murchison River to Eucla and inland to Morawa, Bunketch (40 miles north-east of Wongan Hills), Dowerin and Kellerberrin. It is becoming more plentiful at Lake Grace (I. C. Carnaby). It occurs also on many of the islands of the Archipelago of the Recherche. Inland nesting records are from Mullewa, Dowerin and Lake Grace.

Nesting: The Raven usually builds its nest in a vertical fork high up in a tree, and constructs it of sticks, lining the centre with bark, hair, etc. It is a well-built structure. Ground colour of eggs, pale green, *boldly marked* with dark and olive-brown in the form of frecklings, blotches and spots. Clutch 4 or 5. Average size of 32 eggs from the South-West, 45.1 x 30.4 mm., varying from 42.0-51.7 x 28.9-33.2 mm. When an intruder approaches a Raven's nest the sitting bird does not fly away in a horizontal direction, but slips down towards the ground close to the trunk of a tree in which the nest is situated and then flies quietly away close to the ground.

Australian Crow

Corvus orru

(formerly *Corvus cecilæ*)

Native names: Woggora (Millstream); Wonga-ka (Coolawanyah); Kawk (North-West Cape); Woggora (Lyons River); Karnka (Jigalong, both species); Kaku (Pidong tribe, East Murchison River); Kargoo (Upper Murchison); Carcoo (Wooleen); Gala (Wallawe tribe, Mount Magnet); Kahnga (Warburton Range); Wajigga (Fraser Range).

Similar to the preceding species except that the *concealed bases of the body feathers are snow white* instead of dusky, *contrasting sharply* from the black terminal portions of the feathers. It is this sharply defined colour pattern in the feather which is the important diagnostic factor; in the Raven there is no clear contrast even in those cases where the feather base may have a whitish tinge. The throat hackles also are much shorter. Contrary to popular belief there is no difference in the eye colour—all our members of the genus *Corvus* have brown eyes in immaturity and white in adult life. There is a corresponding change in the colour of the inside of the mouth—flesh-pink in the young and black in the adult. Beak and legs, black. Length, 19½-22 in. Weight, 1 lb. 3 oz.-1 lb. 10 oz. Wing length, 330-380 mm.

Distribution: Found all over the State, except the South-West. The exact limits are not definitely known, but its distribution is probably exclusive of that of the Raven. Known localities at or near the border line are: 15 miles east of Pindar and 50 miles east of Lake King. Nesting has been recorded at Gnow's Nest (south of Yalgoo) and 25 miles south of Mt. Magnet.

Nesting: The nest is built in tall gums, often in inaccessible situations. The 3 to 5 (sometimes as many as 7) eggs of elongated oval shape, are of light blue ground colour *unmarked* or lightly spotted with olive-brown flecks. Western and central Australian eggs of this species are strikingly different from eastern Australian eggs, which resemble Raven's eggs in colour. Average size of 31 eggs, 45.5 x 29.9 mm., varying from 41.2-52.0 x 28.0-32.8 mm. The breeding season is from June to August or September in the North-West, being constant, irrespective of the character of the season.

This is the common large crow of the northern and interior parts of the State. In appearance it is difficult to distinguish from the Raven, but the call-notes have a different quality. The Crow has a shorter note, not so deep as that of the Raven, sounding more like "oh! oh!" rather than the bass, long drawn-out "ah-h-h" of the Raven.

Little Crow

Corvus bennetti

This bird is a smaller edition of the Crow, having *white bases to the body feathers* and short throat hackles. Young birds have a brownish tinge to the plumage, as well as the usual features of immaturity, such as brown instead of white eyes, and flesh-coloured instead of black mouths. The white iris in adult birds differs from that in the Raven and Crow in having a light blue ring around the pupil (A. H. Robinson). Beak and legs, black. Length, 16½-19½ in. Weight, 11 oz. to 1 lb. Wing length, 290-330 mm.

The call is hoarser and shorter-noted than the Raven of the South-West and usually ends in a warble or yodel. The difference in size from the other two species is perceptible on sight. Caution should be used in identifying the Little Crow from single specimens as E. F. Boehm, a South Australian ornithologist, has shown that occasionally dwarf individuals may occur in both of the other species. Furthermore, the small Indian Crow *(Corvus splendens)* is not infrequently brought to Fremantle in ships and specimens have been taken in the port precincts from time to time. From 1950 to the end of 1967 some 30 Indian Crows were destroyed in Western Australia. The Indian Crow differs from all Australian species by *not being uniformly black;* its plumage is mostly glossy black but is variegated by the nape, upper back and breast being ashy-brown. In contrast to all the Australian crows the iris remains dark brown throughout life.

Distribution: The Little Crow is a mid-northern and central Australian species which occurs from the Pilbara and Halls Creek districts, south beyond the Victoria Plains district, seasonally visiting the South-West (outside the heavy jarrah forest zone), and south-east through the northern wheatbelt and eastern goldfields to the Nullarbor Plain.

Although a bird of such wide distribution throughout Australia, its specific distinction from the Raven and Crow was not recognised until 1901, when A. J. North described it as a separate species.

Nesting: The nest is built in scrubby trees, such as mulga and beefwood *(Hakea lorea)* and on telegraph posts and windmill platforms. It never nests in river gums or tall trees as does the Crow. Though not colonial these birds do tend, however, to nest much closer together than is the case with the Crow, and nests may be as near as 50 yards from each other. The Crow is much more solitary and its nests are a considerable distance apart. The Little Crow has a characteristic habit of placing clay between the bottom of the nest and the top lining (of strips of bark, feathers and kangaroos' fur). It lays a small dumpy egg, distinctly smaller than the Crow's and the colour pattern resembles that of the Raven; the ground colour is light blue or green *profusely marked* with olive-brown frecklings, sometimes with spottings and blotches in addition. These may uniformly cover the egg or be mainly over the broad end. Clutch, 4 to 7, varying according to the nature of the season, more eggs in good seasons.

Average size of 48 eggs from the North-West, 39.2 x 27.4 mm. varying from 35.5-43.7 x 25.6-30.2 mm. The time of egg-laying varies with the season and the birds may lay in any month. The usual period in the North-West is March and April, if there are summer and autumn rains and again in the spring if conditions are favourable.

The Little Crow is characteristically the crow of the mulga scrub country, though it occupies the same geographical area as the Crow and the two species may be found together. At Denham, Shark Bay, flocks of predominantly immature members of this species are in constant attendance in the winter months on the waterfront of the little settlement, where they act as the township scavengers. In this generally treeless area they nest on station windmills and telegraph poles.

The occurrence of the Little Crow in the South-West was not recognised until recent years. At first, following its initial identification, it was believed that the species was extending its range and increasing in numbers. This may yet be found to be true, but it is now established that Little Crows had been visiting the South-West since the early part of the century, at least, but had been confused with Ravens. Arising out of trapping and ringing experiments at Coolup Mr. A. H. Robinson has found that the flocks of "crows" which migrate along the scarp of the Darling Range, and previously thought to be Ravens, are really Little Crows. They are more gregarious than the Raven, whose flocks number less than 20 individuals. Along the scarp the birds travel on a very narrow flight line and little or no movement is evident even a few miles out on the coastal plain. E. H. Sedgwick has reported them from Harvey (flocks of up to 100 individuals). The reason for this close hugging of the edge of the tableland apparently is that the birds take advantage of the upward lift derived from the eddy currents always present in the air where there are sudden changes in altitude. The movement is protracted, in the latitude of Perth, to the south in the summer and north in winter. The details of the dispersion in the South-West have not been worked out. In Perth the birds may be seen frequenting municipal rubbish dumps, as at Butler's Swamp, Claremont, and they visit farms in Coolup, in company with Ravens. They have been recorded at Boyup Brook by G. A. Lodge, who has noted that they arrive there in January and February and leave in the autumn, before the winter rains set in.

No record of breeding has been made in the South-West as yet. The present south-western limits of nesting are Yalgoo (P. J. Fuller), Mt. Gibson (I. C. Carnaby) and Yorkrakine (Nicholas Kolichis).

INDEX OF SECTION I

THE HISTORY OF WESTERN AUSTRALIAN ORNITHOLOGY

Alexander, W. B. 45
Amsterdam 16
Armstrong, F. F. 30, 33
Ashby, E. 44
Astrolabe 28
Austin, R. 38
Ayres, F. G. 39
Banda 16
Baudin, N. 23
Bauer, F. 25
Beagle, H.M.S. 32
Beete, J. P. 30
Breton, H. W. 30
Brown, R. 25
Burges, L. C. 32
Bynoe, B. 32
Caen, A. 16
Calvert Expedition 42
Cameron, L. 42
Campbell, A. J. 41
Carter, T. 40
Casuarina 23
Chatham, H.M.S. 19
Clark, W. N. 30, 35
Cockerell, J. 39
Collie, A. 28
Conigrave, C. P. 43
Cunningham, A. 26
Cygnet 16
Dale, R. 28
Dampier, W. 16
De Vergulde Draeck 16
Discovery, H.M.S. 19
Dodd, W. D. 45
Drummond, James 34, 36
Drummond, Johnston 34
Dumont d'Urville 28
Elder Expedition 41
Emeloort 16
Erskine, A. 28, 29
Entrecasteaux, B. d' 21
Esperance 21
Flinders, M. 25
Fraser, C. 28
Fremantle, C. 28
Freycinet, L. C. de 23, 27
Friend, M. 29
Gaimard, J. P. 27
Geelvinck 16
Geographe 23
Gibson, C. G. 44
Gilbert, J. 30, 33
Giles, H. M. 42
Gould, J. 29, 33
Graeff, A. van der 18
Gregory, F. T. 38
Grey, G. 30, 35
Hall, R. 42
Helms, R. 41
Herald, H.M.S. 38
Het Wapen van Hoorn 15
Hillegom, H. C. van 15
Hull, A. F. B. 45
Investigator, H.M.S. 25
Jackson, S. W. 45
Jongh, W. G. de 16
Keartland, G. A. 42
King, P. P. 26
Labillardiere, J. J. H. de 21
Leschenault de la Tour 23
Lesueur, C. A. 23
Levillain, S. 23
Lipfert, H. F. O. 42
Masters, G. 38
Menzies, A. 19
Mermaid, H.M.S. 26
Milligan, A. W. 43
Montague, P. D. 45
Moore, G. F. 31
Napoleon 37
Naturaliste 23
North, M. 39
Ocean Queen 37
Pelsaert, F. 15
Peron, F. 23
Preiss, J. A. L. 31, 34
Quoy, J. R. C. 27
Rayner, F. M. 38
Recherche 21
Riche, C. A. G. 21
Roe, J. S. 28
Roebuck, H.M.S. 17
Roosenbergh, J. van 15
Rossel, E. P. E. de 21
Sanford, W. A. 38
Seewolf 15
Shortridge, G. C. 44
Stokes, J. L. 32
Tunney, J. T. 43
Uranie 27
Vancouver, G. 19
Vlamingh, W. de 16
Webb, W. 39
White, S. A. 41
Whitlock, F. B. L. 43
Wilson, T. B. 28
Wickham, J. C. 32
Wollaston, J. R. 36
Zeewyk 18

INDEX OF VERNACULAR NAMES

(In multiple references the main one is given in black figures. The vernacular names index is a more comprehensive reference to the species mentioned in this book than is the Scientific Names Index).

Albatross, Black-browed 76, **77**
Albatross, Grey headed **79**
Albatross, Light-mantled Sooty **80**
Albatross, Royal **76**
Albatross, Shy **79**
Albatross, Sooty **80**
Albatross, Wandering 25, **75**
Albatross, White-capped .. 5, **79**
Albatross, Yellow-nosed .. 76, **78**
Avocet 11, 17, **226**

Babbler, Grey-crowned 11, **335**
Babbler, Red-breasted 336
Babbler, White-browed 55, 59, 60, **336**
Barking Bird 424
Bee-eater 5, 11, **312**, 410
Bell-bird, Crested 8, 11, 49, 58, **390**, 392
Bell-Magpie, Grey 49, 58, **447**
Berrin-berrin 312
Blue Bonnet 55, **285**
Bittern, Black 49, **129**
Bittern, Brown **129**
Bittern, Chinese Little 5, **131**
Bittern, Green-backed **128**
Bittern, Little **130**
Bittern, Yellow **131**
Booby, Blue-faced **109**
Booby, Brown **110**
Booby, Red-footed 109
Bower-bird, Great 48
Bower-bird, Spotted 11, 38, 49, 55, 63, **454**
Break o'Day Boy 450
Bristle-bird, Eastern 370
Bristle-bird, Rufous 43, 50, 52, **373**
Bristle-bird, Western 34, 38 **370**, 394
Brolga 6, **184**
Bronzewing 24
Bronzewing, Brush 16, 22, 32, 50, **263**
Bronzewing Common 11, 61, **262**
Budgerygah .. 10, 11, 49, 61, **290**
Bush Hen 48
Bushlark, Horsfield **321**
Bushlark, Singing **321**
Bustard, Australian .. 11, 31, **193**
Butcher-bird, Black-throated 11, 60, 61, 294, **450**
Butcher-bird, Grey .. 11, 318, **448**
Butcher-bird, Pied 450
Buzzard, Black-breasted .. 11, **161**

Canary, Native 413
Cape Hen **89**
Caterpillar-eater, White-shouldered 330
Chat, Crimson .. 11, 49, 297, **427**
Chat, Orange 11, **428**
Chat, White-fronted 11, 51, 297, **427**
Chat, Yellow 48
Chock-a-lok 273
Cicada-bird 48
Chough, White-winged **440**
Cockatiel **278**
Cockatoo, Major Mitchell ....
Cockatoo-Parrot 278
Cockatoo, Pink 273
Cockatoo, Red-tailed Black .. **271**
Cockatoo, Sulphur-crested .. 48
Cockatoo,White-tailed Black 5, 27, 50, 54, **269**
Cockatoo, Yellow-tailed Black 50, 270
Coot 10, **192**
Coot, Blue Bald **191**
Copperback 332
Corella, Little 11, 18, **273**
Corella, Long-billed 11, 28, 50, **275**
Cormorant, Black 7, **111**
Cormorant, Black-faced **116**
Cormorant, Little Black 10, 28, **112**
Cormorant, Little Pied 10, **117**
Cormorant, Pied 7, 23, **113**
Corncrake 5, **186**
Crake, Australian Spotted .. **186**
Crake, Malay Banded 5, 186
Crake, Marsh **187**
Crake, Spotless **187**
Crake, White-browed 48
Crow, Australian 11, **456**
Crow, Ceylon 12, 458
Crow, Indian 12, 458
Crow, Little 4, 9, 11, 61, **457**
Cuckoo, Black-eared 8, 11, 49, **295**, 363

Cuckoo, Brush 48
Cuckoo, Channel-bill 48
Cuckoo, Fan-tailed 50, **294**
Cuckoo, Golden Bronze ....
5, 50, **297**
Cuckoo, Horsfield Bronze 296
Cuckoo, Little Bronze 48
Cuckoo Narrow-billed
Bronze 5, 11, **296**, 298
Cuckoo, Oriental **294**
Cuckoo, Pallid 5, **293**, 405
Cuckoo-Shrike, Black-faced 4, 11, **329**
Cuckoo-Shrike, Ground
11, 49, **328**
Cuckoo-Shrike, Papuan 48
Curlew, Eastern 27, 196, **209**
Currawong 447

Dabchick 120, 121
Darter, Australian **118**
Derwent Jackass 448
Desert-bird, Carter's 337
Diamond-bird, Black-headed 48, 401
Diamond-bird, Red-browed ....
11, **401**
Diamond-bird, Red-tipped ....
5, 11, 49, 51, 57, 58, **400**
Diamond-bird, Spotted ....
5, 50, **399**
Diamond-bird, Yellow-rumped 11, **400**
Diving-Petrel, Common 103
Diving-Petrel, Kerguelen **103**
Dollar-bird **314**
Dotterel, Australian 11, 59, **208**
Dotterel, Black-fronted .. 11, **207**
Dotterel, Double-banded .. 5, **206**
Dotterel, Hooded **207**
Dotterel, Large Sand- **203**
Dotterel, Mongolian **204**
Dotterel, Oriental **202**
Dotterel, Red-capped 11, **206**
Dotterel, Red-kneed **208**
Dove, Bar-shouldered **261**
Dove, Diamond 62, **260**
Dove, Indian **257**
Dove, Laughing 258
Dove, Necklace **257**
Dove, Peaceful 11, **259**
Dove, Rock **256**
Dowitcher, Asiatic 217
Duck, Black 10, **145**
Duck, Blue 151
Duck, Blue-billed 10, 29, **153**
Duck, Freckled **143**
Duck, Grey **145**
Duck, Mountain 10, **144**
Duck, Musk 19, 20, **154**
Duck, Pink-eared **149**
Duck, Plumed Tree- **140**
Duck, Spoonbill **148**
Duck, Whistling Tree- **139**
Duck, White-eyed **150**
Duck, White-winged 150
Duck, Wood **151**

Eagle, Australian Little .. 11, **165**
Eagle, Red-backed Sea- **163**
Eagle, Wedge-tailed ....
11, 19, 23, **167**
Eagle, Whistling 10, **162**
Eagle, White-breasted Sea- 23, **169**
Egret, Cattle 125
Egret, Little 124
Egret, Plumed 123
Egret, White 6, **123**
Emu 11, 16, 22, 25, 36, **65**
Emu-Wren, Mallee 353
Emu-Wren, Rufous-crowned 11, 40, 55, 297, **353**
Emu-Wren, Southern ....
33, 50, 54, 60, **352**

Falcon, Black 175
Falcon, Black-cheeked 172
Falcon, Grey 11, 32, **174**
Falcon, Little 11, **173**
Falcon, Peregrine 172
Fantail, Black-and-white ....
Fantail, Grey 5, 378, **381**
Fantail, Northern 48
Fantail, Rufous 48
Faintail-Warbler, Golden-headed 338
Field Wren 11, 297, **365**
Field-Wren, Rufous 365
Finch, Chestnut-eared 432
Finch, Double-bar 410
Finch, Mountain 431
Finch, Painted 11, 33, **431**
Finch, Red-browed 11, **435**
Finch, Red-eared 429
Finch, Red-faced 434
Finch, Star 11, **434**
Finch, Zebra ....
10, 11, 59, 361, 410, **432**
Firetail, Beautiful 50
Firetail, Diamond 52
Firetail, Red-eared ....
28, 33, 50, 62, **429**
Flycatcher, Brown ....
11, 60, 63, **294**

Flycatcher, Lemon-breasted .. 48
Flycatcher, Restless 294, **383**
Flycatcher, Satin 384
Frigate-bird, Greater 5, **106**
Frigate-bird, Lesser 7, **105**
Frogmouth, Tawny 301, **304**
Fulmar, Antarctic 83

Galah 11, 60, 61, **276**
Gallinule 188
Gannet, Australian 5, **108**
Gannet, Brown 7, **110**
Gannet, Masked 7, **109**
Gannet, Red-footed 109
Garganey 149
Gnow 177
Godwit, Bar-tailed ....
195, 196, **211**, 221
Godwit, Black-tailed 195, **210**
Gold-digger 310
Goldfinch 11, **428**
Gold-miner 312
Goose, Cape Barren ....
22, 26, 27, **141**
Goose, Green Pygmy 152
Goose, Magpie 6, **136**
Goose, Maned 10, **151**
Goose, Pied 6, **136**
Goshawk, Australian 164
Goshawk, Grey 174
Goshawk, Red 48
Grape-eater 409
Grass-bird, Little 50, **339**, 406
Grass-bird, Tawny 48
Grass-Parrots 62
Grass-Wren, Black 48
Grass-Wren, Dusky 370
Grass-Wren, Rufous 368
Grass-Wren, Striated 368
Grass-Wren, Thick-billed 369
Grass-Wren, Thin-billed 370
Grass-Wren, Western .. 27, 40, **369**
Grebe, Crested 31, **119**
Grebe, Hoary-headed 121
Grebe, Little 10, **119**
Greenie 402
Greenshank 196, **213**
Grey Peter 409
Ground-bird, Chestnut 332
Ground-bird, Chestnut-breasted 334
Groundlark 327
Ground-Thrush, Australian 51
Ground-Wren, Shy ....
295, 296, 297, **364**
Grundy 327
Gull, Dominican 235
Gull, Franklin's 233

Gull, Kelp 235
Gull, Pacific 7, 22, **234**
Gull, Silver 7, **231**

Hawk, Brown 11, **175**
Hawk, Crested 48
Harrier, Spotted 11, **169**
Harrier, Swamp 170
Heron, Great-billed 48
Heron, Mangrove 17, **128**
Heron, Nankeen Night 128
Heron, Pacific 6, **127**
Heron, Pied 48
Heron, Reef 17, **124**
Heron, White-faced 10, **126**
Heron, White-fronted 10, **126**
Heron, White-necked 6, **127**
Honeyeater, Black 49, **407**
Honeyeater, Blue-faced 48
Honeyeater, Brown ....
11, 25, 49, 341, 386, **404**
Honeyeater, Brown-headed ..
60, 61, 293, **416**
Honeyeater, Golden Backed . 416
Honeyeater, Grey 44, 55, **408**
Honeyeater, Grey-headed ....
11, 55, 294, **412**
Honeyeater, New Holland ....
11, 50, 318, **420**
Honeyeater, Pied 63, **405**
Honeyeater, Purple-gaped ....
11, 55, **411**
Honeyeater, Rufous-throated 48
Honeyeater, Singing ....
11, 49, 60, 61, 293, **409**
Honeyeater, Spiny-cheeked 11, 49, 59, 239, **425**
Honeyeater, Striped 58
Honeyeater,Tawny-crowned ..
50, 60, 293, 297, 318, **418**
Honeyeater, White-cheeked ..
11, 50, **422**
Honeyeater, White-eared ....
60, 63, **414**
Honeyeater, White-fronted ....
49, 55, 59, **419**
Honeyeater, White-gaped 48
Honeyeater, White-naped ....
50, 293, **415**
Honeyeater, White-plumed ....
11, 29, **413**
Honeyeater, White-throated .. 48
Honeyeater, Yellow-fronted ..
21, **411**
Honeyeater, Yellow-plumed ..
55, 60, 61, 63, **410**, 414
Honeyeater, Yellow-throated . 60

Honeyeater, Yellow-tinted 48
Honeyeater, Yellow-winged .. 420
Hookbill 282

Ibis, Glossy 134
Ibis, Straw-necked 6, **132**
Ibis, White 6, **133**

Jabiru 131
Jacana 48
Jacana, Pheasant-tailed 196
Jay, Blue 329

Kestrel, ... 10, **176**
Kingfisher, Azure 48
Kingfisher, Forest 48
Kingfisher, Mangrove 311
Kingfisher, Red-backed .. 11, **310**
Kingfisher, Sacred 5, **309**
Kite, Black 6, **159**
Kite, Black-shouldered .. 62, **157**
Kite, Fork-tailed 159
Kite, Letter-winged 6, **158**
Kite, Square-tailed 160
Kite, Whistling 162
Knot 218
Knot, Great 217
Koel 48, **299**
Kookaburra, Blue-winged 309
Kookaburra, Laughing 11, **308**

Landrail 186
Landrail, Banded 185
Lincoln 330
Lorikeet, Little 51
Lorikeet, Musk 51
Lorikeet, Purple-crowned ....
5, 51, 55, 57, 63, **267**
Lorikeet, Rainbow 268
Lorikeet, Red-collared 48
Lorikeet, Scaly-breasted 48
Lorikeet, Varied 48
Lyre-bird 51

Magpie, Black-backed .. 11, **453**
Magpie, Murray 439
Magpie-Lark .. 9, 11, 61, 294, **439**
Magpie, Western 11, 60, **451**
Major Mitchell 22, 49, **273**
Mallard 12
Mallee-Fowl .. 30, 36, 55, 59, **177**
Martin, Fairy 326
Martin, Tree- 5, 11, **324**
Mickey Miner 423
Miner, Dusky 423
Miner, Noisy 52
Miner, Yellow-throated ....
11, 49, 54, 60, 293, **423**
Mistletoe-bird 398

Moorhen, Dusky 10, **190**
Mudlark 439
Mutton-bird, King 97
Mutton-bird, Tasmanian 95

Native Companion 184
Native-Hen, Black-tailed ....
6, 29, **188**
Night-Heron, Nankeen 128
Nightingale 318, 341
Nightjar, Owlet- 11, **305**
Nightjar, Spotted 306
Nightjar, White-tailed 48
Noddy, Common 7, 17, **251**
Noddy, Lesser 32, **254**

Osprey 26, 27, **171**
Owl, Barking 301
Owl, Barn 302
Owl, Boobook 19, 20, **300**, 305
Owl, Cave 303
Owl, Eastern Grass- 303
Owl, Masked 303
Owl, Rufous 48
Owl, Winking 301
Owlet-Nightjar 305
Oyster-catcher, Pied .. 17, 20, **198**
Oyster-catcher, Sooty ....
19, 20, **198**

Pardolote .. .... see Diamond-bird
Parrakeet, Crested 62
Parrot, Alexandra 55, **280**
Parrot, Black-tailed 279
Parrot, Blue Mountain 268
Parrot, Blue-winged 51
Parrot, Bourke 11, **286**
Parrot, Elegant Grass- 55, **288**
Parrot, Ground 39, 50, **290**
Parrot, King 19, 23, 62, **282**
Parrot, Many-coloured 285
Parrot, Mulga 11, 49, 55, **285**
Parrot, Naretha 285
Parrot, Night 38, 55, **291**
Parrot, Oak 285
Parrot, Plaide-wing 279
Parrot, Port Lincoln 283
Parrot, Red-backed 58
Parrot, Red-capped 282
Parrot, Red-winged 48, **278**
Parrot, Regent 279
Parrot, Ringneck 55, **283**
Parrot, Rock 289
Parrot, Scarlet-chested .. 34, **287**
Parrot, Shell 290
Parrot, Smoker ....
11, 54, 55, 60, 61, **279**
Parrot, Spinifex 291
Parrot, Sundown 286

Parrot, Turquoise 51
Parrot, Twentyeight 11, 24, 28, 51, 53, 55, 56, **283**
Parrot, Varied 285
Parrot, Yellow-cheeked 281
Parrot, Yellow-collared 283
Parrot, Zit 267
Pea-fowl 12
Pee-wee 439
Pediunker 89
Pelican, Australian 6, 20, 24, 27, **107**
Penguin, Adelie 74
Penguin, Crested 72
Penguin, Emperor 7
Penguin, Erect-crested 74
Penguin, Fairy 71
Penguin, Little 19, 21, 22, 26, **71**
Penguin, Rockhopper 72
Penguin, Thick-billed 73
Petrel, Blue 83
Petrel, Cape 24
Petrel, Giant 21, 76
Petrel, Giant, Northern 82
Petrel, Giant, Southern 81
Petrel, Great-winged 7, 21, 29, **97**
Petrel, Grey 25, **89**
Petrel, Kerguelen 98
Petrel, Silver-grey 83
Petrel, Soft-plumaged 99
Petrel, White-chinned 89
Petrel, White-headed 99
Pheasant 12
Pheasant-Coucal 299
"Pigeon, Blue" 329
Pigeon, Box Poison 263
Pigeon, Bronzewing 24
Pigeon, Brush 16, 22, 32, 50, **263**
Pigeon, Bronzewing, Common 11, 61, **262**
Pigeon, Crested 11, 61, **263**
Pigeon, Domestic 11, **256**
Pigeon, Flock 266
Pigeon, Green-winged 48
Pigeon, Partridge 48
Pigeon, Plain 266
Pigeon, Plumed 265
Pigeon, Rose-crowned 48
Pigeon, Red-plumed 11, 33, 38, **264**
Pigeon, Rock 48
Pigeon, Topknot 263
Pigeon, Torres Strait 48
Pilot-bird 52
Pintado 82
Pipit, Australian 11, 294, **327,** 341
Pipit, Richard's 327
Pitta, Blue-winged 5
Pitta, Rainbow 48
Plover, Banded 9, 11, **201**
Plover, Eastern Golden .. 195, 202
Plover, Grey 195, **202**
Plover, Little Ringed 205
Plover, Masked 199
Plover, Prairie 208
Plover, Spur-winged 200
Pratincole, Australian 228
Pratincole, Oriental 229
Prion, Antarctic 87
Prion, Broad-billed 89
Prion, Fairy 85
Prion, Medium-billed 88
Prion, Thin-billed 87

Quail, Brown 181
Quail, Little 10, **183**
Quail, Painted 50, **182**
Quail, Stubble 10, **180**
Quail-Thrush, Chestnut 55, 59, 60, **332**
Quail-Thrush, Cinnamon 11, 55, 334
Quail-Thrush, Nullarbor 333
Quail-Thrush, Spotted 51
Quarrion 278

Rail, Banded Land- 185
Rail, Chestnut 48
Rail, Lewin Water- 185
Rainbird 293
Rainbow-bird 5, **312**
Raven, Australian 50, **456**
Redshank 213
Redthroat 11, 49, 59, 295, 296, **363**
Reed-Warbler 4, 16, 295, 318, **240**, 405
Reeve 223
Robin, Flame 52
Robin, Hooded 11, 61, 294, 297, **377**
Robin, Mangrove 62, **381**
Robin, Pink 52
Robin, Red-capped 49, 58, 61, 297, **375**
Robin, Rose 52
Robin, Scarlet 11, 50, 294, 297, **374**
Robin, Southern Scrub- 55, 59, **332**
Robin, Western Yellow- 28, 50, 51, 53, **377**, 380

Robin, White-breasted 28, 52, 62, 378, **379**
Robin, Yellow 50, 51, 52, 53
Rock Pebbler 279
Roller, ... 48, **314**
Rosella, Crimson 51, 62
Rosella, Eastern 50
Rosella, Northern 48
Rosella, Stanley 281
Rosella, Western 19, 50, **281**
Ruff 223

Sanderling 218
Sandpiper, Broad-billed 222
Sandpiper, Common 196, **212**
Sandpiper, Curlew 195, **218**, 220, 221
Sandpiper, Green 196
Sandpiper, Marsh 213
Sandpiper, Pectoral 195, **222**
Sandpiper, Sharp-tailed 195, **220**, 222
Sandpiper, Terek 196, **214**
Sandpiper, Wood 196, **211**
Scale-bird 293
Scissors-grinder 383
Scrub-bird, Noisy .. 11, 36, 37, 38, 39, 41, 45, 62, **314**, 394
Scrub-bird, Rufous .. 62, 319, 320
Scrub-Fowl 48
Scrub Robin, Southern 55, 59, **332**
Scrub-Wren, Spotted 50, 298, 318, **361**
Shag .... see Cormorant
Shearwater, Fleshy-footed 5, 37, **94**
Shearwater, Fluttering 5, **91**
Shearwater, Hutton's 91
Shearwater, Little 7, **90**
Shearwater, Short-tailed .. 5, **95**
Shearwater, Sooty 97
Shearwater, Wedge-tailed 7, 11, 16, 18, 32, **93**
Shelduck, Chestnut-breasted 144
Shoemaker 25, 89
Shoveler, Blue-winged .. 10, 148
Shrike-Thrush, Brown 48
Shrike-Thrush, Brown-breasted 48
Shrike-Thrush, Grey 55, 389
Shrike-Thrush, Little 48
Shrike Thrush, Western 11, 55, **388**
Shrike-Tit, Crested 29, 50, 60, 61, **389**
Silver-eye, Eastern 50, 403
Silver-eye, Western 11, 50, 54, 295, 297, 298, **402**
Silver-eye, Yellow 403
Sittella, Black-capped 11, 29, 49, 51, 297, **395**
Sittella, White-winged 48, 295
Skua, Arctic 5, 230
Skua, Great 5, 25, **229**
Skua, Parasitic 5, **230**
Skua, Pomarine 5, **231**
Skua, Richardson 5, **230**
Skua, Southern 5, 25, **229**
Skua, Twist-tailed 5, **231**
Skylark, European 321
Sleepy Dick 354
Snipe, Bishop 224
Snipe, Chinese 215
Snipe, Japanese 216
Snipe, Painted 31, **197**
Snipe, Pin-tailed 215
Snipe, Rottnest 224
Songlark, Brown 4, 11, **341**
Songlark, Rufous .. 4, 11, 61, **341**
Song-Thrush, European 417
Sparrow House 11, 403, **436**
Sparrow, Tree 437
Sparrow-Hawk, Collared 165
Spinebill, Eastern 50
Spinebill, Western 11, 29, 50, 253, 297, **417**
Spinifex-bird 40, **337**
Spoonbill, Black-billed 135
Spoonbill, Yellow-billed 136
Spoonbill, Royal 135
Squeaker 49, 51, 55, 57, 57, 63, **447**
Starling, Common 12, **437**
Steamer 154
Stilt, Banded 58, **224**
Stilt, Black-winged 11, **224**
Stilt, White-headed 11, **224**
Stint, Little 219
Stint, Long-toed 195, **220**
Stint, Red-necked .. 195, **219**, 221
Stone-Curlew, Beach 228
Stone-Curlew, Southern 227
Stork, Black-necked 131
Storm-Petrel, Black-bellied .. 101
Storm-Petrel, Leach's 103
Storm-Petrel, Sooty 5, **102**
Storm-Petrel, Wilson 99
Storm-Petrel, White-bellied .. 101
Storm-Petrel, White-faced . 5, **100**
Summer-bird 329
Swallow, Barn 5, **324**
Swallow, Golden 312
Swallow, Welcome 4, **322**

Swallow, White-backed ....
11, 49, 58, 60, **321**
Swamphen 191
Swan, Black 6, 10, 16, 20, 23, 25, 28, 31, **138**
Swan, Mute 11, 139
Swan, White 11, 139
Swift, Fork-tailed 5, **307**
Swift, Spine-tailed 307

Tattler, Grey-tailed 195, **214**
Tattler, Wandering 214
Teal, Chestnut 146
Teal, Grey 8, 10, **147**
Teal, Mountain 146
Teal, Whistling 149
Tern, Arctic 5, 240, **241**
Tern, Bridled 7, 17, 18, **249**
Tern, Brown-winged 249
Tern, Common 5, 242, **243**
Tern, Caspian 7, 17, 25, **238**
Tern, Crested 7, 17, **244**
Tern, Fairy 7, **246**
Tern, Gull-billed 238
Tern, Lesser Crested 245
Tern, Little 247
Tern, Marsh 237
Tern, Panayan 249
Tern, Roseate **240**, 242
Tern, Sooty 7, **247**
Tern, Whiskered 237
Tern, White-winged Black 5, **236**
Thornbill, Broadtailed .. 8, 11, 49, 51, 52, 295, 297, 298, **355**
Thornbill, Brown 51, 356
Thornbill, Buff-rumped 50, **359**
Thornbill, Chestnut-rumped 49, 55, 59, **357**
Thornbill, Little 51
Thornbill, Robust-billed 357
Thornbill, Samphire .. 8, 11, **359**
Thornbill, Slate-backed ....
11, 55, 297, **357**
Thornbill, Slender-billed 359
Thornbill, Striated 51
Thornbill, Western .. 50, 52, **358**
Thornbill, Yellow-rumped ....
9, 11, 33, 296, 297, 298, **360**
Tin Tack 427
Tom-Tit 360
Tree-creeper, Allied 397
Tree-creeper, Black-tailed 397
Tree-creeper, Brown 55, 396
Tree-creeper, Rufous ....
11, 33, 55, 57, 63, **396**
Tree-creeper 11, 63, **396**
Tree-Duck, Plumed 140
Tree-Duck, Whistling 139
Tree-Martin 5, 11, **324**
Triller, Varied 48
Triller, White-winged ....
4, 11, 294, **330**
Tropic-bird 24
Tropic-bird, Red-tailed 21, **104**
Tropic-bird, White-tailed .. 5, **105**
Turkey, Wild 193
Turnstone 196, **214**
Turtledove, Senegal 11, **258**
Turtledove, Spotted 11, **257**
Twentyeight 283

Wagtail, Pied 328
Wagtail, Willy ....
11, 294, 318, **382**
Wagtail, Yellow 5, **328**
Walkingale 279
Warbler, Dusky 355
Warbler, Speckled 51
Warbler, Western ....
49, 51, 56, 297, **354**
Wattle-bird, Little ....
11, 50, 293, **425**
Wattle-bird, Red ....
5, 11, 50, 63, 293, **424**
Weather-bird 293
Wedgebill ....
11, 55, 63, 294, **391**, 393
Weebill 11, **364**
Weero 11, 278
Whimbrel, ... 196, **210**
Whimbrel, Little 210
Whipbird, Western ....
36, 38, 55, **393**
Whistler, Black-tailed 385
Whistler, Brown 48
Whistler, Gilbert 55, **387**
Whistler, Golden .. 50, 294, **385**
Whistler, Mangrove 385
Whistler, Rufous 11, 318, **386**
Whistler, White-breasted 388
Whiteface, Banded .. 55, 297, **367**
Whiteface, Southern ....
8, 49, 55, 57, **366**
White-wings 151
Wideawake 247
Widgeon 149
Willy Wagtail 382
Wockengull 279
Wood-Swallow, Black-faced .. 11, 49, 60, 61, 294, **444**
Wood-Swallow, Dusky ....
50, 63, 294, 388, 406, **445**
Wood-Swallow, Little 11, **446**
Wood-Swallow, Masked Wood-Swallow 442

Wood-Swallow, White-breasted 441
Wood-Swallow, White browed 443
Wren, Banded Blue 11, 28, 51, 55, 56, 297, **343**
Wren, Black-and white 27, 40, 297, **346**
Wren, Black-backed 55
Wren, Blue-and-white 11, 49, 60, 297, **345**, 346
Wren, Blue-breasted 53, **348**
Wren, Field 365
Wren, Lavender-flanked 48
Wren, Marsh 349
Wren, Purple-backed 347
Wren, Purple-crowned 48
Wren, Red-backed 48, **351**
Wren, Red-winged .. 29, 53, **349**
Wren, Rock 369
Wren, Sandplain 365
Wren, Spotted Scrub- 361
Wren, Splendid 343
Wren, Superb Blue 51, 56
Wren, Turquoise 55, **344**
Wren, Variegated 11, 53, 60, 297, 298, **347**
Wren, White-winged 345

INDEX OF NATIVE NAMES

Arranger 107

Baacka-baacka 390
Bagobago 390
Bam-boore 377
Bambun 377
Ban-dene 420, 422
Bandin 420
Banggin 415
Bar-dan-itch 129
Bardanitch 129
Bar-doo-ngoo-ba 148
Bardunguba 148
Barraberry 180
Barrul 193
Bawalilly 400
Bawkadeerna 263
Be-ahlee 276
Be-bil-ya 193
Bee-arkoo 276
Bee-bill-yer 193
Bee-ogu 276
Bee-roo-bee-roo-long 312
Beetunkara 366
Bee-woo-een 444, 445
Beingy 423
Belidita 427
Bel-gar-be-jal 188
Bell 447
Bella 447
Be-lul 401
Beo-dee-lung 107
Ber-ril-ber-ril 415
Berrin-berrin 312
Ber-rit-ber-rit 363
Berulberul 312
Be-wo-wen 444, 445
Biagu 276
Biarka 276
Bibbilya 193
Biddiarra 271
Bildera 271
Billbul 280
Billy deedy 427
Bill-your-ga 423
Bilya-coo 276
Bilyabiet 400
Bil-ya-goo-rong 423
Bilyendi 276
Bimba 107
Binar 302
Bin-birri 176
Bindoon 208
Birr-ear-ra 273
Birringo 312
Birunbirun 312
Biwoen 444, 445
Bo-dal-lang 107
Bogga 126
Boi-jeel 379
Bokanbokan 390
Bokkun 390
Bokkon-bokkun 390
Boodalan 107
Boo-doo 153
Booladoodoo 293
Boolgarra-boolgarra 202
Boolie 273
Bool-jeet 417
Boon boon 423
Boone-yung 332
Boorh-ga 174
Boo-run-dur-dee 290
Booteritch 370
Bootlung 107
Bo-ta 153
Buatu 153
Budibudi 321
Budtallang 107
Bula-doota 278
Bul-bul 299
Buljit 417
Bulli 273
Bun-geen 415
Burabur 193
Bur-den-etch 129
Burridira 193
By-yoo-gool-ye-de 439
By-yu Gulyidi 439

Ca-an-gan 141
Cabbridge 322
Ca-cac-co 335
Ca-go-looloo 263
Cal-cal-goa 263
Calig-yoo-roo 260
Callo-gora 260
Calyung 146, 147
Carakine 164
Carcoo 457
Cardacut 111
Carebumdulah 329
Carra-cal-lu 175
Carrer 271
Chadichadi 321
Chandilla 188
Charlalya 164
Chee-ri-gal 382
Chelyup 282
Cherruck 327
Chidadoo 391
Chiga 432

Chillenbidy 439
Chillian 169
Chillin-birrin 439
Chin-bijee 147
Chindi-chindil 382
Chinin-chinin 397
Chitti-chitti 382
Choogin 285
Chun-din-di 188
Churuna 107
Cullya 65
Coatchuck 154
Cockaraboota 450
Cogadoo 302
Cogalala 276
Combiac 304
Coodardy 65
Coolbardie 451
Coolburra 450
Coolecha 138
Coolie 193
Coolyerrie 290
Cooraloo 138
Cowara 267
Cower 267
Crowbardy 451
Cudgeego 450
Cula-cuga 260
Culerrit 276
Cumbine 304
Curabarty 451
Curraak 271

Dammalak 283
Dara 283
Darandura 409
Dee-will deewill 439
Deldillia 343
Demclick 374
Dia-dia 401
Dilabit 439
Dila-but 439
Dilebert 439
Dil-yurdu 170
Din-gel-yer-ung 290
Dirroon dirroon 312
Djalyup 282
Djang-gang 424, 425
Djanjarak 224
Djan-jar-uk 224
Djardal-ya 370
Djar-dong-garri 290
Djar-doon-gur-ree 290
Djarjilya 343
Djar-rail-bur-tong 282
Djarrybardang 282
Dje-je-nup 231
Dje-laak 447
Djijnak 231
Djillak 447
Djo-bul-djo-bul 358
Djool-be-djool-bung 335
Djoon-dool-goo-roon 389
Djudarran 293
Dju-dur-run 293
Djul-bat-ta 290
Djulbidjulbang 355
Djung-gung 424, 425
Djuritch 296
Djur-jeel-ya 343, 349, 352
Djyr-dal-ya 370
Don-don 175
Dongrruck 424
Doondroon marda 283
Doo-rum-doo-rum 409
Dow-arn 283
Doyndoyn 335
Du-leer 304
Dum-ul-uk 283
Dun-gur-uk 425
Duranduran 414
Du-laar 294
Dwar-den-ngool-gnan-neer .. 429
Dwer-da-ngok ngun-nin 407
Dyular 294

Eelooroo 329
E-role-del 342
E-ro-to 150
Er-roo-doo 150
Errudo 150

Gabbi Kallangorong 324
Gab-by-kal-lan-goo-rong 324
Gaddara 154
Gadjinnak 381
Gala 457
Galiberie 65
Gambigora 304
Gargang-a 111
Geah-ter-but 364
Ge-durn-nul-uk 160
Geetardo ....
Gi-aterbat 364
Gid-jee-broon 192
Girgal 361
Gnal-gan-ning 128
Gnampion 262
Gnimeruk 144
Gnowera 145
Gnow 177
Gnow-un 336
Gnular 269
Gnwool-ye-nug-ger-rang 146
Go-da-ra 154
Go-go-go 117

Gol-jak 138
Goo-ba 374
Goo-ba-mogin 377
Goo-dap 161
Goo-de-lung 388
Good-jee-lum 164
Googagoora 147
Goojuk 154
Goolagoogoo 260
Gool-be-gool-burn 419
Gool-boort 409
Goo-le-ma 191
Gool-ye-der-ung 288
Goo-mul-cul-long 330
Goo-mal-de-dite 395
Gooraba 451
Goo-ra-ga 144
Goor-a-go 144
Goo-ra-ling 328
Goore-bat 451
Goor-de-mul 301
Goore-beet-goore-beet 389
Goor-gal 361
Goorgoor 300
Goor-goor-da 300
Goor-jee-goor-jee 340
Gooroyl 138
Goota-un-goota-un 281
Goran-goran 198
Guddang-uddan 281
Gudilang 388
Gudjelan 164
Gu-gu-mit 300
Guijak 138
Guldanguldan 281
Gullima 191
Gul-yidarang 288
Gumalbidyt 395
Guraga 144
Gurbal 451
Gurbat 451
Gurbitgurbit 389
Gurgurda 300
Gurji-gurji 340
Guroyl 138
Gutuban 296
Gwet-ul-bur 172
Gwetalbur 172
Gwom-na-na 145

I-dal-ya 3
Iligura 454
Indatjarra 413
Indee-narrangee 163
Ir-karka 454
Irodu 150

Jabba-loong 167
Jabba-roongoo 167
Jackal-yackal 273
Jakeburi 65
Jak-kul-yak-kul 273
Jak-kul-ya-ra 273
Jan-doo 162
Jandu 162
Jarbell 447
Jarburn 447
Jarrooroo 309
Jar-jil-ya 352
Jarli 302
Jee-da 360
Jeedadoo 391
Jee-da-mya 3
Jee-mul-uk 314
Jee-ra 369
Jeer-ree 429
Jeranger 231
Jew-la-oola 264
Jida 3, 360
Jida-mya 3
Jilak 447
Jilakarby 107
Jil-le-e-lee 389
Jili-mil-yan 127
Jil-lee-jil-lee 165
Jil-lee-mil-yum 127
Jillijilli 165
Jillimberri 439
Jindangie 390
Jind-ril 382
Jindurka 276
Jindy-jindy 382
Jin-gee 415
Jin-jo-gour 404
Jin-nee 396
Jinni 396
Jiri 429
Jirjil-ya 352
Ji-ro-na 107
Jit-bung 442
Jit-burt 360
Jit-tee-gnut 383
Jit-te-jit-te 382
Jitti-ngat 383
Jockalock 273
Joo-ee 304
Jooleen 285
Jule-we-de-lung 402
Jul-u-up 282
Julwidilang 402
Jungoorna 65

Kahnga 457
Kakalyalya 273
Kaku 456
Kalaya 65

Ka-lee 119
Kal-ga 306
Kali 119
Kalia 65
Kaliat 65
Kal-jeer-gang 244
Kaljirang 244
Kallilia 65
Kandimak 309
Kang-innak 309
Kannamit 322
Kanyinnak 309
Kar-a-dong 183
Karbunga 144
Kargoo 457
Kargyne 175
Karnka 456
Karoo-wolkun 309
Kar-rak 271
Ka: ru 117
Kar-ryne-mun-dyne ....
Kattarr 154
Kawk 456
Keenirri ....
Keer-jan-dee 271
Kerra-jinga 175
Kigaribiloo 145
Kilbaltkilbalt 324
Kijjibrun 192
Kil-lin-gil-lee 164, 165
Kilykilykari 290
Kintillka 324
Kirrkee keen keen 201
Knulumberry 107
Kogga-je 273
Kol-yabbiddy 290
Koo-go-burt 300
Koo-je-lan 431
Kool-brit 264
Koolkillkaree 290
Koon-dah-ree 145
Koon-jere 127
Kooracha 144
Koorburroo 453
Koorark 271
Koor-goo-mah-loo 300
Kooroodor 124
Korlba 300
Korlolba 126
Kor-rar-ra 271
Korrodg-e-cup 271
Kowar 267
Kukubert 305
Kulbardie 451
Kulire 65
Kulla-eedoo 138
Kul-lee 185
Kullin-nunga 193
Kullya 65
Kulyaberuk 439
Kun-na-meet 322
Kun-yee-nuk 309
Kurillthu 138
Kurr-gurr-ee 159
Kurulu 138
Kwaggum 456
Ky-a 65
Ky-lor-ing 290

Labya 273
Lawn 310
Limbee 207
Loorn 310

Mabenga 283
Ma-dee 113
Mah-boo-goo 264
Mal-lee 138
Malpuntari 183
Mal-yi 138
Man-na-brun-ka 260
Man-ga 3
Man-goota 262
Mannich 275
Mannitch 275
Mannite 275
Manpi 262
Manup 275
Manyt 275
Marnbu 107
Mar-ang-an-ner 151
Mar-arl 160
Marlee 138
Marlie 138
Martiet 176
Mat-wel-itch 164
Mawgootagoota 264
Medie 113
Mee-die 113
Meginma 176
Mele 138
Men-de-ete 374
Me-ne-gedang 375
Mennich 275
Miamit 410
Migwin 176
Milka 334
Mimmal 118
Minidang 375
Minijidang 375
Minnar 302
Min-min 165
Moldern 172
Moo-lah 191
Mool-ya-wind-doo 192
Moo-ne-je-tang 398

Moo-reete 181
Moorhait 262
Moo-ro-lum 182
Mow-gra 262
Moyadung 281
Moy-a-dung 281
Muddenburra 290
Mullincoorin 283
Mullin-goora 283
Munnit 275
Murit 181
Murolang 182

Nailoit 167
Nalpintjee 193
Neamoora 432
Ne-dool-ye-rong 207
Nee-an-tee 365
Nee-murri 431
Ne-keit 100
Nelgie 167
Newmerri 432
Nidul-yorong 201, 207
Nimon 147
Nirimba 107
Nga-ga-la 234
Ngalganning 128
Ngampu 3
Nganamara 177
Ngamuti 193
Ngarrawa 145
Ngaruwa 145
Ngo-lak 269
Ngol-ye-nuk 269
Ngoo-nan 120
Ngow-o 177
Ngulor 169
Ngurlawaanda 107
Ngwol-yi-nag-girang 146
Ngwonana 145
Nol-yang 188
Noondeegannan 188
Noor-ruk 65
Nu-lar-go 329
Nurdo 3
Nurrak 65
Nyandi 368
Nye-nye 432
Nyern-de 369
Nyi-nyi 432
Nyinnyinka 432
N-yuneruk 144

Oo-da 262, 263
Oo-lack 269
Oonanah 145
Oonern 145
Orta-goota 305

Padda-moora 290
Paddera 193
Pad-oon 159
Panpanpalala 290
Pan-pan-panella 390
Par-dang 456
Parderua 193
Parr-oor-ra 193
Pa-thara 290
Patilpa 283
Pe-dil-me-dung 385
Peelabong 107
Peel-baal 283
Peen peen 423
Pee-orr-ba 328
Peewee 304
Perg-gee 173
Perina 309
Perinbo 312
Perit Perit 327
Pidilmidang 385
Pillargo 169
Pin-ki-mum 361
Pit-ahn-di 276
Poir-een 429
Pooka 309
Pooker 309
Poo-let 418
Poorajean 121
Poor-een 429
Pooree pooree 327
Por-rut 94
Puer 423
Pullal-pullal 285
Pullong 126
Pun-bun 283
Pun-bunba 283
Punpukara 183
Purinderoo 181
Puttarlung 107

Queg-gum 456
Quodyuk 154
Quoin-kur-dong 306
Qwoolibruk 439

Tantagee 169
Tar-ding 393
Tarkaroo 126
Tarradee 454
Teher-ree 369
Te-by-by 329
Thanni-midi 188
Thudaara 107
Thullundee 188
Thurra-ji 454
Tiru-tiru 396

Tjantji 3
Tjilltjilka 345
Tjitati 3
Tjitta tjalbee 442
Tjittee tjittee 382
Tjoonin 159
Tjoontpjangalankoo 326
Tjootara 107
Tjulily-tjulilpa 375
Tee-luk 447
Tooli 304
Too-low-in 264
Toonat 283
Toorakee 451
Toorir 278
Toorna 113
Toornart 283
Tow-arn 283
Towern 283
Towerrin 283
Toywee 304
Tui-dee 413
Tula 65
Tu-lee 285
Tungaluk 424
Tur-dal-l 425
Turdoo wardoo 335

Wadda-wadda 448
Wad-do-wad-ong 448
Wadje 65
Waid tjulka-tjulka 377
Wait 65
Waitch 65
Wajigga 457
Walada 167
Waladoo 167
Walawuru 167
Walgil 167
Walkinger 279
Wallan 107
Warlick 167
War-lock 167
Waltharrie 263
Wamba 278
Wandoa 448
Wang-anner 147
Wannoo 105
Warlit 167
Warraja 187
War-ra-jah 187
War-ra-joolon 327
Warrajudong 327
Warrangarree 198
Warrdong 456
Warrida 167
Warreedar 167
Waukanga 279
Wauraling 278
Warryl-bardang 354
War-ryle-bur-dang 354
We-dee-we-due 400
Wee-dun-ba 165
We-dup-we-dup 399, 400
Wee-ar-doo 198
Wee-arra 278
Weela-murra 227
Weelawodda 227
Weel-bra 171
Weeler 138
Wee-lo 227, 228
Wee-loo 227
Wee-rirr-eer-ri 345
Weeroo 278
Wegie 65
Welojabbin 227
Widap-widap 400
Wida-wida 400
Widgie 65
Widge-e-doo 312, 413
Widgi-winji 400
Wid-joo-on-ong 209
Widung-gurra 198
Wil-a-ring 382
Willaring 382
Wil-ya 227
Wimbin 149
Windee windee 176
Windoo 227
Winjahna 302
Wirmalo 107
Woggora 456
Wokka Tjalbee 444
Wold-dja 167
Waldja 167
Wollpootatjarra 328
Won-bon-alah 391
Wonga-ka 456
Wongona 145
Woodah 262
Woodta 262
Wool-boo-gle 301
Woo-lock 269
Wool-wool 78
Woordang 456
Woo-ra-ling 278
Woo-re-ning 420
Woorerling 278
Woo-rup 300
Worgener 175
Worle 445
Wouk-un-ga 279
Wow-oo 173
Wulbugli 301
Wulwul 78
Wundoolah 453

Wunnbunnbullalla 401
Wunnbunnpullalla 390
Wuraling 278
Wur-dang 456
Wy-an 126
Wy-ang 126
Wy-uda 121
Wym-bin 149
Wy-ooda 121
Wy-ro-dju-dong 418

Yaba-wilban 427
Ya-et 145
Ya-jin-goo-rong 226
Yalbonjarry 444
Yalla-biddee 65
Yallabiri 65
Yallebiri 65
Yalliberry 65
Yallobiddy 65
Yaralapulpa 263
Yarda 167
Yate 145
Yeellin-berra 439
Ye-jul-lup 341
Yelyel 451
Yerreeyarro 345
Yimbalu 447
Yim-eye 432
Yon-ja 302
Yoon-door-doo 171
Yoyjeddie 448

INDEX OF SCIENTIFIC NAMES

(In multiple references the main one is given in bold type)

Acanthagenys rufogularis 56, **425**
Acanthiza apicalis 49, 51, 52, **355**
A. chrysorrhoa 360
A. inornata 50, 51, 52, **358**
A. iredalei 359
A. lineata 51
A. nana 51
A. pusilla 51
A. reguloides 51, 359
A. robustirostris 357
A. uropygialis 357
Acanthorhynchus superciliosus 50, **417**
A. tenuirostris 50
Accipiter cirrocephalus 165
A. fasciatus 164
A. novæhollandiæ 175
Acrocephalus stentoreus 340
Aegintha temporalis 435
Aegotheles cristatus 305
Alauda arvensis 321
Amaurornis olivaceus 48
Amytornis housei 48
A. purnelli 370
A. striatus 368
A. textilis 369
Anas castanea 146
A. gibberifrons 147
A. platyrhyncha 12
A. querquedula 149
A. rhynchotis 148
A. superciliosa 145
Anhinga rufa 118
Anous stolidus 251
A. tenuirostris 254
Anseranus semipalmata 136
Anthochæra carunculata 50, **424**
A. chrysoptera 50, **525**
Anthus novæseelandiæ 327
Aphelocephala castaneiventris 55
A. leucopsis 55, **366**
A. nigricincta 367
A. whitei 55
Aprosmictus erythropterus .. 48, **278**
Aptenodytes forsteri 7
Apus pacificus 307
Aquila audax 167
Ardea novæhollandiæ 126
A. pacifica 127
A. picata 48
A. sumatrana 48
Arenaria interpres 215
Artamus cinereus 444
A. cyanopterus 445
A. leucorhynchus 441
A. minor 446
A. personatus 442
A. superciliosus 443
Atrichornis clamosus 314
A. rufescens 62, 319, 320
Aviceda subcristata 48
Aythya australis 150

Barnardius 62
B. semitorquatus 53
B. zonarius 53, **283**
Bathilda ruficauda 434
Biziura lobata 154
Botaurus poiciloptilus 129
Bubulcus ibis 125
Burhinus grallarius 227
Butorides striatus 128

Cacatua galerita 48
C. leadbeateri 273
C. roseicapilla 276
C. sanguinea 273
C. tenuirostris 50, **275**
Cacomantis flabelliformis 294
C. pyrrhophanus 50, **294**
C. variolosus 48
Calamanthus compestris 365
C. fuliginosus 365
C. montanellus 365
Calidris acuminata 220
C. alba 218
C. canutus 218
C. ferruginea 218
C. melanotos 222
C. subminuta 220
C. ruficollis 219
C. tenuirostris 217
Calyptorhynchus baudini 50, **269**
C. funereus 50, 270
C. magnificus 271
Caprimulgus macrurus 48
Cardeulis cardeulis 428
Catharacta antarctica 229
Centropus phasianinus 229
Cereopsis novæhollandiæ 141
Certhionyx variegatus 56, **405**
Ceyx azureus 48

Chætura caudacuta 307
Chalcophaps indica 48
Charadrius bicinctus 206
C. cinctus 208
C. cucullatus 207
C. dubius 205
C. leschenaultii 203
C. melanops 207
C. mongolus 204
C. ruficapillus 206
C. veredus 202
Chenonetta jubata 151
Cheramœca leucosternum 321
Chlamydera guttata 55
C. maculata 55, **454**
C. nuchalis 48
Chlidonias hybrida 237
C. leucoptera 236
Chrysococcyx basalis 296
C. lucidus 50, **297**
C. minutillus 48
C. plagosus 50, **297**
C. osculans 295
Chthonicola sagittata 51
Cincloramphus cruralis 341
C. mathewsi 342
Cinclosoma alisteri 333
C. castaneothorax 55
C. castanotum 332
C. cinnamomeum 55, **334**
C. punctata 51
Circus approximans 170
C. assimilis 169
Cisticola exilis 338
Cladorhynchus leucocephalus 224
Climacteris affinis 396
C. melanura 397
C. picumnus 55, 396
C. rufa 55, **396**
C. wellsi 397
Colluricincla brunnea 48
C. harmonica 55, 389
C. parvula 48
C. rufiventris 55, **389**
C. woodwardi 48
Columba livia 256
Conopophila rufogularis 48
Coracina novæhollandiæ 329
C. papuensis 48
Corcorax melanorhamphus .. 440
Corvus bennetti 457
C. ceciliæ 456
C. coronoides 456
C. orru 456
C. splendens 12, 458
Coturnix pectoralis 180
Cracticus nigrogularis 450
C. torquatus 448
Crex crex 186
Cuculus pallidus **293**, 405
C. saturatus 294
Cygnus atratus 138
C. olor 139

Dacelo gigas 308
D. leachii 309
Daption capensis 82
Dasyornis brachypterus 371
D. broadbenti 373
D. longirostris 370
Demigretta sacra 124
Dendrocygna arcuata 139
D. eytoni 140
Dicæum hirundinaceum 398
Diomedea cauta 79
D. chlororhynchos 78
D. chrysostoma 79
D. epomophora 76
D. exulans 25, **75**
D. melanophris 77
Dromaius novæhollandiæ 65
Drymodes brunneopygia 332
Ducula spilorrhoa 48
Dupetor flavicollis 129

Edoliisoma tenuirostre 48
Egretta alba 123
E. garzetta 124
E. intermedia 123
E. sacra 124
Elanus notatus 157
E. scriptus 158
Emblema oculata 429
E. pictus 431
Entomyzon cyanotis 48
Eopsaltria australis 50, 51
E. georgiana 52, 62, **379**
E. griseogularis .. 50, 51, 52, **377**
E. pulverulenta 62, **381**
Ephthianura albifrons 427
E. aurifrons 428
E. crocea 48
E. tricolor 427
Eremiornis carteri 337
Erythrotriorchis radiatus 48
Esacus magnirostris 228
Eudynamys scolopacea .. 48, **299**
Eudyptes chrysocome 72
E. cristatus 72
E. pachyrhynchus 73
E. sclateri 74
Eudyptula minor 71
Eulabeornis castaneoventris .. 48
Eunymphicus cornutus 62
Eupodotis australis 193

Eurostopodus guttatus 306
Eurystomus orientalis 48, 314

Falco berigora 175
F. cenchroides 176
F. hypoleucus 174
F. longipennis 173
F. peregrinus 172
F. subniger 175
Falcunculus frontatus 50, **389**
Fregata ariel 105
F. minor 106
Fregatta grallaria 101
F. tropica 101
Fulica atra 192
Fulmarus glacialoides 83

Gabianus pacificus 234
Gallinago hardwickii 216
G. megala 215
Gallinula tenebrosa 190
Gelochelidon nilotica 238
Geopelia cuneata 260
G. humeralis 261
G. striata 259
Geophaps smithi 48
Geopsittacus occidentalis 291
Gerygone fusca 354
G. tenebrosa 355
Glareola maldivarum 229
Gliciphila albifrons 419
G. indistincta 404
G. melanops 50, **418**
Glossopsitta concinna 51
G. porphyrocephala 267
G. pusilla 51
Grallina cyanoleuca 439
Grus rubicunda 184
Gymnorhina dorsalis 451
G. tibicen 453

Hæmatopus fuliginosus 198
H. longirostrus 198
H. ostralegus 198
Halcyon chloris 311
H. macleayi 48
H. pyrrhopygia 310
H. sancta 309
Haliæetus leucogaster 169
Haliastur indus 163
H. sphenurus 162
Halobæna cœulea 73
Hamirostra melanosternon 161
Hieraætus morphnoides 165
Himantopus himantopus 224
Hirundo neoxena 322
H. rustica 5, **324**
Histriophaps histrionica 266

Hydrophasianus chirurgus 196
Hydroprogne caspia 239
Hylacola cauta 364

Ixobrychus minutus 130
I. sinensis 131

Jacana gallinacea 48

Lacustroica whitei 408
Lalage leucomela 48
L. sueurii 330
Larus dominicanus 235
L. novæhollandiæ 231
L. pacificus 234
L. pipixcan 233
Leipoa ocellata 177
Leptolophus hollandicus 278
Lichmera indistincta 404
Limicola falcinellus 222
Limnodromus semipalmatus .. 217
Limosa lapponica 211
L. Limosa 210
Lophoictinia isura 160
Lophophaps ferruginea 264
L. plumifera 265

Macronectes giganteus 81
M. halli 82
Malacorhynchus membranaceus 149
Malurus callainus 55, **344**
M. coronotus 48
M. cyaneus 51, 56
M. dulcis 48
M. elegans 52, **349**
M. lamberti 53, **347**, 350
M. leuconotus 345
M. leucopterus 346
M. melanocephalus 48, **351**
M. melanotus 55
M. pulcherrimus 53, **348**
M. splendens 55, **343**
Manorhina flavigula .. 54, 56, **423**
M. melanocephala 52
Megalurus gramineus 50, **339**
M. timoriensis 48
Megapodius freycinet 48
Melanodryas cucullata 377
Meliornis niger 422
M. novæhollandiæ 420
Meliphaga cratitia 411
M. flavescens 48
M. flavicollis 60
M. keartlandi 412
Meliphaga leucotis 56, **414**
M. ornata 56, **410**
M. penicillata 56, **413**

Meliphaga plumula 56, **411**
M. virescens 409
M. unicolor 48
Melithreptus albogularis 48
M. brevirostris 56, **416**
M. gularis 416
M. lunatus 50, **415**
Melopsittacus undulatus 290
Menura 319, 320
Menura novæhollandiæ 51
Merops ornatus 312
Microeca flavigaster 48
M. leucophæa 373
Milvus migrans 159
Mirafra javanica 321
Misocalius osculans 295
Morus bassana 109
M. capensis 109
M. serrator 108
Motacilla alba 328
M. flava 5, **328**
Myiagra cyanoleuca 384
M. inquieta 383
Myzantha flavigula 423
M. obscura 423

Neochmia ruficauda 434
Neophema bourkii 286
N. chrysostoma 51
N. elegans 288
N. petrophila 289
N. pulchella 51
N. splendida 34, **287**
Neositta leucoptera 48, 395
N. pileata 395
Nestor notabilis 7
Nettapus pulchellus 152
Ninox connivens 301
Ninox novæhollandiæ 300
N. rufa 48
Numenius madagascariensis . 209
N. minutus 210
N. phæopus 210
Nycticorax caledonicus 128

Oceanites oceanicus 99
Oceanodroma leucorhoa 103
O. matsudeiræ 102
Ocyphaps lophotes 263
Oreocincla lunulata 51
Oreoica gutturalis 390
Oxyura australis 153

Pachycephala inornata 387
P. lanioides 388
P. melanura 385
P. pectoralis 50, 385
P. rufiventris 386
P. simplex 48
Pachyptila belcheri 87
P. desolata 87
P. salvini 88
P. turtur 85
P. vittata 89
Pandion haliætus 171
Pardalotus assimilis 58
P. melanocephalus 48, 401
P. ornatus 58
P. punctatus 50, **399**
P. rubricatus 401
P. striatus 58
P. substriatus 51, 58, **400**
P. xanthopygus 400
Passer domesticus 403, **436**
P. montanus 437
Pavo cristatus 12
Pelagodroma marina 100
Pelecanoides exsul 103
P. urinatrix 103
Pelecanus conspicillatus 107
Peltohyas australis 208
Peneœnanthe pulverulenta 381
Petrochelidon ariel 326
P. nigricans 324
Petroica cucullata 377
P. goodenovii 375
P. multicolor 50, **374**
P. phœnicea 52
P. rhodinogaster 52
P. rosea 52
Petrophassa albipennis 48
Pezoporus occidentalis 291
P. wallicus 290
Phæthon lepturus 105
P. rubricauda 104
Phalacrocorax carbo 111
P. fuscescens 116
P. melanoleucos 117
P. sulcirostris 112
P. varius 113
Phaps chalcoptera 262
P. elegans 263
Phasianus colchicus 12
Philomachus pugnax 223
Phœbetria fusca 25, **80**
P. palpebrata 80
Phylidonyris albifrons 56, **419**
P. nigra 50, **422**
P. novæhollandiæ 50, **420**
Pitta iris 48
P. moluccensis 5
Platalea flavipes 136
P. regia 135
Platycercus 62
Platycercus elegans 51
P. eximius 50

Platycercus icterotis 50, **281**
P. venustus 48
Plectorhynchus lanceolata 58
Plegadis falcinellus 134
Pluvialis dominica 202
P. squatarola 202
Podargus strigoides 304
Podiceps cristatus 119
P. novæhollandiæ 120
P. poliocephalus 121
Pœcilodryas superciliosa 48
Polytelis alexandræ 280
P. anthopeplus 54, **279**
Pomatostomus superciliosus .. 336
P. temporalis 335
Porphyrio porphyrio 191
Porzana cinerea 48
P. fluminea 186
P. pusilla 187
P. tabuensis 187
Procellaria æquinoctialis 25, **89**
P. cinera 25, **89**
Psephotus hæmatogaster 285
P. hæmatonotus 58
P. varius 285
Psitteuteles versicolor 48
Psophodes nigrogularis 393
Pterodroma brevirostris 98
P. lessoni 99
P. macroptera 97
P. mollis 99
Pteropodocys maxima 328
Ptilinopus regina 48
Puffinus assimilis 90
P. carneipes 94
P. gavia 91
P. griseus 97
P. huttoni 92
Puffinus pacificus 93
P. tenuirostris 95
Purpureicephalus spurius 62, **282**
Pygoscelis adeliæ 74
Pycnoptilus floccosus 52
Pyrrholæmus brunneus 363

Q*uoyornis leucurus* 381

R*allina fasciata* 5, 186
Rallus pectoralis 185
R. philippensis 185
Recurvirostra novæhollandiæ 226
Rhipidura flabellifera 381
R. fuliginosa 381
R. leucophrys 382
R. rufifrons 48
R. rufiventris 48
Rostratula bengalensis 197

S*cythrops novæhollandiæ* 48
Seisura inquieta 383
Sericornis maculatus 361
Smicrornis brevirostris 364
Sphenostoma cristatum 391
Stercorarius parasiticus 230
S. pomarinus 231
S. skua 229
Sterna albifrons 247
S. anæthetus 249
S. bengalensis 245
S. bergii 244
S. dougallii 240
S. fuscata 247
S. hirundo 243
S. macrura 241
S. nereis 246
S. paradisæa 241
Stictonetta nævosa 143
Stiltia isabella 228
Stipiturus malachurus 50, **352**
S. mallee 353
S. ruficeps 353
Strepera versicolor 58, **447**
Streptopelia chinensis 257
S. senegalensis 258
Sturnus vulgaris 12, **437**
Sula dactylatra 109
S. leucogaster 110
S. sula 109
Synoicus ypsilophorus 181

T*adorna tadornoides* 144
T. radjah 48
Taeniopygia castanotis 432
Threskiornis molucca 133
T. spinicollis 132
Tribonyx ventralis 188
Trichoglossus chlorolepidotus 48
T. hæmatodus 48
T. moluccanus 268
Tringa brevipes 214
T. glareola 211
T. hypoleucos 212
T. incana 214
T. nebularia 213
T. stagnatalis 213
T. totanus 213
Turdus ericetorum 417
Turnix varia 182
T. velox 183
Tyto alba 302

MAP OF
WESTERN AUSTRALIA
SHOWING PLACES REFERRED TO IN TEXT
SCALE
MILES 50 40 30 20 10 0 50 100 150 200 MILES
115°
117°
119°
121°
123°
125°
127°
16°
18°
20°
22°
24°
KIMBERLEY DIVISION
DERBY
YEEDA HSTD
UPPER LIVERYNGA HSTD
NOONKANBAH HSTD
HALLS CREEK
FITZROY RIV.
MT ALEXANDER
Mungi Rock Hole
BROOME
Joanna Sp.
EIGHTY MILE BEACH
Bedout I.
CONDON
PORT HEDLAND
DAMPIER ARCHIPELAGO
Legendre I.
NICKOL B.
Depuch I.
Lewis I.
Hermite I.
Montebello Is.
COSSACK
ROEBOURNE
Barrow I.
DE GREY RIV.
SHAW RIV.
PILBARA
MARBLE BAR
PILBARA
GOLDFIELD
NULLAGINE
FORTESCUE RIV.
HAMERSLEY RANGE
Thevenard I.
N.W. Cape
EXMOUTH GULF
ONSLOW
ASHBURTON RIV.
Pt Cloates
OPHTHALMIA RA.
Tropic of Capricorn
LYNDON RIV
C. Farquhar
MINILYA HSTD
MINILYA RIV.
SALT LAKE
BARLEE RA
RIV.
LYONS
MT AUGUSTUS
Nicholl Sps
BOOLATHANA HSTD
CARNARVON
GASCOYNE
Bernier I.
Dorre I.
SHARK
FENCE
CANNING STOCK ROUTE
Well 51
Godfreys Tank (Well 48)
Well 45
Well 43
Well 37
Well 33
Well 29
Separation Well
Well 20
The Weld Sps (Well 9)